Soil Erosion and Sediment Redistribution in River Catchments

Measurement, Modelling and Management

Soil Erosion and Sediment Redistribution in River Catchments

Measurement, Modelling and Management

Edited by

P.N. Owens

National Soil Resources Institute
Cranfield University
North Wyke Research Station
Okehampton, Devon, UK

and

A.J. Collins

Landcare Research
Palmerston North
New Zealand

www.cabi.org

CABI is a trading name of CAB International

CABI Head Office
Nosworthy Way
Wallingford
Oxfordshire OX10 8DE
UK

Tel: +44 (0)1491 832111
Fax: +44 (0)1491 833508
E-mail: cabi@cabi.org
Website: www.cabi.org

CABI North American Office
875 Massachusetts Avenue
7th Floor
Cambridge, MA 02139
USA

Tel: +1 617 395 4056
Fax: +1 617 354 6875
E-mail: cabi-nao@cabi.org

A catalogue record for this book is available from the British Library, London, UK.

Library of Congress Cataloging-in-Publication Data

Soil erosion and sediment redistribution in river catchments / edited by P.N. Owens and A.J. Collins.
p. cm.
Includes bibliographical references.
ISBN-10: 0-85199-050-9 (alk. paper)
ISBN-13: 978-0-85199-050-7 (alk. paper)
1. Soil erosion. 2. River sediments. 3. Watershed management. I. Owens, Philip N. (Philip Neil), 1966- II. Collins, A.J. (Alison J.) III. Title.

S623.S5747 2006
551.3'02--dc22

2005014354

Typeset by AMA DataSet Ltd, UK.
Printed and bound in the UK by Biddles Ltd, King's Lynn.

Contents

Contributors ix

Preface xiii

PART I: INTRODUCTION

1 Introduction to Soil Erosion and Sediment Redistribution in River Catchments: Measurement, Modelling and Management in the 21st Century 3
A.J. Collins and P.N. Owens

PART II: MEASUREMENT

2 Tracing versus Monitoring: New Challenges and Opportunities in Erosion and Sediment Delivery Research 13
D.E. Walling

3 A Comparison of Caesium-137 and Erosion Pin Data from Tai To Yan, Hong Kong 28
M.R. Peart, M.E. Ruse and R.D. Hill

4 Assessing the Contribution of Different Processes to Soil Degradation within an Arable Catchment of the Stavropol Upland, Southern European Russia 40
V.R. Belyaev, A.Yu. Sidorchuk, V.N. Golosov, P.J. Wallbrink and A.S. Murray

5 Hillslope Soil Erosion and Bioturbation after the Christmas 2001 Forest Fires near Sydney, Australia 51
R.A. Shakesby, W.H. Blake, S.H. Doerr, G.S. Humphreys, P.J. Wallbrink and C.J. Chafer

6 Tracing Eroded Soil in a Burnt Water Supply Catchment, Sydney, Australia: Linking Magnetic Enhancement to Soil Water Repellency 62
W.H. Blake, S.H. Doerr, R.A. Shakesby, P.J. Wallbrink, G.S. Humphreys and C.J. Chafer

7 **Land Use, Sediment Delivery and Sediment Yield in England and Wales** 70
R. Evans

8 **Seasonal Trends of Suspended Sediment Concentration in a Mediterranean Basin (Anoia River, NE Spain)** 85
J. Farguell and M. Sala

9 **Suspended Sediment Transport during Rainfall and Snowmelt–Rainfall Floods in a Small Lowland Catchment, Central Poland** 94
L. Hejduk, A. Hejduk and K. Banasik

10 **Sediment in the River Bush, Northern Ireland: Transport, Sources and Management Implications** 101
D.J. Evans and C.E. Gibson

11 **The Physical and Biological Influence of Spawning Fish on Fine Sediment Transport and Storage** 112
E.L. Petticrew

12 **Lakes and Reservoirs in the Sediment Delivery System: Reconstructing Sediment Yields** 128
I.D.L. Foster

PART III: MODELLING

13 **Can Soil Erosion be Predicted?** 145
M.A. Nearing

14 **Erodibility Assessment in Dynamic Event-based Erosion Models** 153
N.J. Kuhn

15 **Double-averaging Methodology in Stochastic Modelling of Soil Erosion** 162
A.Yu. Sidorchuk, A. Smith and V. Nikora

16 **Runoff and Predicting Erosion on Hillslopes within Catchments** 170
P.I.A. Kinnell

17 **The Roles of Natural and Human Disturbances in Forest Soil Erosion** 177
W.J. Elliot

18 **Runoff and Erosion Modelling by WEPP in an Experimental Mediterranean Watershed** 186
F. Licciardello, E. Amore, M.A. Nearing and S.M. Zimbone

19 **Spatial Modelling of Ephemeral Gully Incision: a Combined Empirical and Physical Approach** 195
V. Jetten, J. Poesen, J. Nachtergaele and D. van de Vlag

20 **Simulating Fine Sediment Delivery in Lowland Catchments: Model Development and Application of INCA-Sed** 207
N.P. Jarritt and D.S.L. Lawrence

PART IV: MANAGEMENT

21 Estimating Sediment Generation from Hillslopes in England and Wales: Development of a Management Planning Tool 217
G.A. Wood, M. McHugh, R.P.C. Morgan and A. Williamson

22 Management of Sediment Production and Prevention in River Catchments: a Matter of Scale? 228
R.J. Rickson

23 Changes in the Spatial Distribution of Erosion within a Selectively Logged Rainforest Catchment in Borneo 1988–2003 239
R.P.D. Walsh, M.A. Clarke, K. Bidin, W.H. Blake, N.A. Chappell, I. Douglas, N. Ramli, A.M. Sayer, W. Sinun, J. Larenus and J. Hanapi

24 Erosion and Deposition Rates on 'Headlands' in Low-gradient Sugarcane Land in Australia 254
F. Visser

25 Land Use Change, Sediment Fluxes and Reef Conservation in Belize, Central America 264
R.S. Nunny, P. Stone and D.E. Walling

26 Understanding the Distribution, Structure and Behaviour of Urban Sediments and Associated Metals Towards Improving Water Management Strategies 272
I.G. Droppo, K.N. Irvine, K.J. Curran, E. Carrigan, S. Mayo, C. Jaskot and B. Trapp

27 Managing Sediment in the Landscape: Current Practices and Future Vision 287
R.P.C. Morgan

PART V: SUMMARY AND OUTLOOK

28 Soil Erosion and Sediment Redistribution in River Catchments: Summary, Outlook and Future Requirements 297
P.N. Owens and A.J. Collins

Index 319

Contributors

Amore, E., *Department of Civil and Environmental Engineering, University of Catania, Via le A. Doria, 6-95125 Catania, Italy.*

Banasik, K., *Department of Hydraulic Engineering and Environmental Recultivation, Warsaw Agricultural University SGGW, 02-787 Warszawa, ul. Nowoursynowska 166, Warsaw, Poland.*

Belyaev, V.R., *Faculty of Geography, Moscow State University, GSP-2, Moscow 119992, Russia.*

Bidin, K., *School of Science and Technology, Universiti Malaysia Sabah, Sepanggar Bay, 88502 Kota Kinabalu, Malaysia.*

Blake, W.H., *School of Geography, University of Plymouth, Drake Circus, Plymouth PL4 8AA, UK.*

Carrigan, E., *School of Geography and Geology, McMaster University, Hamilton, Ontario L8S 4L8, Canada.*

Chafer, C.J., *Sydney Catchment Authority, Cordeaux Office, PO Box 323, Penrith, NSW 2751, Australia.*

Chappell, N.A., *Lancaster Environment Centre, Lancaster University, Lancaster LA1 4YW, UK.*

Clarke, M.A., *National Soil Resources Institute, Cranfield University, Ty Cambria, 29 Newport Road, Cardiff CF24 0TP, UK.*

Collins, A.J., *National Soil Resources Institute, Cranfield University, Silsoe, Bedfordshire MK45 4DT, UK.* Present address: *Landcare Research, Private Bag 11052, Palmerston North, New Zealand.*

Curran, K.J., *Dalhousie University, Halifax, Nova Scotia B3H 4R2, Canada.*

Doerr, S.H., *Department of Geography, University of Wales Swansea, Singleton Park, Swansea SA2 8PP, UK.*

Douglas, I., *Department of Geography, Manchester University, Oxford Road, Manchester M13 9PL, UK.*

Droppo, I.G., *National Water Research Institute, Environment Canada, PO Box 5050, Burlington, Ontario L7R 4A6, Canada.*

Elliot, W.J., *USDA Forest Service Rocky Mountain Research Station, 1221 South Main, Moscow, ID 83843, USA.*

Evans, D.J., *Agriculture and Environmental Sciences Department, Queen's University Belfast, Newforge Lane, Belfast BT9 5PX, UK.* Present address: *Environmental Consultant, Atkins Limited, Chadwick House, Birchwood Science Park, Warrington WA3 6AE, UK.*

Evans, R., *Department of Geography, Anglia Ruskin University, East Road, Cambridge CB1 1PT, UK.*

Farguell, J., *Departament de Geografia Física, Facultat de Geografia i Història, Universitat de Barcelona, Baldiri Reixac s/n, 08028 Barcelona, Spain.*

Foster, I.D.L., *Centre for Environmental Research and Consultancy, School of Science and the Environment (Geography and Environmental Science), Coventry University, Priory Street, Coventry CV1 5FB, UK and Geography Department, Rhodes University, Grahamstown 6140, South Africa.*

Gibson, C.E., *Aquatic Science Section, Department of Agriculture and Rural Development, New Forge Lane, Belfast BT9 5PX, UK.*

Golosov, V.N., *Faculty of Geography, Moscow State University, GSP-2, Moscow 119992, Russia.*

Hanapi, J., *Danum Valley Field Centre, PO Box 60282, 91112 Lahad Datu, Sabah, Malaysia.*

Hejduk, A., *Department of Hydraulic Engineering and Environmental Recultivation, Warsaw Agricultural University SGGW, 02-787 Warszawa, ul. Nowoursynowska 166, Warsaw, Poland.*

Hejduk, L., *Department of Hydraulic Engineering and Environmental Recultivation, Warsaw Agricultural University SGGW, 02-787 Warszawa, ul. Nowoursynowska 166, Warsaw, Poland.*

Hill, R.D., *Department of Ecology and Biodiversity, University of Hong Kong, Pokfulam Road, Hong Kong, PR China.*

Humphreys, G.S., *Department of Physical Geography, Macquarie University, North Ryde, NSW 2109, Australia.*

Irvine, K.N., *Department of Geography and Planning, Buffalo State, State University of New York, Buffalo, NY 14222, USA.*

Jarritt, N.P., *Department of Geography, School of Human and Environmental Sciences, University of Reading, Reading, Berkshire RG6 6AB, UK.* Present address: *Entec UK Ltd, 17 Angel Gate, City Road, London EC1V 2SH, UK.*

Jaskot, C., *National Water Research Institute, Environment Canada, PO Box 5050, Burlington, Ontario L7R 4A6, Canada.*

Jetten, V., *Department of Physical Geography, Faculty of Geosciences, Utrecht University, PO Box 80115, 3508 TC Utrecht, The Netherlands.*

Kinnell, P.I.A., *School of Resource, Environmental and Heritage Sciences, University of Canberra, Canberra, ACT 2601, Australia.*

Kuhn, N.J., *Department of Geography, University of Exeter, Amory Building, Rennes Drive, Exeter, Devon EX4 4RJ, UK.*

Larenus, J., *Danum Valley Field Centre, PO Box 60282, 91112 Lahad Datu, Sabah, Malaysia.*

Lawrence, D.S.L., *Department of Geography, School of Human and Environmental Sciences, University of Reading, Reading, Berkshire RG6 6AB, UK.*

Licciardello, F., *Department of Agricultural Engineering, University of Catania, Via Santa Sofia, 100-95123 Catania, Italy.*

Mayo, S., *School of Geography and Geology, McMaster University, Hamilton, Ontario L8S 4L8, Canada.*

McHugh, M., *National Soil Resources Institute, Cranfield University, North Wyke Research Station, Okehampton, Devon EX20 2SB, UK.*

Morgan, R.P.C., *National Soil Resources Institute, Cranfield University, Silsoe, Bedfordshire MK45 4DT, UK.*

Murray, A.S., *Nordic Laboratory for Luminescence Dating, Department of Earth Sciences, Aarhus University, Riso National Laboratory, DK-4000 Roskilde, Denmark.*

Nachtergaele, J., *Physical and Regional Geography Research Group, Department of Geography-Geology, Katholieke Universiteit Leuven, Redingenstraat 16, B-3000 Leuven, Belgium.*

Nearing, M.A., *USDA-ARS Southwest Watershed Research Center, 2000 E Allen Road, Tucson, AZ 85719, USA.*

Nikora, V., *National Institute of Water and Atmospheric Research, PO Box 8602, Christchurch, New Zealand.*

Nunny, R.S., *Ambios Ltd, The Ferns, Kilkenny Avenue, Taunton, Somerset TA2 7PJ, UK.*

Owens, P.N., *National Soil Resources Institute, Cranfield University, North Wyke Research Station, Okehampton, Devon EX20 2SB, UK.*

Peart, M.R., *Department of Geography, University of Hong Kong, Pokfulam Road, Hong Kong, PR China.*

Petticrew, E.L., *Geography Program, University of Northern British Columbia, Prince George, British Columbia, Canada V2N 4Z9.* Present address: *School of Geography, University of Plymouth, Drake Circus, Plymouth, Devon PL4 8AA, UK.*

Poesen, J., *Physical and Regional Geography Research Group, Katholieke Universiteit Leuven Geo-Institute, Celestijnenlaan 200E, B-3001 Heverlee, Belgium.*

Ramli, N., *School of Science and Technology, Universiti Malaysia Sabah, Sepanggar Bay, 88502 Kota Kinabalu, Malaysia.*

Rickson, R.J., *National Soil Resources Institute, Cranfield University, Silsoe, Bedfordshire MK45 4DT, UK.*

Ruse, M.E., *Department of Geography, University of Hong Kong, Pokfulam Road, Hong Kong, PR China.*

Sala, M., *Departament de Geografia Física, Facultat de Geografia i Història, Universitat de Barcelona, Baldiri Reixac s/n, 08028 Barcelona, Spain.*

Sayer, A.M., *Department of Geography, University of Wales Swansea, Singleton Park, Swansea SA2 8PP, UK.*

Shakesby, R.A., *Department of Geography, University of Wales Swansea, Singleton Park, Swansea SA2 8PP, UK.*

Sidorchuk, A.Yu., *Faculty of Geography, Moscow State University, GSP-2, Moscow 119992, Russia.* Present address: *Landcare Research, Private Bag 11052, Palmerston North, New Zealand.*

Sinun, W., *Innoprise Corporation Sdn. Bhd., 9th Floor, Wisma Innoprise, Teluk Likas, PO Box 11622, 88817 Kota Kinabula, Sabah, Malaysia.*

Smith, A., *National Institute of Water and Atmospheric Research, PO Box 8602, Christchurch, New Zealand.*

Stone, P., *PM Atkins Water, 280 Aztec West, Park Avenue, Almondsbury, Bristol BS32 4 SY, UK.*

Trapp, B., *National Water Research Institute, Environment Canada, PO Box 5050, Burlington, Ontario L7R 4A6, Canada.*

Van de Vlag, D., *Department of Earth Observation Systems, International Institute for Geoinformation Science and Earth Observation, PO Box 6, 7500 AA Enschede, The Netherlands.*

Visser, F., *Physical and Regional Geography Research Group, Department of Geography-Geology, Katholieke Universiteit Leuven, Redingenstraat 16, B-3000 Leuven, Belgium.*

Wallbrink, P.J., *CSIRO Land and Water, Christian Building, Clunies Ross Street, Acton, ACT 2600, Australia.*

Walling, D.E., *Department of Geography, University of Exeter, Amory Building, Rennes Drive, Exeter, Devon EX4 4RJ, UK.*

Walsh, R.P.D., *Department of Geography, University of Wales Swansea, Singleton Park, Swansea SA2 8PP, UK.*

Williamson, A., *Environment Agency, King's Meadow House, Reading, Berkshire RG1 8DG, UK.*

Wood, G.A., *National Soil Resources Institute, Cranfield University, Silsoe, Bedfordshire MK45 4DT, UK.*

Zimbone, S.M., *Department of Agro-forestry and Environmental Sciences and Technologies, Mediterranean University of Reggio Calabria, Piazza S. Francesco 4-89061, Reggio Calabria, Italy.*

Preface

While research on soil erosion and on sediment redistribution in rivers both have long histories, each tends to be conducted in relative isolation and by specialist teams. Recent years have, however, seen a move towards more integrated management of soil–water systems, particularly at the scale of the river basin, which in turn has led to a greater need for more integrated research so as to inform management and assist with the decision-making process. The link between soil erosion and redistribution on land and sediment transport and deposition within rivers and lakes is clear to many, but a culture persists whereby the two groups of scientists rarely interact. This book is in part an attempt to get experts in soil erosion and experts in sediment transport and deposition to cooperate, so that it may be possible to understand the movement of soil and sediment particles from source to sink. The book stems from a conference held at the National Soil Resources Institute (NSRI) at the Silsoe campus of Cranfield University, UK, between 9th and 11th September 2003. The conference was attended by over 80 delegates from more than 15 countries. The chapters in this book represent a selection of oral and poster presentations given during the conference, in addition to an invited contribution. All chapters were peer-reviewed.

We are grateful to a number of individuals who provided help with the conference and book. We would like to thank Prof. Mark Kibblewhite, the Director of NSRI, for supporting us (both personally and financially) in these activities, and several members of staff at NSRI who have helped along the way, including Michelle Clarke, Julia Duzant, Inga Wells and the admin group at Silsoe. We would also like to acknowledge the support and assistance of CABI, in particular Tim Hardwick and Rebecca Stubbs, and also Alison Foskett for her copy-editing work. Finally, we would like to thank the following referees:

Ramon Batalla
Vladimir Belyaev
Will Blake
John Boardman
Michelle Clarke
Adrian Collins
Lynda Deeks
Julia Duzant
William Elliot
Bob Evans
Dan Evans
Joaquim Farguell
Peter Fiener
Ian Foster
Wojciech Froehlich
Carlos Garcia
Mathew Heywood
David Higgitt
Anton Imeson
Victor Jetten
Paula Jones
Klaus Kuhn
Feliciana Licciardello
Lee MacDonald
Kit McCleod
Marianne McHugh
Roy Morgan
Mark Nearing
Jon Olley
Jean Poesen
Paolo Porto
Ellen Petticrew
John Quinton
Jane Rickson
Jerry Ritchie
Mark Russell
David Sear
Mike Stone
Ross Sutherland

Fred Swanson
Kevin Taylor
Marcel Van Der Perk
Kristof Van Oost
Gert Verstraeten
Fleur Visser
John Walden
Des Walling
Rory Walsh
Sue White
Gavin Wood

The success of this book and associated conference will be measured by the level of research that they inspire and, ideally, by an increased level of integration between those working in soil erosion and in sediment redistribution within rivers. We also hope that this book will encourage further collaboration between those studying measurement and modelling and those concerned with the implications of these for management. We must not forget that as scientists and researchers our ultimate responsibility is to provide the information and knowledge base needed for informed management of soil–sediment–water systems.

Phil Owens, North Wyke, UK
Alison Collins, Palmerston North, New Zealand

1

Introduction

Soil erosion, mass movements and sediment deposition on grazing land after heavy rain in February 2004, Manawatu, New Zealand (photo: Landcare Research, New Zealand).

1 Introduction to Soil Erosion and Sediment Redistribution in River Catchments: Measurement, Modelling and Management in the 21st Century

A.J. Collins[1] and P.N. Owens[2]
[1]*National Soil Resources Institute, UK.* Present address *Landcare Research, New Zealand;* [2]*National Soil Resources Institute, UK*

The Importance of Soil Erosion and Sediment Redistribution

There is increasing awareness of the need to protect our natural resources in order to meet present and future requirements. Since economies and environments are dependent on healthy soil and water, it is essential to ensure the sustainable use of the resource base in the face of growing demand.

Excessive or enhanced soil erosion due to poor land management can result in both on- and off-site impacts that are detrimental to a whole range of receptors. Where soil erosion occurs, the soil resource can be severely depleted if the rate of erosion exceeds the rate of natural soil formation. This loss often corresponds to the most agriculturally important topsoil and any fertilizer or pesticide application, causing subsequent reductions in agricultural productivity. Soil erosion is a hazard traditionally associated with agriculture, and often occurs in tropical and semiarid areas (Morgan, 1986). The topic of soil erosion was never so emotive as during the 1930s 'Dust Bowl' in the USA. Whilst soil erosion is still a significant issue in tropical and semiarid areas, it is being increasingly recognized as a hazard in temperate countries. In the UK, for example, the annual present-day value of lost production due to soil erosion is estimated to be £700 million (of the order of €1000 million) (Evans, 1996). It is estimated that erosion affects 40% of arable land, with these soils losing more than 25% of their agricultural productivity (Evans, 1996). The redistribution of eroded soil material within the field, and accompanying changes in soil structure, can also result in habitat damage, reduced crop yields and changes in flood storage capacity.

Significantly, recent interest in soil erosion has been triggered by a growing awareness of the off-site impacts. These impacts are predominantly associated with the movement of eroded soil and sediment particles, and changes in water flows (both through and across the soil). The off-site problems are often more evident, and include the loading and sedimentation of watercourses and reservoirs, and increases in stream turbidity, all of which can disturb aquatic ecosystems and upset the geomorphological functioning of river systems (Owens *et al.*, 2005). In China, soil erosion and the off-site effects of sediment deposition have resulted in multiple impacts. For example, since the 1950s, over 90,000 reservoirs have been built in

China, with a cumulative storage capacity of over 400×10^9 m^3. However, increased soil erosion has caused annual sedimentation levels to rise to as much as 10×10^6 m^3, reducing storage capacity by approximately 10%. Sedimentation resulting from soil erosion has adversely affected hydroelectric output, availability of irrigation water, flood control potential and the navigation of waterways (from 172,000 km in the 1960s to 108,000 km today). Furthermore, increased sedimentation along the lower reaches of the Yellow River has caused a 0.10 m annual rise of the river bed, with the result that the Yellow River is in many places 'suspended' above the plains over which it flows, precariously controlled by dikes constructed along its course (Yuqian and Ning, 1986).

If eroded soil carries with it nutrients, contaminants and pathogens, it can present serious problems in terms of watercourse and groundwater pollution and eutrophication, threatening habitat and human health (Owens *et al.*, 2005). The cost of eutrophication (including the loss of drinking water supplies, reduction in the value of waterside property, decline in the recreational and amenity value of water-bodies, and impacts on water for industrial uses and commercial fisheries), mainly due to phosphate enrichment, is estimated at £128.5 million per annum in the UK alone (Pretty *et al.*, 2000).

Since the 1990s, environmental management legislation in both the USA and Europe has switched primarily to the control of off-site impacts, as mentioned above. For example, the European Union has implemented several directives (including the Water Framework Directive, Habitats Directive, Fisheries Directive, and Bathing Water Directive) that aim to prevent off-site impacts on economics, environment and people caused by waste, water and soil (see Owens and Collins, Chapter 28, this volume). That said, emerging EU policy under development within the Thematic Strategy for Soil Protection is likely to be more soil-centric. Should this succeed, it is likely to recognize not only the impact of soil erosion and sediment transfer on other receptors (Owens, 2004) but also draw attention to the intrinsic value of the soil as a resource and the need to prevent soil erosion and sediment transfer for soil protection (Van-Camp *et al.*, 2004).

The Significance of the River Basin Scale

Since soil erosion and sediment redistribution have implications for both soil and water resources, and scientists have established that the movements of soil, sediment and water are intrinsically linked, it is critical to implement integrated resource protection strategies. Soil erosion and sediment redistribution (transport, storage and remobilization) are controlled by hydrological and geomorphological processes, which operate within the context of a river basin. The river basin therefore represents a convenient and meaningful unit for the management of soil erosion and sediment redistribution, since the shape and characteristics of the river basin control the pathways and fluxes of soil, water and sediment (Owens *et al.*, 2004). It is, therefore, encouraging that policy-makers and managers are now opting to manage soil erosion and sediment transfers at a catchment or river basin scale, as has been proposed in the EU Water Framework Directive, for example. This issue is discussed further below and in Owens and Collins (Chapter 28, this volume).

The Three Themes of this Book: Measurement, Modelling and Management

In order to effectively protect and manage natural resources there is a need to develop the science and to assemble the necessary information on which to base decision-making. There are several *measurement* and *modelling* tools available to scientists and managers for use in *management* of soil erosion and sediment redistribution in river basins. Here, these terms are defined as:

- *Measurement tools*: tools for identifying and quantifying the magnitude, rate, severity and timescale of erosion and sediment sources, pathways, sinks and impacts.
- *Modelling tools*: tools, including physically based, conceptual, statistical and stochastic models, to predict or understand spatial patterns and trends in soil erosion and sediment transfers. Models require sufficient data, both spatial and temporal, for

development and validation, and hence there is a feedback loop between measurement and modelling, as shown in Fig. 1.1 below.

- *Management*: an all-embracing term, referring to identifying the problem, quantifying its importance, planning and implementing a strategy to control or mitigate it, and evaluating the effectiveness of the solution. Management strategies are informed by the data collected, or trends predicted within measurement and modelling. Also, once a management strategy has been implemented, further measurement or modelling is required in order to evaluate or appraise success, and where necessary plan modifications.

Management can relate to any stage in the soil erosion–sediment redistribution continuum and may include the following generic range of management options, all of which have the potential to manipulate either soil erosion or sediment transport within the landscape:

- *Source control*: includes soil conservation techniques, reducing runoff risk or protecting against erosion.
- *Control of sediment pathways and delivery to receptors*: includes interception and retention methods.
- *Remediation at the sink*: includes removing sediment through filtering and dredging, or immobilizing and cleaning polluted material.

Whilst soil erosion and sediment redistribution are influenced by the characteristics of the river basin, and should be managed at such a scale (Owens *et al.*, 2004; Owens, 2005), generation and delivery processes vary spatially with different kinds of processes dominating at various locations (Verstraeten *et al.*, 2002). Often a different soil conservation or sediment control measure is required to combat each specific type of soil erosion or sediment transport process, and therefore the 'broad-brush' application of a conservation technique across an entire catchment may not be feasible. An effective conservation strategy should therefore integrate a variety of suitably located control techniques into a catchment or river basin management plan (Verstraeten *et al.*, 2002).

In all three themes, measurement, modelling and management, the appropriate mantra is *integration*: between various soil, sediment

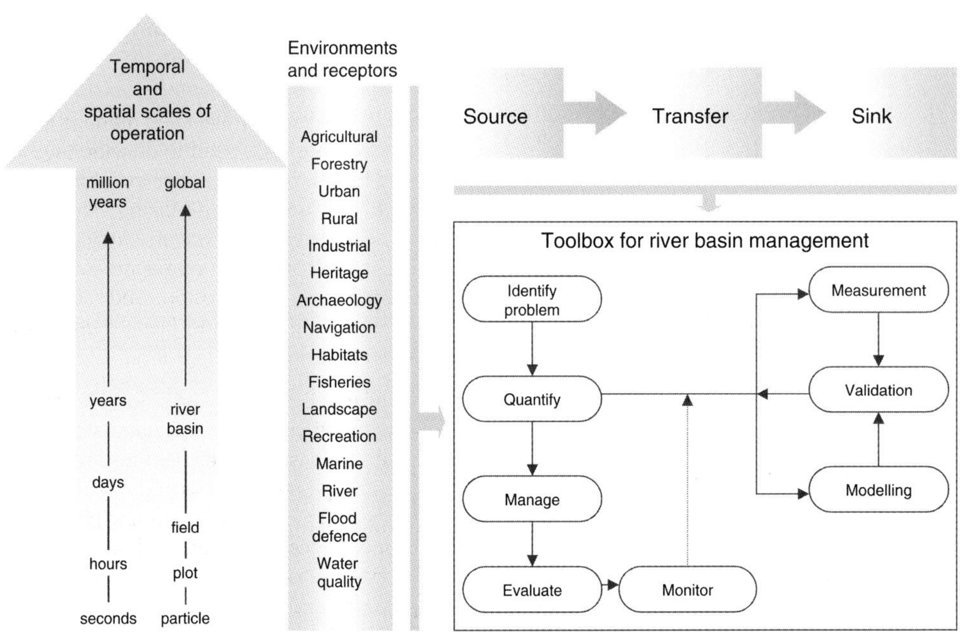

Fig. 1.1. Overview of the river basin management of soil erosion and sediment redistribution.

and water disciplines to allow holistic land–river management, and between scientists and managers to ensure that the science is producing tools as well as management strategies that are accessible and practical to implement.

Overview of the Following Chapters

Many of the chapters that follow reiterate the key themes presented in this chapter – the need for integration in our science and management of natural resources. Where possible, the chapters have been chosen to represent the variety of temporal and spatial scales of operation, from the modelling of dynamic event-based processes to the annual monitoring of a site to determine longer term trends. The chapters also explore some of the many environments in which erosion and sediment redistribution occur, and indeed the unique controls that these environments may exert on geomorphological functioning. These environments may also qualify as the main receptors affected by erosion and/or sediment redistribution. Figure 1.1 demonstrates these varying issues as well as the toolbox available for river basin scale management of both erosion and sediment redistribution, and how the chapters that follow may contribute to the scientific rationale and basis for that management.

Measurement

Chapters in Part II of the book use a variety of contrasting techniques (both traditional and novel) to measure soil erosion on land and sediment fluxes in rivers, including deposition within lakes and reservoirs. They have been broadly arranged so as to illustrate the differences and similarities in the techniques used to measure soil–sediment transfers from land to rivers to lakes.

Walling (Chapter 2) presents examples of methods for linking erosion with sediment delivery in a review of traditional measurement techniques in the research area and the new challenges and opportunities we can look to within the science. This chapter also traces the changes in policy that have directed concerns from on-site problems associated with soil erosion to off-site problems resulting from sedimentation. This, he argues, has resulted in a shift from traditional soil erosion monitoring methods to more sophisticated sediment tracing techniques. The comparison of traditional and recent (i.e. tracing) soil erosion assessment techniques is continued by Peart *et al.* (Chapter 3), who explore the usefulness of caesium-137 (^{137}Cs) and erosion plot data from a hillslope region in Hong Kong. Whilst the two techniques were in broad agreement, there were problems associated with calibrating the ^{137}Cs measurements. Similarly, Belyaev *et al.* (Chapter 4) describe some of the various measurement techniques available to assess the contribution of sheet, rill and ephemeral gully erosion, as well as tillage translocation, for two arable catchments in Southern Russia, and what factors may be causing observed local differences.

The implications of land use and climate change on soil erosion and sediment redistribution are considered in several chapters. Shakesby *et al.* (Chapter 5) present hillslope erosion response following forest fires in Australia, and explore the link between fire-induced soil water repellency and erodibility. The paper is an interesting insight into the effects of acute and catastrophic triggers on soil erosion, and has important implications for environments where climate change may result in an increase in such events. Blake *et al.* (Chapter 6) continue the theme of soil water repellency, attempting to link soil magnetic signatures to water repellency in order to quantify soil and sediment export from catchments of varying wildfire burn severity in Australia. Evans (Chapter 7) examines the link between land use and sediment delivery. He stresses the importance of understanding and quantifying the respective contributions of various sources in order to effectively manage sediment redistribution. In Chapter 8, Farguell and Sala explore the impact of severe rainfall events on resulting suspended sediment loads for a semiarid catchment in the Iberian Peninsula. The same theme is applied to a study area with a contrasting climate in the chapter by Hejduk *et al.* (Chapter 9). These authors present data on discharge and suspended sediment transport in a small lowland catchment in central Poland during rainfall and snowmelt events.

The final set of chapters in Part II present measurement techniques and data that link soil erosion and sediment redistribution with water quality and aquatic ecosystems. The survival of salmonid populations in County Antrim, Northern Ireland, was the driver for a study on sediment transport dynamics by Evans and Gibson (Chapter 10). The work illustrates techniques available for elucidating the link between soil erosion and downstream sediment delivery, and for informing sediment management. In contrast, Petticrew (Chapter 11) describes how salmonids influence both the amount and composition (including organic matter content and type, and aggregate structure) of fine-grained sediment stored in gravel bed streams in British Columbia, Canada. In the final chapter of the measurement theme (Chapter 12), Foster presents lakes and reservoirs as the ultimate components of the sediment transfer story. The UK case studies presented provide a history of sediment and associated phosphorus concentrations in reservoir catchments and the impact of land management, including land drainage, on sediment sources.

Modelling

The chapters in Part III explore some of the common advantages and problems associated with modelling approaches, and new developments within the science. The chapters recommend refinements and methods to improve the availability and reliability of data, parameterization of key factors and approaches for modelling components previously neglected in soil erosion and sediment research. As with the chapters in Part II, the chapters follow the progression from soil erosion to sediment redistribution.

Nearing (Chapter 13) examines the critical link between measurement and modelling, by focusing on the importance of having data that can be fed into modelling and prediction tools to support management decisions and inform policy. He argues this is best achieved by collecting and managing data effectively and consistently to avoid temporal and spatial variability. Kuhn (Chapter 14) examines methods for the assessment of erodibility for inclusion in dynamic event-based models. In this study he assesses the suitability of techniques based on soils from Canada and Mexico, and how appropriately characterizing erosion processes can improve dynamic process-based prediction. Sidorchuk *et al.* (Chapter 15) argue for a third generation of erosion models to account for the stochastic nature of soil erosion, and outline a method called 'double averaging'. Kinnell (Chapter 16) also focuses on the refinement of models to improve prediction. He presents a modified version of the Universal Soil Loss Equation (USLE) to include runoff as a factor in accounting for event erosivity.

Several of the chapters discuss the development and application of models that consider gully erosion and landslides in addition to rill and interrill erosion. Recent research quantifying erosion rates due to roads, fires, landslides and harvesting is presented by Elliot (Chapter 17). These data are used to explore the performance of GeoWEPP (Water Erosion Prediction Project) modelling to understand the source, production and attenuation of sediment from upland forested catchments. Elliot also makes suggestions to improve model performance, especially in sediment and flood routing components. The WEPP model is also applied by Licciardello *et al.* (Chapter 18) to a small Sicilian watershed in order to assess its performance compared with other physically based erosion models in Mediterranean areas. The authors also suggest possible improvements in the WEPP (and GeoWEPP) model. Jetten *et al.* (Chapter 19) discuss the lack of explicit modelling for gully incision and formation. The authors propose combining landscape indicators with process modelling in order to improve the simulation of ephemeral gully incision, presenting a method requiring little additional data above basic erosion modelling.

The final chapter in this section, by Jarritt and Lawrence (Chapter 20), investigates the application of a new model, INCA-Sed, to simulate sediment delivery processes at the catchment scale. The model is applied to catchments in southern England to demonstrate the effectiveness of the approach in reproducing supply- and transport-limited conditions.

Management

In Part IV the chapters look at the effect of various land management practices on the generation of erosion and sediment and the implications for receptors in forested, agricultural and urban areas in temperate and tropical environments. In attempting to control these effects, the case studies highlight the dependency of management strategies on the outputs of measurement and modelling tools. The scale and accuracy of the data on which management decisions are based, and the need for correct characterization of processes and parameters, are identified as being of critical importance.

Wood *et al.* (Chapter 21) open the management part by demonstrating the link between the three themes of this book. They illustrate how the data from measurement and monitoring studies can be used to model and then map at larger spatial scales the delivery of eroded sediment from land to water, and how this could be used as an effective management tool. Spatial scale issues are considered further by Rickson (Chapter 22), who presents an interesting debate on the importance of scale when assessing the effectiveness of erosion and sediment control practices. Her work illustrates how the performance of control practices can vary depending on the scale of the data and, reiterating the sentiments of Nearing (Chapter 13), suggests that accurate and reliable data are critical for effective implementation and policy-making.

Collectively, the subsequent four chapters (Chapters 23–26) underline how detailed and appropriate scientific information on soil erosion and sediment redistribution can be used for effective and targeted management in contrasting river catchments. Walsh *et al.* (Chapter 23) examine changes in the spatial distribution of erosion within a catchment in the rainforests of Borneo. The changes over a 10-year monitoring period are shown to be attributable to the management of the catchment and the effect of practices such as selective logging. Continuing with the tropical environment, albeit in a lowland setting, Visser (Chapter 24) offers an insight into practical erosion management on sugarcane plantations on tropical floodplains. Her data have implications for controlling sediment delivery to rivers, and particularly illustrate the importance of the connectivity between land and rivers. Nunny *et al.* (Chapter 25) examine the impact of land clearance to create plantations on sediment delivery to the barrier reef in Belize, Central America. They argue that whilst the land management practices appear to be loading rivers with sediment, turbidity at the reef is decoupled from such effects by wave and current action.

Most of the chapters in this book focus on natural and agricultural environments, but it is important to also address sediment redistribution in urban environments, especially given that the proportion of people living in urban areas, and therefore the amount of land that is urbanized, is expected to increase dramatically this century. Droppo *et al.* (Chapter 26) present methods for determining the distribution, structure and behaviour of urban sediments in watersheds within Ontario, Canada. The authors demonstrate how this information can be used to improve water management strategies in the urban environment.

The final words in Part IV go to Morgan (Chapter 27), who provides a personal perspective on current practices and future visions for managing sediment in the landscape. He highlights the importance of integrated and holistic approaches to land and river management at the catchment or river basin scale. Morgan suggests that as more countries establish legislation for soil protection, the true measure of success will be the ability to connect a variety of disciplines and involve all stakeholders, including the local community.

Conclusion

Soil and water resource protection are clearly crucial for productive and sustainable economies and environments. Both soil and water resources can be threatened by processes of soil erosion and sediment redistribution, and the chapters in this volume illustrate some of the main forces driving research in this area, such as concerns with aquatic ecosystems, tropical forests and urban systems.

It is evident from the work presented here that great strides are being made within the soil erosion and sediment redistribution research area. Both science and policy communities are raising soil and water protection up the political agenda to encourage future research in this key topic (see Owens and Collins, Chapter 28). However, it is critical that before we embark on further research we learn from the conclusions and recommendations of existing research, attempt to plug the gaps and meet the requirements of end-user groups. These lessons include, first, the recognition that measurement and modelling are the basic tools necessary to inform and allow the implementation and evaluation of management strategies and practices. Secondly, that the data derived or generated from these tools must be at the appropriate scale and level of complexity for end-users. Thirdly, that the management of soil erosion and sediment redistribution is integrated and focused at the river basin scale. Several of the chapters that follow outline some of the tools and approaches available to do this.

Finally, without doubt the true value of our science will only be realized through communication of the key messages to stakeholders, including catchment managers, policy-makers and the local community, and this is discussed further by Morgan (Chapter 27) and Owens and Collins (Chapter 28). Our task as a scientific community must be, therefore, not only to continue to build the science and the tools as described in this book, but to develop effective language and dissemination skills to communicate them.

References

Evans, R. (1996) *Soil Erosion and its Impacts in England and Wales.* Friends of the Earth Trust, London.

Morgan, R.P.C. (1986) *Soil Erosion and Conservation.* Longman, London.

Owens, P.N. (2004) The link between soil erosion and the diffuse contamination of water and air. In: Van-Camp, L., Bujarrabal, B., Gentile, A.R., Jones, R.J.A., Montanarella, L., Olazabal, C. and Selvaradjou, S.K. (eds) *Reports of the Technical Working Groups Established Under the Thematic Strategy for Soil Protection*, EUR 21319 EN/1. Office for the Official Publications of the European Commission, Luxembourg, pp. 263–266.

Owens, P.N. (2005) Conceptual models and budgets for sediment management at the river basin scale. *Journal of Soils and Sediments* 5, 201–212.

Owens, P.N., Apitz, S., Batalla, R., Collins, A.J., Eisma, M., Glindemann, H., Hoonstra, S., Köthe, H., Quinton, J., Taylor, K., Westrich, B., White, S. and Wilkinson, H. (2004) Sediment management at the river basin scale: synthesis of SedNet Working Group 2 outcomes. *Journal of Soils and Sediments* 4, 219–222.

Owens, P.N., Batalla, R.J., Collins, A.J., Gomez, B., Hicks, D.M., Horowitz, A.J., Kondolf, G.M., Marden, M., Page, M.J., Peacock, D.H., Petticrew, E.L., Salomons, W. and Trustrum, N.A. (2005) Fine-grained sediment in river systems: environmental significance and management issues. *River Research and Applications* 21, 693–717.

Pretty, J.N., Brett, C., Gee, D., Hine, R., Mason, C.F., Morison, J.I.L., Raven, H., Rayment, M. and van der Bijl, G. (2000) An assessment of the total external costs of UK agriculture. *Agricultural Systems* 65, 113–136.

Van-Camp, L., Bujarrabal, B., Gentile, A.-R., Jones, R.J.A., Montanarella, L., Olazabal, C. and Selvaradjou, S.-K. (eds) (2004) *Reports of the Technical Working Groups Established Under the Thematic Strategy for Soil Protection*, EUR 21319 EN/1. Office for the Official Publications of the European Commission, Luxembourg.

Verstraeten, G., Van Oost, K., Van Rompaey, A., Poesen, J. and Govers, G. (2002) Evaluating an integrated approach to catchment management to reduce soil loss and sediment pollution through modelling. *Soil Use and Management* 19, 386–394.

Yuqian, L. and Ning, C. (1986) Erosion and transportation of sediment in the Yellow River basin. *International Journal of Sediment Research* 1, 1–38.

II

Measurement

Collecting soil samples for caesium-137 analysis, Chinamora, Zimbabwe (photo: P.N. Owens).

2 Tracing versus Monitoring: New Challenges and Opportunities in Erosion and Sediment Delivery Research

D.E. Walling
University of Exeter, UK

Introduction

Soil erosion and associated increases in sediment mobilization and transport have long been recognized as a major problem in those regions of the world where erosion rates and suspended sediment yields are high. There, soil erosion can result in destruction of agricultural land or a marked reduction in agricultural productivity (Lal, 2001) and these 'on-site' problems are frequently coupled with 'off-site' problems associated with increased sediment mobilization and transport (Sundborg, 1982). Such off-site problems include reductions in reservoir storage resulting from sedimentation and the siltation of canals and water distribution networks, river channels and harbours. These can in turn threaten the sustainable development of water resources and hydropower production, as well as having serious consequences for the maintenance of navigable waterways. More recently, a growing recognition of the important role of fine sediment in the transfer, storage and fate of sediment-associated nutrients and contaminants, including pesticides, heavy metals and persistent organic pollutants (Allan, 1986; Walling *et al.*, 1997; Stone, 2000; Warren *et al.*, 2003) and in the more general degradation of aquatic habitats, including the siltation of salmonid spawning gravels and clogging of aquatic vegetation (Clark *et al.*, 1985; Wood and Armitage, 1997, 1999; Soulsby *et al.*, 2001), has emphasized the wider incidence of sediment problems in many other areas of the world where erosion rates and sediment yields are substantially lower. These problems reflect the wide-ranging environmental and ecological significance of fine sediment and have highlighted the wider need to incorporate effective sediment control strategies into catchment management programmes. In the UK, for example, the transport of fine sediment by rivers was essentially ignored as a potential problem until fairly recently, because reservoir sedimentation was not seen as representing a significant constraint on water resource development. Recent concern for the improvement of river water quality and aquatic habitats prompted by the EC Water Framework and Habitats Directives has, however, now identified fine sediment as a key contributor to diffuse source pollution and the degradation of aquatic habitats and emphasized the need to control sediment mobilization and delivery to water courses, even though rates of soil loss and specific suspended sediment yields are relatively low by world standards.

Against this background, sediment control strategies now represent a vital component of catchment management in many areas of the world, including both those with high sediment yields, which have traditionally been accepted as experiencing sediment problems, and those where specific sediment yields are an order of magnitude or more lower, but fine sediment represents a key source of diffuse source pollution and an important environmental problem. The design and implementation of effective sediment management and control strategies is, however, frequently hampered by a lack of data on erosion rates and sediment yields, as well as limited understanding of the transfer and storage of fine sediment within drainage basin systems. The linkages between sediment mobilization, transport, deposition, storage and sediment yield at the basin outlet can be highly complex, especially in situations where sediment storage equals or exceeds sediment export (Trimble, 1983; Phillips, 1992; Walling, 2000). Paradoxically, the need for data and a better understanding of catchment sediment dynamics is frequently greatest in those areas of the world where erosion rates and sediment yields are lowest. This situation is, in part, a reflection of a lack of past monitoring activity in such areas, but, perhaps more importantly, it also reflects the greater difficulty of identifying sediment sources, pathways and sinks in areas where erosion rates and sediment fluxes are relatively low.

The sediment budget concept affords a valuable framework for assembling the detailed information required to elucidate and characterize the drainage basin sediment delivery system (Golosov *et al.*, 1992; Reid and Dunne, 1996; Walling, 2000; Walling *et al.*, 2001). By identifying and quantifying the dominant sources, the transfer pathways and the sinks, as well as the output of sediment, for a drainage basin, it is possible to identify and quantify the key areas of sediment mobilization and storage and to assess the efficiency of the sediment delivery system and its sensitivity to change. However, catchment monitoring programmes traditionally have placed emphasis on monitoring sediment loads and yields, particularly at the catchment outlet, and the need to establish and elucidate catchment sediment budgets has introduced a requirement for information on other components of the sediment budget, including sediment mobilization, transfer and storage, which can be highly variable spatially and thus considerably more difficult to document. Again, it can be suggested that the problems and uncertainties associated with constructing a meaningful sediment budget for a catchment frequently increase in those areas of the world where erosion rates and sediment yields are relatively low. In such areas, the primary sediment sources within a drainage basin may not be clearly evident, significant soil erosion is frequently restricted to small areas, and sediment deposition and storage may be difficult to identify and document.

Faced with demands for new information to underpin the development of sediment control and management strategies, particularly in those locations where sediment problems have previously attracted little attention, there is a need for new approaches to assembling the data required for establishing catchment sediment budgets. The potential for using environmental radionuclides as sediment tracers, both as an alternative to traditional monitoring techniques and to complement these existing approaches, has been increasingly recognized and exploited. As a result, it is arguably possible to identify a general shift in emphasis away from the traditional reliance on *monitoring* towards a growing emphasis on *tracing*. Key advantages of the use of environmental radionuclides as sediment tracers include:

1. The potential for assembling retrospective (medium-term) information on the basis of a limited programme of contemporary measurements, thereby avoiding the need for expensive long-term monitoring.
2. The possibility of using essentially the same measurements within different components of the sediment budget and thus tracing the movement of sediment through the delivery system.
3. The provision of spatially distributed point estimates of sediment mobilization and deposition that are directly compatible with the current generation of spatially distributed numerical models.
4. The ability to apply the approach at a range of spatial scales.

This contribution aims to demonstrate further the potential for using environmental radionuclides as tracers in catchment sediment budget investigations, in order to assemble the data required to underpin the design and implementation of sediment control and management strategies.

Environmental Radionuclides

The term *environmental radionuclide* is frequently used to refer to those radionuclides which are commonly occurring and widely distributed in the environment or landscape and, whilst found at relatively low levels, are readily measurable. In most cases they are of natural ori gin. For applications relating to sediment tracing, most work to date has focused on a particular group of environmental radionuclides, namely, *fallout radionuclides*, or radionuclides that reach the land surface as fallout from the atmosphere. In this case, the fallout input can generally be assumed to be spatially uniform, at least over a relatively small area. Because the radionuclides used are rapidly and strongly adsorbed by the soil on reaching the catchment surface as fallout, they accumulate at or near the surface and afford a means of tracing sediment mobilization and deposition by documenting the post fallout redistribution of the radionuclide tracer, which will directly reflect the mobilization, transport and deposition of soil and sediment particles. In essence, therefore, it is possible to view the fallout as being analogous to the artificial application of a sediment tracer to the land surface of a study area. Observation of the subsequent redistribution of the radionuclide provides a basis for establishing rates and patterns of sediment transfer and establishing the magnitude and relative importance of sediment storage within the landscape.

The radionuclide that has been most widely used as a sediment tracer is caesium-137 (^{137}Cs) (Ritchie and McHenry, 1990; Zapata, 2002). Caesium-137 is a man-made radionuclide, with a half-life of 30.2 years, which was released into the stratosphere by the atmospheric testing of thermonuclear weapons during the period extending from the mid 1950s to the 1960s. Fallout of ^{137}Cs began in 1954, peaked in the early 1960s and subsequently decreased, reaching near zero levels in the mid 1980s. Fallout levels were globally variable, reflecting both annual precipitation amount and location relative to the main weapons tests (Walling, 2002). Smaller amounts of ^{137}Cs have also been released into the atmosphere by accidents at nuclear power stations, notably the Chernobyl disaster in 1986, which resulted in additional inputs of ^{137}Cs fallout over large areas of Europe and adjacent regions.

Use of other fallout radionuclides as sediment tracers has primarily focused on unsupported or excess lead-210 (^{210}Pb) and beryllium-7 (^{7}Be). These two radionuclides differ from ^{137}Cs in two important respects. First, they are both of natural origin and, secondly, their fallout input can be treated as essentially constant over time. Lead-210 is a naturally occurring product of the ^{238}U decay series, with a half-life of 22.2 years, that is derived from the decay of gaseous ^{222}Rn, the daughter of ^{226}Ra. Radium-226 exists naturally in soils and rocks and the ^{210}Pb in soils generated *in situ* by the decay of ^{226}Ra is termed *supported* ^{210}Pb and is in equilibrium with ^{226}Ra. However, upward diffusion of a small portion of the ^{222}Rn produced in the soil and rock introduces ^{210}Pb into the atmosphere and its subsequent fallout provides an input of this radionuclide to surface soils and sediments that will not be in equilibrium with its parent ^{226}Ra. Fallout ^{210}Pb is commonly termed *unsupported* or *excess* ^{210}Pb, when incorporated into soils and sediments, to distinguish it from the ^{210}Pb produced *in situ* by the decay of ^{226}Ra. In contrast to ^{137}Cs and ^{210}Pb, ^{7}Be has a very short half-life (53 days). It is produced by the bombardment of the Earth's atmosphere by cosmic rays and is subsequently deposited as fallout.

The different half-lives of the three fallout radionuclides considered above and the different temporal distributions of their fallout mean that their inventories (i.e. the total amount of radionuclide contained within a soil or sediment profile (Bq/m^2)) will exhibit different temporal behaviour. In the case of unsupported ^{210}Pb, the essentially constant fallout means that the inventory of a stable soil, unaffected by erosion or deposition, will also remain essentially constant and in steady state, with loss by decay being balanced by new fallout input. In contrast, the ^{137}Cs inventory of a stable soil would have been zero prior to the onset of fallout in the mid 1950s. It will then have increased through to the late 1960s, in response to the main period of fallout input, and subsequently it will have decreased as the rate of decay exceeded the rate of replenishment by new fallout. However, because of its relatively long half-life (30.2 years), significant amounts of ^{137}Cs will still remain some 40 years after the main period of fallout input. As a result of its short half-life, the ^{7}Be inventory of a stable soil will evidence considerable short-term variability. During periods of dry weather, when fallout is limited, the inventory will rapidly decline due to decay, only to increase again as a result of rainfall and associated fallout.

Caesium-137, ^{7}Be and ^{210}Pb activities can be readily measured by gamma spectrometry (see Wallbrink *et al.*, 2002). By using appropriate high purity germanium (HPGe) detectors, it is possible to measure all three radionuclides simultaneously. The long count times (e.g. 6–24 h) commonly required for accurate measurements must, nevertheless, be seen as a significant limitation, in that the number of samples that can be analysed may be restricted.

Using Fallout Radionuclides to Trace Sediment Mobilization and Delivery

Figure 2.1 illustrates typical distributions of ^{137}Cs, unsupported ^{210}Pb and ^{7}Be in adjacent permanent pasture and cultivated soils at a site near Crediton in Devon, UK. At undisturbed pasture sites, the radionuclides are typically concentrated close to the surface, and concentrations decline exponentially with depth. The minor differences between the vertical distributions of ^{137}Cs and unsupported ^{210}Pb primarily reflect the different temporal patterns of fallout input associated with the two radionuclides over the past few decades. The vertical distribution of ^{7}Be differs significantly from that of the other two radionuclides and this directly reflects its very much shorter half-life. Beryllium-7 is only found at or very near the surface, where it is continually replenished by fallout. If an undisturbed site that has been influenced by neither erosion nor deposition can be identified, measurement of the total inventory of the individual radionuclides at that site (Bq/m^2) can be used to provide an estimate of the local fallout input. Such sites are commonly referred to as *reference sites* (Loughran *et al.*, 2002) and are normally located in areas with limited relief, and particularly on interfluves.

The ^{137}Cs and unsupported ^{210}Pb depth profiles from adjacent cultivated areas, shown in Fig. 2.1, clearly demonstrate the effects of cultivation or tillage in mixing the soil contained within the plough layer to produce near

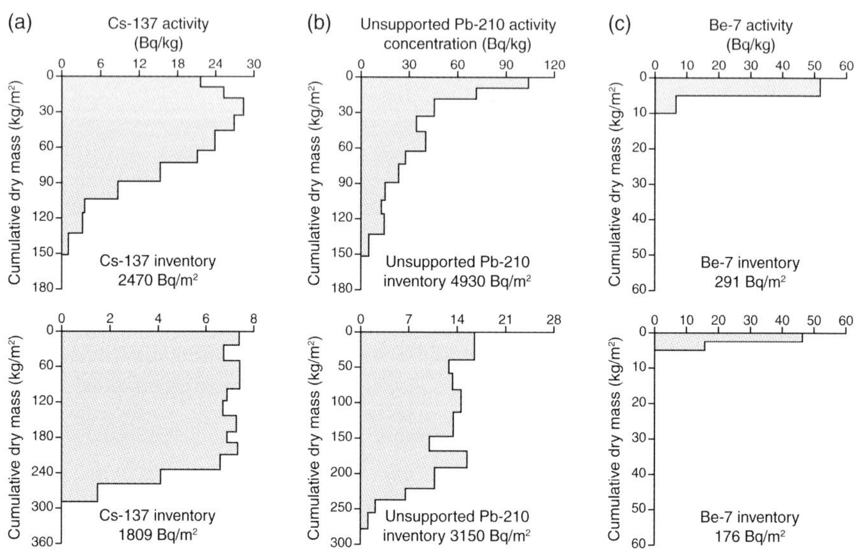

Fig. 2.1. Typical depth distributions of ^{137}Cs, unsupported ^{210}Pb and ^{7}Be in undisturbed pasture (top) and cultivated soils (bottom) in Devon, UK.

delivery from agricultural land has involved measurements of ^{137}Cs (Ritchie and McHenry, 1990; Walling, 1998). However, both unsupported ^{210}Pb and ^{7}Be have also been used in similar applications (Blake *et al.*, 1999, 2002; Walling and He, 1999a; Walling *et al.*, 1999). By collecting cores from a study site, measuring the ^{137}Cs, unsupported ^{210}Pb or ^{7}Be inventories, and applying a conversion model, it is possible to derive point estimates of the erosion and deposition rates associated with the individual cores and, by integrating these values across the study site, the relative importance of erosion and deposition and thus the gross and net erosion and the sediment delivery ratio, can be established. Figure 2.3 presents the results of an investigation of soil redistribution within a 6.7 ha field at Higher Walton Farm near Crediton, Devon, UK (see Walling *et al.*, 1999). In this study, measurements of both ^{137}Cs and ^{7}Be activities were undertaken, with the former providing estimates of average rates of soil redistribution over the past *c.* 40 years and the latter estimates of the erosion rates associated with a particular period of heavy rainfall (69 mm in 7 days) occurring in early January 1998. The soil cores used for the ^{137}Cs and ^{7}Be measurements were collected in two separate sampling campaigns, although they could have been collected at the same time. In both cases, the coring points were located at the intersections of a 20 m × 20 m grid, resulting in a suite of approximately 140 cores. Cores used to establish the local reference inventory were also obtained from adjacent areas of undisturbed land. For the ^{137}Cs measurements, the cores were collected in August 1996, using a motorized percussion corer equipped with a 6.9 cm

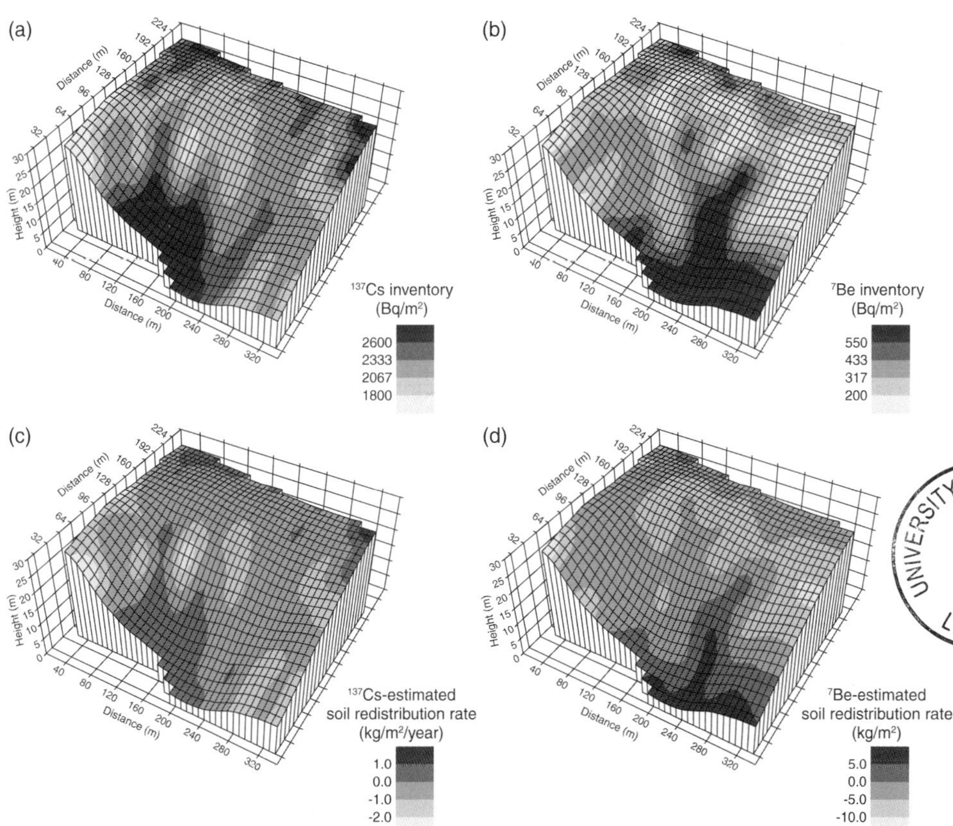

Fig. 2.3. The spatial distribution of ^{137}Cs and ^{7}Be inventories within a field at Higher Walton Farm, near Crediton, Devon, UK (a, b) and of the estimates of soil redistribution rates derived from these measurements (c, d).

internal diameter steel core tube, which was inserted into the soil to a depth of about 60 cm. The cores used for the ^{7}Be measurements were, in contrast, much shallower and were collected using a 15 cm diameter plastic core tube pushed manually into the soil to depths of 3–5 cm, in January 1998. During the preceding spring/summer of 1997, the field had been cultivated and sown to maize and the crop was harvested in early November 1997, when the soil was compacted by the harvesting equipment. After harvesting, the field was left bare and uncultivated over the winter and the period of heavy rainfall in early January 1998 resulted in substantial surface runoff and soil erosion.

The pattern of ^{137}Cs inventories within the study field documented by the cores collected in August 1996 is presented in Fig. 2.3a. The local reference inventory was estimated to be approx. 2500 Bq/m^2 and the pattern of ^{137}Cs inventories therefore shows clear evidence of erosion (reduced inventories) as well as deposition (increased inventories). Use of a conversion model enables estimates of the mean annual soil redistribution rates over the past *c.* 40 years to be derived from the measured inventories. The resulting pattern has been mapped in Fig. 2.3c and the data have been summarized in Table 2.1, which presents values for the range of soil redistribution rates, the mean erosion rate for the eroding areas, the mean deposition rate for the depositional areas, the net soil loss from the field and the sediment delivery ratio. The latter measure is of considerable importance, since it provides an estimate of the relative proportions of the mobilized sediment which have been transported *beyond* the field or redeposited *within* the field. Such information is extremely difficult to obtain using conventional monitoring techniques.

The spatial pattern of ^{7}Be inventories within the study field measured at the end of the period of heavy rainfall in early January 1998 is presented in Fig. 2.3b. The equivalent value for the local reference inventory was estimated to be 533 Bq/m^2 and the pattern shown in Fig. 2.3b again provides clear evidence of areas with both reduced and increased inventories and thus of both erosion and deposition within the field. In order to interpret this pattern in terms of soil redistribution rates associated with the period of heavy rainfall in early January, it is important to consider the extent to which it may reflect spatial variability inherited from previous erosion events. In this case, however, the preceding autumn and early winter had been relatively dry and there was no evidence of surface erosion having occurred during the previous 6 months. It is therefore reasonable to assume that the spatial variability in ^{7}Be inventories within the study field evident in Fig. 2.3b reflects soil redistribution associated with the period of heavy rainfall in early January 1998. By calculating the increase or decrease in inventory relative to the reference inventory and knowing the depth distribution of ^{7}Be at uneroded points within the field, it is possible to estimate the soil redistribution rates (Walling *et al.*, 1999). The resulting pattern of soil redistribution rates is presented in Fig. 2.3d and summary data, equivalent to those provided for the ^{137}Cs measurements, are also presented in Table 2.1.

The soil redistribution rates (kg/m^2) associated with the short period of heavy rainfall in

Table 2.1. A comparison of rates of soil redistribution within the study field at Higher Walton Farm estimated from ^{137}Cs and ^{7}Be measurements on the soil cores collected from the field. Based on data presented in Walling *et al.* (1999).

Measure	^{137}Cs (kg/m^2/year)	^{7}Be (kg/m^2)
Range	−4.5 to +2	−11.9 to +9.8
Mean erosion rate for eroding area	−1.1	−5.3
Mean deposition rate for depositional areas	0.69	4.0
Net soil loss	−0.48	−2.0
Sediment delivery ratio	0.83	0.8

early January 1998 estimated from the ^{7}Be measurements are substantially greater than the equivalent longer-term mean annual redistribution rates estimated using the ^{137}Cs measurements. However, the sediment delivery ratios are very similar, indicating that about 80% of the eroded sediment was transported out of the field. The high sediment redistribution rates associated with the period of heavy rainfall in early January 1998 reflect both the extreme nature of this period of rainfall and, perhaps more importantly, the condition of the field, which having been compacted by the maize harvesting machinery and left bare after the harvest, was particularly susceptible to surface runoff and erosion. Such results underscore the potential significance of a small number of extreme events and the incidence of particular land use conditions in controlling erosion from the study field.

In the example presented above, a large number of cores were used to establish the pattern of soil redistribution within the study field. It is clearly impossible to extend sampling at this intensity to more than a few fields and, if a sediment budget is to be constructed for a larger area, it will be necessary to design a sampling strategy that focuses on representative areas and permits extrapolation of the results to a wider area (e.g. Walling *et al.*, 2001, 2002, 2003b,c).

Tracing or Fingerprinting Sediment Sources

Sediment source tracing or fingerprinting techniques can be used to provide information on the relative importance of a number of potential sources contributing to the sediment load of a river. Such information can clearly be of considerable value when designing sediment control strategies, since it will assist in identifying those sources which should be targeted for application of control measures. Although the approach can be used to establish the relative importance of different parts of a river basin (i.e. *spatial sources*), information on the relative importance of different *source types* (e.g. sheet and rill erosion, gully erosion and channel erosion) is frequently more useful in a management context. In the latter case, fallout radionuclides will commonly provide a key component of the composite fingerprint used to discriminate potential sources. Although most work of this type has focused on fingerprinting the source of the suspended sediment load transported by a river (e.g. Walling and Woodward, 1995; Collins *et al.*, 1997), it can, for example, also be used to trace the source of fine sediment deposited on floodplains or accumulating within river gravels (e.g. Bottrill *et al.*, 2000; Walling *et al.*, 2003a). An example of the latter application is provided below.

The siltation of spawning gravels has frequently been identified as a key factor contributing to the declining success of salmon fisheries in British rivers. Concern for this problem has focused attention on the need to reduce gravel siltation and thus to reduce fine sediment mobilization and transport in impacted catchments, through the establishment of sediment control programmes. The development of effective sediment control programmes requires information on the likely sources of the fine sediment accumulating in spawning gravels, since these sources must be targeted if the control programme is to prove effective. In an attempt to provide such information, a reconnaissance source fingerprinting survey of several representative rivers, located in different parts of Britain, was undertaken by the author and his co-workers in collaboration with the Environment Agency (Walling *et al.*, 2003a).

In this study, samples of interstitial fine sediment were recovered from salmonid spawning gravels for a representative selection of rivers in England and Wales (Fig. 2.4a), by means of a national fieldwork programme conducted by the Environment Agency over the period 1999–2000. Sample collection involved the use of retrievable basket samplers, which were installed in artificial redds constructed in spawning gravels at representative locations. Between one and five samplers were installed in each of the rivers identified in Fig. 2.4a. The basket samplers were filled with clean framework gravel (> 6.4 mm) prior to their installation and they were retrieved about 3 months later. The gravel contained within the basket was subsequently wet sieved to recover the fine (< 0.125 mm) interstitial sediment that had accumulated within the gravel during the period of deployment and this fraction was used for

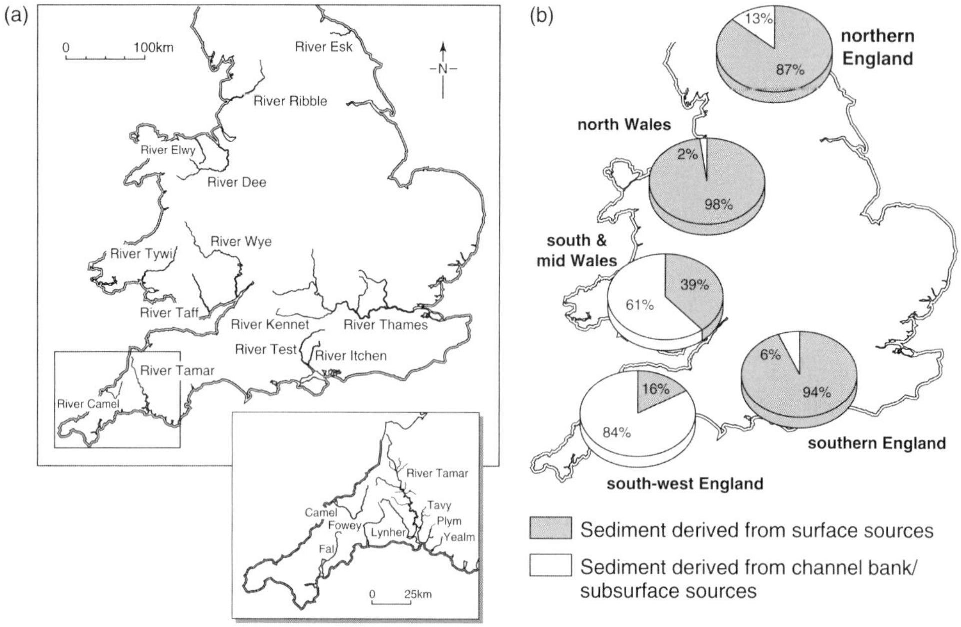

Fig. 2.4. Fingerprinting the source of fine sediment accumulating in salmon spawning gravels in UK rivers, showing (a) the location of the sampled rivers and (b) regional contrasts in the source of the fine sediment.

sediment source fingerprinting. By virtue of the reconnaissance nature of the study, which included 18 catchments, sampling of potential source materials focused on the broad distinction between surface and channel bank/subsurface sources, and a total of 672 source material samples were collected from the different study areas. These samples were sieved to < 0.125 mm to facilitate direct comparison with the samples of fine interstitial sediment. The limited resources available to the study also precluded the use of an extensive range of fingerprint properties to discriminate the two potential sources and emphasis was placed on the use of radiometric (^{137}Cs, unsupported ^{210}Pb, ^{226}Ra) measurements, coupled with information on the organic matter content (C and N) of the potential sources. Mean values of these properties were used to characterize the two potential sources in each of the river basins investigated.

A multicomponent mixing model, incorporating correction for the effects of contrasts in particle size and organic matter content between the samples of interstitial sediment and the source material, was used to estimate the relative contribution of surface and channel bank/subsurface sources to the samples of fine sediment recovered from the spawning gravels. The results of these computations for different regions of England and Wales are presented in Fig. 2.4b. Appreciable contrasts in the relative importance of surface and channel bank/subsurface sources between the regions are apparent in Fig. 2.4b, with, for example, surface sources accounting for > 90% of the fine interstitial sediment in north Wales and southern England, but only 16% and 39% in south-west England and south and mid Wales, respectively. These regional contrasts reflect the interaction of land use with both erosion processes and the efficiency of sediment transfer to the channel network. In south-west England, where stocking densities are high and river channels are frequently quite deeply incised, the combination of livestock trampling of channel margins and erosion of unstable channel banks means that channel and subsurface sources are the dominant source of fine interstitial sediment. In contrast, the greater importance of arable farming, and more specifically soil erosion, on large cultivated fields with few boundaries to interrupt slope–channel connectivity, combined with the relative stability of the well-vegetated and relatively low channel banks, result in surface soils providing the dominant source of the fine interstitial sediment recovered from spawning gravels in southern England. Surface sources are also important in the upland areas of northern England and north Wales, where high rainfall and grazing pressure promote the erosion of surface soils under upland pasture or moorland and the steep topography combined with an absence of field boundaries results in the efficient routing of sediment to the river channel.

The contrasts in the relative importance of surface and channel bank/subsurface sources in different regions of England and Wales outlined above have important implications for the design and implementation of effective sediment control strategies. Where bank erosion is the dominant source, attention should clearly focus on reduction of livestock trampling of channel margins and improvement of bank stability (e.g. by fencing and revegetation of channel margins). However, such measures are likely to be of little value in areas where surface sources are dominant, where emphasis should be placed on controlling sediment mobilization and transfer from the catchment surface more generally.

Investigating Floodplain Sedimentation

Overbank deposition on river floodplains during flood events can represent an important sink for suspended sediment transported through a river system, and recent studies have demonstrated that such transmission losses can account for as much as 40–50% of the suspended sediment load delivered to the main channel system (Walling and Owens, 2002). Where the nutrient and contaminant content of the sediment is high, floodplains can represent significant nutrient and contaminant sinks, posing problems for their longer term sustainable use. Equally, the progressive aggradation of river floodplains can result in reduced floodwater conveyance capacity and thus an increasing flood risk. Information on rates of overbank sedimentation on river floodplains may be needed to investigate further their role as sediment sinks and, in view of the difficulties of obtaining such information using conventional approaches, the use of environmental radionuclides to establish deposition rates has been shown to offer considerable potential.

As an example, Fig. 2.5 shows how ^{137}Cs and unsupported ^{210}Pb measurements have been used to document overbank sedimentation rates along a short reach of the floodplain of the River Severn near Buildwas in Shropshire, UK. In this study, 124 sediment cores were collected at the intersections of a 25 m $\times$ 25 m grid, using a motorized percussion corer equipped with a 6.9 cm internal diameter core tube. Cores were collected to a depth of about 70 cm to ensure that they included the complete ^{137}Cs and unsupported ^{210}Pb profiles. Measurements of the ^{137}Cs and unsupported ^{210}Pb inventories of the individual cores were used to estimate the mean annual sedimentation rates at the coring points, using the procedures documented by He and Walling (1996) and Walling and He (1997). These estimates have been used to map the patterns of sedimentation within the reach shown in Fig. 2.5. By comparing the estimates of sedimentation rate derived from the ^{137}Cs measurements, which relate to the past *c.* 40 years, with those based on the unsupported ^{210}Pb measurements, which relate to the past *c.* 100 years, it is possible to assess longer term changes in sedimentation at this location. The mean annual sedimentation rate at this site over the past 40 years is 0.28 g/cm^2/year and the equivalent rate for the past 100 years is 0.33 g/cm^2/year. This suggests that rates of overbank sedimentation have changed little over the past 100 years.

Although the example presented in Fig. 2.5 represents a detailed investigation of an individual reach, it is equally possible to use the approach to obtain representative information on overbank sedimentation rates for a range of rivers within a region (e.g. Walling and He, 1999c) or to establish the magnitude of the longer term transmission losses associated with overbank deposition on the floodplains bordering the main channels of a river basin. In the latter case there will be a need to extrapolate measurements from representative transects or small reaches to the entire floodplain area, in order to calculate the mass of sediment involved and to compare this with the suspended sediment flux at the basin outlet

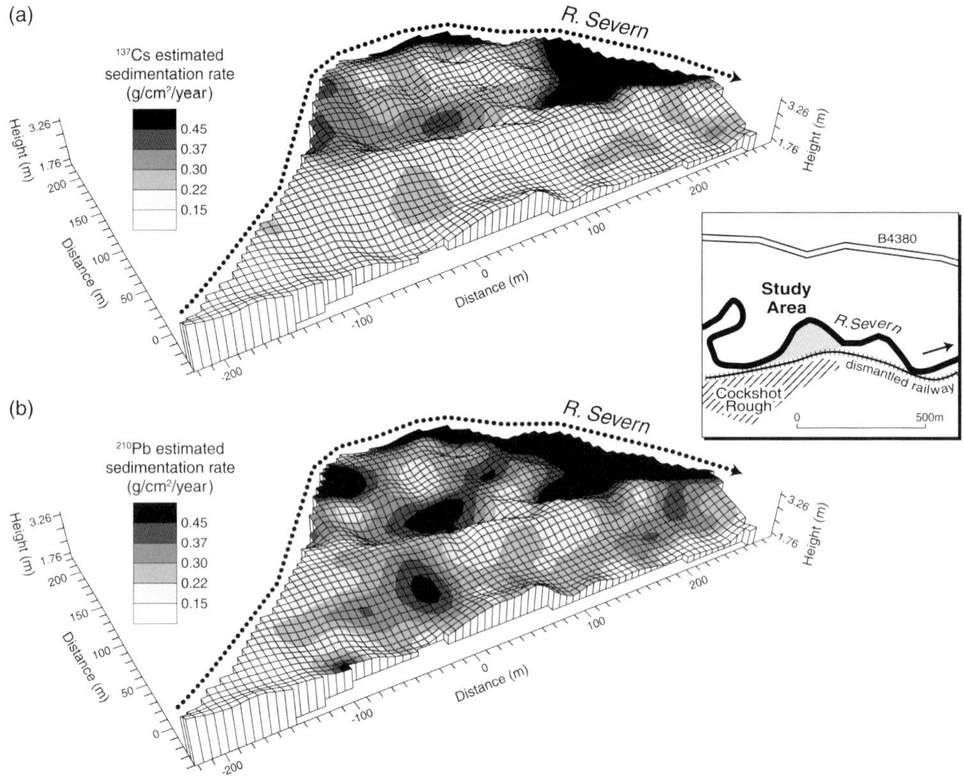

Fig. 2.5. The spatial distribution of overbank sedimentation rates within a small reach of the floodplain of the River Severn near Buildwas, Shropshire, UK derived from ^{137}Cs and unsupported ^{210}Pb measurements undertaken on floodplain cores.

(e.g. Walling *et al.*, 1998). The potential for using measurements of both ^{137}Cs and unsupported ^{210}Pb to assess changes in overbank sedimentation rates over the past 100 years also offers considerable scope to investigate recent changes in floodplain sedimentation in response to changes in catchment land use (Walling and He, 1999c).

Perspective

The case studies described above provide examples of the potential for using fallout radionuclides as tracers in order to obtain information on the functioning of catchment sediment budgets. However, each of the case studies focuses on a particular component of the sediment budget. In many investigations, the ultimate aim will be to establish the overall catchment sediment budget and it is important to recognize that the results obtained for the individual components of a sediment budget can be combined to establish the overall sediment budget of a catchment. The use of the same radionuclide tracer in studies of the individual components will clearly facilitate this exercise.

The work of Walling *et al.* (2001) in the Upper Kaleya catchment in southern Zambia provides an example of how the results from investigations of several components of the catchment sediment budget can be combined. Caesium-137 measurements on cores collected from representative slope transects under commercial cultivation, communal cultivation and bush grazing were used to establish gross and net erosion rates under these three land use types. These results were combined with measurements of the suspended sediment flux at the catchment outlet, source fingerprinting of the sediment load at the

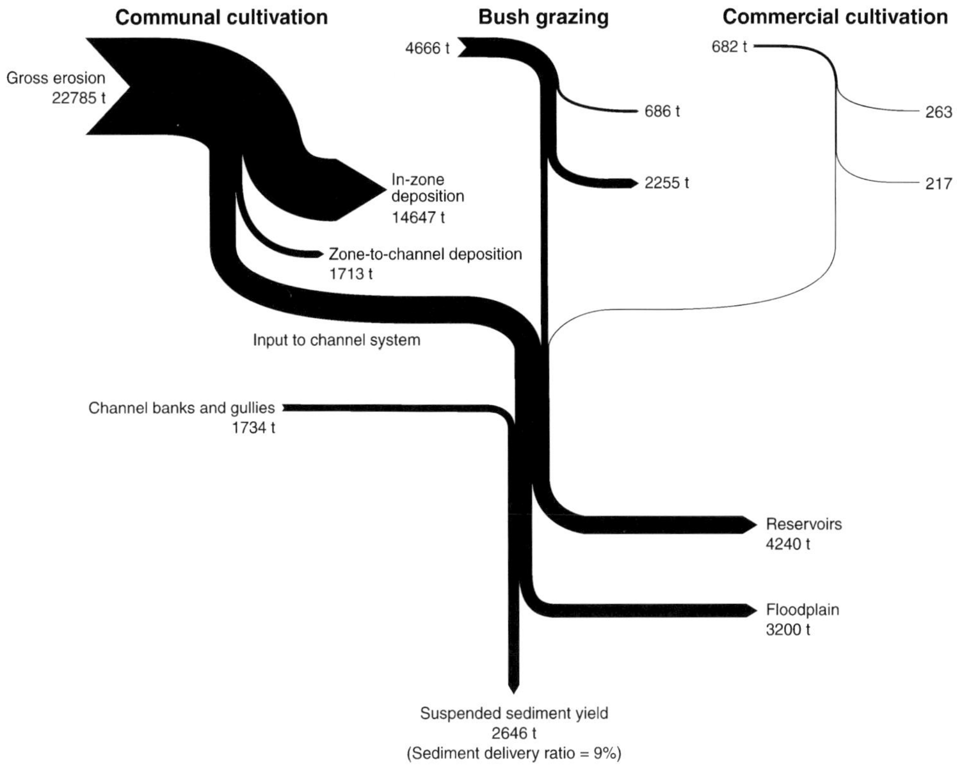

Fig. 2.6. The catchment sediment budget derived for the Upper Kaleya catchment, near Mazabuka, southern Zambia, by Walling *et al.* (2001).

catchment outlet, estimation of conveyance losses associated with overbank sedimentation on the river floodplain and measurements of sediment accumulation in a number of small reservoirs to construct the sediment budget shown in Fig. 2.6. By summing the estimates of mean annual suspended sediment load at the catchment outlet, floodplain conveyance loss and reservoir sedimentation, it was possible to estimate the total amount of sediment delivered to the main channel system. The source fingerprinting results were used to apportion this load to surface erosion within the three main land use types and to channel erosion. Comparison of these load components with the estimates of net soil loss from the three land use types provided an estimate of the conveyance losses associated with sediment delivery from the slopes to the channel network.

Figure 2.6 provides a good example of the potential of tracing techniques to provide an improved understanding of erosion and sediment delivery processes operating within a drainage basin, by extending and complementing the data provided by traditional monitoring techniques. In this case, traditional monitoring techniques have been used to quantify the suspended sediment flux at the catchment outlet, but tracing techniques have made it possible to quantify sediment mobilization and storage within the basin and thereby establish the overall catchment sediment budget. This budget provides valuable information on the functioning of the catchment in terms of its sediment dynamics. Equally importantly, it provides a sound basis for developing an effective sediment management or control programme, in that it identifies the key sources that would need to be targeted and highlights the importance of within-catchment storage in attenuating the transfer of sediment between its source and the catchment outlet. In the latter context, any reduction in conveyance losses associated with reduced storage could readily offset any reduction in sediment mobilization through improved land management.

Acknowledgements

The important contribution of a number of my co-workers, including Dr Qingping He, Dr Will Blake, Dr Adrian Collins and Dr Henry Sichingabula, to the work described in the various case study examples cited and the financial support provided for that work by the Natural Environment Research Council, the Environment Agency and the UK Department for International Development are gratefully acknowledged.

References

Allan, R.J. (1986) *The Role of Particulate Matter in the Fate of Contaminants in Aquatic Ecosystems.* Environment Canada Scientific Series No. 142. Inland Waters Directorate, Burlington, Ontario.

Blake, W.H., Walling, D.E. and He, Q. (1999) Fallout beryllium-7 as a tracer in soil erosion investigations. *Applied Radiation and Isotopes* 51, 599–605.

Blake, W.H., Walling, D.E. and He, Q. (2002) Using cosmogenic beryllium-7 as a tracer in sediment budget investigations. *Geografiska Annaler* 84A, 89–102.

Bottrill, L.J., Walling, D.E. and Leeks, G.J.L. (2000) Using recent overbank deposits to investigate contemporary sediment sources in large river systems. In: Foster, I.D.L. (ed.) *Tracers in Geomorphology.* Wiley, Chichester, UK, pp. 369–387.

Clark, E.H., Haverkamp, J.A. and Chapman, W. (1985) *Eroding Soils: The Off-farm Impacts.* The Conservation Foundation, Washington, DC.

Collins, A.L., Walling, D.E. and Leeks, G.J.L. (1997) Sediment sources in the Upper Severn catchment: a fingerprinting approach. *Hydrological and Earth System Sciences* 1, 509–521.

Golosov, V.N., Ivanova, N.N., Litvin, L.F. and Sidorchuk, A.Yu. (1992) Sediment budget in river basins and small river aggradation. *Geomorphologiya* 14, 62–71 (in Russian).

He, Q. and Walling, D.E. (1996) Use of fallout Pb-210 measurements to investigate longer-term rates and patterns of overbank sediment deposition on the floodplains of lowland rivers. *Earth Surface Processes and Landforms* 21, 141–154.

Lal, R. (2001) Soil degradation by erosion. *Land Degradation and Development* 12, 519–539.

Loughran, R.J., Wallbrink, P.J., Walling, D.E. and Appleby, P.G. (2002) Sampling methods. In: Zapata, F. (ed.) *Handbook for the Assessment of Soil Erosion and Sedimentation Using Environmental Radionuclides.* Kluwer, Dordrecht, The Netherlands, pp. 41–57.

Phillips, J.D. (1992) Delivery of upper-basin sediment to the lower Neuse River, North Carolina, USA. *Earth Surface Processes and Landforms* 17, 699–709.

Reid, L.M. and Dunne, T. (1996) *Rapid Evaluation of Sediment Budgets.* GeoEcology Paperbacks. Catena, Germany.

Ritchie, J.C. and McHenry, J.R. (1990) Application of radioactive fallout cesium-137 for measuring soil erosion and sediment accumulation rates and patterns: a review. *Journal of Environmental Quality* 19, 215–233.

Soulsby, C., Moir, H.J., Youngson, A. and Malcolm, I. (2001) Fine sediment influence on salmonid spawning in a lowland stream: a preliminary assessment. *The Science of the Total Environment* 265, 295–307.

Stone, M. (ed.) (2000) *The Role of Erosion and Sediment Transport in Nutrient and Contaminant Transfer* (Proceedings of the Waterloo Symposium, July 2000). IAHS Publication 263. IAHS Press, Wallingford, UK, pp. 69–81.

Sundborg, A. (1982) *Sedimentation Problems in River Basins.* UNESCO Studies and Reports in Hydrology No. 35. UNESCO Division of Water Sciences, Paris.

Trimble, S.W. (1983) A sediment budget for Coon Creek basin in the Driftless Area, Wisconsin. 1853–1977. *American Journal of Science* 283, 454–474.

Wallbrink, P.J., Walling, D.E. and He, Q. (2002) Radionuclide measurement using HPGe gamma spectrometry. In: Zapata, F. (ed.) *Handbook for the Assessment of Soil Erosion and Sedimentation Using Environmental Radionuclides.* Kluwer, Dordrecht, The Netherlands, pp. 67–96.

Walling, D.E. (1998) Use of ^{137}Cs and other fallout radionuclides in soil erosion investigations: progress, problems and prospects. In: *Use of ^{137}Cs in the Study of Soil Erosion and Sedimentation.* International Atomic Energy Agency Publication IAEA-TECDOC-1028. IAEA, Vienna, pp. 39–64.

Walling, D.E. (2000) Linking land use, erosion and sediment yields in river basins. *Hydrobiologia* 410, 223–240.

Walling, D.E. (2002) Recent advances in the use of environmental radionuclides in soil erosion investigations. In: *Nuclear Techniques in Integrated Plant Nutrient, Water and Soil Management.* IAEA C&S Papers Series 11/C. IAEA, Vienna, pp. 279–301.

Walling, D.E. and He, Q. (1997) Use of fallout ^{137}Cs in investigations of overbank sediment deposition on river floodplains. *Catena* 29, 263–282.

Walling, D.E. and He, Q. (1999a) Using fallout lead-210 measurements to estimate soil erosion on cultivated land. *Soil Science Society of America Journal* 63, 1404–1412.

Walling, D.E. and He, Q. (1999b) Improved models for estimating soil erosion rates from ^{137}Cs measurements. *Journal of Environmental Quality* 28, 611–622.

Walling, D.E. and He, Q. (1999c) Changing rates of overbank sedimentation on the floodplains of British rivers during the past 100 years. In: Brown, A.G. and Quine, T.A. (eds) *Fluvial Processes and Environmental Change.* Wiley, Chichester, UK, pp. 207–222.

Walling, D.E. and Owens, P.N. (2002) The role of floodplain sedimentation in catchment sediment and contaminant budgets. In: *The Structure and Management Implications of Fluvial Sedimentary Systems* (Proceedings of the Alice Springs Symposium, September 2002). IAHS Publication 276. IAHS Press, Wallingford, UK, pp. 407–416.

Walling, D.E. and Woodward, J.C. (1992) Use of radiometric fingerprints to derive information on suspended sediment sources. In: *Erosion and Sediment Transport Monitoring Programmes in River Basins* (Proceedings of the Oslo Symposium, August 1992). IAHS Publication 210. IAHS Press, Wallingford, UK, pp. 153–164.

Walling, D.E. and Woodward, J.C. (1995) Tracing sources of suspended sediment in river basins. A case study of the River Culm, Devon, UK. *Marine and Freshwater Research* 46, 327–336.

Walling, D.E., Webb, B.W. and Russell, M.A. (1997) Sediment-associated nutrient transport in British rivers. In: *Freshwater Contamination* (Proceedings of the Rabat Symposium). IAHS Publication 243. IAHS Press, Wallingford, UK, pp. 69–81.

Walling, D.E., Owens, P.N. and Leeks, G.J.L. (1998) The role of channel and floodplain storage in the suspended sediment budget of the River Ouse, Yorkshire, UK. *Geomorphology* 22, 225–242.

Walling, D.E., He, Q. and Blake, W.H. (1999) Use of ^{7}Be and ^{137}Cs measurements to document short- and medium-term rates of water-induced soil erosion on agricultural land. *Water Resources Research* 35, 3865–3874.

Walling, D.E., Collins, A.L., Sichingabula, H.M. and Leeks, G.J.L. (2001) Integrated assessment of catchment sediment budgets. *Land Degradation and Development* 12, 387–415.

Walling, D.E., Russell, M.A., Hodgkinson, R.A. and Zhang, Y. (2002) Establishing sediment budgets for two small lowland agricultural catchments in the UK. *Catena* 47, 323–353.

Walling, D.E., Collins, A.L. and McMellin, G.K. (2003a) A reconnaissance survey of the source of interstitial fine sediment recovered from salmonid spawning gravels in England and Wales. *Hydrobiologia* 497, 91–108.

Walling, D.E., Collins, A.L. and Sichingabula, H.M. (2003b) Using unsupported lead-210 measurements to investigate soil erosion and sediment delivery in a small Zambian catchment. *Geomorphology* 52, 193–213.

Walling, D.E., Collins, A.L., Sichingabula, H.M. and Leeks, G.J.L. (2003c) Use of reconnaissance measurements to establish catchment sediment budgets: a Zambian example. In: *Erosion Prediction in Ungauged Basins: Integrating Methods and Techniques* (Proceedings of the Sapporo Symposium, July 2003). IAHS Publication 279. IAHS Press, Wallingford, UK, pp. 1–12.

Warren, N., Allan, I.J., Carter, J.E., House, W.A. and Parker, A. (2003) Pesticides and other micro-organic contaminants in freshwater sedimentary environments – a review. *Applied Geochemistry* 18, 159–194.

Wood, P.J. and Armitage, P.D. (1997) Biological effects of fine sediment in the lotic environment. *Environmental Management* 21, 203–217.

Wood, P.J. and Armitage, P.D. (1999) Sediment deposition in a small lowland stream: management implications. *Regulated Rivers* 15, 199–210.

Zapata, F. (ed.) (2002) *Handbook for the Assessment of Soil Erosion and Sedimentation Using Environmental Radionuclides.* Kluwer, Dordrecht, The Netherlands.

3 A Comparison of Caesium-137 and Erosion Pin Data from Tai To Yan, Hong Kong

M.R. Peart, M.E. Ruse and R.D. Hill
University of Hong Kong, PR China

Introduction

Techniques for deriving soil erosion data are often constrained to site-specific, labour-intensive monitoring of erosion plots, erosion pins and suspended sediment yields. An alternative approach is provided by the measurement of caesium-137 (^{137}Cs) in the soil. The loss or gain of ^{137}Cs in the soil at a particular point, determined by comparison to a reference site, may indicate erosion or deposition of soil. Empirical or theoretical relationships can be used to convert ^{137}Cs inventories to an estimate of soil loss (e.g. Ritchie and McHenry, 1990; Walling and He, 1999). Measurement of ^{137}Cs to examine soil erosion has been applied in a range of different environments and has a number of advantages over other techniques (Ritchie and McHenry, 1990; Zapata, 2003). A key advantage of the ^{137}Cs technique is its ability to provide medium-term estimates (~ 40 years) from a single site visit (Loughran, 1989), a scale 'more appropriate to studies of hillslope processes than monitoring methods such as the use of small plots' (Wise, 1980, p. 117). Perhaps surprisingly, the simple issue of how ^{137}Cs inventories compare with erosion quantities reported from other monitoring techniques has rarely been addressed in the literature, and never in Asia. Notable exceptions include Campbell *et al.* (1986), Saynor *et al.* (1994) and Higgitt and Haigh (1995).

An erosion pin process monitoring study conducted at Tai To Yan from 1992 to the present day provides an opportunity to compare ^{137}Cs and erosion pin data. This allows a test of the ^{137}Cs technique as a way to develop information on soil erosion in Hong Kong over time-periods that reflect management practices. Given significant changes in land use practices since 1945, particularly afforestation in catchment areas for water supply (Corlett, 1999) and the decline in the use of phytofuel (Chen *et al.*, 1997), such data would be invaluable in identifying the erosional sources of drainage network sedimentation. Furthermore, while ^{137}Cs tracing has been investigated in many areas of the world, and proven to have distinct advantages (Zapata, 2003), there have been no attempts to examine the potential of the ^{137}Cs approach in Hong Kong and the region.

Study Area Characteristics

The study area, located on the footslopes of Tai To Yan, has a vegetation cover comprising grass, fern and some shrubs. This fire-climax community covers over 26% of Hong Kong's total land area. Hong Kong has a subtropical monsoonal climate, with a high mean annual rainfall (2214 mm; 1961–1990) and a considerable range (901–3330 mm; from 1884) recorded by the Hong Kong Observatory. A rain gauge near the study site at the Kadoorie Agricultural Research Centre (KARC) gave a mean annual rainfall of 2403 mm for the period 1991–1997, with a range of 1972–3111 mm. Around 80% of annual rainfall occurred between May and September inclusive.

Methods

Tai To Yan erosion pin study

The soil erosion study at Tai To Yan started in 1992 using six bounded plots of 6 m × 20 m (Fig. 3.1). Plot size was dictated by manageability and the need to minimize edge effects that might distort erosion data. Plot Nos 1 and 3 are control plots, vegetated predominantly by ferns with shrubs and grasses, against which the effects of vegetation clearance could be gauged. They have occasionally experienced hillfires, but vegetation recovery is rapid. Two additional plots (Nos 7 and 8), established in January 1994 and 1995 respectively, were maintained as unvegetated plots by initial clearance of vegetation with ongoing application of herbicide and hand weeding as required. Mean slope angles of the plots are given in Table 3.1. Although the site is located across a lithological contact between granodiorite and tuff (Geotechnical Control Office, 1988), field and laboratory textural analyses show that the soils are very similar sandy clay loams, with some boulders in Plot No. 1. The study area is between 120 and 160 m above sea level.

Erosion pins of mild steel, 1 m long and 10 mm diameter (slightly thinner on Plot No. 8), were inserted vertically to a minimum pin depth of 40 cm in order to achieve a stable datum. The erosion pins were spaced at regular intervals in three rows of 10 pins. The rows were 2 m apart, as was downslope pin spacing. The vertical ground surface change data reported in this paper refer to a comparison of pin heights measured at the time of plot establishment and measurements made in October 1997, at the end of the wet season when erosion rates reduce considerably. An increase in pin exposure signifies soil erosion, whilst a decrease in pin length may identify deposition.

Measurement of soil ^{137}Cs inventories

Ten regularly spaced soil cores from between the three rows of pins were sampled from each of Plot Nos 1, 7 and 8 in January 1998, prior to the start of the wet season. The steel core tubes had 7 cm diameters and were inserted to depths of 25 cm for Plot No. 1 and 22 cm for Plot Nos 7 and 8. These depths were chosen to avoid diluting the ^{137}Cs too much by incorporating unlabelled soil from lower in the soil profile (see Ruse, 1999; Ruse and Peart, 2000).

After collection, samples were dried, sieved through a 2 mm mesh, in order to remove clasts and ensure a constant sample density and geometry, and 1 kg samples were sent to Westlakes Scientific Consulting, UK, to determine soil ^{137}Cs content. Samples were counted on an HPGe gamma spectrometer of 30–75% relative efficiency in a calibrated counting geometry for periods ranging between 1 and 3 days.

The ^{137}Cs-loading with depth has also been obtained in 2 cm depth increments at two flat non-eroding and undisturbed input reference sites, A and B, using a 300 mm × 200 mm scraper frame (Campbell *et al.*, 1988) and the preparation detailed above. The reference sites were in close proximity to the Tai To Yan erosion plots (Ruse, 1999; see Fig. 3.1) with a vegetation cover of grasses, ferns and shrubs that was better developed at Site B. Aerial

Table 3.1. ^{137}Cs content and soil surface change.

	Mean ^{137}Cs (Bq/m^2)	^{137}Cs as % of Plot 1	^{137}Cs CV (%)	Average vertical soil surface change (mm)	Pin CV (%)	Average slope angle
Plot 1	1574.39 ± 391	–	24.86	+3.2 ± 8.13	254.1	15°
Plot 8	988.50 ± 359	62.8	36.35	–24.4 ± 8.30	34.0	17°
Plot 7	711.05 ± 298	45.2	41.92	–57.1 ± 17.43	30.5	26°
Plot 3	no data	no data	no data	–4.5 ± 9.81	218.0	26°

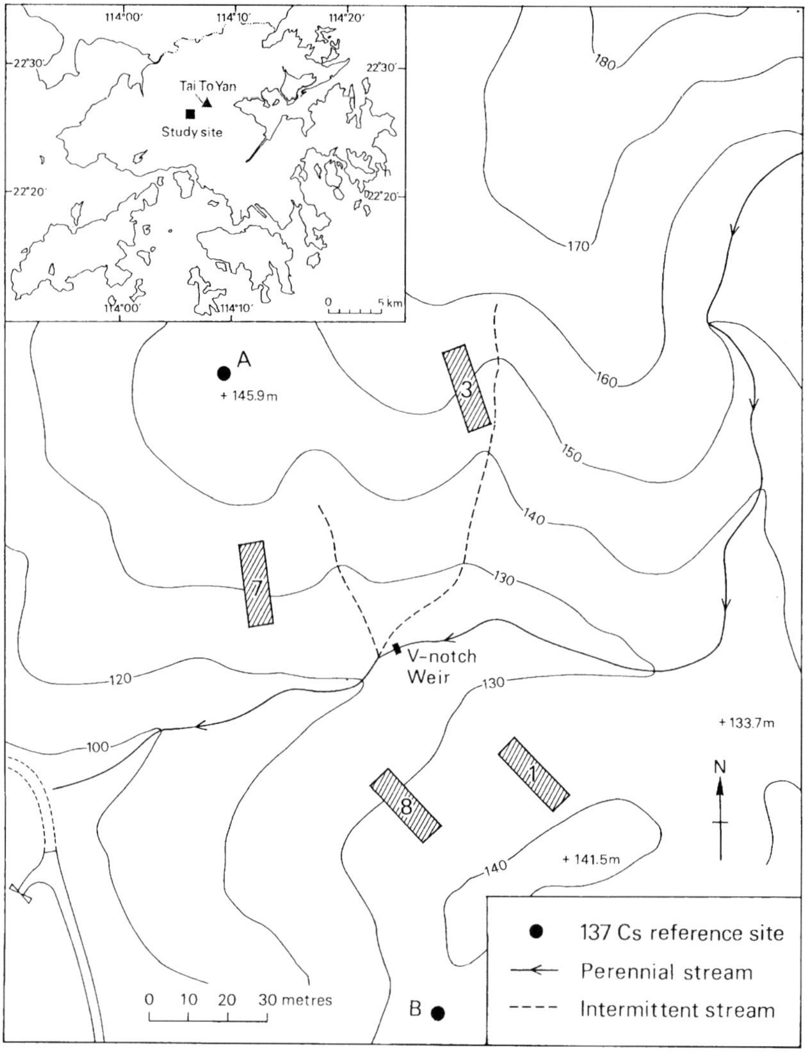

Fig. 3.1. Location of study plots.

photographs from 1956 evidence a vegetative cover and no obvious signs of disturbance. The data were obtained at the Institute of Limnology and Geography, Nanjing, and Hong Kong University.

Results

Erosion pin monitoring results

Erosion pin monitoring results are given in Table 3.1. Despite limitations of sample size and replications in the present study, the results appear to show absolute differences and offer reasonable indications of erosion rates. Control Plot No. 1 exhibited + 3.2 mm (± 8.1 mm) mean vertical increase of the ground surface over the 70 month monitoring period with a very high coefficient of variation (CV) of 254%. Taking a mean dry bulk density value of 0.98 g/cm^3, the Plot No. 1 values equate to between 19.1 t/ha/year of soil formation and 8.2 t/ha/year of soil erosion. Control Plot No. 3 also shows general stability, with a small mean ground surface decline of –4.5 mm (± 9.8 mm) and a CV of 218% (Table 3.1). Plot Nos 1 and 3 evidence the protective effects of vegetation.

Bare Plot Nos 8 and 7 exhibited for the 33 and 45 month measurement periods respectively: mean erosion of –24.4 mm (± 8.3 mm) and –57.1 mm (± 17.4 mm), with CVs of 34.0% and 30.5% (Table 3.1). The estimated erosion rates are –42.4 to –85.1 t/ha/year for Plot No. 8 and –82.5 to –153.0 t/ha/year for Plot No. 7 (Table 3.2). Mann-Whitney test results indicate statistically significant differences in the erosion noted at the three plots at a probability of 0.01. Differences between Plot Nos 7 and 8 arise primarily from the time periods for vegetation clearance (45 months and 33 months for Plot Nos 7 and 8, respectively), though differences in slope angle are noted (26° and 17° for Plots 7 and 8, respectively). They are valuable here because the differences in erosion totals based on the erosion pin measurements can be tested against the measured ^{137}Cs inventories.

Figure 3.2a presents isoline maps of erosion based upon the pin data for Plot Nos 1, 8 and 7 for which ^{137}Cs data are available. They clearly demonstrate differences in erosion totals and the large degree of spatial variability, reflected in the CVs in Table 3.1.

Soil ^{137}Cs results

The mean ^{137}Cs inventory of the ten soil cores recorded on each plot were 1574.4 Bq/m^2 (± 391.5), 988.5 Bq/m^2 (± 359.4) and 711.1 Bq/m^2 (± 298.1) for Plot Nos 1, 8 and 7, respectively (Table 3.1). These exhibit the same ranking as the erosion totals recorded by erosion pins (Table 3.2). Mann-Whitney U-test results indicate that the ^{137}Cs inventories of all three plots are significantly different to each other, at the 0.05 confidence level. This is reflected in the fact that none of the measured inventories on Plot No. 8, which recorded the least erosion, exceeded the mean ^{137}Cs inventory for Plot No. 1. As the mean ^{137}Cs inventory declines, the ^{137}Cs variability increases, with CVs of 24.9%, 36.4% and 41.9% for Plot Nos 1, 8 and 7, respectively. The extent of the ^{137}Cs spatial variability at each plot is reflected in Fig. 3.2b.

The median ^{137}Cs inventories were 993 and 1003 Bq/m^2 respectively for Reference Sites A and B, based upon 10 core samples. Both reference site profiles have similar ^{137}Cs distributions with depth in that their peak values are located in the 2–4 cm layer and for both profiles the ^{137}Cs concentration starts to decline markedly from 10 cm. The depth distribution of ^{137}Cs for Site A is shown in Fig. 3.3 and approximates that of a typical reference profile. These inventories are lower than reported for Plot No. 1, the stable site.

Comparison of ^{137}Cs Inventories to Erosion Totals

This investigation shows that ^{137}Cs inventories in the grass and fernlands of Tai To Yan decline as soil is removed from the surface by erosion. This work provides independent verification of the use of ^{137}Cs loadings to evaluate soil erosion. It also supports the use of ^{137}Cs as a tracer of soil movement.

The relationship between ^{137}Cs depletion and ground lowering by soil erosion is not directly proportional. A mean value of 24 mm of vertical erosion on Plot No. 8 reduced the ^{137}Cs loading by around 37% compared to the control plot, but a further 33 mm of erosion on Plot No. 7 only lowered the soil ^{137}Cs loading by a further 18% (Table 3.1). This point, central to the 'calibration, or conversion, question' (see Walling and Quine, 1991; Quine, 1995), arises from the differential loading of ^{137}Cs with depth from the soil surface.

Conversion of ^{137}Cs Inventories to Erosion Totals

The depth distribution of ^{137}Cs in the soil profile can be used to derive an 'independent' estimate of erosion rate, using a simple variant of the profile distribution model (Walling and He, 1999) for converting the ^{137}Cs inventories to erosion. Similar work was conducted by Chappell *et al.* (1998).

The reference total reported for Plot No. 1 was taken to equal 100%, having experienced no erosion. Using the standard deviation about the mean of the ^{137}Cs values, Plot Nos 8 and 7 equal 53.2–68.6% and 34.9–51.3% of

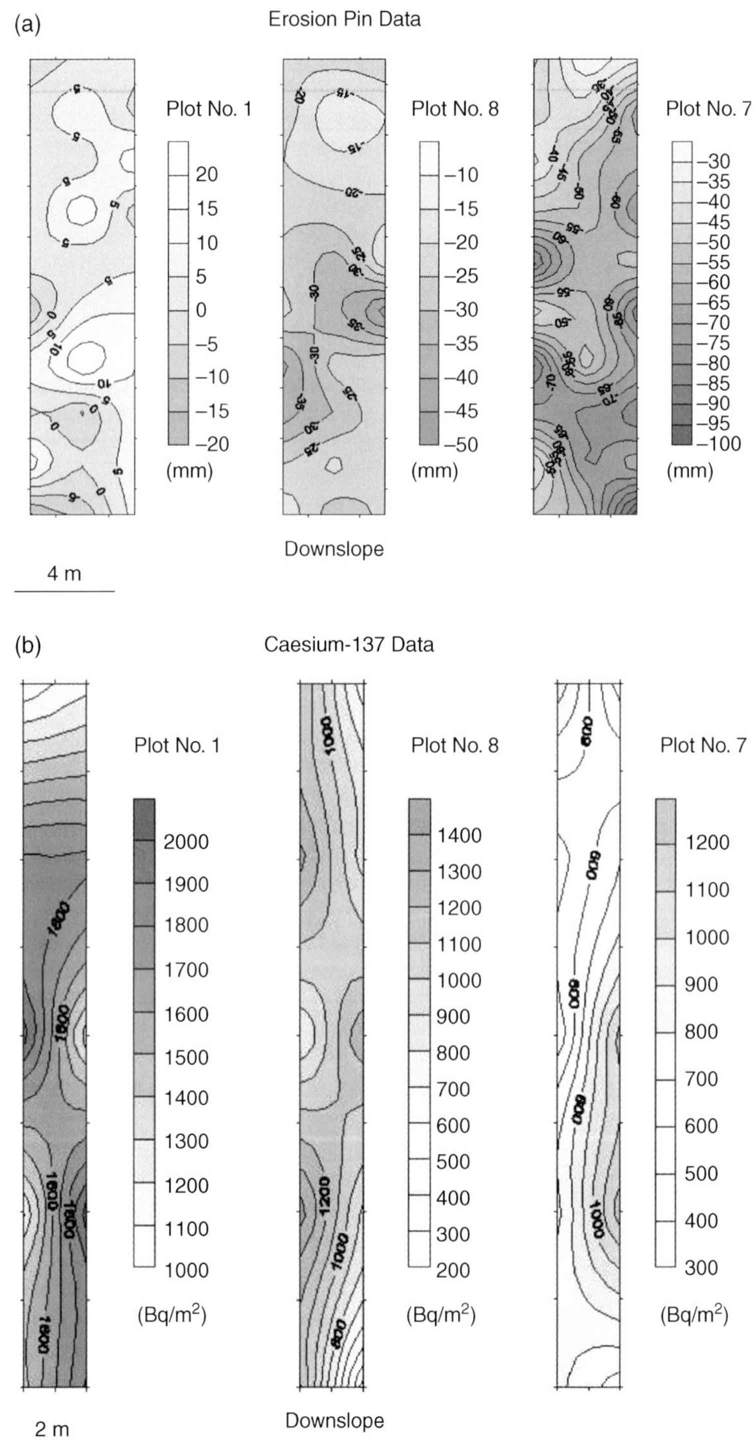

Fig. 3.2. Spatial variation of (a) erosion and (b) ^{137}Cs on Plots 1, 7 and 8. The different plot size is due to the relative locations of the erosion pins and the ^{137}Cs core samples.

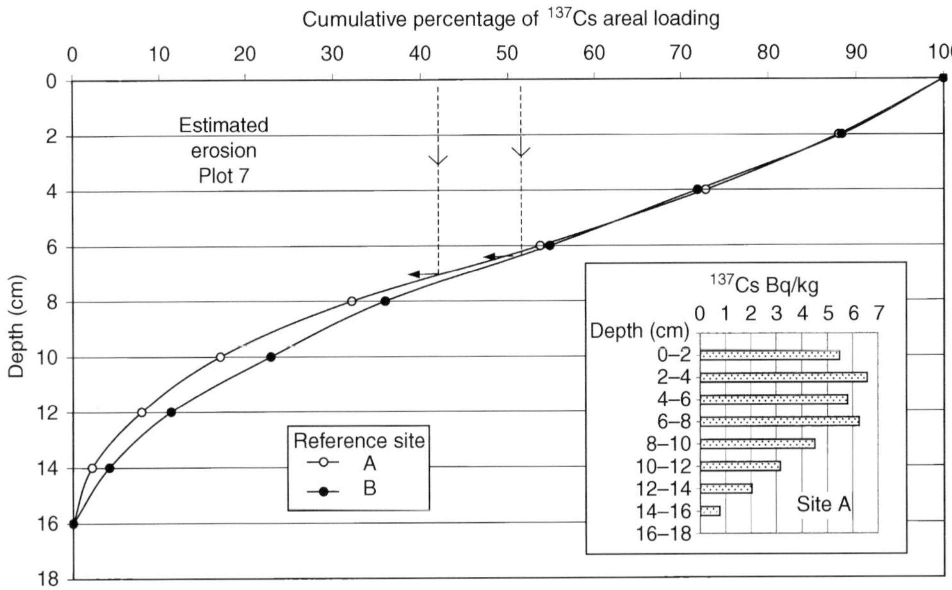

Fig. 3.3. Cumulative percentage ^{137}Cs areal loading by depth at two reference sites and ^{137}Cs concentration with depth at Site A.

Table 3.2. Erosion rates comparison.

Erosion pins (erosion (t/ha/year))		^{137}Cs (erosion (t/ha/year))		Small basin studies (erosion (t/ha/year))		
Plot 1	+19.1 to –8.2	Plot 1	Control: No erosion	Fok (2001)		0.42
Plot 8	–42.4 to –85.1	Plot 8	–117.6 to –159.4	Lam (1978)	A	85.9*
Plot 7	–82.5 to –153.0	Plot 7	–130.0 to –163.0		B	63.0*
					C	2.30*

*15 months' erosion data.

the total ^{137}Cs loading respectively. This equates to ground lowering totals, using the ^{137}Cs depth distribution of Fig. 3.3, of approximately 45–61 mm for Plot No. 8 and 63–79 mm for Plot No. 7 when derived from Fig. 3.3. For Plot No. 7, this amounts to –130.0 to –163.0 t/ha/year when expressed as estimated erosion rates from the ^{137}Cs inventories, which is similar to the erosion pin estimates of –82.5 to –153.0 t/ha/year (Table 3.2). Plot No. 8 estimates are less similar, with –117.6 to –159.4 t/ha/year estimated from ^{137}Cs data compared with –42.4 to –85.1 t/ha/year estimated from the erosion pins (Table 3.2). This may be consequent upon the fact that, given the shape of the ^{137}Cs reference profile in Fig. 3.3, initial relatively large losses of ^{137}Cs do not result in large surface lowering. A minimum amount of erosion may be necessary before the techniques show close agreement, unsurprising given the variability of data.

A proportional approach comparing the soil profile and plot ^{137}Cs data is appropriate given that the ^{137}Cs values were recorded on more than one machine: the use of percentage and relative values removes any potential influence of machine variability. The ^{137}Cs depth distributions of two adjacent sites were utilized because the soil erosion pin study is on-going and excavating soil pits from the plots themselves to obtain depth profiles would result in considerable disturbance.

Methods for relating ^{137}Cs losses to actual rates of soil erosion include those based upon theoretical models or accounting procedures and those using empirical data from erosion plots (Walling and Quine, 1991). Most empirical calibrations utilize erosion plots to develop simple functions relating ^{137}Cs loss to soil erosion rates (e.g. Loughran and Campbell, 1995) and they are attractive because they are easy to apply. The more sophisticated models, such as those incorporating diffusion and migration (Walling and He, 1999), were not considered for this exploratory study. Given the limited examples of the latter approach (Campbell *et al.*, 1986; Higgitt and Haigh, 1995), the proportionality, though not direct, observed between ^{173}Cs inventories and soil erosion in Table 3.1 is encouraging. In the future, repeat sampling of Plot Nos 7 and 8 may permit a simple calibration equation of surface erosion and ^{137}Cs inventory similar to that derived by Loughran and Campbell (1995).

Discussion

Quine (1995) suggested that the simplest approach to converting ^{137}Cs loss to soil loss is the application of an empirically derived relationship. Limitations with this approach include the variability of the ^{137}Cs and erosion pin data (Sutherland, 1991), the difficulties of obtaining sites where quantitative records of erosion extend back to 1954, potential measurement errors and the representativeness of empirical conversion models (Quine, 1995), which may be limited both spatially and in terms of process.

Variability of data

Table 3.1 evidences the variability of the ^{137}Cs and the erosion pin measurements at the three plots in this study. This is consistent with other studies of ^{137}Cs or erosion pins (Lam, 1977; Statham, 1990; Sutherland, 1991; Saynor *et al.*, 1994; Owens and Walling, 1996; Basher, 2000; Theocharopoulos *et al.*, 2003). The coefficients of variation for the data in Table 3.1 indicate a need to consider spatial variability in monitoring erosion and when comparing different measurement techniques. Part of the variability for both techniques may arise from the number of samples taken (see Sutherland, 1991; Higgitt, 1995), but the present study is necessarily limited by practical concerns. Furthermore, part of this variability may also reflect measurement error.

The increasing variability of ^{137}Cs for the three soil erosion plots as erosion proceeds suggests that eroded sites will generally require a larger sample size to estimate ^{137}Cs content to a specified level of accuracy (Table 3.1). This ^{137}Cs variability may reflect the spatial variation of erosion superimposed upon the original variation arising from ^{137}Cs input. Figure 3.2a shows the spatial variability of erosion measured using the erosion pins, with no obvious pattern evident. Interestingly, the variability of erosion obtained from the pin data is considerably higher for the vegetated control Plot No. 1 than the eroded plots. This may arise from the variation of process and material caused by the vegetation, with rainsplash, stemflow or organic matter accumulation varying with proximity of pins to individual plants.

Temporal representativeness of measurements

A difficulty with ^{137}Cs conversion is that the ^{137}Cs inventory is a measure of soil movement since 1954 (e.g. Loughran, 1989; Higgitt, 1995), while the erosion pin study only recorded soil movement over a 6-year period for Plot No. 1 and much shorter timescales (around 3–4 years) for the two bare plots. Consequently, low frequency, high magnitude erosion events (e.g. very large rainstorms) may have occurred outside the monitoring period (Loughran, 1989; Quine and Walling, 1993; Walling and He, 1999), and these short-term erosion rate estimates may not be directly comparable to those predicted by ^{137}Cs. Similar problems were identified by Schuller *et al.* (2000) and Fulajtar (2003), who also comments on the problems that may be raised in comparing different slope angles and lengths.

It is suggested that the stability afforded to the soil by the vegetation cover on the control plots and the artificial clearance of Plot Nos 7 and 8 mean that the data on pin exposure and the relative changes in ^{137}Cs inventories are measuring the same net quantities of erosion. As noted previously, the general stability of the site prior to the commencement of the study has been independently checked by the examination of aerial photographs dating back to 1956 and by careful field observations. The absence of erosion on the control plots for the duration of the study is further supportive evidence.

The results of soil erosion reported in Tables 3.1 and 3.2 derived from ^{137}Cs in the case of bare Plot Nos 7 and 8 indicate that, whilst ^{137}Cs can be used to derive estimates of erosion over the medium term, the technique is also capable of providing information of soil erosion over the short term. It therefore has potential to be used as an indicator of soil movement in systems that have recently undergone land use changes. In Hong Kong and the region, an area experiencing rapid change, this ability may prove very useful.

Spatial representativeness of ^{137}Cs reference inventories in Hong Kong

The spatial variability of ^{137}Cs inventories in Hong Kong is particularly influenced by rainfall variability. Figure 3.4 shows considerable spatial variation in mean annual rainfall of Hong Kong for the period 1961–1990, in part reflecting topography. Given evidence of rainfall-dependent ^{137}Cs loadings in Hong Kong (Ruse, 1999), the reference or control plot must have restricted spatial representativeness. A number of empirical calibrations may need to be developed for different altitudinal zones. Consequently, despite the encouraging results that ^{137}Cs inventories decrease as soil loss increases (Table 3.1), it may be necessary to investigate theoretically derived relationships, such as those described by Owens and Walling (1998) and Walling and He (1999), to convert ^{137}Cs to soil loss across larger areas of Hong Kong.

The enhanced ^{137}Cs levels on Plot No. 1 in comparison to the reference sites may also be due to deposition of ^{137}Cs-rich sediment or ^{137}Cs derived from run-on water. Overland flow has been observed during intense rainfall near Plot No. 1. However, overland flow has not contained noticeable amounts of sediment during the years of monitoring, and Plot Nos 1, 7 and 8 have similar topographical positions and morphology.

Using the shape of the ^{137}Cs profile to convert per cent ^{137}Cs loss to actual soil surface loss assumes that the similar reference ^{137}Cs profiles of Sites A and B approximate the unmeasured soil ^{137}Cs profiles of the rather steeper erosion plot sites and that they have not experienced sediment deposition. It also assumes that the depth distributions of ^{137}Cs have no significant differences in post-depositional redistribution of ^{137}Cs.

Measurement errors

The existence of potential measurement errors in the variation of ^{137}Cs and surface change recorded by the erosion pins and reported in Table 3.1 must be recognized. A key question is how accurately pin height above the ground surface can be assessed. Haigh (1977) examined some of the methods proposed to minimize the measurement problem, indicating that changes in exposure of the erosion pin may result not from erosion or deposition of soil but from, for example, freezing and thawing, soil creep or compaction. The comparison of mean values of ^{137}Cs and pin exposure assume an accurate measurement and, in the case of pins, also assumes that a change in exposure reflects only erosion or deposition. The use of distributed erosion pin and ^{137}Cs measurements removes a common difficulty with erosion pin studies that not all soil displaced on the plot may be fully removed from the study plot.

Krause *et al.* (2003) provide evidence of ^{137}Cs detector errors of around 3–8% at six reference sites in New South Wales, Australia. The possibility of detector error may be particularly important when interpreting ^{137}Cs data which give erosion rates that are marginally above zero. Under such circumstances Krause *et al.* (2003) suggest it may be better to accept that the region is not affected by surface erosion to any significantly detrimental extent. Table 3.2 reveals that such may be the case in the comparison of Plot Nos 1 and 8.

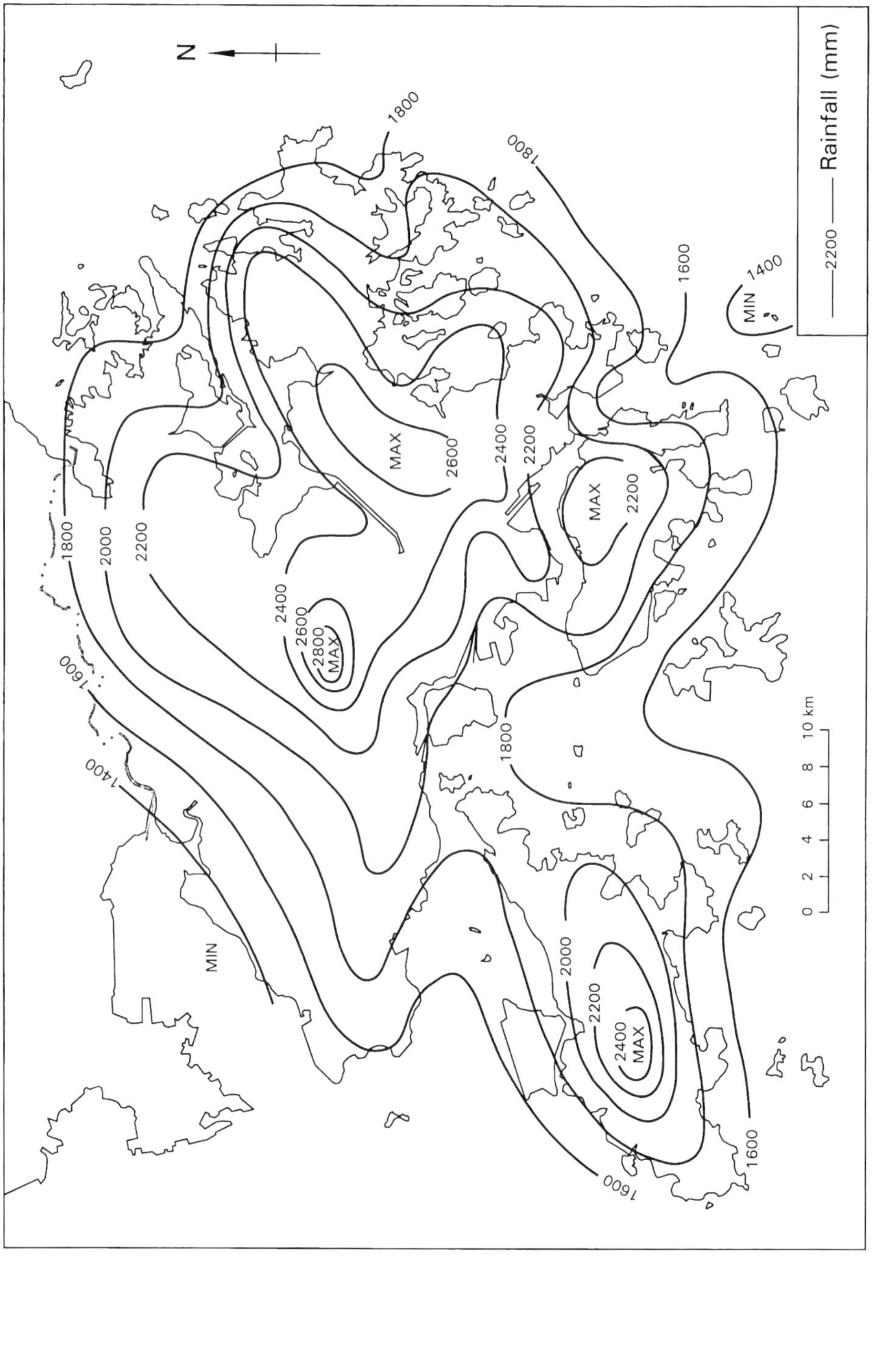

Fig. 3.4. Spatial variation of annual rainfall 1961–1990 in Hong Kong.

Comparison with other erosion values

Table 3.2 compares the erosion data from this study with that found by others in Hong Kong. Fok (2001) and Lam's (1978) drainage basin studies confirm the stability afforded by vegetation in small catchments with erosion rates of 0.42 and 2.3 t/ha/year respectively, values which are similar to the vegetated erosion Plot No. 1 in this study (Table 3.2). Lam's (1978) data for two small basins (A and B, Table 3.2) containing 'badlands', or bare soil surfaces, indicate accelerated rates of erosion when vegetation is absent. Moreover, the erosion rates reported by Lam (1978) for his drainage basins containing 'badlands' are not dissimilar to those recorded on the two erosion plots from which vegetation has been removed (Plot Nos 7 and 8, Table 3.2). It should also be remembered that, when comparing erosion plot, or at-a-point, estimates of erosion with those derived from drainage basin studies, the effects of sediment delivery need to be considered. It is well known that only a fraction of the sediment eroded within a drainage basin will find its way to the outlet and be represented in the sediment yield and erosion rate. Indeed, Lam (1978) reports that the observed suspended sediment yields for Catchments A and B are respectively only 59% and 69% of the estimated sediment production from badland surfaces in the basins. The present study recorded lower rates, of around 24 mm and 57 mm of erosion over 33 and 45 months of observations, than Lam's (1977) results, which gave mean vertical ground lowering of 23.9 mm for a 15-month sampling period. However, the bare plots of the present study retained plant root systems that were absent from Lam's true 'badland' soils.

Conclusion

Conventional techniques for monitoring sediment redistribution have a number of limitations (see Loughran, 1989; Zapata, 2003). These include the representativeness of the data, the necessity for repeated visits/measurement to the site and their limited potential for providing medium to long-term records. Some of these difficulties have been overcome through the application of the ^{137}Cs technique. This study has observed statistically significant differences in ^{137}Cs content on three plots where differences of erosion have been monitored using erosion pins. While ^{137}Cs inventories were not proportional to erosion pin data, converted ^{137}Cs erosion estimates showed encouraging agreement with erosion pin data as erosion proceeds. Further, repeat sampling of the soil erosion plots for ^{137}Cs and pin height change may develop an empirical calibration between ^{137}Cs and soil erosion in the area. However, because of the rugged terrain, the association between ^{137}Cs input and precipitation and the intrinsic variability of erosion data, the representativeness of such a calibration may be spatially limited. As the bare erosion plots of the present study did not develop rills or gullies, the ^{137}Cs inventory appears to be detecting sheet erosion. As Loughran noted, this ability to measure sheet erosion may be 'one reason for the accelerated use of ^{137}Cs in soil erosion studies' (1989, p. 227). The present work indicates the value of ^{137}Cs to identify areas that have experienced sheetwash and splash erosion in Hong Kong. However, other processes act to remove ^{137}Cs from hillslopes. Soil piping, tunnelling, eluviation and particularly landslides, common in the climatic regime of Hong Kong (e.g. Ruse *et al.*, 2002), could remove, entrain or bury the entire ^{137}Cs profile. This fact supports the ongoing calls for integrated approaches to assessment of geomorphological change within watersheds (Wallbrink *et al.*, 2002).

Quine (1995) observed: 'the full potential of the (^{137}Cs) technique will only be realised when there is general acceptance that the derived rates are reliable' (p. 308). The comparison of soil ^{137}Cs levels to known erosion rates, such as those derived in the present study, indicates that the ^{137}Cs technique offers potential for erosion studies in Hong Kong and the region. Moreover, the data from this study also suggest that, where a temporal record of land use is available, the ^{137}Cs inventory may identify the effects of land use change upon soil erosion and redistribution over the short term.

Acknowledgements

Financial support from the CRCG and the Stephen S.F. Hui Fund is gratefully acknowledged. We also wish to express our thanks to the two anonymous reviewers and Phil Owens and Alison Collins for their helpful comments.

References

Basher, L.R. (2000) Surface erosion assessment using ^{137}Cs: examples from New Zealand. *Acta Geologica Hispanica* 35, 219–228.

Campbell, B.L., Loughran, R.J., Elliott, G.L. and Shelly, D.J. (1986) Mapping drainage basin sediment sources using caesium-137. In: Hadley, R.F. (ed.) *Drainage Basin Sediment Delivery*. IAHS Publication 159, IAHS Press, Wallingford, UK, pp. 437–446.

Campbell, B.L., Loughran, R.J. and Elliott, G.L. (1988) A method for determining sediment budgets using caesium-137. In: Bordas, M.P. and Walling, D.E. (eds) *Sediment Budgets*. IAHS Publication 174. IAHS Press, Wallingford, UK, pp. 171–179.

Chappell, A., Warren, A., Oliver, M.A. and Charlton, M. (1998) The utility of ^{137}Cs for measuring soil redistribution rates in southwest Niger. *Geoderma* 81, 313–337.

Chen, R., Hill, R.D. and Corlett, R.T. (1997) Energy and nutrient flow through the storage and consumption of upland phytomass fuel. *Forest Ecology and Management* 93, 63–71.

Corlett, R.T. (1999) Environmental forestry in Hong Kong: 1871–1997. *Forest Ecology and Management* 116, 93–105.

Fok, L. (2001) Sediment and solute loads in a small Hong Kong stream. Unpublished MPhil thesis. University of Hong Kong, Hong Kong.

Fulajtar, E. (2003) Assessment of soil erosion on arable land using ^{137}Cs measurements: a case study from Jaslovske Bohunice, Slovakia. *Soil and Tillage Research* 69, 139–152.

Geotechnical Control Office (1988) *Yuen Long: Solid and Superficial Geology*, 1 : 20,000 map. Map HGM20 Sheet 6, 1st edn. Geotechnical Control Office, Civil Engineering Department, Hong Kong.

Haigh, M.J. (1977) The use of erosion pins in the study of slope evolution. *Shorter Technical Methods II*, Technical Bulletin 18, British Geomorphological Research Group. Geo Books, Norwich, UK, pp. 31–49.

Higgitt, D.L. (1995) The development and application of caesium-137 measurements in erosion investigations. In: Foster, I.D.L., Gurnell, A.M. and Webb, B.W. (eds) *Sediment and Water Quality in River Catchments*. Wiley, Chichester, UK, pp. 287–305.

Higgitt, D.L. and Haigh, M.J. (1995) Rapid appraisal of erosion rates on mining spoils using caesium-137 measurements. *International Journal of Surface Mining, Reclamation and Environment* 9, 141–148.

Krause, A.K., Loughran, R.J. and Kalma, J.D. (2003) The use of caesium-137 to assess surface soil erosion status in a water-supply catchment in the Hunter Valley, New South Wales, Australia. *Australian Geographical Studies* 41, 73–84.

Lam, K.C. (1977) Patterns and rates of slopewash on the badlands of Hong Kong. *Earth Surface Processes* 2, 319–332.

Lam, K.C. (1978) Soil erosion, suspended sediment and solute production in three Hong Kong catchments. *Journal of Tropical Geography* 47, 51–62.

Loughran, R.J. (1989) The measurement of soil erosion. *Progress in Physical Geography* 13, 216–233.

Loughran, R.J. and Campbell, B.L. (1995) The identification of catchment sediment sources. In: Foster, I.D.L., Gurnell, A.M. and Webb, B.W. (eds) *Sediment and Water Quality in River Catchments*. Wiley, Chichester, UK, pp. 189–205.

Owens, P.N. and Walling, D.E. (1996) Spatial variability of caesium-137 inventories at reference sites: an example from two contrasting sites in England and Zimbabwe. *Applied Radiation and Isotopes* 47, 699–707.

Owens, P.N. and Walling, D.E. (1998) The use of a numerical mass balance model to estimate soil redistribution on uncultivated land from ^{137}Cs measurements. *Journal of Environmental Radioactivity* 40, 185–203.

Quine, T.A. (1995) Estimation of erosion rates from caesium-137 data: the calibration question. In: Foster, I.D.L., Gurnell, A.M. and Webb, B.W. (eds) *Sediment and Water Quality in River Catchments*. Wiley, Chichester, UK, pp. 307–329.

Quine, T.A. and Walling, D.E. (1993) Use of caesium-137 measurements to investigate relationships between erosion rates and topography. In: Thomas, D.S.G. and Allison, R.G. (eds) *Landscape Sensitivity*. Wiley, Chichester, UK, pp. 31–48.

Ritchie, J.C. and McHenry, J.R. (1990) Application of radioactive fallout caesium-137 for measuring soil erosion and sediment accumulation rates and patterns: a review. *Journal of Environmental Quality* 19, 215–233.

Ruse, M.E. (1999) An investigation of Caesium-137 at reference locations in Hong Kong. Unpublished PhD thesis. University of Hong Kong, Hong Kong.

Ruse, M.E. and Peart, M.R. (2000) Intrasite sampling of Hong Kong soils contaminated by caesium-137. *Chemosphere* 41, 45–51.

Ruse, M.E., Waring, D.P., Kaldy, A., Chan, S.K. and Ng, K.C. (2002) Initiation and runout characteristics of a swarm of 121 landslides in the Tsing Shan Foothills, Hong Kong. In: Franks, C., Hall, J., Martin, D., Ng, S., Parry, S., Pinches, G. and Thorn, M. (eds) *Natural Terrain – a Constraint to Development?* Institute of Mining and Metallurgy, Hong Kong Branch, Hong Kong, pp. 77–87.

Saynor, M.J., Loughran, R.J., Erskine, W.D. and Scott, P.F. (1994) Sediment movement on hillslopes measured by caesium-137 and erosion pins. In: Olive, L.J., Loughran, R.J. and Kesby, J.A. (eds) *Variability in Stream Erosion and Sediment Transport*. IAHS Publication 224. IAHS Press, Wallingford, UK, pp. 87–93.

Schuller, P., Sepulveda, A., Trumper, R.E. and Castillo, A. (2000) Application of the ^{137}Cs technique to quantify soil redistribution rates in paleohumults from Central-South Chile. *Acta Geologica Hispanica* 35, 285–290.

Statham, I. (1990) Slope processes. In: Goudie, A., Anderson, M., Burt, T., Lewin, J., Richards, K., Whalley, B. and Worsley, P. (eds) *Geomorphological Techniques*, 2nd edn. Unwin Hyman, London, pp. 225–259.

Sutherland, R.A. (1991) Examination of caesium-137 areal activities in control (uneroded) locations. *Soil Technology* 4, 33–50.

Theocharopoulos, S.P., Florou, H., Walling, D.E., Kalantzakos, H., Christou, M., Tountas, P. and Nikolaou, T. (2003) Soil erosion and deposition rates in a cultivated catchment area in central Greece, estimated using the ^{137}Cs technique. *Soil and Tillage Research* 69, 153–162.

Wallbrink, P.J., Belyaev, V., Golosov, V.N., Sidorchuk, A.S. and Murray, A.S. (2002) *Use of radionuclide, field based and erosion modelling methods for quantifying rates and amounts of soil erosion processes*. Consultancy report. CSIRO Land and Water, Canberra.

Walling, D.E. and He, Q. (1999) Improved models for estimating soil erosion rates from caesium-137 measurements. *Journal of Environmental Quality* 28, 611–622.

Walling, D.E. and Quine, T.A. (1991) The use of caesium-137 measurements to investigate soil erosion on arable fields in the UK: potential applications and limitations. *Journal of Soil Science* 42, 147–165.

Wise, S.M. (1980) Caesium-137 and lead-210: a review of the techniques and some applications in geomorphology. In: Cullingford, R.A., Davidson, D.A. and Lewin, J. (eds) *Timescales in Geomorphology*. John Wiley, Chichester, UK, pp. 109–127.

Zapata, F. (2003) The use of environmental radionuclides as tracers in soil erosion and sedimentation investigations: recent advances and future developments. *Soil and Tillage Research* 69, 3–13.

4 Assessing the Contribution of Different Processes to Soil Degradation within an Arable Catchment of the Stavropol Upland, Southern European Russia

V.R. Belyaev,[1] A.Yu. Sidorchuk,[2] V.N. Golosov,[1] P.J. Wallbrink[3] and A.S. Murray[4]

[1]*Moscow State University, Russia;* [2]*Moscow State University, Russia.* Present address *Landcare Research, New Zealand;* [3]*CSIRO Land and Water, Australia;* [4]*Aarhus University, Denmark*

Introduction

A number of different methods and techniques are used to quantify rates of soil redistribution on arable land. However, the opportunities for comparison and integration of their results are limited, and such examples are rare in the literature. In this study we applied a suite of methods to quantify soil redistribution and evaluate the relative importance of particular processes and their contribution to a sediment budget for the entire period of cultivation in a study area, located within the Stavropol Upland, southern European Russia. Four different approaches are used in this study (Table 4.1): (i) soil survey; (ii) ephemeral gully volume measurements; (iii) tracing soil redistribution using caesium-137 (^{137}Cs) measurements; and (iv) Universal Soil Loss Equation (USLE)-based model modified for Russian conditions.

Study Area

General description

The study area is located within the eastern part of the Stavropol Upland (Fig. 4.1a). The regional climate is characterized by relatively cold winters (< 0°C for 3–4 months) and hot summers (up to 40°C). The mean annual precipitation is about 400 mm, mostly associated with heavy summer rainstorms > 30–60 mm for one event.

The study area occupies the southern aspect slope of the Dolgaya balka (a dry valley) 3.5 km north of Gofitskoe village (Fig. 4.1a). The range in height between the balka bottom and the ridge top is about 150 m. The selected slope unit is approximately 2.5 km long and has a complex shape (Fig. 4.1b). The upper part is a relatively steep, uncultivated convex escarpment 600–700 m long

Table 4.1. Summary of approaches used in this study for quantifying soil redistribution processes and rates.

Methods	References	Temporal resolution	Soil redistribution processes accounted for
(1) Soil profile comparison	Larionov *et al.* (1973), Kiryukhina and Serkova (2000)	Period of cultivation	All
(2) ^{137}Cs proportional model	Kachanoski (1987)	*c.* 50 years	Sheet erosion and tillage translocation/deposition
(3) ^{137}Cs mass balance model	Walling and He (1999)	*c.* 50 years	Sheet erosion and tillage translocation/deposition
(4) ^{137}Cs depth profiles	Belyaev *et al.* (2004)	*c.* 50 years with possible additional time marks (1957, 1963, 1986)	Deposition
(5) Ephemeral gully volume measurements	Boardman and Evans (1994), Poesen *et al.* (1996)	Single runoff event or several runoff events between tillage operations	Linear erosion
(6) USLE-based model modified for Russian conditions	Larionov (1993), Larionov *et al.* (1998), Krasnov *et al.* (2001)	From single runoff event up to entire cultivation period; limited by availability of meteorological data	Sheet erosion

(unit 1, Fig. 4.1b) with an average gradient of 0.1–0.2. Downslope it changes into a midslope concavity (unit 2, 300–400 m long), which connects the upper steep section to the lower gentle straight slope. This is about 1500 m long (comprising units 4–7) with an average gradient of 0.04–0.05 and an undulating cross profile. The cultivated part of the study area (comprising units 2 and 4–7) is about 1800 m long, with an average width of 700 m and area of 1.3 km^2. The cultivated field is separated from the Dolgaya balka bottom by a grassed buffer strip less than 100 m wide. A forest shelter belt (approximate width 35 m, unit 3), designed to protect the field from wind erosion, runs along the entire length of the western side of the field. As it has been a permanent feature since the beginning of cultivation (around 1930), we chose it as an undisturbed reference site for the soil survey and ^{137}Cs methods. The study field is characterized by Chernozem soils. Visual observation suggests that the soil surface in the cultivated area is 20–30 cm lower than the adjacent undisturbed surface of the forest shelter belt (unit 3), most likely due to severe erosion.

Erosion and deposition features of the study site

The major features of the study area are a series of semi-parallel hollows running downhill almost the entire length of the arable slope. Ephemeral gullies develop in the hollow bottoms after high-magnitude runoff events; the length of the main branches can exceed 1 km. Sediment from the gullies is deposited as a series of fans on the grassed balka slope below the lower field boundary (Fig. 4.1b).

Observations over successive years have shown that ephemeral gullies are semi-permanent features and reappear almost at the same place, even after their removal by ploughing. The largest branches are completely filled only by harrowing during cereal treatment, but the hollows remain prominent. A schematic representation of the interaction of linear erosion, mechanical translocation and sheet erosion is presented in Fig 4.2. Between ephemeral gully infill and incision events, soil redistribution occurs as a result of moderate water erosion and mechanical translocation, mainly limited to within the arable field. The ephemeral gully network is the main route of

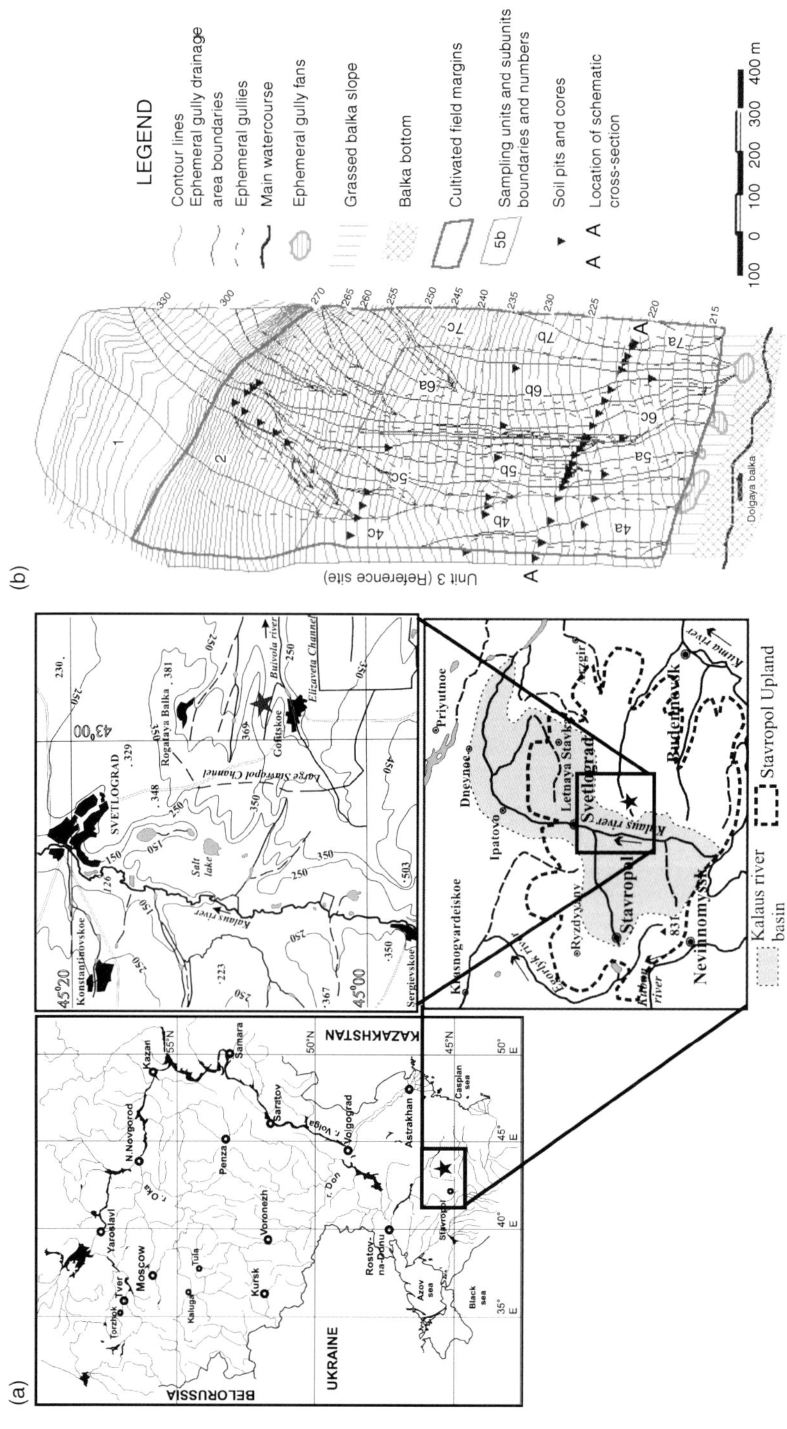

Fig. 4.1. The study area location (a), landscape units and sampling locations within the study site (b).

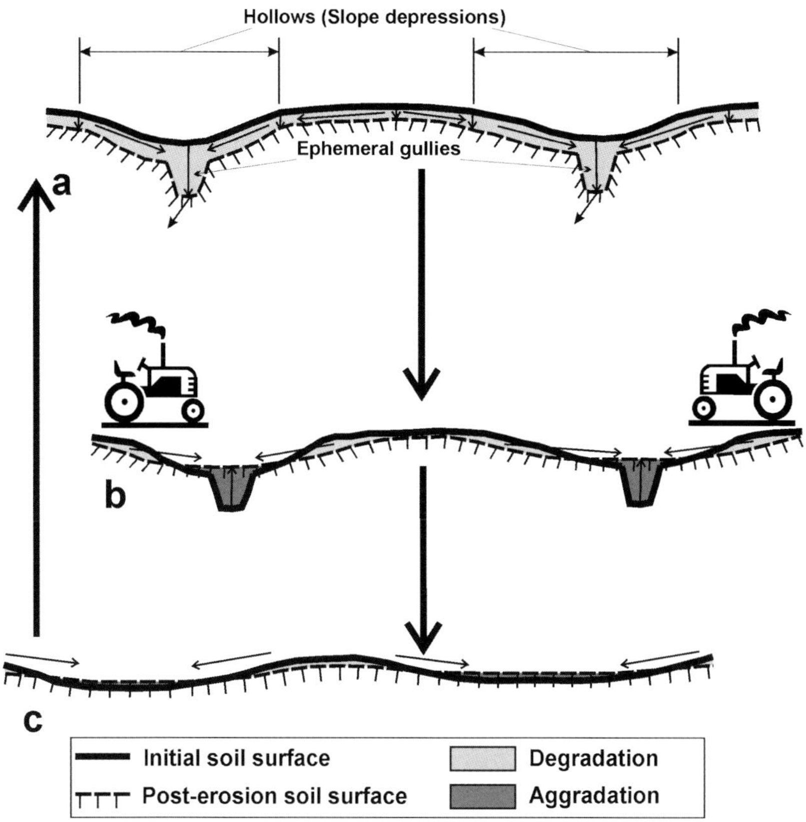

Fig. 4.2. A schematic representation of soil redistribution processes within the study field. (a) High-magnitude runoff event results in ephemeral gully incision in the bottoms of hollows, delivering material from their former infill stage as well as sediment produced by erosion from inter-hollow areas out of the arable field. (b) Ephemeral gullies are infilled by harrowing procedure during cereal treatment, hollows remain prominent, providing routes for runoff concentration in the future. (c) Moderate soil redistribution by water erosion and mechanical translocation occurs between stages (b) and (a), mainly limited to within the arable field.

sediment delivery from the study field to the adjacent dry valley bottom.

Experimental Design and Sample Analysis

Fieldwork was undertaken during the summers of August 1993 and August–September 1994. The observation and sampling programme included:

1. The organic topsoil layer thickness was measured in 31 soil pits and 19 soil cores (Fig. 4.1b). The locations of the pits and cores within the cultivated field were chosen to represent (a) hollow bottoms and (b) the flat areas between them. Soil profile differences between these two geomorphic locations were analysed to determine differences in soil degradation rates (Fig. 4.3).

2. The cross-sectional areas of ephemeral gullies were measured by tape at 50 representative locations.

3. A topographic survey by an optical theodolite was undertaken to build a digital elevation model of the study site (Fig. 4.1b).

4. The soil sampling for ^{137}Cs inventories involved taking (a) depth increment samples from pits in the reference area of the forest-shelter

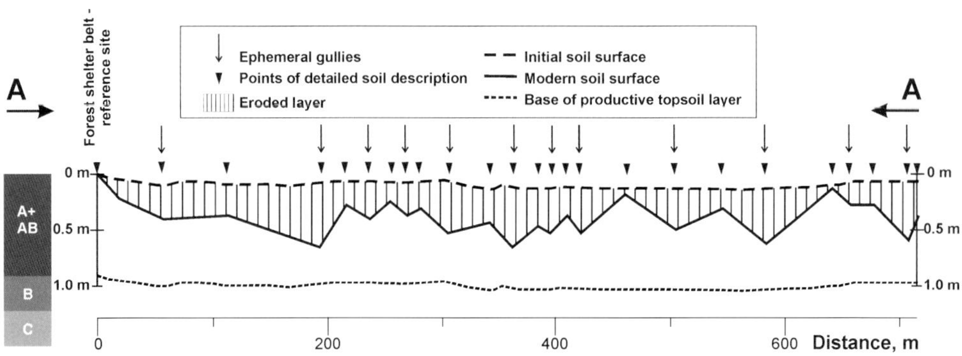

Fig. 4.3. A schematic representation of soil thickness variation along the cross-section A–A (for location see Fig. 4.1b) and method of erosion layer reconstruction, assuming initially uniform soil thickness.

belt (one pit), cultivated field (three pits), sheet deposition zone (one pit) and ephemeral gully fan (one pit), and (b) a series of bulked cores from the different landscape units (Fig. 4.1b). For bulk sampling, each of the cultivated field units (2, 4–7) was divided into three roughly equal areas (subunits a, b and c, Fig. 4.1b). Ten cores were taken randomly from each area, weighed and then mixed together. One subsample (250 g) was then extracted for subsequent gamma analysis. The mixing method does not allow a point-based analysis of ^{137}Cs distribution, but provides a cost-effective method to quantify differences in mean ^{137}Cs inventories between landscape units (Wallbrink *et al.*, 2002).

5. A USLE-based model was employed to independently calculate sheet erosion rates for the cultivated field. The version we used was developed for use in Russia using a large spatially distributed data set of coefficients. Modifications include an improved set of equations for determining topographic factors (Larionov, 1993; Larionov *et al.*, 1998), calculating and mapping a rainfall erosivity index for European Russia (Krasnov *et al.*, 2001), as well as adaptation of land use factors and soil protection techniques specific to the Russian agricultural system. The component for calculating soil erosion by snowmelt runoff, based on an empirical model developed by the Russian State Hydrological Institute, was also added (Larionov, 1993). The input data, including topography, geology, climate, soil, vegetation properties and land use, were defined from a combination of field observations, laboratory analyses and published and unpublished sources.

All ^{137}Cs and grain size analyses were undertaken at the CSIRO Land and Water radionuclide facility in ACT, Australia. The soil and sediment samples were oven-dried, ashed at 400°C, ground in a ring mill and then analysed by gamma spectrometry according to methods outlined in Murray *et al.* (1987).

Results

Estimates of soil redistribution rates obtained by all the methods described in Table 4.1 were first analysed separately, Table 4.2 and Fig. 4.3 presenting results from method (1), Table 4.3 methods (2), (3), (5) and their combination, and Table 4.4 method (6). By integrating these and analysing all data in combination, a provisional sediment budget for the study slope was constructed (Table 4.5 and Fig. 4.4).

Estimates of soil loss based on soil survey and pit data

Quantifying soil degradation rates from soil profile analysis involves measuring the difference of the organic topsoil thickness between undisturbed and cultivated locations. With information on the soil bulk density, average values of soil degradation or aggradation rates can be estimated. These values must be attributed to the

Table 4.2. Soil erosion rates and volumes estimated from soil survey pit data for the cultivated part of the study area (1.3 km²).

	Flat areas between hollows	Hollows	Total erosion (sheet and linear erosion)*
Area (ha)	117.0	10.0	127.0
Mean soil loss (mm/year)	3.5	5.7	4.0
Mean soil loss (kg/m^2/year)	5.2	8.5	5.4
Mean annual soil loss (t/year)	6,230	850	7,080
Total soil loss since 1930 (t)	399,000	54,300	453,300
Total eroded layer since 1930 (mm)	225	365	255

*Values have been obtained by summing erosion rates from both geomorphic units considered, recalculated for the entire arable area.
Data have been rounded.

Table 4.3. Estimates of ephemeral gully volumes, sheet erosion and tillage translocation rates from the ^{137}Cs conversion models and total soil erosion from the entire arable field by combining data from these two approaches.

	Ephemeral gully volumes*	Proportional model		Mass-balance model	
Method	Direct soil loss from linear erosion	Sheet erosion and mechanical translocation	Ephemeral gully volumes added	Sheet erosion and mechanical translocation	Ephemeral gully volumes added
Mean soil loss (mm/year)	0.4	3.9	4.1	3.7	3.9
Mean soil loss (kg/m^2/year)	0.6	5.5	5.6	5.3	5.4
Total eroded layer since 1930 (mm)	27	260	270	240	250
Mean amount of soil eroded (t/year)	740	6,950	7,100	6,680	6,800
Total soil loss since 1930 (t)	47,400	437,700	447,300	421,000	428,400

*Assuming re-incision cycle recurrence of 10 years.
Data have been rounded.

Table 4.4. Estimates of soil loss by sheet erosion from the study field obtained using the USLE model modified for Russia.

	Mean (1960–1980)	Maximum (1968)	Minimum (1978)
Rainfall erosivity index	13.9	34.4	0.7
Mean soil loss (kg/m^2/year)	0.9	2.7	0.1
Annual eroded layer (mm)	0.6	2.0	0.1
Mean amount of soil eroded (t/year)	1,175	3,490	81
Extrapolated for 1930–1994 period, assuming constant erosion rates:			
Total eroded layer since 1930 (mm)	42	125	3.0
Total soil loss since 1930 (t)	75,200	223,360	5,184

Data have been rounded.

Table 4.5. Summary of mean annual soil losses and gains (t/year and kg/m^2/year) from the different landscape units in the study area.

	Landscape units				
	Main arable slope (erosion)		Grassed lower slope and balka bank (deposition)		Sediment delivery to the balka bottom (output)
Methods	Sheet erosion areas	Ephemeral gullies	Sheet deposition areas	Ephemeral gully fans	
Soil profile comparison	−6230/−4.7	−850/−0.7		+150/+10.7	−6930/−5.2
^{137}Cs proportional model	−6950/−5.5		+80/+1.5	+95/+7.0	−6775/−5.1
^{137}Cs mass balance model	−6680/−5.3		+80/+1.5	+86/+6.3	−6514/−4.9
^{137}Cs depth profiles			+180/+3.4	+490/+35.9	
Direct gully volume measurement		−740/−0.6			
USLE	−1175/−0.9				
Preferred values (methods used)	−6620/−5.3 (1, 2, 3)	−795/−0.6 (1, 5)	+180/+3.4 (4)	+320/+23.3 (1, 4)	−6915/−5.2

Data have been rounded. Reasons for selection of methods described in text.

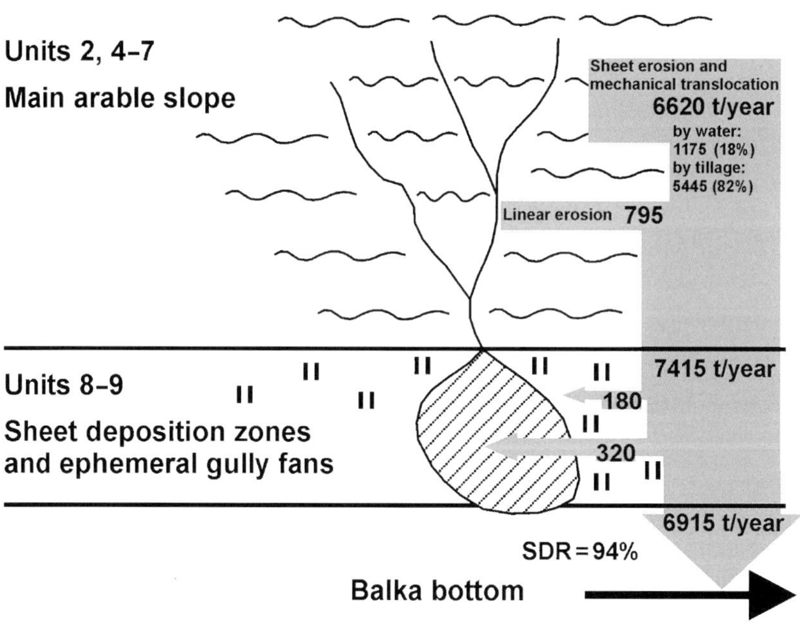

Fig. 4.4. A sediment budget for the study area based on erosion rate data from Table 4.5.

entire cultivation period and integrate the impact of all soil redistribution processes. By analysing soil profiles from hollows and flat areas separately, it is possible to evaluate increases in soil degradation rates at locations directly affected by ephemeral gully development (Figs 4.2 and 4.3).

Two pits from an undisturbed soil surface under the forest shelter belt (Fig. 4.1a) were used as a reference. Total thickness of the upper organic soil layer (A and AB horizons) was 78 cm for the upper pit and 91 cm for the lower pit, thus proving an expected increase of the undisturbed soil thickness downslope. Both of these values are within the normal range of variation (70–120 cm, with coefficient of variation (C_V) 7–15% for individual soil horizons) obtained from the analysis of 40 statistically significant populations of soil pits for undisturbed Chernozem soil profiles in North Caucasian and Stavropol regions (Valkov, 1977). For 48 observation points from the cultivated area, the mean organic layer thickness was estimated as 56.5 cm, with a C_V of 38% and skewness towards lower values. The difference between the mean value from the cultivated area and reference pits (21.5–34.5 cm) is comparable to the observed excess of forest shelter belt soil surface above the field (20–30 cm). Of 48 observation points, 16 had an A + AB thickness below 50 cm, showing a degradation of 35% to 45% of the topsoil (Fig. 4.3). Clear evidence of accelerated soil degradation within the study field suggests that our reference pit data are suitable for comparing with values from the arable field.

The area of hollows occupied by ephemeral gullies is 10.0 ha. The mean annual soil loss from these hollows (850 t/year or 8.5 kg/m^2/year) represents 12% of the total value obtained for the arable field (7080 t/year or 5.4 kg/m^2/year) and can be considered as a direct contribution of linear erosion (Table 4.2). About 6230 t/year or 5.2 kg/m^2/year (88% of the total erosion from the field) is estimated to result from water erosion and mechanical translocation. However, the cumulative surface lowering by ephemeral gullies concentrated in hollows, even in the situation of repeated infill, is more prominent than that of sheet erosion and tillage translocation in inter-hollow areas (Fig. 4.3).

Soil pit evidence from one of the gully fans at the base of the field showed that the original soil surface was buried by about 1 m of deposited sediment. The surveyed area of gully fans (Fig. 4.1b) was 1.4 ha. Using this information we estimate the total amount of sediment deposited on the fans to be 9540 t (150 t/year or 10.7 kg/m^2/year) or 2% of the total soil eroded from the field since the beginning of cultivation (Table 4.5).

Depth distributions and inventories of ^{137}Cs

The ^{137}Cs depth profile for the undisturbed soil within the forest shelter belt has an approximately exponential shape, with most ^{137}Cs retained within the upper 10 cm of the soil. The total inventory agrees well with the integral samples taken nearby. The reference ^{137}Cs inventory was calculated as 5440 ± 380 Bq/m^2, representing the mean inventory of the forest shelter belt integral samples (30 cores, three mixed samples counted) and the reference pit.

The depth profile from the sheet deposition zone on the grassed balka slope shows an increase in total ^{137}Cs inventory (20%) and depth penetration (10 cm) relative to the reference site. There are no additional ^{137}Cs peaks in the depth profile other than that at the surface. This implies that accumulation here is gradual and peaks from the recent sediment deposits overlie one another in the top 10 cm of the profile. Using these data, the mean rate of deposition for the entire area of grassed balka slope, except gully fans (5.2 ha), is estimated as 180 t/year (3.4 kg/m^2/year) (Table 4.5).

Substantial sediment accumulation can be inferred from the ^{137}Cs depth distribution in the sampled ephemeral gully fan. The 1 m thick sediment layer was deposited above the supposed location of the 1963 year peak, which coincided with the location of the buried original soil surface. The mean sedimentation rate estimated for the entire area of gully fans (1.4 ha) is 490 t/year (35.9 kg/m^2/year) (Table 4.5).

Assessing soil redistribution using ^{137}Cs conversion models

Soil redistribution rates can be derived from the ^{137}Cs data by using various conversion models. Two of the most widely used models were applied in this study: (i) the proportional model (Kachanoski, 1987), and (ii) the mass balance model (Walling and He, 1999). Erosion values estimated by both models are similar (Table 4.3). The good agreement is explained by the extremely high soil degradation rates, as seen by the > 50% total depletion of ^{137}Cs from the cultivated part of the study area. Under such circumstances the uncertainties associated with the ^{137}Cs technique are substantially decreased.

Erosion rates within the cultivated area obtained from both conversion models (5.5 and 5.3 kg/m^2/year, Table 4.3) are comparable to the soil survey data for flat inter-hollow zones (5.2 kg/m^2/year, Table 4.2). This implies that no significant increase in soil redistribution rates occurred between 1930–1953 and 1953–1994. We conclude that the total eroded soil depth and volume for the entire period of cultivation can be calculated from the ^{137}Cs data by simple extrapolation (Table 4.3) in this case.

The calculated deposition rates for the balka bank grassed buffer zone from the conversion models are lower than those from the soil pit and ^{137}Cs depth distribution profiles (Table 4.5). However, these ^{137}Cs conversion models were designed for application to arable fields. Therefore, it is unlikely that their application to uncultivated areas such as grassed slopes will yield reliable results. We believe that the best approach to quantify sediment accumulation in such uncultivated areas is either to compare ^{137}Cs depth distributions with undisturbed reference sites or to use a conversion model developed for uncultivated soils (Walling and He, 1999).

Combining ^{137}Cs conversion model data with gully volume measurements

The estimation of ephemeral gully erosion volumes was based on the assumption that their observed spatial extent and temporal pattern of development has remained constant for the period of cultivation since 1930. The total volume of ephemeral gullies for the one incision cycle has been estimated by multiplying their total surveyed length (Fig. 4.1b) by the mean value of representative cross-section areas measured. From the mean cross-section area of 0.6 m^2 and total surveyed length of 18,700 m, a volume of 11,220 m^3 (equivalent to 7900 t) has been obtained for a single ephemeral gully incision cycle.

Complete infill of ephemeral gullies by cultivation operations can take place only once every 3 years, during cereal treatment. However, major re-incision is likely to be associated with extreme summer rainstorms, which have a recurrence period of 8–10 years (Belyaev et al., 2004). On this basis, we suggest a hypothetical recurrence interval of the complete cycle of ephemeral gully formation of approximately 10 years. Total direct soil loss for the cultivation period is estimated as 47,400 t or 740 t/year (Table 4.3). The cycle may alter due to changes in land management practices and extreme runoff events. Consequently, these values should be treated as an approximation only. These data have been further combined with the erosion rates from the inter-hollow areas estimated by the ^{137}Cs conversion models to derive total soil erosion rates for the arable field (Table 4.3). Comparison of Tables 4.2 and 4.3 shows that the soil survey method and the ^{137}Cs conversion models (combined with gully volume measurements) provide similar estimates of soil loss amounts and rates.

Estimates of soil loss using USLE-based model

Estimates of sheet erosion rates for the arable field from the USLE method are given in Table 4.4. The model was initially run for the period 1960–1980, determined by the availability of input data. The data obtained were then extrapolated for the entire cultivation period (1930–1994), assuming constant erosion rates. This is supported by comparison of soil profile and ^{137}Cs data as discussed above.

Calculations were initially undertaken on an annual basis with inputs of rainfall erosivity index, R, estimated separately for each year. Spatially distributed estimates of annual soil losses from each DEM grid cell were averaged to obtain a single representative value over the 21-year period of 0.9 kg/m^2/year (Table 4.4). It is also possible to analyse each year individually. The highest number of heavy rainstorms occurred in 1968, the year with lowest R value was 1978 (a drought year) and there is nearly a factor of 30 difference in estimated soil losses between those years (Table 4.4).

Soil loss estimates from the model are lower than those from field-based methods (Tables 4.2–4.4). The mean value (0.9 kg/m^2/year) is a factor of six below the mean soil survey and ^{137}Cs-based approaches. However, this does not mean that the USLE-based model applied is completely inappropriate for our study area. The strength of the model may be in its ability to assess erosion rates from areas between ephemeral gully systems, where redistribution occurs by sheet and rainsplash erosion. The discrepancy between model and field-based techniques may be attributed to tillage translocation not accounted for by the USLE model. If this assumption is correct, soil translocation by cultivation operations may be responsible for > 80% (5.6 – 0.9 = 4.7 kg/m^2/year) of the sediment delivered into and further redistributed via the ephemeral gully network.

Discussion

All the methods employed suggest severe soil losses from the arable part of the study field. The estimates from the physically based methods were in very good agreement (Tables 4.2, 4.3 and 4.5). Estimates of the total eroded layer for the entire cultivation period (255, 270 and 250 mm from soil survey, proportional model with gully volumes and mass-balance model with gully volumes, respectively; Tables 4.2 and 4.3) are consistent with one another, and with visual observations of the relative height difference between the forest shelter belt and the arable field (20–30 cm).

Within the cultivated field it is necessary to consider the complex mechanisms of the spatial and temporal interaction of sheet erosion, ephemeral gully erosion and mechanical translocation (Fig. 4.2). The mean value of total soil redistribution rates from inter-hollow zones from methods 1, 2 and 3 is 6620 t/year (5.3 kg/m^2/year) (Table 4.5 and Fig. 4.4). The USLE-based (method 6) value (1175 t/year or 0.9 kg/m^2/year) is treated as an estimate of water erosion in inter-hollow zones, thus allowing us to assess a contribution of mechanical translocation to overall soil redistribution. It has been estimated that it contributes as much as 82% of the total soil loss from interrill zones (5445 t/year or 4.7 kg/m^2/year) (Fig. 4.4). This is consistent with observations elsewhere (Quine *et al.*, 1999).

Soil translocation by cultivation operations does not deliver soil out of the arable field, but serves as a main sediment source for the ephemeral gully network. The latter, in turn, directly produces a relatively low amount of sediment (795 t/year or 0.6 kg/m^2/year – the mean value from methods 1 and 5) (Table 4.5 and Fig. 4.4). This finding agrees well with studies cited in Poesen *et al.* (1996). However, the role of ephemeral gullies as a very efficient transportation pathway for sediment mobilized by other processes is of much higher importance for total soil redistribution within the study area, as almost all of the 6620 t/year (5.3 kg/m^2/ year) of material delivered from inter-hollow areas passes through them on the way to the system outlet.

Deposition within the grassed slope between the ephemeral gully fans was estimated by three methods, but the value from the ^{137}Cs depth profile distribution was taken as the most reliable. For deposition on ephemeral gully fans we took the mean of the soil profile description and ^{137}Cs depth distribution methods. Total losses from the system were estimated by three different methods and there is reasonably good agreement between them. Combining all these data we calculated the sediment yield from the study slope to the adjacent balka bottom as 6915 t/year 5.2 kg/m^2/year) (Table 4.5 and Fig. 4.4).

A provisional sediment budget constructed using the data from Table 4.5 reflects the long-term annual redistribution of soil within and from the field (averaged over 1930–1994) (Fig. 4.4). It shows that severe erosion occurs from the arable field. Sedimentation in fans and on the grassed buffer strip is very limited (6% of the total soil loss from the field). This implies that about 94% of eroded sediment is delivered to the valley system. This is consistent with the findings of Golosov (1996) that many low order streams and rivers in European Russia have aggraded severely following intensive cultivation. The high sediment delivery ratio estimated for the study field can be attributed to specific mechanisms of soil redistribution (Figs 4.2 and 4.4).

Conclusions

Several independent techniques were employed to evaluate soil redistribution for a study area containing a 1.3 km^2 arable field with a semi-permanent ephemeral gully network and a downslope buffer zone of grassed dry valley (balka) bank with depositional fans. In the arable field, severe erosion occurs at a rate of about 5 kg/m^2/year. The contribution of sheet erosion by water (calculated with the USLE) is about 1 kg/m^2/year, whereas mechanical translocation contributed around 4 kg/m^2/year (or 80% of the total). The ephemeral gully network is a transport pathway for practically all eroded and mechanically translocated soil material as well as a source of fresh material itself (at 1 kg/m^2/year). Sediment deposition within the buffer zone is low, representing only 6% of total eroded material. High sediment losses from this system may have significant downstream impacts. The data obtained were combined in a provisional sediment budget showing that inputs from erosion far exceed storage capacity (Table 4.5 and Fig. 4.4).

Our attempt to use multiple independent methods to evaluate soil redistribution rates has highlighted the importance of: (i) comparing such techniques; (ii) validating the results from them; and (iii) the value of combining or integrating them. Best estimates of soil redistribution rates have been obtained by integrating the soil survey and the ^{137}Cs conversion model (combined with gully volume measurement) approaches.

Acknowledgements

The authors would like to acknowledge funding from IAEA technical contract no. 11406/RO, the Australian Department of Trade, the Russian Foundation for Basic Research (project 04-05-64215) and CSIRO Land and Water.

References

Belyaev, V.R., Wallbrink, P.J., Golosov, V.N., Murray, A.S. and Sidorchuk, A.Yu. (2004) Reconstructing the development stages of a gully in the upper Kalaus River basin. *Earth Surface Processes and Landforms* 29, 323–341.

Boardman, J. and Evans, R. (1994) Soil erosion in Britain: a review. In: Rickson, R.J. (ed.) *Conserving Soil Resources: European Perspective*. CAB International, Wallingford, UK, pp. 3–12.

Golosov, V.N. (1996) Redistribution of sediments in small river catchments of temperate zone. In: Walling, D.E. and Webb, B.W. (eds) *Erosion and Sediment Yield: Global and Regional Perspectives*. IAHS Publication 236. IAHS Press, Wallingford, UK, pp. 339–346.

Kachanoski, R.G. (1987) Comparison of measured soil 137-cesium losses and erosion rates. *Canadian Journal of Soil Science* 67, 199–203.

Kiryukhina, Z.P. and Serkova, Yu.V. (2000) Podsolic soils morphometric characteristics variability and the diagnostics of soil erosion. *Soil Erosion and Channel Processes* 12, 63–70 (in Russian).

Krasnov, S.F., Dobrovolskaya, N.G. and Litvin, L.F. (2001) Spatial and temporal aspects of the rainfall erosivity evaluation. *Soil Erosion and Channel Processes* 13, 8–17 (in Russian).

Larionov, G.A. (1993) *Soil Erosion and Deflation*. Moscow University Publication, Moscow (in Russian).

Larionov, G.A., Kiryukhina, Z.P. and Samodurova, L.S. (1973) Determination of slope wash rates by the method of paired soil pits descriptions. *Soil Erosion and Channel Processes* 3, 162–167 (in Russian).

Larionov, G.A., Dobrovolskaya, N.G., Krasnov, S.F., Liu, B.Y. and Nearing, M.A. (1998) Theoretical-empirical equation of topography factor for a statistical model of soil erosion by water. *Soil Erosion and Channel Processes* 11, 25–44 (in Russian).

Murray, A.S., Marten, R., Johnston, A. and Martin, P. (1987) Analysis for naturally occurring radionuclides at environmental concentrations by gamma spectrometry. *Journal of Radioactivity and Nuclear Chemistry, Articles* 115, 263–288.

Poesen, J.W., Vandaele, K. and Van Wesemael, B. (1996) Contribution of gully erosion to sediment production on cultivated lands and rangelands. In: Walling, D.E. and Webb, B.W. (eds) *Erosion and Sediment Yield: Global and Regional Perspectives*. IAHS Publication 236. IAHS Press, Wallingford, UK, pp. 251–266.

Quine, T.A., Walling, D.E. and Zhang, X.X. (1999) Tillage erosion, water erosion and soil quality on cultivated terraces near Xifeng in the Loess Plateau, China. *Land Degradation and Development* 10, 251–274.

Valkov, V.F. (1977) *Genesis of Soils of the North Caucasian Region*. Rostov University Publication, Rostov (in Russian).

Wallbrink, P.J., Roddy, B.P. and Olley, J.M. (2002) A tracer budget quantifying soil redistribution on hillslopes after forest harvesting. *Catena* 47, 179–201.

Walling, D.E. and He, Q. (1999) Improved models for estimating soil erosion rates from cesium-137 measurements. *Journal of Environmental Quality* 28, 611–622.

5 Hillslope Soil Erosion and Bioturbation after the Christmas 2001 Forest Fires near Sydney, Australia

R.A. Shakesby,[1] W.H. Blake,[2] S.H. Doerr,[1] G.S. Humphreys,[3] P.J. Wallbrink[4] and C.J. Chafer[5]

[1]University of Wales Swansea, UK; [2]University of Plymouth, UK; [3]Macquarie University, Australia; [4]CSIRO Land and Water, Australia; [5]Sydney Catchment Authority, Australia

Introduction

Hillslope soil losses can often be extremely high after forest fire (e.g. Robichaud *et al.*, 2000). High losses can result from, first, damage to or destruction of tree canopy, ground vegetation and litter layer. This reduces interception and storage of rainfall and exposes the soil to rainsplash detachment and overland flow transport (e.g. Shakesby *et al.*, 1993; Germanoski and Miller, 1995; Zierholz *et al.*, 1995; Prosser and Williams, 1998). Second, burning usually generates large amounts of easily removed ash and organic debris, and can reduce the stability of soil aggregates (e.g. Giovannini and Lucchesi, 1983). Third, where soil water repellency is fire-induced or enhanced, this will tend to reduce infiltration and enhance overland flow (e.g. Burch *et al.*, 1989). Bioturbation is not often given prominence in the context of post-fire hillslope erosion as post-fire numbers of birds, mammals and insects tend to be substantially reduced (e.g. Newsome *et al.*, 1975), but in some circumstances it can have an important influence on soil erosion after fire by providing readily eroded surface material and affecting infiltration, soil bulk density, overland flow and downslope transfer of sediment (e.g. Humphreys, 1981; Adamson *et al.*, 1983; Humphreys and Mitchell, 1983; Humphreys 1994; Dragovich and Morris, 2002).

Between Christmas 2001 and early January 2002, fires near Sydney affected 225,000 ha of eucalypt forest covering heavily dissected sandstone tablelands (Chafer *et al.*, 2004). The fires were followed by periods of prolonged and intense rainfall so that widespread serious soil erosion of the predominantly highly water repellent, erodible soils on steep slopes would be anticipated. This paper focuses on post-fire hillslope erosion in steep forested terrain in Nattai National Park, *c.* 80 km south-west of Sydney, using as evidence soil pedestal heights, ground-level change monitoring data and measurements of bioturbation impacts on soil translocation.

Background

The study was carried out mainly in two sub-catchments (63 and 89 ha) of Blue Gum Creek (150° 29.5′E, 34° 13.3′S) in Nattai National Park (Fig. 5.1), selected because they were of similar size and relief, represented an assemblage of slope and vegetation type units

 Soil Erosion and Sediment Redistribution in River Catchments (eds P.N. Owens and A.J. Collins)

typical of this region, but had experienced contrasting fire severities. The smaller sub-catchment (H) was affected mainly by high to extreme severity fire and the larger one (L) mainly by low to moderate severity fire (for details see Shakesby *et al.*, 2003). Blue Gum Creek is incised into Hawkesbury Sandstone and Narrabeen Group sediments leaving *c.* 200-m high valley sides dominated by sandy materials (sand to sandy loams) on a characteristic sequence of slope units (Table 5.1) (Tomkins *et al.*, 2004). After moderate–extreme fire, burnt, dark organic-rich topsoil up to 1–2 cm thick overlies sandy, minerogenic, light-coloured, sub-surface material. Relatively sheltered, damp valley floors and deep tributary valleys are characterized especially by Deanes blue gum (*Eucalyptus deanei*) and yellow bloodwood (*Corymbia gummifera*) on foot- and mid-slopes. On ridge tops, tree species typical of drier conditions occur (e.g. *E. creba*, *E. punctata*).

By the first visit in early May 2002, post-fire activity by ants (especially funnel ants) (Fig. 5.2) was widespread except where fire had been most severe. Evidence of large animals was scarce, but small mammal activity was apparent during the first year after fire.

Table 5.1. Typical valley-side characteristics, Blue Gum Creek study site.

Slope unit	Description	Slope angles
Ridge top	Sets of low-relief exposed sandstone (Hawkesbury Sandstone) benches with small downslope depositional aprons of shallow sandy material	2–10°
Cliff	Steep, near vertical bluffs of Hawkesbury Sandstone commonly up to 20 m (occasionally up to 60 m) high	70–90°
Mid-slope	Steep, 100–300 m planar to gently concave slopes comprising gravelly, sandy material interspersed with clay loams	15–35°
Foot-slope	Low-angled, typically 100–150 m wide concave slopes comprising deep sandy material (sand to sandy loams) with variable gravel content	5–10°
Valley floor	Alluvial terrain (terraces, floodplain and channels) comprising organic and fine-textured material (clay loam) > 1 m thick	< 5°

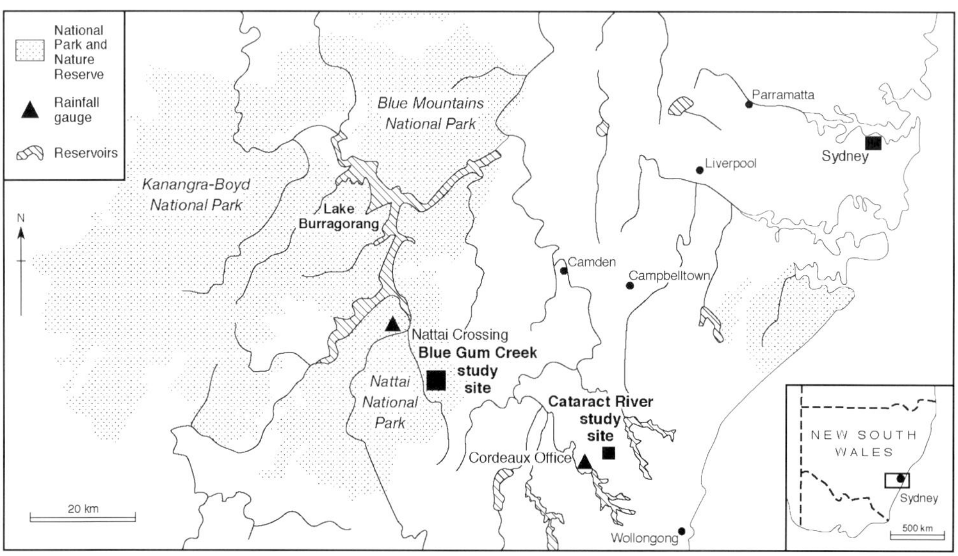

Fig. 5.1. Location of Blue Gum Creek and Cataract River basin study sites and rainfall gauges.

Fig. 5.2. Nests of funnel ants (*Aphaenogaster longiceps*), Blue Gum Creek, May 2002.

Fig. 5.3. Litter dam, Blue Gum Creek, May 2002.

In addition, litter dams were common on low-angled slopes (Mitchell and Humphreys, 1987; Fig. 5.3). A severe fire in January 2003 in Cataract River catchment (Fig. 5.1) provided the opportunity to monitor erosion at a second site (150° 46′E, 34° 18.2′S) with similar terrain to the Blue Gum Creek site prior to any post-fire rainfall. The climate at both sites is humid temperate with moist summers and cool winters. Average rainfall is 900–1000 mm, with large year-to-year variations and no marked dry season.

Methods

Measurement of post-fire hillslope soil redistribution

Two main techniques for estimating soil erosion were adopted. First, in early May 2002, following 440 mm of rainfall since fire, the heights of soil pedestals beneath stone and root caps (Fig. 5.4), together with the heights of live roots above the soil surface, were measured in fifteen 0.16 m^2 plots at representative locations in each of the four slope units (Table 5.1) in the two Blue Gum Creek sub-catchments.

Second, 27 ground-level change monitoring sites were installed during the period 11–21 May 2002 within the different slope units (ridge-top: 6 sites; upper mid-slope: 6 sites; lower mid-slope: 6 sites; and foot-slope: 9 sites) in the two Blue Gum Creek sub-catchments. The sites were located mainly on sub-surface sandy material. Some may have been installed on redistributed soil material but none was on *in situ* topsoil remnants. Only 153–166 mm of rain fell between May 2002 and 30 January 2003, when the sites were remeasured. They were measured again on 20 February 2003 after 40 mm of rainfall, and lastly during 20–24 February 2004 following an additional 811 mm of rain. To gain a better understanding of post-fire erosion in the early weeks and months after fire, four ground-level change monitoring sites were installed in January 2003 in the Cataract River basin immediately after a moderate–high severity fire. They were remeasured on 20 February 2003 after 29 mm of rain and on 23 February 2004 after 924 mm of rain. No sites were installed in unburnt locations because of known measurement problems in mature forest using this technique (Shakesby, 1993). Erosion rates are, however, extremely low in such locations. Prosser (1990), for example, estimated an average erosion rate of as little as 0.01 t/ha/year in similar unburnt eucalypt forest.

A modified version of the erosion bridge described by Shakesby (1993) was used to measure ground-level changes (Fig. 5.5). This type of soil micro-profiling device provides a stable datum for measuring soil level changes at 70 points spaced 25 mm apart, providing a transect length of 1.75 m. At each selected site, 1-m steel stakes were installed parallel to the contour, the c. 2-m long bridge mounted on them and stake heights adjusted until the bridge was horizontal and near the ground surface. At each point, any loose litter was temporarily removed and

Fig. 5.4. Soil pedestals, Blue Gum Creek, May 2002.

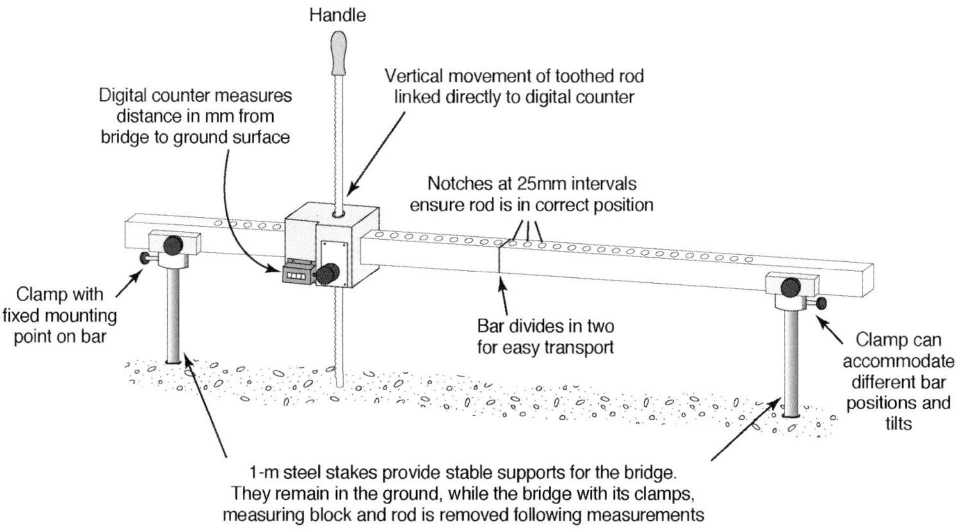

Fig. 5.5. Annotated diagram of the soil erosion bridge.

features of interest, such as charcoal, stones and evidence of bioturbation recorded.

Soil translocation by surface bioturbation

Plots 5 m × 1 m in size were marked out near the erosion bridge sites in each of the four slope units in the Blue Gum Creek sub-catchments and bioturbation observations made during the period 30 January–28 February 2003, some 14 months after the fire. There were three plots in each of the ridge top and mid-slope units and six plots in the foot-slope units of both sub-catchments. Ant-mound material above the newly burnt surface was carefully removed, oven-dried at 105°C and sieved through a 2-mm mesh to remove the negligible amount (< 1% by weight) of coarse litter, which was discarded. Bulk density measurements (ring volume 211.3 cm^3) of the upper 6 cm at 15 undisturbed soil sites and material in seven ant mounds enabled calculation of mass and volume from sample weights. The means and standard deviations of bulk density for surface soil were 0.98 ± 0.13 g/cm^3 and 0.85 ± 0.08 g/cm^3 for ant mounds. Mammal scrape volumes were determined from their dimensions. Surface gravel content, a possible influence on soil quantities moved particularly by ants, was assessed by measuring the percentage length of the 5 m sides of the plots coinciding with stones > 1 cm in size.

Results

Measurements of post-fire hillslope soil redistribution

The mean heights of soil pedestals and exposed roots above the soil surface recorded in May 2002 at the Blue Gum Creek site ranged from 6.8 to 14.9 mm with relatively high standard deviations (4.3–8.1) (Table 5.2). The means are statistically indistinguishable (Student's *t*-test; $P > 0.05$) for the two sub-catchments. Based on a bulk density of 1 g/cm^3, this range represents losses of soil of the order of 70–150 t/ha. However, the stone content in the soils is typically about 30% (Shakesby *et al.*, 2003), so that lower, but still substantial estimated losses of about 50–100 t/ha are more appropriate. All those pedestals inspected were formed of minerogenic, sandy, friable, unscorched sediment, and contained little if any organic matter in contrast to the topsoil material. In addition, stones capping the pedestals tended to have scorch lines on their sides or near their tops, suggesting that the stones had been embedded in the soil during fire. By implication, therefore, the soil losses indicated by the pedestal heights are primarily of the sub-surface soil rather than of topsoil.

Table 5.3 presents the erosion bridge results for three measurement periods between May 2002 and February 2004 and the whole period at the Blue Gum Creek sites, and two measurement periods between January 2003 and February 2004 and the entire period at the Cataract River basin site. Ground lowering (loss) is recorded as a

Table 5.2. Means and standard deviations of combined pedestal and exposed root heights (mm) measured in May 2002 (data from Shakesby *et al.*, 2003).

Sub-catchment L (mainly low to moderate fire severity)	
Foot-slope	9.1 ± 5.1
Lower mid-slope	12.2 ± 8.1
Upper mid-slope	14.9 ± 8.0
Ridge top	7.7 ± 4.9
Sub-catchment H (mainly high to extreme fire severity)	
Foot-slope	9.9 ± 4.9
Lower mid-slope	9.8 ± 4.7
Upper mid-slope	13.7 ± 7.7
Ridge top	6.8 ± 4.3

Table 5.3. Ground-level change results from Blue Gum Creek slopes and Cataract River basin.

			11–21 May 2002–30 January 2003					31 January 2003–20 February 2003				
			Rainfall					Rainfall				
Location and fire severity	No. of sites	Slope angle(s) (°)	Total[b] (mm)	Days with > 20 mm	Mean loss/gain (mm)	Percentage of points recording losses	Mean abs. change (mm)	Total (mm)	Days with > 20 mm	Mean loss/gain (mm)	Percentage of points recording losses	Mean abs. change (mm)
Blue Gum Creek												
Foot-slope												
High severity	6	6–8	166	8	−1.8	65	3.1	40	1	−2.8	54	2.2
Low severity	3	5–8	166	8	−3.3	62	6.9	40	1	−0.9	50	3.8
Mid-slope												
V. high severity	6	15–32	166	8	−3.3	65	6.0	40	1	2.0	36	5.4
Low severity	5[a]	19–32	161	8	−4.2	66	12.5	40	1	−1.1	58	10.2
Ridge top												
Extreme severity	3	8	166	8	−2.0	65	4.0	40	1	−1.7	67	4.3
High severity	3	8	153	8	−1.8	53	4.3	40	1	−1.0	58	4.0
Cataract River basin												
Severely burnt January 2003	4	8	–		–		–	29	0	−1.8	74	2.3

Negative mean loss/gain values represent loss and positive values represent gain.
The rainfall figures are obtained from Nattai Crossing for Blue Gum Creek and from Cordeaux Office for Cataract River basin (see Fig. 5.1).
[a]One mid-slope site abandoned because of a fallen tree across the transect.
[b]Variations reflect rainfall differences over the installation period.

negative value and gain as a positive value. At the Blue Gum Creek site, all mean changes show loss (–1.0 to –10.2 mm) for all but one period of measurement (January–February 2003) which recorded one positive mean value. Based on a bulk density of 1 g/cm^3 and a 30% stone cover, this range would represent soil losses of about 7–70 t/ha. Mean absolute changes (all ground height changes recorded as positive values) range from similar values to about ten times those of corresponding losses, indicating that between about a half and most ground-level changes were losses rather than gains. This is broadly confirmed by the percentages of losses recorded which, with the exception of one slope and fire severity category, range from 50% to 86%. In the final, comparatively wet measurement period (14–26 months after fire), the overall mean ground-level changes recorded for the different slope and fire severity categories (–0.6 to –7.3 mm) were higher than the values recorded for the period May 2002–January 2003 in three out of six cases, with one showing no difference. The relatively brief second measurement period (20–21 days) with 40 mm of rain had similar ground-level changes to those in the preceding 9-month period when there was about 160 mm of rain. Although the number of sites was small, it is nevertheless noteworthy that there is no clear link between fire severity and the magnitude of soil losses in Table 5.3, which is supported by the evidence of soil magnetic signatures described in Blake *et al.* (Chapter 6 this volume). At four erosion bridge sites at the Blue Gum Creek study site, a total of 42 points representing 1.1% of all measurement points coincided with active bioturbation. Excluding these points would have a minimal effect on the results in Table 5.3, except for the high and low severity foot-slope sites for the May 2002–January 2003 period where mean losses would be reduced by 0.8 mm and 0.4 mm, respectively.

The mean elevation changes for the four sites in Cataract River basin (Table 5.3) ranged

21 February 2003–20–24 February 2004					Entire period				
Rainfall					Rainfall				
Total (mm)	Days with > 20 mm	Mean loss/gain (mm)	Percentage of points recording losses	Mean abs. change (mm)	Total (mm)	Days with > 20 mm	Mean loss/gain (mm)	Percentage of points recording losses	Mean abs. change (mm)
811	11	–1.8	62	4.1	1018	20	–4.5	78	6.1
811	11	–0.6	50	4.9	1018	20	–4.6	69	7.7
811	11	–6.0	70	9.3	1018	20	–6.9	73	12.1
811	11	–7.3	54	13.9	1013	20	–7.6	70	16.1
811	11	–6.6	82	7.6	1018	20	–10.2	86	11.3
811	11	–1.4	56	7.5	1005	20	–4.2	66	11.1
924	10	–9.7	94	10.3	953	10	–11.8	96	12.5

Table 5.4. Summary of bioturbation quantities and rates measured between 30 January 2003 and 28 February 2003 for different slope units in sub-catchments H and L, Blue Gum Creek study site.

Slope position[a]	Surface gravel[b] (%)	Ant mounding[b] (g)	Small mammal scrapes[b] (g)	Total bioturbation[c] (t/ha/year)	Total bioturbation[c] (mm/year)
Sub-catchment L (mainly low to moderate fire severity)					
Foot-slope	< 1 (< 1)	1042 (20–2016)	2819 (0–8881)	6.5	0.6
Lower mid-slope	17.3 (9.2–24.3)	218 (108–397)	4445 (0–11,672)	7.8	0.8
Upper mid-slope	33.8 (30.8–39.4)	104 (0–158)	1698 (0–2934)	3.0	0.3
Ridge top	8.0 (4.5–14.7)	777 (0–2146)	456 (0–1368)	2.1	0.2
Sub-catchment H (mainly high to extreme fire severity)					
Foot-slope	4.2 (< 1–9.7)	2134 (563–3852)	950 (0–2313)	5.2	0.5
Lower mid-slope	24.3 (19.5–29.7)	26 (13–39)	288 (0–864)	0.5	0.1
Upper mid-slope	27.4 (24.3–32.4)	384 (254–634)	3560 (0–7748)	6.6	0.7
Ridge top	6.4 (0–9.8)	1196 (473–2317)	329 (0–988)	2.6	0.2

[a]Slope units as in Table 5.1.
[b]Mean value followed by range in parentheses.
[c]Ant mounding and mammal scraping combined and expressed as an estimated annual rate.

from −1.8 mm for the period 31 January–20 February 2003, following 29 mm of rain on the newly burnt surface, to −9.7 mm for the much longer and wetter period from February 2003 to February 2004. Corresponding mean absolute changes for these periods are similar to mean loss/gain values, indicating that most individual point changes were losses, and this is confirmed by the high percentages of measurement points recording losses for each period (74–96%).

Soil translocated by surface bioturbation

Table 5.4 gives post-fire rates of sediment translocation by ants and small mammals. The combined rates range from 0.5 to 7.8 t/ha/year. Considerable between-plot variations were found in each slope unit. There are no clear differences in bioturbation quantities between the sub-catchments. Ant mounding generally led to more soil translocation in ridge-top and foot-slope than in mid-slope locations, whereas the opposite was the case for mammal scraping. Regression analysis of ant mounding and surface gravel content produced a statistically significant negative correlation ($r^2 = 0.61$; $P < 0.05$), but there was no significant correlation between mammal scraping and surface gravel content, indicating that the ability of ants to move sediment is affected by gravel content, but this was not the case for mammals.

Discussion

Delicate pedestals like those observed in May 2002 in the Blue Gum Creek sub-catchments (Fig. 5.4) are typically interpreted predominantly as the product of rainsplash erosion (Stocking and Murnaghan, 2001), as any substantial overland flow would tend to destroy them. Their delicacy indicates a post-Christmas 2001 fire origin and their sizes and lack of organic material suggest that they represent considerable erosion of the sandy sub-surface soil, possibly helped by its highly water repellent nature (Terry and Shakesby, 1993). There were extensive deposits of redistributed soil that had been transported into the stream system, but these comprised burnt, fines- and organic-rich topsoil as shown by mineral magnetic (Blake *et al.*, Chapter 6 this volume, Blake *et al.*, 2006) and radionuclide (Wallbrink *et al.*, 2005) evidence. Comparatively little of the sub-surface sandy material, however, had apparently been transported beyond the foot-slope zones by May 2002: only very small quantities of such material, and certainly only a small fraction of the amounts implied by the pedestal heights, were found in the drainage ditch on the upslope side of the forest road aligned along the base of the foot-slope or in the fans of sediment spread across the road at the mouths of the gullies draining the sub-catchments. Reconnaissance of Blue Gum Creek indicated that this situation was typical.

Little burnt topsoil and ash remained *in situ* on the slopes of the sub-catchments by May 2002. The highly erodible nature of this fine-grained, organic-rich soil and easy transportation of charred debris are thought to have been more important than differences in wettability caused by soil heating differences (Doerr *et al.*, 2006) in leading to its large-scale export from the sub-catchments (Shakesby *et al.*, 2003). Observations of widespread removal of burnt, organic-rich topsoil sediment have been made on similar forested terrain in south-east Australia (Blong *et al.*, 1982; Leitch *et al.*, 1983; Atkinson, 1984; Zierholz *et al.*, 1995). The predominance of ground-level lowering throughout the period of monitoring indicates continued, though declining erosion of this material after May 2002. This is supported by the tendency for greater mean lowering during the wetter February 2003–February 2004 than in the much drier May 2002–January 2003 period, despite the more complete vegetation and litter cover.

The implication of erosion of the sub-surface material shown by the pedestal and erosion bridge evidence and limited export of this material from foot-slope areas is that redistribution must have been largely local. Pedestals would not have survived large quantities of overland flow so that substantial infiltration must have occurred despite the water repellent sub-surface material. Possible explanations are that, although the soil was commonly repellent, its intensity varied, leading to the development of zones of preferential flow in the soil (Doerr *et al.*, 2006), and that many infiltration pathways would have been provided by bioturbation, particularly by ants. Nests of the main ant species, *Aphaenogaster longiceps*, comprise mounds of wettable soil around entrances up to 4 cm in diameter and 30 cm deep leading to extensive gallery systems (Humphreys, 1981, 1994; Australian Museum Online, 2004). The nest mounds would have increased surface roughness reducing overland flow velocity early during rainfall events. Also, the litter dams would have trapped much of the mobilized sediment (Mitchell and Humphreys, 1987).

In addition to its probable impact on overland flow, post-fire bioturbation provides erodible material for redistribution by rainsplash and overland flow. Preliminary results during the second year after fire (up to early 2004) indicate similar rates of soil translocation by ants but declining rates of mammal activity to those reported here. The decline in the latter is probably explained by a post-fire decline in small mammal numbers due to habitat destruction and reduced food availability (e.g. Fox, 1990). Whilst not unimportant, the translocation figures represent a relatively small proportion of the estimated rates of soil loss indicated by the pedestal and ground-level change evidence, which do not take into account the additional widespread redistribution of the topsoil material. The role of post-fire bioturbation in providing erodible surface material for removal by slopewash or in the direct downslope transfer of soil would, therefore, seem to be comparatively minor (Dragovich and Morris, 2002).

The modest ground-level changes following the first post-fire rainfall events between 31 January and 20 February 2003 at the Cataract catchment site might seem surprising given the extremely erodible, newly burnt surface material and the ubiquitous presence of near-surface repellency (Doerr *et al.*, 2006). The most likely explanation is that the maximum daily rainfall recorded at the nearby Cordeaux Office (Fig. 5.1) of 9 mm over this period did not exceed the storage capacity of the surface 1–2 cm of soil in which water repellency had been destroyed by fire (Doerr *et al.*, 2006). Substantial soil losses (mean –9.7 mm) between February 2003 and February 2004 can be attributed to the high rainfall amount and large daily totals during this period.

Conclusion

Following severe fire in the Christmas 2001 period which affected eucalypt forest near Sydney, Australia, erosion of sandy sub-surface material did not repeat the widespread removal and redistribution into the stream system of the overlying burnt, fines- and organic-rich topsoil. Instead, erosion seems to have been largely restricted to localized redistribution. Likely reasons for this limited transport despite intense rainfall and the highly erodible and water repellent nature of this material include litter dams acting as sediment traps and bioturbation (particularly by ants) increasing surface roughness and providing bypass drainage routes for overland flow. This role of bioturbation seems to have been more important than that of enhancing the downslope transfer of soil.

Acknowledgements

The support of a NERC Research Grant (NER/A/S/2002/00143), the NERC Geophysical Equipment Pool and a NERC Advanced Fellowship (NER/J/S/2002/00662) are gratefully acknowledged. Thanks are also due to: Macquarie University for a financial contribution (GSH); Nicola Jones for drawing Figs 5.1 and 5.5; V. Farwig, R. Field, Z. Hardiman and B. Harrington for logistical and fieldwork help; R. Walsh for design modifications to the erosion bridge; and Sydney Catchment Authority personnel, especially G. Capararo, for granting access and in-kind support. The comments of two anonymous referees were helpful in improving the chapter.

References

Adamson, D., Selkirk, P.M. and Mitchell, P. (1983) The role of fire and lyrebirds in the sandstone landscape of the Sydney Basin. In: Young, R.W. and Nanson, G.C. (eds) *Aspects of Australian Sandstone Landscapes.* Australian and New Zealand Geomorphology Group, Special Publication No. 1. University of Wollongong, New South Wales, pp. 81–93.

Atkinson, G. (1984) Erosion following bushfires. *Journal of Soil Conservation NSW* 40, 4–9.

Australian Museum Online (2004) Funnel ant fact file. http: //www.faunanet.gov.au/wos/factfile.cfm?Fact_ID=253 (accessed November 2004).

Blake, W.H., Wallbrink, P.J., Doerr, S.H., Shakesby, R.A. and Humphreys, G.S. (2006) Magnetic enhancement in fire-affected soil and its potential for sediment source ascription. *Earth Surface Processes and Landforms* 31, 249–264.

Blong, R.J., Riley, S.J. and Crozier, P.J. (1982) Sediment yield from runoff plots following bushfire near Narrabeen Lagoon, NSW. *Search* 13, 36–38.

Burch, G.J., Moore, I.D. and Burns, J. (1989) Soil hydrophobic effects on infiltration and catchment runoff. *Hydrological Processes* 3, 211–222.

Chafer, C.J., Noonan, M. and McNaught, E. (2004) The post-fire measurement of fire severity and intensity in the Christmas 2001 Sydney wildfires. *International Journal of Wildland Fire* 13, 227–240.

Doerr, S.H., Shakesby, R.A., Blake, W.H., Chafer, C.J., Humphreys, G.S. and Wallbrink, P.J. (2006) Effects of differing wildfire severities on soil wettability and implications for hydrological response. *Journal of Hydrology* doi: 10.1016/j.jhydrol.2005.06.038.

Dragovich, D. and Morris, R. (2002) Fire intensity, slopewash and bio-transfer of sediment in eucalypt forest, Australia. *Earth Surface Processes and Landforms* 27, 1309–1319.

Fox, B.J. (1990) Changes in the structure of mammal communities over successional time scales. *Oikos* 59, 321–329.

Germanoski, D. and Miller, J.R. (1995) Geomorphic response to wildfire in an arid watershed, Crow Canyon, Nevada. *Physical Geography* 16, 243–256.

Giovannini, G. and Lucchesi, S. (1983) Effect of fire on hydrophobic and cementing substances of soil aggregates. *Soil Science* 137, 457–463.

Humphreys, G.S. (1981) The rate of ant mounding and earthworm casting near Sydney, New South Wales. *Search* 12, 129–131.

Humphreys, G.S. (1994) Bioturbation, biofabrics and the biomantle. In: Ringrose-Voase, A.J. and Humphreys, G.S. (eds) *Soil Micromorphology: Studies in Management and Genesis.* Developments in Soil Science 22. Elsevier, Amsterdam, pp. 421–436.

Humphreys, G.S. and Mitchell, P.B. (1983) A preliminary assessment of the role of bioturbation and rainwash on sandstone hillslopes in the Sydney basin. In: Young, R.W. and Nanson, G.C. (eds) *Aspects of Australian Sandstone Landscapes.* Australian and New Zealand Geomorphology Group, Special Publication No. 1. University of Wollongong, New South Wales, pp. 66–80.

Leitch, C.J., Flinn, D.W. and Van De Graaf, R.H.M. (1983) Erosion and nutrient loss resulting from Ash Wednesday (February 1983) wildfires: a case study. *Australian Forestry* 46, 173–180.
Mitchell, P.B. and Humphreys, G.S. (1987) Litter dams and microterraces formed on hillslopes subject to rainfall. *Geoderma* 39, 331–357.
Newsome, A.E., McIlroy, J. and Catling, P. (1975) The effects of an extensive wildfire on populations of twenty ground vertebrates in south-east Australia. *Proceedings of the Ecological Society of Australia* 9, 107–123.
Prosser, I.P. (1990) Fire, humans and denudation at Wangrah Creek, Southern Tablelands, New South Wales. *Australian Geographical Studies* 28, 77–95.
Prosser, I.P. and Williams, L. (1998) The effect of wildfire on runoff and erosion in native *Eucalyptus* forest. *Hydrological Processes* 12, 251–265.
Robichaud, P.R., Beyers, J.L. and Neary, D.G. (2000) *Evaluating the Effectiveness of Post-fire Rehabilitation Treatments.* General Technical Report RMRS-GTR-63, September. United States Department of Agriculture, Forest Service.
Shakesby, R.A. (1993) The soil erosion bridge: a device for micro-profiling soil surfaces. *Earth Surface Processes and Landforms* 18, 823–827.
Shakesby, R.A., Coelho, C. de O.A., Ferreira A.D., Terry, J.P. and Walsh, R.P.D. (1993) Wildfire impacts on soil erosion and hydrology in wet Mediterranean forest, Portugal. *International Journal of Wildland Fire* 3, 95–110.
Shakesby, R.A., Chafer, C.J., Doerr, S.H., Blake, W.H., Wallbrink, P., Humphreys, G.S. and Harrington, B.A. (2003) Fire severity, water repellency characteristics and hydrogeomorphological changes following the Christmas 2001 Sydney forest fires. *Australian Geographer* 34, 147–175.
Stocking, M.A. and Murnaghan, N. (2001) *Handbook for the Field Assessment of Land Degradation.* Earthscan, London.
Terry, J.P. and Shakesby, R.A. (1993) Soil hydrophobicity effects on rainsplash: simulated rainfall and photographic evidence. *Earth Surface Processes and Landforms* 18, 519–526.
Tomkins, K.M., Humphreys, G.S., Skeen, H.J., Taylor, G.M., Farwig, V.J., Shakesby, R.A., Doerr, S.H., Blake, W.H., Wallbrink, P. and Chafer, C.J. (2004) Deciphering a colluvial mantle: Nattai catchment. In: Singh, B. (ed.) *SuperSoil 2004: Program and Abstracts for the 3rd Australian–New Zealand Soil Conference*, University of Sydney, 5–9 December 2004. http://www.regional.org.au/au/asssi/supersoil2004 (accessed June 2005).
Wallbrink, P., Blake, W., Doerr, S., Shakesby, R., Humphreys, G. and English, P. (2005) Using tracer based sediment budgets to assess redistribution of soil and organic material after severe bush fires. In: Walling, D.E. and Horowitz, A.J. (eds) *Sediment Budgets 1.* IAHS Publication 292. IAHS Press, Wallingford, UK, pp. 223–230.
Zierholz, C., Hairsine, P. and Booker, F. (1995) Runoff and soil erosion in bushland, following the Sydney bushfires. *Australian Journal of Soil and Water Conservation* 8, 28–37.

6 Tracing Eroded Soil in a Burnt Water Supply Catchment, Sydney, Australia: Linking Magnetic Enhancement to Soil Water Repellency

W.H. Blake,[1] S.H. Doerr,[2] R.A. Shakesby,[2] P.J. Wallbrink,[3] G.S. Humphreys[4] and C.J. Chafer[5]

[1]*University of Plymouth, UK;* [2]*University of Wales Swansea, UK;* [3]*CSIRO Land and Water, Australia;* [4]*Macquarie University, Australia;* [5]*Sydney Catchment Authority, Australia*

Introduction

Substantial redistribution of soil and sediment can occur within river basins following wildfire events, particularly when vegetation destruction by fire is followed by intense rainstorm events (Elliot and Parker, 2001). Whilst soil erosion in the slope environment has received some attention (see Shakesby *et al.*, Chapter 5, this volume), less work has focused on the downstream impacts of these significant but infrequent events. Considering the affinity of nutrients and other trace elements for fine sediment particles, a fuller understanding of post-fire sediment dynamics will assist in the monitoring, modelling and mitigation of post-fire nutrient fluxes downstream.

Sediment tracing approaches offer a means to link downstream sediment deposits to specific or generic source area types and can further offer information on slope and channel processes that act to transport mobilized material (Walling *et al.*, 1979; Yu and Oldfield, 1993; Wallbrink *et al.*, 1999). The work described in this contribution explores a potential link between temperature-sensitive, fire-induced changes in soil hydrology associated with soil water repellency, and soil Fe mineralogy, which can be utilized for tracing. Linking downstream sediment to sources of contrasting soil hydrology would allow the role of fire-modified water repellency in post-fire sediment generation to be assessed more thoroughly than has been possible to date. The work described below explores this possibility.

Fire-induced changes in soil surface hydrology

Although the degree of ground cover destruction in response to varying fire severities is one of the key factors determining post-fire soil erodibility (Shakesby *et al.*, 2000; De Luis *et al.*, 2003), changes to its infiltration behaviour associated with fire-induced alterations in water repellency are also considered to be critical (Scott and van Wyk 1990; Shakesby *et al.*, 2000). Thus, in many forested catchments, post-fire erosion processes are complicated by the transformation of soil surface hydrological processes through modification of the soil's water repellency status (Burch *et al.*, 1989; Doerr *et al.*, 2000; Shakesby *et al.*, 2003,

Chapter 5, this volume). During burning, soil water repellency can be enhanced (*T* = 270–330°C; depending on heating duration) or abruptly destroyed (T > ~ 330°C) (DeBano and Krammes, 1966; Doerr *et al.*, 2004), strongly affecting runoff and erosion dynamics. For example, recent work in the USA by Gabet (2003) shows that the presence of a water repellent layer beneath a thin wettable surface layer can lead to the development of a 'perched water table' which may cause thin hillslope failures or, as suggested by Shakesby *et al.* (2003), widespread 'rafting' of the organic rich topsoil following saturation.

Fire-induced changes in soil Fe mineralogy

It is well known that soil temperatures > 400°C, as caused by excessive heating during severe wildfires, can generate large quantities of secondary magnetic minerals. These changes alter the magnetic signature of soil and derived sediment (Rummery *et al.*, 1979; Gedye *et al.*, 2000) which can be detected using sensitive equipment in the field and laboratory (Walden *et al.*, 1999). Changes in the geophysical properties of soil (linked to reduction of Fe-bearing minerals and the formation of new ultra-fine grained Fe minerals) can be used to trace downstream sediment back to source areas of contrasting burn severity (see Blake *et al.*, 2004).

Given that the reported destruction of soil water repellency and enhancement in magnetic properties occur, in the natural environment, above a broadly similar temperature threshold (300–400°C), this contribution aims to explore the link between burn severity, presence of soil water repellency and mineral magnetic properties.

Study Area

The study was undertaken in the Nattai River drainage basin which flows into the Warragamba River, which in turn forms part of Lake Burragorang, Sydney's principal water supply reservoir (formed in 1960 with completion of the Warragamba dam). The area is underlain by Hawkesbury Sandstone, which is dissected by the local drainage network into a characteristic gorge–plateau landscape. Soils are sandy to sandy loam and forested by a variety of native eucalypt species (Shakesby *et al.*, 2003, Chapter 5 this volume). Soil and sediment characterization work was carried out primarily in two small (< 1 km^2) sub-catchments of Blue Gum Creek, a tributary of Little River which in turn flows into Nattai River (Fig. 6.1).

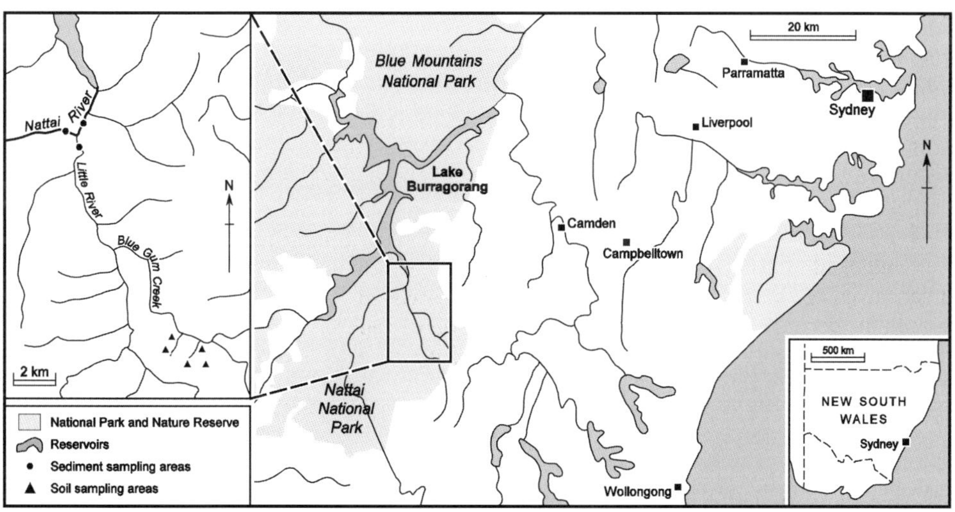

Fig. 6.1. Study area and location of soil source area and downstream sediment samples.

Methods

Field sampling for characterization of source area soils

Samples of burnt surface soil were collected from ridge-top and footslope units (see Table 5.1 in Shakesby *et al.*, Chapter 5, this volume) in areas representative of 'severe', 'moderate' and 'unburnt' burn regimes. Unburnt slope units were not available within the study region and hence a total of only five slope unit types were sampled. The degree of fire severity experienced at each site was determined using an index of vegetation-consumed-by-fire (supplied by the Sydney Catchment Authority and based on comparison of SPOT satellite imagery before and after the wildfire event – for further details see Shakesby *et al.*, 2003; Chafer *et al.*, 2004) and corroboration of the index through ground-based observation. At each site, a 0.5 m^2 area was carefully cleared of ash and other debris before a thin layer (~ 100 g) of mineral surface soil was sampled. A sample of subsurface soil (denoted by an uncharred appearance) was also taken. The soils were air-dried at ambient temperature before being sieved to < 2 mm ready for analysis.

Water repellency status

The soil surface at each sampling site was tested *in situ* for water repellency using the Water Drop Penetration Time (WDPT) method. Based on placing three droplets (80 µl) of water on to separate positions on the dry soil surface, soils were classed as wettable if all droplets penetrated within 5 s, and as water repellent for longer penetration times (Bisdom *et al.*, 1993). In a few cases where infiltration behaviour of the three drops was inconsistent (e.g. damp patches), measurements were excluded from analysis.

Field sampling for characterization of downstream sediment

Bulk (*c.* 5 kg) sediment samples were collected from deposits in and around the channels of Little River and Nattai River and also from the Nattai Arm of Lake Burragorang where grab and core samples were recovered (see Fig. 6.1 inset). Samples were air-dried and disaggregated before carrying out chemical dispersion and particle size separation as described below.

Sample preparation

Sub-samples of soil and sediment samples were packed into 10 cm^3 plastic pots ready for mineral magnetic analysis. Also, sub-samples of each sample population were combined into spatially averaged samples for each landscape unit (following the principles of Wallbrink *et al.*, 2003). These, along with the downstream sediment samples, were subjected to sieving and particle settling (see Gibbs *et al.*, 1971) providing particle size separates (< 10, 10–20, 20–40, 40–63, 63–125, 125–250, 250–500, 500–1000 and 1000–2000 µm), which in turn were packed as above. These allowed particle size controls on sediment signatures to be explored whilst the < 10 µm fraction could be utilized for direct comparison of sample properties without the need to consider particle size effects.

Mineral magnetic analysis

Low- (4.6 kHz) and high-frequency (0.46 kHz) magnetic susceptibility (χ_{lf} and χ_{hf}) were measured on each soil and sediment sample using a Bartington MS2 dual frequency susceptibility meter, allowing frequency-dependent susceptibility to be calculated on both a mass specific (χ_{fd}) and percentage ($\chi_{fd\%}$) basis. Mass-specific magnetic susceptibility is an indicator of the 'magnetizability' of the sample and generally increases with production of fine-grained pyrogenic minerals. Values are expressed in units of 10^{-6} m^3/kg. Frequency-dependent susceptibility indicates the proportion of super paramagnetic (SP) minerals present in the assemblage with values ranging from 0 to 14% (Dearing, 1999). Pyrogenic minerals are typically fine grained. Anhysteretic remanent magnetism (ARM) was grown using a Molspin AF demagnetizer, measured using a Molspin flux-gate spinner magnetometer and is expressed as a susceptibility of anhysteretic remanent magnetization (χARM) with units of 10^{-5} m^3/kg. Susceptibility of anhysteretic remanent magnetization is an indicator of the concentration of magnetic minerals of the 0.02–0.4 μm fraction (Maher, 1988) and contributes to characterization signatures. Forward- and back-field isothermal remanent magnetization (IRM) measurements were also made using a Magnetic Measurements MMPM10 pulse magnetizer and measured using a Molspin fluxgate spinner magnetometer. The rapidity and magnitude of remanence acquisition indicate the relative proportions of 'magnetically soft' (e.g. magnetite or maghaemite, which can be produced by fire) and 'magnetically hard' (e.g. haematite and goethite) magnetic minerals. Some haematite may be produced at high temperature where levels of organic reducing agents are depleted. Again, these differences contribute to source characterization signatures.

Results and Discussion

Magnetic enhancement of soil during burning

Soil experiencing both moderate and severe burn severity in the study area is magnetically enhanced as found elsewhere (e.g. Rummery *et al.*, 1979), although the relationship between burning and magnetic signature is shown not to be entirely straightforward (Blake *et al.*, 2006). Table 6.1 gives values for the mineral magnetic properties of the bulk surface soil samples for each slope unit type (< 10 μm fraction). In general, the magnetic properties of the soil become increasingly enhanced with increasing fire severity. The relationship between two mineral magnetic properties enhanced by fire (χ_{lf} and χARM) for the most severely burnt ridge-top unit, which forms the focus of this investigation into water repellency and magnetic enhancement, is shown in Fig. 6.2. The ridge-top soil has been selected to eliminate possible effects of slope storage of previously burnt and mobilized soil on surface soil magnetic characteristics. The strong linear relationship demonstrates a similarity in the response of these properties, which tie into both the bulk magnetic properties of the soil (χ_{lf}) and magnetic-grain-size-specific properties (χARM).

In situ water repellency status

Table 6.1 shows the percentages of samples classified as water repellent for each slope unit.

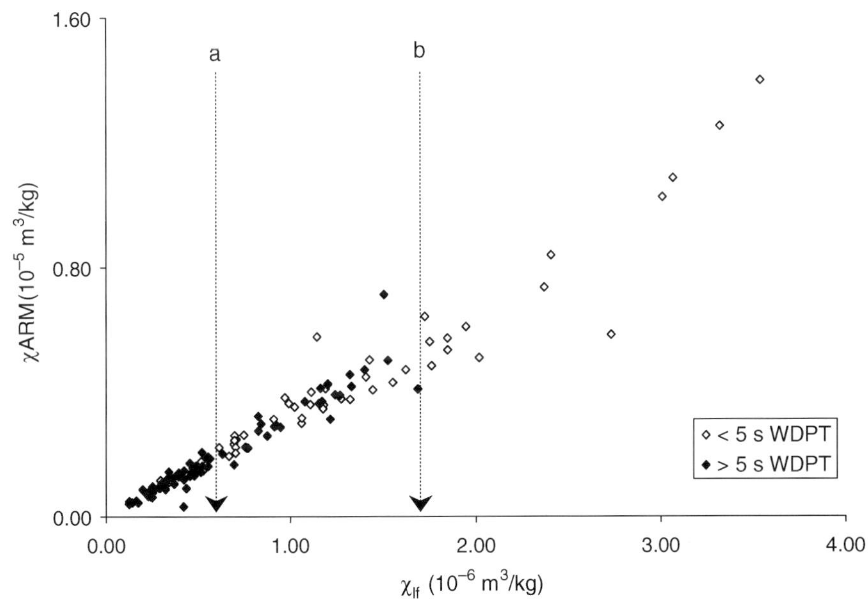

Fig. 6.2. Plot of magnetic susceptibility (horizontal axis) against susceptibility of anhysteretic remanent magnetization (vertical axis) showing threshold values of magnetic susceptibility 'a' and 'b' – see text for discussion.

Table 6.1. Bulk soil magnetic properties for each landscape unit burn type (< 10 μm fraction) alongside % of soil samples from each population classified as water repellent (during field tests).

Surface soil location (in order of bulk magnetic enhancement; χ_{lf})	χ_{lf} (10^{-6} m^3/kg)	χ_{fd}%	χARM (10^{-5} m^3/kg)	SIRM (10^{-5} Am^2/kg)	% samples water repellent
Unburnt surface	0.77	8.84	0.35	829	100
Moderately burnt footslope	4.06	10.65	1.82	2368	68
Moderately burnt ridge-top	4.59	11.18	1.21	2104	74
Severely burnt ridge-top	8.03	11.24	2.92	4751	12
Severely burnt footslope	10.39	10.68	3.36	6700	12

In the unburnt unit, 100% of samples were water repellent. This figure declines in the moderately burnt units to 68–74% and drops further to 12% in the severely burnt units. In addition to the linear relationship between χ_{lf} and χARM for the soils of the severely burnt ridge-top unit, Fig. 6.2 also shows the proportions of these samples classified as water repellent (solid diamonds) or wettable (open diamonds). Two thresholds in magnetic susceptibility are marked as 'a' and 'b'. Below threshold 'a', which represents the magnetic susceptibility of the unburnt background soil, the majority of soil samples are water repellent. These samples are presumed to have been obtained from areas where soil temperatures were insufficient either to enhance magnetically the soil and/or, in most cases, to destroy its water repellency. Threshold 'b' marks the value of magnetic susceptibility above which all soil samples were wettable. At this threshold, soil is believed to have attained temperatures sufficiently high for both magnetic enhancement and destruction of water repellency.

Linking magnetic enhancement and soil wettability

At the two extremes of magnetic susceptibility (i.e. above threshold 'b' and below threshold 'a'), the link between magnetic enhancement and water repellency status is convincing. However, between the two magnetic susceptibility thresholds, the repellency status of the soil samples is mixed. Considering individual soil samples, this overlap is most likely due to the 50–100°C difference between the temperature thresholds at which transformation of the two properties reportedly take place. Poor correlation between laboratory WDPT times (not presented) and magnetic enhancement further suggests that the probability of transformation of each property at specific temperature thresholds is not consistent and that each property can be modified within a range of temperature values. Other factors, for example soil organic content or soil moisture, could influence the response of soil properties to temperature. Observed heterogeneity in the response of both properties within a relatively narrow temperature range suggests that a firm link between downstream sediment and upstream sources classified according to water repellency status is not appropriate here. However, comparison of the bulk magnetic properties and the overall repellency status of the slope units, represented by the percentage of samples classified as water repellent in each unit (Table 6.1), indicates that the overall magnetic response is linked to the overall repellency response when the effects of spatial heterogeneity in the distribution of both properties are overcome. At the individual soil sample scale, the link between magnetic enhancement and water repellency status is complicated by different temperature (and temperature duration) thresholds for transformation and variable response to temperature due to other factors (e.g. soil organic content). At the slope unit scale, a broad link between the dominant repellency status across the unit and the spatially averaged magnetic signal can be seen.

Comparing downstream sediment to slope unit soil

Figure 6.3 gives a plot of χ_{lf} and χARM for the < 10 μm fractions of bulk slope soil samples and recent sediment deposits sampled from the river channel downstream. The location of the downstream sediment within the plot suggests, on the basis of these two properties alone, that the material could be derived from both severely burnt and moderately burnt sources. Considering the broad link between repellency status and the slope unit signatures, a minor impact of altered repellency status is implied, i.e. there is no apparent dominance of the downstream signal by severely burnt soil sources. If this is the case, the intensity of post-fire rainstorm events is likely to have overcome any changes in soil surface hydrology (i.e. increased surface storage where repellency is destroyed), an issue that is discussed in further detail by Shakesby *et al.* (2004). The issue of tracer 'linear additivity' (see Walden *et al.*, 1997) also requires attention, and for more robust discrimination of catchment source types additional geochemical evidence is required (Blake *et al.*, 2004). Analysis of subaqueously stored sediment further downstream suggests extension of the above interpretation to lacustrine sedimentary deposits, to provide a longer term perspective, is complicated by the development of anoxic bottom-water conditions leading to distortion of the magnetic signature in this environment (Blake *et al.*, 2004).

Conclusions

1. Soil magnetic signatures within the burnt water supply catchment of Sydney become enhanced during burning with an apparent discrimination between areas of contrasting fire severity and overall surface soil water repellency.
2. At the individual soil sample scale, the link between magnetic enhancement and water repellency status is complicated by different temperature thresholds for alteration of magnetic properties and possible variable response to temperature and heating duration. At the slope unit scale, a broad link between the dominant repellency status across the unit and the spatially averaged magnetic signal is evident.
3. Simple comparison of slope soil magnetic signatures and downstream sediment properties

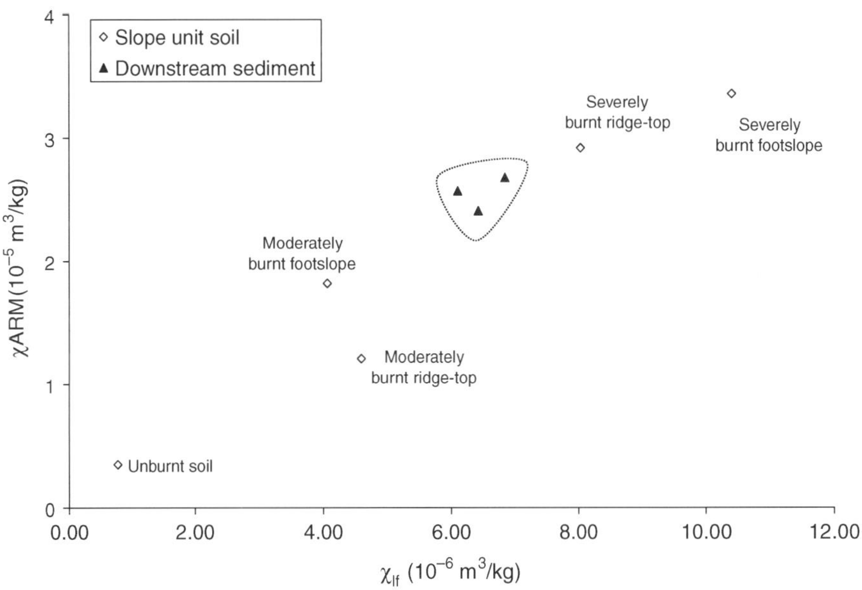

Fig. 6.3. Relationship between magnetic properties of slope unit surface soils (diamonds) and downstream sediment deposits in Little River (triangles).

suggests that fire-modification of surface repellency has had a limited effect on soil surface erodibility. This could be linked to the intensity of post-fire rainstorm events which might have exceeded any increase in surface storage capacity. More robust tracer signatures using additional geochemical evidence are required to confirm this inference.

Acknowledgements

The support of a NERC Research Grant (NER/A/S/2002/00143), Advanced NERC Fellowship (SHD) (NER/J/S/2002/00662) and the NERC Geophysical Equipment Pool is acknowledged. Thanks are also due to: Macquarie University (MU) for a financial contribution (GSH); Nicola Jones and Jamie Quinn for drawing Fig. 6.1; Russell Field, Ben Harrington (MU) and Danny Hunt (CSIRO) for logistical and fieldwork help; Gareth James for laboratory analysis; and Sydney Catchment Authority personnel, especially Glen Capararo. John Walden, Fleur Visser and Phil Owens are gratefully acknowledged for constructive reviews.

References

Bisdom, E.B.A., Dekker, L.W. and Schoute, J.F.Th. (1993) Water repellency of sieve fractions from sandy soils and relationships with organic material and soil structure. *Geoderma* 56, 105–118.

Blake, W.H., Wallbrink, P.J., Doerr, S.H., Shakesby, R.A. and Humphreys, G.S. (2004) Sediment redistribution following wildfire in the Sydney region, Australia: a mineral magnetic tracing approach. In: Golosov, V., Belyaev, V. and Walling, D.E. (eds) *Sediment Transfer Through the Fluvial System*. IAHS Publication 288. IAHS Press, Wallingford, UK, pp. 52–59.

Blake, W.H., Wallbrink, P.J., Doerr, S.H., Shakesby, R.A. and Humphreys, G.S. (2006) Magnetic enhancement in fire-affected soil and its potential for sediment source ascription. *Earth Surface Processes and Landforms* 31, 249–264.

Burch, G.J., Moore, I.D. and Burns, J. (1989) Soil hydrophobic effects on infiltration and catchment runoff. *Hydrological Processes* 3, 211–222.

Chafer, C.J., Noonan, M. and McNaught, E. (2004) The post-fire measurement of fire severity and intensity in the Christmas 2001 Sydney wildfires. *International Journal of Wildland Fire* 13, 227–240.

Dearing, J. (1999) Magnetic susceptibility. In: Walden J., Oldfield, F. and Smith, J. (eds) *Environmental Magnetism: A Practical Guide*. Technical Guide 6. Quaternary Research Association, Cambridge/London.

DeBano, L.F. and Krammes, J.S. (1966) Water repellent soils and their relation to wildfire temperatures. *International Association of Hydrological Sciences Bulletin* 2, 14–19.

De Luis, M., Gonzalez-Hidalgo, J.C. and Raventos, J. (2003) Effects of fire and torrential rainfall on erosion in a Mediterranean gorse community. *Land Degradation and Development* 14, 203–213.

Doerr, S.H., Shakesby, R.A. and Walsh, R.P.D. (2000) Soil hydrophobicity: its causes, characteristics and hydro-geomorphological significance. *Earth-Science Reviews* 51, 33–65.

Doerr, S.H., Blake, W.H., Humphreys, G.S., Shakesby, R.A., Stagnitti, F., Vuurens, S.H. and Wallbrink, P. (2004) Post-fire water repellency: an indicator of soil temperature reached during burning? *International Journal of Wildland Fire* 13, 157–163.

Elliot, J.G. and Parker, R.S. (2001) Developing a post-fire chronology and recurrence probability from alluvial stratigraphy in the Buffalo Creek watershed, Colorado, USA. *Hydrological Processes* 15, 3039–3051.

Gabet, E.J. (2003) Post-fire thin debris flows: sediment transport and numerical modelling. *Earth Surface Processes and Landforms* 28, 1341–1348.

Gedye, S.J., Jones, R.T., Tinner, W., Ammann, B. and Oldfield, F. (2000) The use of mineral magnetism in the reconstruction of fire history: a case study from Lago di Origlio, Swiss Alps. *Paleogeography, Paleoclimatology, Paleoecology* 164, 101–110.

Gibbs, R.J., Matthews, M.D. and Link, D.A. (1971) The relationship between sphere size and settling velocity. *Journal of Sedimentary Petrology* 41, 7–18.

Maher, B. (1988) Magnetic properties of some synthetic sub-micron magnetites. *Geophysical Journal of the Royal Astronomical Society* 94, 83–96.

Rummery, T.A., Bloemendal, J., Dearing, J.A., Oldfield, F. and Thompson, R. (1979) The persistence of fire-induced magnetic oxides in soil and lake sediments. *Annales de Géophysique* 35, 103–107.

Scott, D.F. and Van Wyk, D.B. (1990) The effects of wildfire on soil wettability and hydrological behaviour of an afforested catchment. *Journal of Hydrology* 121, 239–256.

Shakesby, R.A., Doerr, S.H. and Walsh, R.P.D. (2000) The erosional impact of soil hydrophobicity: current problems and future research directions. *Journal of Hydrology* 231–232, 178–191.

Shakesby, R.A., Chafer, C.J., Doerr, S.H., Blake, W.H., Wallbrink, P., Humphreys, G.S. and Harrington, B.A. (2003) Fire severity, water repellency characteristics and hydrogeomorphological changes following the Christmas 2001 Sydney forest fires. *Australian Geographer* 34, 147–175.

Shakesby, R.A., Doerr, S.H., Blake, W.H., Humphreys, G.S., Wallbrink, P. and Chafer, C.J. (2004) Hillslope soil redistribution following severe wildfire in eucalypt stands in the Sydney water supply catchment, Australia. *Geophysical Research Abstracts* 6, 04730.

Walden, J., Slattery, M.C. and Burt, T.P. (1997) Use of mineral magnetic measurements to fingerprint suspended sediment sources: approaches and techniques for data analysis. *Journal of Hydrology* 202, 353–372.

Walden, J., Oldfield, F. and Smith, J. (eds) (1999) *Environmental Magnetism: A Practical Guide*. Technical Guide 6. Quaternary Research Association, Cambridge/London.

Wallbrink, P.J., Murray, A.S. and Olley, J.M. (1999) Relating suspended sediment to its original soil depth using fallout radionuclides. *Soil Science Society of America Journal* 63, 369–378.

Wallbrink, P.J., Roddy, B.P. and Olley, J.M. (2003) A tracer budget quantifying soil redistribution on hillslopes after forest harvesting. *Catena* 47, 129–201.

Walling, D.E., Peart, M.R., Oldfield, F. and Thompson, R. (1979) Suspended sediment sources identified by magnetic measurements. *Nature* 281, 110–113.

Yu, L. and Oldfield, F. (1993) Quantitative sediment source ascription using magnetic measurements in a reservoir-catchment system near Nijar, S.E. Spain. *Earth Surface Processes and Landforms* 18, 441–454.

7 Land Use, Sediment Delivery and Sediment Yield in England and Wales

R. Evans
Anglia Ruskin University, UK

Introduction

Runoff transports pollutants from the land to watercourses both in solution and in suspension. Sediment is the most visible pollutant. It is deposited in reservoirs (Butcher *et al.*, 1993) and can impact severely on freshwater fisheries (Stewart, 1963). But to sediment may be attached pesticides (Brown *et al.*, 1995) and nutrients, especially phosphorus (Frost, 1996; Hodgkinson and Withers, 1996; Fraser *et al.*, 1999; Quinton *et al.*, 2001). Water enriched by phosphate, mostly attached to soil particles, can support the formation of algal blooms, which can be toxic, in reservoirs and slow flowing streams and rivers as well as enhance the growth of unwanted plants, which reduces the biodiversity of the aquatic environment (Lund, 1970; RCEP, 1992; DETR, 1998; Environment Agency, 1998, 1999, 2000). At great cost, the water industry may have to remove sediment and its attached pollutants, and treat the enriched water, with its algal blooms, to make it fit to drink (Evans, 1995; Pretty *et al.*, 2001; Environment Agency, 2002). This is especially a problem in southern and eastern England where water is pumped from high-flowing rivers in winter to be stored in reservoirs to be released in the drier half of the year. Sediment can be expensive to remove from rivers and impoundments (Jarman, personal communication, with reference to National Trust property). Sediment-laden flood water can also cause great damage to property, as the autumn floods of 2001 showed (Environment Agency, 2002).

Here I will deal only with the transfer by water of sediment from the land into water courses. There are many sources of sediment both in upland and pastoral England and Wales and in the cultivated lowlands and there is not scope here to describe them in detail. I will give information on sediment yields in catchments and how those may have changed in the recent past and the reasons behind those changes. I will deal more specifically with where the sediment comes from in lowland arable landscapes, and then explain the apparent discrepancy between sediment yields measured in rivers and the amounts of soil carried from cropped fields by runoff and erosion. Some conclusions are then drawn. The work takes further that on sediment yields and land use outlined by Foster (1995).

Sediment Yield and Land Use

There is now sufficient evidence to make reasonable inferences about sediment yield in catchments in England and Wales and its relation to land use. Suspended sediment yields in some larger English and Welsh catchments range from 0.1 to 1 t/ha/year (Mitchell, 1990; Collins *et al.*, 1998; Walling *et al.*, 1999). With regard to much of the rest of the world these sediment yields are very low (Walling, 1988).

 Soil Erosion and Sediment Redistribution in River Catchments (eds P.N. Owens and A.J. Collins)

Large parts of these catchments are in the uplands and the pastoral areas flanking them, and arable land covers relatively small portions of the catchments. Higher yields may be associated with upland sub-catchments with more peat cover, or lower lying catchments which are more heavily stocked (Mitchell, 1990) or contain more cultivated land (Foster in Collins *et al.*, 1998).

Information given by Walling (1990) and Butcher *et al.* (1993) allows the dominant land use of smaller catchments to be identified (Table 7.1a). The positions of the catchments are portrayed on maps and these can be related to topographic maps and to the National Soil Map, the legend of which (Mackney *et al.*, 1983) describes the dominant land use of the soil associations found in these catchments.

Table 7.1. Suspended sediment yields (t/ha/year) in some English and Welsh river catchments.

Land use	Mean	Range	No.	Source
(a) Land use implied from National Soil Map Legend (Mackney *et al.*, 1983)				
Moorland	1.1	0.01–4.9	16	Walling (1990)
Pasture	0.4	0.01–1.6	30	Walling (1990)
Arable	0.2	0.02–0.8	17	Walling (1990)
Moorland with peat	1.22	0.22–2.89	14	Butcher *et al.* (1993)
Moorland, no peat	0.51	0.11–0.74	3	Butcher *et al.* (1993)
Pastoral	1.62	0.51–2.86	4	Butcher *et al.* (1993)
Mostly forest	0.04			Butcher *et al.* (1993)
(b) Land use described in article				
Forested	0.05			Oxley (1974)
Pasture	0.01			Oxley (1974)
Dominantly forested	0.10			Foster *et al.* (1985)
Not forested	0.09	0.01–0.38	6	Soutar (1989)
Mature forest	0.24	0.12–0.35	3	Soutar (1989)
Afforestation/felling	0.41	0.09–1.31	6	Soutar (1989)
Mixed land uses	0.44			Mitchell (1990)
Mixed	0.5 (0.2 rising to 0.9)			Foster and Walling (1994)
Lowland agriculture	0.36 and 0.17			Foster (1995)
Mixed	0.80			Foster *et al.* (1998)
Moorland	0.19 and 0.24			Foster and Lees (1999)
Forested	0.12 and 0.20			Foster and Lees (1999)
Pasture	0.09 and 0.19			Foster and Lees (1999)
Mixed	0.52			Foster and Lees (1999)
Arable	0.16 and 0.22			Foster and Lees (1999)
Mixed	0.09–0.12			Walling and Amos (1999)
Forested	0.42	(0.24–0.57)	3 Increase with replanting	Leeks (2000)
Moorland/grassland	0.12	(0.05–0.20)	4 Increasing with road construction	Leeks (2000)
Arable	0.77–1.22			Russell *et al.* (2001)
Mixed	0.82			Russell *et al.* (2001)

Moorland sheep-grazed catchments, especially those with a component of peat soils, have generally higher yields than have the pastoral catchments of lower ground, although those grazed intensively by cattle and sheep can have high yields, as in the southern Pennines (Butcher *et al.*, 1993). Some forested catchments may also have high yields. The arable catchments in the drier parts of England have low yields.

More detailed information on land use in catchments where sediment yield has been measured is available (Table 7.1b); some catchments for which data are given by Walling (1990) are also included in Table 7.1b. The highest sediment yield was in a moorland catchment where soil exposed by fire was eroding (Imeson, 1971). Other high yields were found in catchments containing recently planted or cut down forests, 'mixed' farms with both grazed and cultivated fields, as well as catchments that are dominantly cultivated. The ploughing of ground to aid afforestation, a technique brought into use in the late 1940s as surplus heavy war machinery became available, led to severe erosion and sedimentation/filtration problems in reservoirs (Evans, 1996). Increased yields in upland lakes in Wales since the 1950s have been attributed to an increase in tourist activity (George *et al.*, 2000). There is other evidence to suggest that sediment yields in small catchments have risen in the recent past, mainly related to more intensive grazing in catchments (Foster *et al.*, 1990; Van der Post, 1997). Foster and Walling (1994) show that in a small catchment in Devon sediment yields have risen markedly over the last 50 years or so as farming has intensified, thus stocking intensities have increased and a greater proportion of the land is now cultivated. Water erosion of arable land is much more widespread than it was 50 years ago (Evans, 1996; and below) and will have led to more sediment reaching rivers. Foster (1995) has shown that sediment yield increases as cereal pollen in the sediments increase and as field boundaries are removed. The increase in sedimentation rates in small catchments does not appear to be reflected in increased rates in larger catchments. Thus, overbank sedimentation rates essentially have been constant over the last 100 years (Owens *et al.*, 1999) in the Ouse catchment in Yorkshire. From around 1900 more sediment has come from the Vale of York rather than the Yorkshire Dales, and before the Second World War could be associated with increased grassland and grazing in the Vale. Prior to 30 years ago sediment came largely from topsoils, but since then more has come from subsoils and river banks. The changes in sediment yields could be related to changes in land use and possibly to climate.

The picture is a complex one, therefore. Although sediment yields are generally higher in the uplands, especially where eroding peat moors occur, where moors and mountains are relatively undisturbed by mankind's activities yields are low. Yields are also low in pastoral landscapes where stocking intensities have not increased greatly and where forests are long established. Sediment yields may have changed little in arable landscapes where the relief is low and heavy textured soils predominate.

Crude threshold values of suspended sediment yield may be defined that separate what may be termed a 'background' level for a particular land use from a higher level caused by intensification of that land use. The general background level of suspended sediment yield in 'stable' landscapes is less than *c.* 0.2 t/ha/year, whereas landscapes known to be being farmed or forested more intensively and suffering more runoff (and drainflow) and erosion have sediment yields greater than *c.* 0.4 t/ha/year. It is suggested here that as land use has intensified since the Second World War, sediment yields have increased and are often more than double what they were.

Sediment Delivery from Eroded Cropped Fields

There is a little information on the amounts of sediment transported from eroded cultivated fields into water courses. However, information on sediment delivery from eroded cultivated fields in England and Wales allows estimates to be made of the amounts of soil reaching lowland rivers which (see following section) further allows comparison with river suspended sediment yields. Between 1982 and 1986 an average of 700 km^2 of farm land spread across 17 localities in lowland England

and Wales (Fig. 7.1) was surveyed each year to locate fields eroded by rills and gullies. The localities were selected not only to sample soils considered to be at risk of erosion, but also to be representative of the different soils found in the lowlands. The large dataset (*c.* 1700 eroded fields) provides information to estimate rates of erosion and deposition within fields and hence sediment delivery to watercourses (Evans, 2002). Mean erosion rates have been estimated for each monitored transect (Table 7.2).

Sediment delivery, as a percentage, was estimated for 367 very small catchments within eroded fields. This was done by relating the estimated volumes of rills and gullies to the estimated volumes of sediment deposited within the field for those fields where it was considered that all the sediment that had been deposited within the field had been accounted for. Fine textured deposits were often not found for they had been carried into fieldside drainage ditches or on to farm tracks and roads and then into roadside ditches and so into streams.

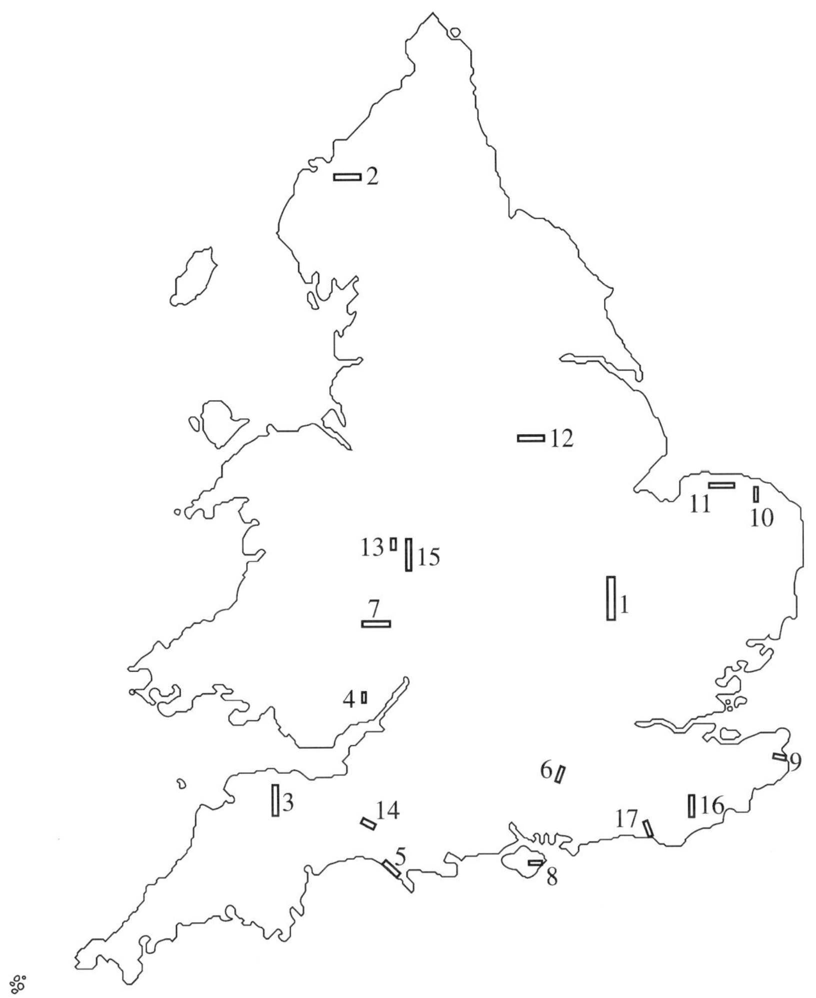

Fig. 7.1. Transects monitored for rill and gully erosion 1982–1986. Key: 1 Bedfordshire/Cambridgeshire; 2 Cumbria; 3 Devon; 4 Gwent; 5 Dorset; 6 Hampshire; 7 Herefordshire; 8 Isle of Wight; 9 Kent; 10 Norfolk East; 11 Norfolk West; 12 Nottinghamshire; 13 Shropshire; 14 Somerset; 15 Staffordshire; 16 Sussex East; 17 Sussex West.

Table 7.2. Rates of erosion and sediment delivery to watercourses for 17 monitored transects.

Locality	Erosion rate (t/ha/year)	Sediment delivery (t/ha/year)
Isle of Wight	0.47	0.19
Shropshire	0.33	0.09
Nottinghamshire	0.25	0.06
Somerset	0.22	0.14
Staffordshire	0.21	0.06
Hampshire	0.13	0.09
Norfolk East	0.11	0.03
Sussex West	0.11	0.04
Kent	0.08	0.06
Norfolk West	0.07	0.03
Dorset	0.06	0.05
Gwent	0.04	0.03
Cumbria	0.03	0.01
Herefordshire	0.02	0.01
Bedfordshire	0.01	0.01
Sussex East	0.01	0.01
Devon	0.01	0.01
Mean	0.10	0.05

There is great variability in the data, related probably to crop cover, storm intensity and the amount and velocity of runoff generated.

$$\% \text{ sediment delivery} = 100 - (19.6 + 0.69\% \text{ sand} + 0.06\% \text{ silt} - 0.25\% \text{ clay})$$
$$r^2 = 27.8\% \ (P = 0.000)$$

For each of the 17 localities the mean particle size of the topsoil in the eroded fields was estimated. This was done by summing the sand, silt and clay fractions of the topsoils of the dominant soil series within the soil associations (SSEW, 1983, 1984) in which the eroded fields were located and dividing the sums for each fraction by the number of eroded fields. Using the mean topsoil particle size determination, the sediment transported out of the field and which is mostly delivered to watercourses was estimated (Table 7.2). The mean rate of sediment delivery for all transects is estimated to be 0.05 t/ha/year, ranging from 0.01 to 0.19 t/ha/year. These values may be on the high side, because not all fine soil particles may reach watercourses, but observation suggests much does. On the other hand, the 5-year monitoring period did not include erosion events of large magnitude, say of > 1 in 30 year occurrence, in which delivery of sediment to watercourses would be greater than normal, and so the estimates may err on the conservative side.

Based on the data (extent of erosion and rate of erosion) obtained from the national monitoring scheme, other *ad hoc* survey data and anecdotal evidence, Evans (1990) determined five classes of risk of a soil association being eroded by water. Risk of erosion is primarily defined by the estimated mean annual area of eroded fields likely to occur within an association, which is largely governed by the extent of arable land within a soil landscape, the texture of the topsoil and the extent of slopes steeper than *c.* 3°. To a much lesser extent risk is defined by its likely erosion rate, which is mainly a function of topsoil texture.

Erosion rates of the 'very small' and 'small' risk classes are considered to be similar, the major difference between the classes being the extent of erosion, < 0.5% and 0.5–1.0% respectively. Six of the transects (Bedfordshire, Cumbria, Devon, Dorset, Gwent and Sussex East; Fig. 7.1) have low erosion rates, the mean volumetric rate of erosion per field being 1.14 m^3/ha. Assuming a topsoil bulk density of 1300 kg/m^3, this equates to an erosion rate of 1.48 t/ha/field. For soil associations at moderate, high and very high risk, the extent of land covered by eroding fields is 1–5%, 5–10% and > 10% respectively, and the mean erosion rate based on eroded fields in 11 monitored transects is 3.22 t/ha/field. Taking into account the respective areas covered by the soil associations in each risk category, the delivery of suspended sediment to watercourses is estimated for associations at very low risk of rill and gully erosion to be 0.005 t/ha/year, for small risk 0.01 t/ha/year, moderate risk 0.04 t/ha/year, high risk 0.17 t/ha/year and very high risk 0.19 t/ha/year.

The 17 transects cover a large number (61) of soil associations. Of these, in only 17 associations were more than 30 eroded fields located. Three associations contained more than 30 eroded fields in each of two transects. The fields covered 2–14% of the soil associations. Three of these associations are considered at little risk of erosion (Evans, 1990), except in localities where the association is dominantly cultivated land or its relief is more conducive to runoff and channel erosion, for example at the edge of the chalky boulder clay plateau in eastern England. The sediment delivery in the 20 soil associations ranges from 0.01 to 0.29 t/ha/year, with a mean of 0.09 t/ha/year. For the three associations where more than 30 fields were located in two transects, sediment deliveries are estimated to be respectively 0.03 and 0.03 t/ha/year, 0.03 and 0.09 t/ha/year and 0.07 and 0.08 t/ha/year.

The above estimates of sediment yield are, in the main, for large areas of countryside. However, within one soil association erosion can differ in its extent. Thus, erosion may vary in extent because in some localities within the association proportions vary of arable to pasture, or rainfall regimes or topography differ. Estimates of sediment delivery were made therefore for 10 km^2 blocks of land lying within different soil associations and suffering varying extents of erosion (Table 7.3). In the mid-1980s water erosion of fields along valley sides could occur extensively on Hanslope soils (soil association 4lld; Mackney *et al.*, 1983) of the chalky boulder clay in eastern England. The most extensive erosion recorded in the mid-1980s was on the sandy soils of the Cuckney 1 association in Nottinghamshire, but erosion was also extensive in parts of the South Downs, Sussex (Andover 1 association), the sand lands of the Midlands (Bridgnorth), the silty soils of Somerset (South Petherton) and soils on the Cretaceous lower greensand of the Isle of Wight (Fyfield 4). On clayey (Hanslope) soils, because erosion rates are low, and on sandy (Cuckney 1) soils where erosion is more severe but delivery rates are low, sediment delivery to watercourses is high only when erosion is extensive. Soils with high erosion rates and high sediment delivery (South Petherton and Fyfield 4) provide much material to be transported to watercourses even when erosion is not extensive.

Sediment Supply and Yield in the Lowlands

Estimated yields of sediment from eroded fields underestimate suspended sediment yields measured in lowland catchments that are considered to be more at risk of erosion (Table 7.4). However, in smaller blocks of land where erosion

Table 7.3. Estimated sediment delivery (t/ha/year) in 10 km^2 catchments – as extent of erosion increases.

	Extent of erosion				
	10%	20%	30%	40%	50%
Clayey soils of the Hanslope association	0.05	0.10	0.15	0.20	0.25
Sandy soils of the Cuckney association	0.07	0.13	0.20	0.27	0.34
Silty soils of the Andover association	0.10	0.21	0.31	0.42	0.52
Sandy and coarse loamy soils of the Bridgnorth association	0.18	0.36	0.53	0.71	0.89
Silty soils of the South Petherton association	0.38	0.76	1.15	1.53	1.91
Sandy and coarse loamy soil of the Fyfield association	0.43	0.86	1.29	1.72	2.15

Table 7.4. Comparison of suspended sediment yields in rivers and estimated sediment deliveries from eroded fields.

	Suspended sediment (t/ha/year)	Sediment delivery (t/ha/year)		
		Transect	Risk class	Erodible soil association
Less erodible	< 0.2	0.01–0.03	0.005–0.01	–
More erodible	> 0.4	0.04–0.19	0.04–0.19	0.01–0.29

can be more extensive, yields of > 0.4 t/ha/year are estimated where soils are erodible and have a high delivery ratio (Table 7.3). Instances are known of eroded fields covering more than 20% of areas less than about 10 km^2 in extent (Evans and Boardman, 2003; Evans unpublished, data from national erosion monitoring survey).

It is unlikely that many rilled or gullied fields were missed when monitoring erosion in the mid-1980s, certainly not double the number those actually found. However, a doubling of the number of fields recorded would, in localities more at risk of erosion, give sediment yields nearer to those considered representative of catchments suffering more than 'background' levels of erosion. Is there, therefore, a discrepancy between estimated amounts of sediment delivered from fields and measured amounts yielded in lowland catchments? Or, can the 'gap' between sediment delivered and sediment yielded be explained? A number of explanations can be put forward.

First, the discrepancy partly may be a result of a mismatch of the locations where sediment yield has been measured and the localities where erosion has been surveyed, for there is little overlap between the two. Many of the more recently recorded higher sediment yields (Table 7.1) are in the wetter south and west of the country, whereas many of the localities monitored for erosion in the mid-1980s are to the north and east. In the wetter south and west of England, land use changes and intensification of use have been more marked in the last two decades or so than elsewhere (see below). There is also a temporal aspect to be considered. More monitoring of erosion has been carried out in western and southern England (Chambers *et al.*, 1992; Harrod, 1998; Chambers and Garwood, 2000) since the original national monitoring scheme was carried out.

Secondly, many of the high sediment yields are associated with what may be termed 'mixed' farming, where there is a fair proportion of the land in the catchment grazed by livestock, rather than being dominantly arable land. Trampling and the breaking down of channel banks can markedly change sediment loads in streams, accounting for between 3% and 54% of suspended sediment loads (Foster *et al.*, 1990; Russell *et al.*, 2001), and has been put forward to explain high sediment yields in a number of catchments (Foster *et al.*, 1990; Mitchell, 1990; Van der Post, 1997).

Thirdly, if there is a discrepancy between sediment delivery from the land to watercourses and sediment yields, it is not large. A background level of sediment yield will pertain related to sources other than from eroded fields. River bank and stream channel erosion can contribute between 6% and 45% of suspended sediment load (Foster *et al.*, 1990; Walling *et al.*, 1999; Russell *et al.*, 2001). Runoff from roads and especially unsurfaced tracks can explain much of the suspended sediment load in a stream, for example 98% in the Polish Carpathian Mountains (Froehlich, 1995). Material transported through field drains installed from the mid-1800s to when subsidies on drainage were stopped in the mid-1980s (Robinson and Armstrong, 1988), a source of sediment too little studied, can explain 31–55% of suspended sediment loads in lowland catchments (Russell *et al.*, 2001). If there is a background level of sediment yield of 0.2 t/ha/year, the increase in soil erosion recorded over the last four decades or so (see below) could explain a good proportion of the higher yields recorded in smaller lowland catchments.

Fourthly, runoff from cultivated land, as from heavily trampled ground (see above), will carry fine soil particles splashed into the water by raindrops or by slaking of the saturated bare soil surface. In fields where no water-cut channels can be found, evidence of flow such as 'flow lines' of deposited organic material or very small sandy fans below slight incisions in the soil surface can often be found in bare cultivated fields, especially in tractor wheelings. In the erosion survey of the mid-1980s such features were often seen but discounted, whereas in a survey in the later 1990s more notice was taken of such features, which may account for an apparent increase in erosion recorded in that survey compared to the earlier one (Evans, 2005). Such 'sheetflow' or 'sheetwash' may be important in transporting fine sediment to water courses.

Harrod (1994) suggests that runoff from unrilled parts of cultivated fields (= sheetwash and splash erosion) typically carries a sediment load of 0.1 t/ha/year, though it can be up to 3 t/ha/year depending on the number of rainfall events, their duration and intensity. Sand grains

(> 0.06 mm) are often left on the surface of soils after the soil aggregates have been broken down by raindrops and the fines have been splashed and washed away. If 10% of the surface is left covered by sand grains, up to 0.7 t/ha may have been washed off the land. This figure is not greatly different to those (0.6–0.9 t/ha/year) measured by Morgan *et al.* (1987) on a range of soils, although the use of troughs to trap the soil may have created a 'driver' and so overstated erosion somewhat. The magnitude/frequency curve drawn from the mid-1980s survey data (Fig. 7.2) is very similar in shape to those constructed by Boardman and Favis-Mortlock (1999), and implies that frequent events such as those associated with a small rill found in a large field and which is often associated with evidence of flow as outlined above, suggest rates of sheet erosion of < 0.1–0.3 t/ha/year. Quinton *et al.* (2001) show that much of this fine material leaves the erosion plot often in small but frequently occurring storms. This relates well to the findings of Evans (2002) outlined above, i.e. it is the fine material which moves from the field.

As land use and intensity of use has changed in lowland England and Wales since the late 1940s, especially since joining the European Union in 1973 (see below), more sediment will have been washed from the land. The background level of sediment delivery and yield in those parts of the country where land use changes and intensification have been greatest will have increased therefore, probably from *c.* 0.2 t/h/year to 0.3–0.5 t/ha/year. These minor changes in sediment yields, or so they will be perceived in world terms (Walling, 1988), appear from well-grounded as well as anecdotal evidence (Evans, 1996) and from policy actions (Environment Agency, 2000, 2002) to have had marked impacts on water quality, the water environment generally (its biodiversity) and the wider environment (impacts of flooding).

Changes in Agricultural Land Use Since 1945 and their Likely Impacts on Sediment Yield

Changes in agricultural land use in England (MAFF, 1945–2002) can only be described briefly here. The trends in Wales are similar. The changes in proportions of cultivated land to grassland and the changes in cropping (Fig. 7.3) were, prior to 1973, driven by UK government policies and after that date by European Union policies. These policy shifts were in response to economic and technological changes. Similar reasons explain the fluctuations in cattle and sheep numbers (Fig. 7.3), but foot-and-mouth disease also played a part in 1966 and 2001.

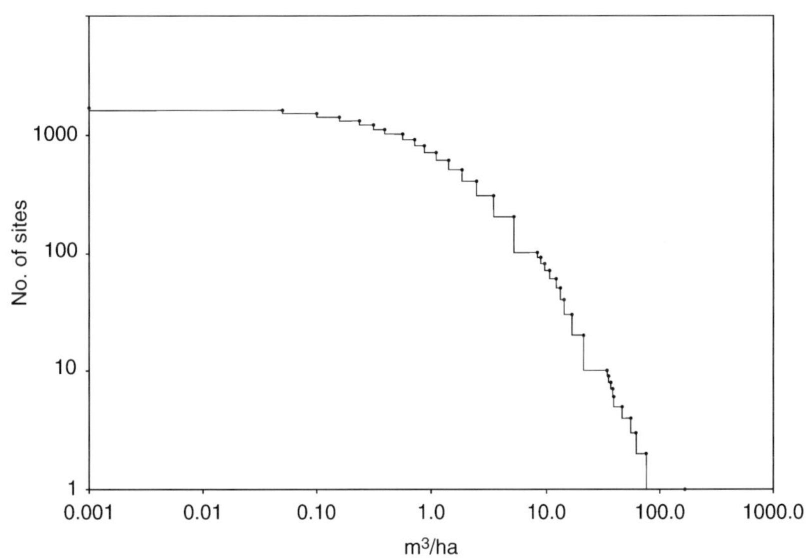

Fig. 7.2. Occurrence and rate of erosion in eroded fields, 1982–1986 – magnitude/frequency curve.

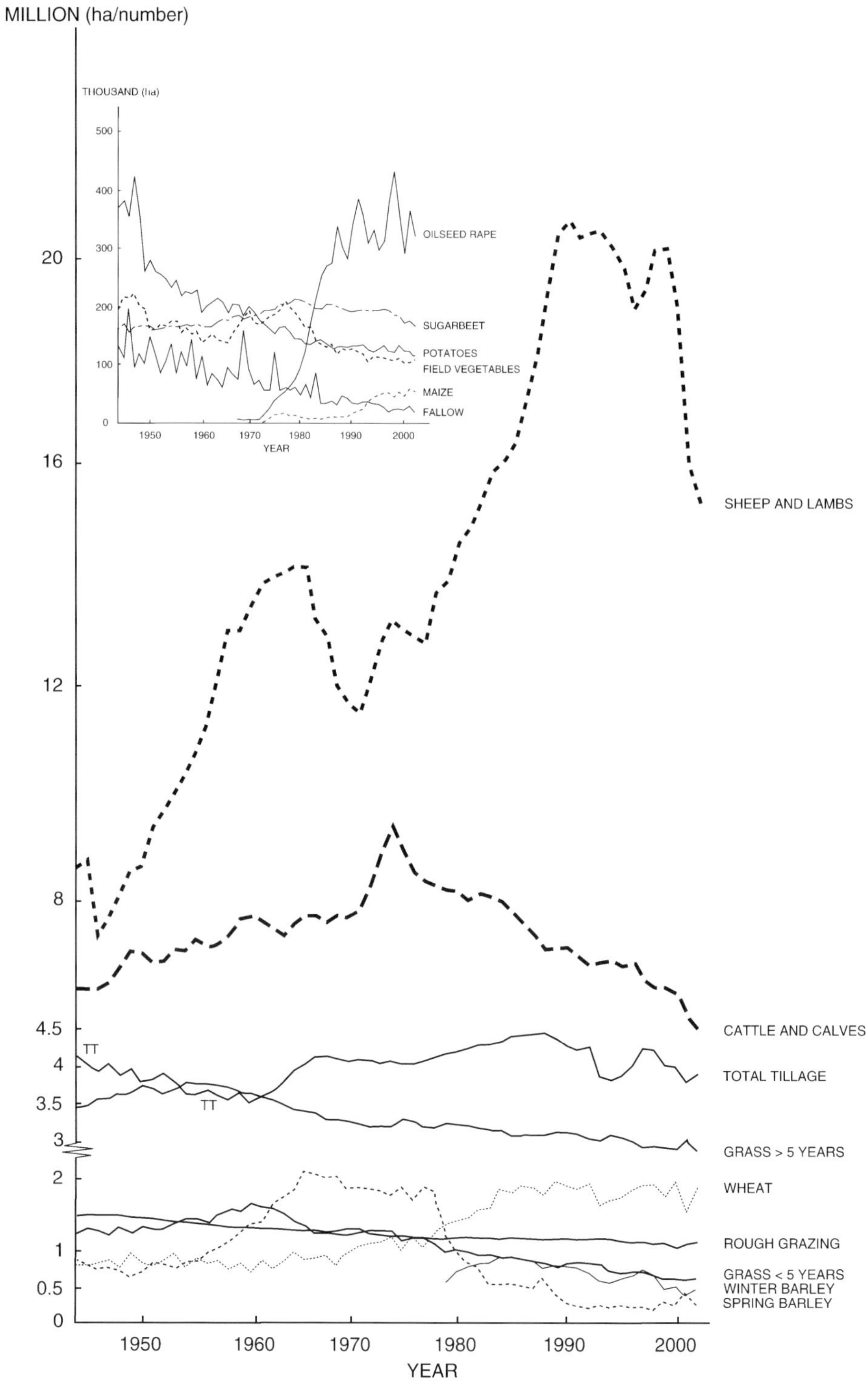

Fig. 7.3. Change in agricultural land use, crops and number of sheep and cattle in England 1945–2002.

Large-scale planting of coniferous trees ceased in the late 1980s, when tax incentives were removed.

The impacts of afforestation, moorland drainage and overgrazing with regard to soil erosion and sediment availability in the uplands have been summarized by Evans (1996, in press). The major impacts of afforestation on sedimentation should have declined since the early 1990s with the introduction of better working practices (Forestry Commission, 1993). In pastoral areas it is known, both in the uplands (Van der Post, 1997) and lowlands (Foster *et al.*, 1990) that stocking of livestock intensively when the ground is wet can release sediment to watercourses, by overland flow or via field drains. Bare soil is created where land is heavily trampled and tracked, especially around feeding places and gateways. Such overstocking is by no means uncommon, I have seen instances throughout England and Wales. In some localities where sheep, especially, and cattle were slaughtered during the foot-and-mouth epidemic of 2001 pastures (may) have recovered somewhat, as they had in early June 2002 near Settle in the Yorkshire Dales, so inhibiting transport of sediment to streams.

In arable England and Wales, in recent decades most soil erosion has occurred in winter cereals, but erosion occurs disproportionately in relation to their area in fields of hops, sugarbeet, maize, potatoes and field vegetables (Table 7.5) because these crops are grown on more vulnerable soils and in ways which promote runoff. Over time the areas sown to different crops have differed (Fig. 7.3), particularly with regard to spring-sown barley, autumn-sown wheat and barley, and oilseed rape. Assuming for the sake of simplicity that the average size of cropped field is 7.5 ha, the size of the average eroded field located in the mid-1980s survey, the change in extent of erosion in different crops can be traced over time. This assumes the risk of erosion occurring in a particular crop has not changed over time.

The minimum extent of tilled ground was around 1960 and the maximum extent in the 1980s, coinciding with the period when the national water erosion monitoring scheme was carried out, peaking in 1988 when set-aside was brought in to curb production. Cereals are the dominant crops grown in England and Wales, but the proportions of spring- and autumn-sown cereals have changed markedly over time (Fig. 7.3). The area planted to winter barley has declined greatly in recent times due to unfavourable economic conditions. The number of fields affected by rill and gully erosion is estimated for the major crops for three periods: 1958–1962, 1984–1988 and 1998–2002 (Table 7.6). As noted above the estimate assumes the average field size is 7.5 ha, the average size of the eroded field in the mid-1980s. It is likely mean field size was smaller in 1960 than in 1984, but will not have changed much since then. Such estimates are very crude but give some information about the likely trend of erosion over time.

It is estimated that the number of eroded fields, excluding those of maize and outdoor pigs, increased by 13% between around 1960 and the 1980s, and has declined by 17% since then (Table 7.6). The estimated increase in number of eroded fields up to the 1980s seems small, especially in the light of the attitude by agricultural advisers throughout the 1970s that although erosion was known and accepted to occur in fields of sugarbeet and field vegetables it was not considered a problem in other crops. It may be that the assumption made here, that

Table 7.5. Risk of erosion according to crop type (Evans, 2005).

Crop	Risk
Outdoor pigs	1 field in 3
Hops	1 field in 6
Sugarbeet	1 field in 7
Maize	1 field in 7
Potatoes	1 field in 10
Other crops	1 field in 11
Field vegetables	1 field in 14
Bare soil	1 field in 21
Kale	1 field in 24
Ley grasses	1 field in 32
Spring barley	1 field in 34
Peas	1 field in 38
Winter wheat and barley	1 field in 42
Field beans	1 field in 71
Oilseed rape	1 field in 100

Table 7.6. Estimated number of fields suffering channel erosion in selected crops in England, 1958–1962, 1984–1988 and 1998–2002.

Crop	1958–1962	1984–1988	1998–2002
Winter wheat and barley	2,505	8,585	7,335
Sugarbeet	3,205	3,855	3,405
Spring barley	4,785	2,135	1,120
Potatoes	2,840	1,810	1,690
Field vegetables	1,440	1,250	1,045
Ley grasses	660	370	265
Bare soil	885	230	160
Oilseed rape	–	385	475
Hops	185	100	55
Total	16,505	18,720	15,550
Maize	–	325	2,085
Total		19,045	17,635
Outdoor pigs	–	< 250	> 1,250
Total		< 19,295	> 18,885

around 1960 erosion occurred widely in cereals, is incorrect.

Since the 1960s, agriculture has changed greatly. It has become more mechanized, and machines for cultivating the land and for spraying and harvesting crops have all got bigger; there is more spraying of the land and more tractor wheelings; smoother seedbeds are now prepared for winter cereals so that herbicides can act efficiently; fields to be cropped by potatoes are de-stoned; and many fields have been enlarged. There is some evidence, mostly anecdotal, that topsoils have become more compact, which may be related to declines in amounts of organic matter in continuously cropped fields, so enhancing the instability of soil structure. In the recent past, bulk densities of topsoils in England have often been considered to be *c.* 1300 kg/m^3 and this figure has been used to convert volumes eroded to mass (Evans, 2002). However, coarse loamy topsoils in a field in north Norfolk from which there was rapid runoff and erosion in a large storm had a mean ($n = 10$) bulk density of 1560 kg/m^3, and in tractor wheelings 1690 kg/m^3. The supposition that topsoils may be becoming more compact may also be supported by the data showing the declining extent of fallow land over the years (Fig. 7.3). Thus, bigger machinery which allows the land to be worked at more-or-less any time will lead to land being worked when it is too wet, so causing compaction. All these changes have encouraged runoff and erosion. It seems likely therefore that erosion was less extensive around 1960 than the estimate implies.

The great expansion in area of autumn-sown cereals, by a factor of about three, explains a large proportion of the increase in erosion since 1960. Winter cereals are grown because they out-yield spring cereals and are more profitable. The ground under winter cereals is saturated for a large part of the winter, often crop cover is thin and the cereals have been sprayed, leaving tractor wheelings (tramlines) down which runoff and erosion can take place. Also, since 1960 autumns may have got wetter and storms may have become more intensive (Foster, personal communication). The onset of set-aside will have reduced the number of eroding fields, and also will have reduced the 'connectivity' between eroding fields (Evans and Boardman, 2003), so there is less chance of sediment being carried far down valleys and into watercourses. Erosion will also be less extensive in other localities where crops vulnerable to runoff, hops for example, have declined in area.

However, since the mid-1980s two 'crops' (maize: MGA/EA, no date; outdoor pigs: Evans, 2004) that are vulnerable to runoff and erosion have greatly expanded in area, although it is not easy to obtain a precise figure for the area of land used to rear outdoor pigs. Only about 3.5% of the total arable area is covered by fields of maize and outdoor pigs, but because of the vulnerability of that land the decline in estimated total number of eroded fields from the mid-1980s until now is not 17% but 2%. Much of the expansion in area of maize and outdoor pigs has taken place in southern and western England, where rainfall is higher, and on soils that are vulnerable to runoff.

Conclusions

Changes in land use and intensivity of use in England and Wales driven by political, economic and social factors, and not by environmental considerations, have led over the last 40 years to more sediment reaching watercourses in many small catchments. Although sediment yields in England and Wales are (still) low, with regard to those in much of the rest of the world, they have severely affected water quality.

Sediment delivery to water courses probably peaked nationally in the 1980s, when soil erosion was most extensive. However, in some regions and localities sediment delivery will have remained high and may even have increased in those areas where maize or outdoor pigs recently have been introduced. To people living in these areas, especially in the south west of England, the perception may be that erosion and sedimentation have increased or become more severe only in the last two decades or so, whereas they have been happening for much longer in other parts of the country. The 'push' to tackle sedimentation is largely coming from these localities.

In the lowlands, sediment is coming from sources other than eroding cultivated fields. There is probably a 'background' level of sedimentation, with particles being transported by sheetwash from cultivated fields, from roads and tracks and from eroding river channels. Locally these sources may be the most important. The intensification of grazing in pastoral regions and the draining of heavy textured fields (Foster *et al.*, 2003) may both be major but largely un-investigated sources of sediment that have become important over the last few decades.

It is easier to identify sources of sediment than to measure sediment loads. Thus, measuring 'sheetwash' is difficult and it may be that indirect methods of assessment, such as those discussed here, will have to suffice for the present. It may well be that small-scale but frequent (say once or twice per year) erosion events are the source of much of the sediment reaching rivers – probably accounting for yields of 0.1–0.3 t/ha/year.

This raises a paradox. Water erosion monitoring schemes were set up to locate fields with rills and gullies, for it was considered that in terms of volumes of soil moved 'sheetwash' was not important, because its impacts on soil productivity over the short term would be negligible. At that time it was not realized that water quality would become, because of European Union legislation, an issue.

Although crude, the analysis presented here suggests that we know the major drivers of sedimentation – they are economic, and brought about by political decisions – and unless these drivers are tackled sedimentation will continue to be a problem for water quality and fisheries. The recent proposals emanating from the Curry Report (PCFFF, 2002) may lead to a more sustainable way of farming the land and result in lower sediment loads in rivers.

The implication of much of this work is that land use and intensivity of use are the most important drivers of sedimentation. However, changing climate may also be having an impact. There is a need to disentangle these factors.

Acknowledgements

John Walsh drew the diagrams. I thank Ian Foster for stimulating discussions about sources of sediment and sediment yields. The Department for Environment, Food and Rural Affairs at York kindly supplied to me, via the Internet, agricultural statistics for England for the years 1989–2002. The two referees, and Phil Owens and Alison Collins made most helpful comments on an earlier draft.

References

Boardman, J. and Favis-Mortlock, D. (1999) Frequency-magnitude distributions for soil erosion, runoff and rainfall – a comparative analysis. *Zeitschrift für Geomorphologie* NF 115, 51–70.

Brown, C.D., Hodgkinson, R.A., Rose, D.A., Syers, J.K. and Wilcockson, S.J. (1995) Movement of pesticides to surface waters from a heavy clay soil. *Pesticide Science* 43, 131–140.

Butcher, D.P., Labadz, J.C., Potter, A.W.R. and White, A.P. (1993) Reservoir sedimentation rates in the southern Pennine region, UK. In: McManus, J. and Duck, R.W. (eds) *Geomorphology and Sedimentology of Lakes and Reservoirs*. John Wiley & Sons, Chichester, UK, pp. 73–92.

Chambers, B.J. and Garwood, T.W.D. (2000) Monitoring of water erosion on arable farms in England and Wales, 1990–94. *Soil Use and Management* 16, 93–99.

Chambers, B.J., Davies, D.B. and Holmes, S. (1992) Monitoring of water erosion on arable farms in England and Wales, 1989–90. *Soil Use and Management* 8, 163–170.

Collins, A.L., Walling, D.E. and Leeks, G.J.L. (1998) Use of composite fingerprints to determine the provenance of the contemporary suspended sediment load transported by rivers. *Earth Surface Processes and Landforms* 23, 31–52.

DETR (1998) *Environmental Effects of Agriculture. Final Report*. Department of the Environment, Transport and the Regions, London.

Environment Agency (1998) *The State of the Environment of England and Wales: Fresh Waters*. HMSO, London.

Environment Agency (1999) *The State of the Environment of England and Wales: Coasts*. HMSO, London.

Environment Agency (2000) *Aquatic Eutrophication in England and Wales: A Management Strategy*. Environment Agency, Bristol, UK.

Environment Agency (2002) *Agriculture and Natural Resources: Benefits, Costs and Potential Solutions*. Environment Agency, Bristol, UK.

Evans, R. (1990) Soils at risk of accelerated erosion in England and Wales. *Soil Use and Management* 6, 125–131.

Evans, R. (1995) Soil erosion and land use: towards a sustainable policy. In: Evans, R. (ed.) *Soil Erosion and Land Use: Towards a Sustainable Policy*. Proceedings 7th Professional Environmental Seminar, Cambridge Committee for Interdisciplinary Environmental Studies. White Horse Press, Knapwell/Cambridge, UK, pp. 14–26.

Evans, R. (1996) *Soil Erosion and its Impacts in England and Wales*. Friends of the Earth Trust, London.

Evans, R. (2002) An alternative way to assess water erosion of cultivated land – field-based measurements and analysis of some results. *Applied Geography* 22, 187–208.

Evans, R. (2004) Outdoor pigs and flooding – an English case study. *Soil Use and Management* 20, 178–181.

Evans, R. (2005) Monitoring water erosion in lowland England and Wales – a personal view of its history and outcomes. *Catena* 64, 142–161.

Evans, R. (in press) Erosion of uncultivated land. In: Boardman, J. and Poesen, J. (eds) *Soil Erosion in Europe*. John Wiley & Sons, Chichester, UK.

Evans, R. and Boardman, J. (2003) Curtailment of muddy floods in the Sompting catchment, South Downs, West Sussex, southern England. *Soil Use and Management* 19, 223–231.

Forestry Commission (1993) *Forests and Water Guidelines*. Forestry Commission, Edinburgh, UK.

Foster, I.D.L. (1995) Lake and reservoir bottom sediments as a source of soil erosion and sediment transport data in the UK. In: Foster, I.D.L., Gurnell, A.M. and Webb, B.W. (eds) *Sediment and Water Quality in River Catchments*. John Wiley & Sons, Chichester, UK, pp. 265–283.

Foster, I.D.L. and Lees, J.A. (1999) Changing headwater suspended sediment yields in the LOIS catchments over the last century: a palaeo-limnological approach. *Hydrological Processes* 13, 1137–1153.

Foster, I.D.L. and Walling, D.E. (1994) Using reservoir sediments to reconstruct changing sediment yields and sources in the catchment of the Old Mill Reservoir, South Devon, UK, over the past 50 years. *Hydrological Sciences Journal* 39, 347–368.

Foster, I.D.L., Dearing, J.A., Simpson, A., Carter, A.D. and Appleby, P.G. (1985) Lake catchment based studies of erosion and denudation in the Merevale catchment, Warwickshire, UK. *Earth Surface Processes and Landforms* 10, 45–68.

Foster, I., Grew, R. and Dearing, J. (1990) Magnitude and frequency of sediment transport in agricultural catchments: a paired lake-catchment study in Midland England. In: Boardman, J., Foster, I.D.L. and Dearing, J.A. (eds) *Soil Erosion on Agricultural Land*. John Wiley & Sons, Chichester, UK, pp. 153–171.

Foster, I.D.L., Lees, J.A., Owens, P.N. and Walling, D.E. (1998) Mineral magnetic characterization of sediment sources from an analysis of lake and floodplain sediments in the catchments of the Old Mill Reservoir and Slapton Ley, South Devon, UK. *Earth Surface Processes and Landforms* 23, 685–703.

Foster, I.D.L., Chapman, A.S., Hodgkinson, R.M., Jones, A.R., Lees, J.A., Turner, S.E. and Scott, M. (2003) Changing suspended sediment load and particulate phosphorus loads and pathways in underdrained lowland agricultural catchments: Herefordshire and Worcestershire, UK. *Hydrobiologia* 494, 119–126.

Fraser, A.I., Harrod, T.R and Haygarth, P.M. (1999) The effect of rainfall intensity on soil erosion and particulate phosphorus transfer from arable soils. *Water Science and Technology* 39, 41–45.

Froehlich, W. (1995) Sediment dynamics in the Polish Flysch Carpathians. In: Foster, I.D.L., Gurnell, A.M. and Webb, B.W. (eds) *Sediment and Water Quality in River Catchments*. John Wiley & Sons, Chichester, UK, pp. 453–461.

Frost, A. (1996) Loch Leven and diffuse pollution. In: Petchey, A.M., D'Arcy, B.J. and Frost, C.A. (eds) *Diffuse Pollution and Agriculture*. The Scottish Agricultural College, Aberdeen, UK, pp. 174–182.

George, D.G., Hewitt, D.P. and Appleby, P.G. (2000) Long-term trends in the rate of sediment accumulation in some selected Welsh lakes. In: Harrod, T.R., McHugh, M., Appleby, P.G., Evans, R., George, D.G., Haworth, E.Y., Hewitt, D., Hornung, M., Housen, G., Leeks, G., Morgan, R.P.C. and Tipping, E. *Research on the Quantification and Causes of Upland Erosion*. Study Number JX4118E. Final Project Report to MAFF, Appendix B2 – Sediment in Lakes. Soil Survey and Land Research Centre, Cranfield, UK, pp. 59–70.

Harrod, T.R. (1994) Runoff, soil erosion and pesticide pollution in Cornwall. In: Rickson, R.J. (ed.) *Conserving Soil Resources. European Perspectives*. CAB International, Wallingford, UK, pp. 105–115.

Harrod, T.R. (1998) *A Systematic Approach to National Budgets of Phosphorus Loss Through Soil Erosion and Surface Runoff at National Soil Inventory (NSI) Nodes*. Final Project Report to Ministry of Agriculture, Fisheries and Food. Soil Survey and Land Research Centre, Cranfield, UK.

Hodgkinson, R.A. and Withers, P.J.A. (1996) Non-point source phosphorus loss from small agricultural catchments. In: Petchey, A.M., D'Arcy, B.J. and Frost, C.A. (eds) *Diffuse Pollution and Agriculture*. The Scottish Agricultural College, Aberdeen, UK, pp. 120–126.

Imeson, A.C. (1971) Heather burning and soil erosion on the North Yorkshire Moors. *Journal of Applied Ecology* 8, 537–542.

Leeks, G. (2000) Comparison of turbidity and flow from moorland and forest catchments at Plynlimon. In: Harrod, T.R., McHugh, M., Appleby, P.G., Evans, R., George, D.G., Haworth, E.Y., Hewitt, D., Hornung, M., Housen, G., Leeks, G., Morgan, R.P.C. and Tipping, E. *Research on the Quantification and Causes of Upland Erosion*. Study Number JX4118E. Final Project Report to MAFF, Appendix A2. Soil Survey and Land Research Centre, Cranfield, UK.

Lund, J.W.G. (1970) Primary production. *Water Treatment and Examination* 19, 332–348.

Mackney, D., Hodgson, J.M., Hollis, J.M. and Staines, S.J. (1983) *Legend for the 1: 250,000 Soil Map of England and Wales*. Soil Survey of England and Wales, Harpenden, UK.

MAFF (1945–2002) *Agricultural Statistics* (to 1989) HMSO, London; (1989–2002) Department for Environment, Food and Rural Affairs, York, UK.

MGA/EA (no date) *Managing Maize. Environmental Protection with Profit*. Maize Growers Association/ Environment Agency, Bath/Bristol, UK.

Mitchell, D.J. (1990) The use of vegetation and land use parameters in modelling catchment sediment yields. In: Thornes, J.B. (ed.) *Vegetation and Erosion*. John Wiley & Sons, Chichester, UK, pp. 289–316.

Morgan, R.P.C., Martin, L. and Noble, C.A. (1987) *Soil Erosion in the United Kingdom: A Case Study from Mid-Bedfordshire*. Occasional Paper No. 14. Silsoe College, Cranfield Institute of Technology, Silsoe, UK.

Owens, P.N., Walling, D.E. and Leeks, G.J.L. (1999) Use of floodplain sediment cores to investigate recent historical changes in overbank sedimentation rates and sediment sources in the catchment of the River Ouse, Yorkshire, UK. *Catena* 36, 21–47.

Oxley, N.C. (1974) Suspended sediment delivery rates and the solute concentration of stream discharge in two Welsh catchments. In: Gregory, K.J. and Walling, D.E. (eds) *Fluvial Processes in Instrumented Watersheds*. Institute of British Geographers, London, pp. 141–154.

PCFFF (2002) *Farming & Food. A Sustainable Future*. Policy Commission on the Future of Farming and Food, London.

Pretty, J., Brett, C., Gee, D., Hine, R., Mason, C., Morison, J., Rayment, M., van der Rijl, G. and Dobbs, T. (2001) Policy challenges and priorities for internalising the externalities of modern agriculture. *Journal of Environmental Planning and Management* 44, 263–283.

Quinton, J.N., Catt, J.A. and Hess, T.M. (2001) The selective removal of phosphorus from soil: is event size important? *Journal of Environmental Quality* 30, 538–545.

RCEP (1992) *Freshwater Quality*. 16th Report, Royal Commission on Environmental Pollution, Command 1966. HMSO, London.

Robinson, M. and Armstrong, A.C. (1988) The extent of agricultural field drainage in England and Wales 1971–80. *Transactions of the Institute of British Geographers New Series* 13, 19–28.

Russell, M.A., Walling, D.E. and Hodgkinson, R.A. (2001) Suspended sediment sources in two small lowland agricultural catchments in the UK. *Journal of Hydrology* 252, 1–24.

Soutar, R.G. (1989) Afforestation and sediment yields in British waters. *Soil Use and Management* 5, 82–86.

SSEW (1983) *Soil Map of England and Wales. Scale 1 : 250,000*. Soil Survey of England and Wales, Harpenden, UK.

SSEW (1984) *Soils and Their Uses in Northern England (and Wales, Midlands and Western England, Eastern England, South West England, and South East England)*. Technical Bulletins 10 (and 11–15). Soil Survey of England and Wales, Harpenden, UK.

Stewart, L. (1963) *Investigation into Migratory Fish Propagation in the Area of the Lancashire River Board*. Report, Lancashire River Board, UK.

Van Der Post, K.D., Oldfield, F., Haworth, E.Y., Crooks, P.R.J. and Appleby, P.G. (1997) A record of accelerated erosion in the recent sediments of Blelham Tarn in the English Lake District. *Journal of Paleolimnology* 18, 103–120.

Walling, D.E. (1988) Measuring sediment yield from river basins. In: Lal, R. (ed.) *Soil Erosion Research Methods*. Soil and Water Conservation Society, Ankeny, Iowa, pp. 39–73.

Walling, D.E. (1990) Linking the field to the river: sediment delivery from agricultural land. In: Boardman, J., Foster, I.D.L. and Dearing, J.A. (eds) *Soil Erosion on Agricultural Land*. John Wiley & Sons, Chichester, UK, pp. 129–152.

Walling, D.E. and Amos, C.M. (1999) Source, storage and mobilisation of fine sediment in a chalk stream system. *Hydrological Processes* 13, 323–340.

Walling, D.E., Owens, P.N. and Leeks, G.J.L. (1999) Fingerprinting suspended sediment sources in the catchments of the River Ouse, Yorkshire, UK. *Hydrological Processes* 13, 955–975.

8 Seasonal Trends of Suspended Sediment Concentration in a Mediterranean Basin (Anoia River, NE Spain)

J. Farguell and M. Sala
Universitat de Barcelona, Spain

Introduction

Suspended sediment concentration is a key parameter to compute sediment loads and yields in rivers, and it is essential to understand its variability in time and space in order to assess the sediment dynamics of a river for a given period of time. In Mediterranean systems, seasonality plays an important role in erosion and sediment transport processes. Experimental catchments in Vallcebre (Pre-Pyrenees mountains), a subhumid Mediterranean environment draining 4.5 km^2, show that depending upon the season weathering is more important than sediment transport and vice versa (Clotet-Perarnau *et al.*, 1986; Gallart *et al.*, 2002). Specific studies, undertaken within this basin, show different patterns of sediment transport according to seasonal rainfall events (Regüés *et al.*, 2000). Furthermore, in Les Gavarres massif, an ephemeral low Mediterranean mountain experimental catchment draining 2.5 km^2, events take place during winter and early spring, when the water table has reached the maximum level after recovering from the dry season by rainfall events during autumn (Sala and Farguell, 2002).

Seasonality has an important role in sediment transport in larger Mediterranean basins, such as the Tordera River (898 km^2), in which flow is not always continuous throughout the year in its lower part, resulting in an accumulation of sediment in upstream reaches. The sediment is then flushed during the wet season at different rates depending upon the number and the magnitude of floods and the amount of sediment accumulated upstream (Rovira *et al.*, 2004).

The purpose of this work is to compare the mean annual trends of suspended sediment concentration with the seasonal trends at two different sampling sites in the Anoia River basin, to assess the variability of this parameter in a medium-sized Mediterranean basin (926 km^2) with sedimentary rock type and with agriculture as the main land use.

Study Area

The Anoia River is the second major tributary of the Llobregat River and is located in the north-eastern part of the Iberian Peninsula, draining an area of 926 km^2 (Fig. 8.1). The upper part of the basin is mainly formed of sedimentary rocks, such as sandstones, mudstones and clays from the Oligocene and Eocene periods. In the lower part, large deposits of silts and clays from the Miocene are the main rock types. The Catalan Coastal Ranges, formed of limestone and small areas of granites and schist, divide the upper and lower part of the basin from the southwest to the northeast. The river flows from the upper

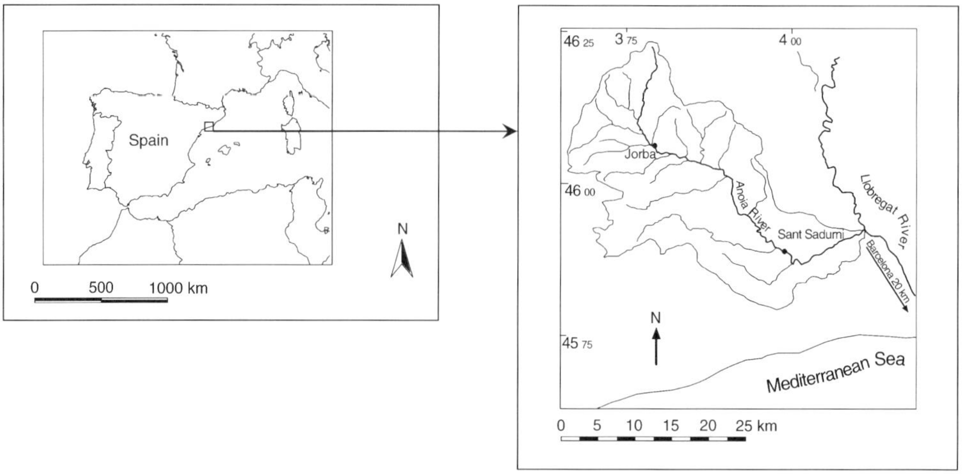

Fig. 8.1. Location of the Anoia River basin.

to the lower part following a fault through the ranges. The maximum height is 900 m above sea level and the lowest is 60 m above sea level at the junction with the Llobregat River. Agriculture, which is based on winter cereal crops and vineyards, occupies 45% of the basin's surface area. Forests and shrub lands take up 51%, and urban areas the remaining 4% of the basin's surface area.

Methodology

The hydrological data have been obtained from the Water Authorities of Catalunya and have been statistically processed using SPSS, Statgraphics and Excel packages. Suspended sediment samples were collected during the water years 2001–2002 and 2002–2003 at Jorba, which drains an area of 220 km^2 and at Sant Sadurní, which drains an area of 726 km^2. During low flows (< 1 m^3/s) samples were taken at a weekly rate by means of a manual, integrated USDH48 depth-height sampler at both sampling sites. Two automatic ISCO 3700 type samplers were installed in the study area, one at Jorba and the other at Sant Sadurní, to collect samples at a higher temporal resolution during events, in addition to the manual sampling. The samplers were able to collect up to 24 samples of 1000 ml each, following a pre-defined sampling programme.

Two different sampling programmes were carried out:

1. Hourly interval sampling: samples were taken at hourly intervals from the first sample taken. This procedure was applied during the beginning and the main part of an event in order to record the rise and peak or peaks of suspended sediment concentration.
2. Multi-interval sampling: samples were taken according to different pre-defined time intervals. The first six samples out of the set of 24 bottles were taken at hourly intervals. The following six samples were taken every 2 h; the following six were taken every 3 h and, finally, the last six samples were taken every 4 h. This sampling procedure was used during the recession of the event in order to record the decreasing rate of suspended sediment concentration.

The ISCO intake was installed at 25 cm above the river bed at Sant Sadurní gauging station, which represented a discharge of 1.2 m^3/s, equalled or exceeded 9% of the time during the study period. At Jorba, the intake was placed at 10 cm above the river bed, which represented 0.08 m^3/s, equalled or exceeded 53% of the time. Previous to the sample collection, a rinse cycle was run in order to clean out the pipe so as to avoid sample contamination.

A comparison between suspended sediment concentrations obtained by means of the manual sampler and the automatic sampler was performed in order to determine the ISCO's efficiency and reliability. At both sampling sites the variation is small and the regression coefficient is 0.99 and 0.98 at Jorba and at Sant Sadurní, respectively. Both regressions are statistically significant ($P < 0.01$).

To determine suspended sediment concentration all samples were filtered using a Millipore vacuum-pump. The amount of filtered sample was 250 ml and the filters used were of 0.45 μm pore size, previously weighted on a precision scale to four decimal places. The samples were air-dried and reweighed again after a week.

Hydrology of the Study Area

Table 8.1 summarizes mean annual hydrological values at the Jorba and Sant Sadurní sites during the periods 1990–2003 and 2001–2003.

Upstream site: Jorba

Mean annual rainfall was 450 mm and the monthly distribution shows a maximum from October to December, a period which accounts for 37% of the total annual rainfall. A secondary maximum is registered from April to June, which accounts for 26% of the total annual rainfall. Winter and summer are the dry seasons, accounting for 14% and 23%, respectively. For 7.8% of the time, discharge equals or exceeds 1 m^3/s, and 0.6% of the time it equals or exceeds 5 m^3/s. Likewise, there is no discharge for 0.6% of the time, which represents an average of 2 days per year. The maximum instantaneous peak discharge recorded was 94 m^3/s. The historical data reveal that 25% of the maximum instantaneous peak discharges recorded within a year occur during September and 20% take place in October.

The mean annual values recorded during the study period show that despite the fact that the rainfall was greater than the mean, discharge, total runoff and water yield were lower than their means. This indicates that for the same rainfall input, the water yield has been reduced. The maximum instantaneous peak discharge recorded was 8 m^3/s, which has a recurrence interval of 1.5 years and took place on 1 September 2003, representing a specific discharge of 0.04 $m^3/s\ km^2$.

Downstream site: Sant Sadurní

Mean annual rainfall was 550 mm and the monthly distribution is also seasonal, with a maximum during autumn, which accounts for 35% of the total annual rainfall and a secondary maximum during spring, accounting for 21%. Winter and summer are the dry periods and their contributions are 17% and 19%, respectively. A discharge equalling or exceeding 1 m^3/s takes place 45% of the time, while a flow of 10 m^3/s is equalled or exceeded 1.8% of the time. A discharge equalling or exceeding 20 m^3/s only occurs 0.5% of the time. On the contrary, a discharge of 0.5 m^3/s is equalled or exceeded 60% of the time. The highest instantaneous peak discharges recorded within

Table 8.1. Hydrological mean annual values at Jorba and Sant Sadurní (a) 1990–2003 and (b) 2001–2003.

	Rainfall (mm)	Discharge (m^3/s)	Specific discharge ($10^{-3}\ m^3/s/km^2$)	Runoff ($m^3 \times 10^6$)	Water yield (mm)	Maximum instantaneous discharge (m^3/s)
(a) 1990–2003						
Jorba	450	0.4	2	12	56	94
Sant Sadurní	550	2.2	3	68	94	400
(b) 2001–2003						
Jorba	480	0.15	1	4.6	21	8
Sant Sadurní	640	0.8	2	24	33	72

a year occur during October and September. The maximum instantaneous peak discharge was 72 m^3/s, which took place on 22 August 2002 with a recurrence interval of 1.8 years and represented a specific discharge of 0.1 $m^3/s/km^2$. During the study period, rainfall was greater than the annual mean value. However, as occurred at the upstream sampling site, discharge, specific discharge, runoff and water yield were lower than the mean values (Table 8.1b).

Comparison of both sampling sites

Figure 8.2 shows the specific discharge of both sampling sites, indicating similar behaviour during the study period, except for the late summer and early autumn months. Specific discharge increases in the downstream site due to storm events, which mostly affect the lower part of the basin, producing greater discharges. Despite this, the trends of discharge are similar at both sites: during autumn the basin recovers from the summer drought due to strong rainfall events. However, it is not until spring that the maximum discharge takes place. Winter and summer are dry periods and a steep decrease in discharge takes place, especially in summer, during which the evapotranspiration rates are higher.

General Trends of Suspended Sediment

Upstream site: Jorba

Table 8.2 shows basic statistics for Jorba during the study period. Suspended sediment concentration ranged from 5 to 4400 mg/l and the associated discharges were 0.02 m^3/s and 1.2 m^3/s. Mean suspended sediment concentration was 300 mg/l and the median was 128 mg/l. The coefficient of variation was about 150%, with concentration equal or greater than 320 mg/l for 25% of the time and equal or greater than 1300 mg/l for 5% of the time. The relationship established between the suspended sediment concentration and specific discharge is shown in Fig. 8.3a. The scatter of the data points, although a common feature in these kinds of relationships (Walling, 1977), is high at this site as the regression coefficient shows (Table 8.3). The r^2 value only explains 17% of the variability of the scatter, despite the fact that it is a statistically significant relationship ($P < 0.01$).

Downstream site: Sant Sadurní

Table 8.2 also shows basic statistics for the Sant Sadurní sampling site. Suspended sediment concentration measured ranged from 3 to 15,000 mg/l and the discharges associated with

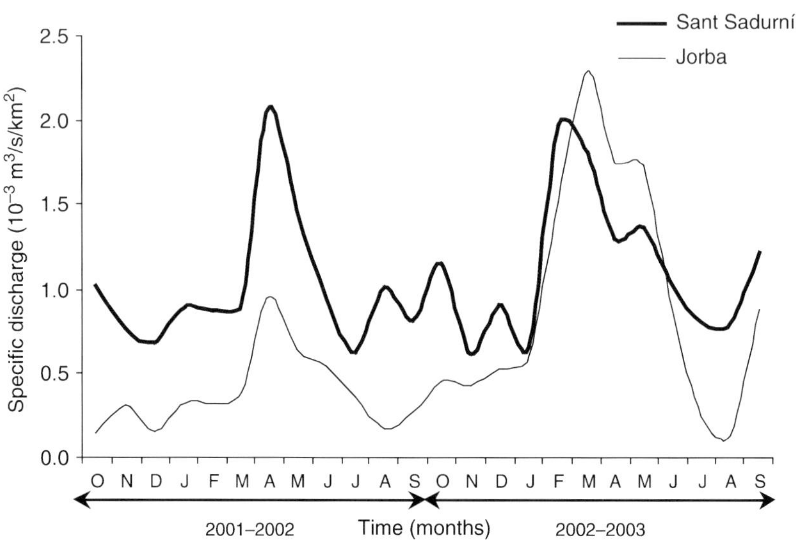

Fig. 8.2. Discharge at Jorba and Sant Sadurní during the study period.

Table 8.2. Mean suspended sediment concentration values at Jorba and Sant Sadurní during the study period.

	Jorba			Sant Sadurní		
	All data	Autumn	Spring	All data	Autumn	Spring
Number of samples	561	251	310	437	148	289
Mean concentration (mg/l)	300	480	200	1,280	2,490	690
Standard deviation	475	608.8	281.9	2,840	4,140	1,530
Median concentration (mg/l)	128	226	75.5	169	275	140
Maximum concentration (mg/l)	4,400	4,400	1,960	15,400	15,400	10,700
Coefficient of variation (%)	158	137	132	221	166	222
Mean specific discharge (10^{-3} m^3/s/km^2)	2.5	1.2	1	4	4	4

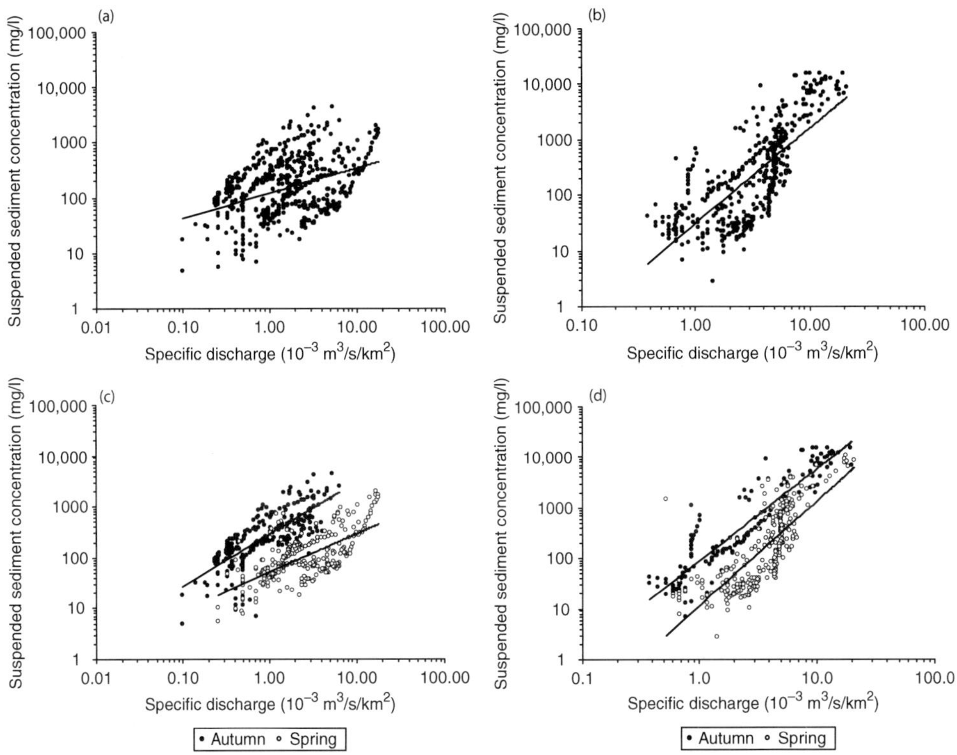

Fig. 8.3. Suspended sediment rating curves at Jorba (a) and at Sant Sadurní (b); and seasonal rating curves at Jorba (c) and at Sant Sadurní (d).

these concentrations were 1.04 m^3/s and 14.2 m^3/s, respectively. Mean and median suspended sediment concentration were 1280 mg/l and 169 mg/l, respectively. The coefficient of variation was 221%, with the concentration equal or exceeding 750 mg/l for 25% of the time and 8340 mg/l for 5% of the time. Furthermore, 1% of the time the concentration equalled or exceeded 14,700 mg/l. The relationship between concentration and specific

Table 8.3. Rating curve equations and coefficients at Jorba and Sant Sadurní for all data and seasonal data. SSC is suspended sediment concentration in mg/l and Q is discharge in m^3/s.

	Upstream (Jorba)			Downstream (Sant Sadurní)		
	No. samples	Equation	r^2	No. samples	Equation	r^2
All data	561	$SSC = 240.8Q^{0.5}$	0.17	436	$SSC = 53.7Q^{1.7}$	0.56
Autumn	251	$SSC = 1342.5Q^{1.0}$	0.60	147	$SSC = 156.3Q^{1.8}$	0.85
Spring	310	$SSC = 160.8Q^{0.75}$	0.46	289	$SSC = 22.2Q^{2.1}$	0.61

discharge is presented in Fig. 8.3b and shows a degree of scatter, although it explains 56% of the variance and is statistically significant ($P < 0.01$) (Table 8.3).

Comparison between sampling sites

Figure 8.4 presents the relationship at both sampling sites and the basic analysis is shown in Table 8.2. Mean and maximum suspended sediment concentrations are greater at Sant Sadurní than at Jorba by one order of magnitude. However, the median value is very similar at both sampling sites (Table 8.2), which demonstrates that, despite the fact that Sant Sadurní registers greater concentrations, these take place occasionally while lower values occur more often. Thus the variability in suspended sediment concentration is high, as shown by the coefficient of variation, which is larger than 100% at both sampling sites, especially at Sant Sadurní which is larger than 200%. Some studies reveal coefficients of variation around 150%, as shown by Batalla and Sala (1994) in the Arbúcies River, or 123% in Vallcebre (Llorens *et al.*, 1997). Both sites are sub-humid Mediterranean forested catchments, with catchment areas of 114 km^2 and 4.5 km^2, respectively.

On the other hand, both sampling sites show an increasing pattern of suspended sediment concentration with increasing discharge. However, the rate of increase is greater at Sant Sadurní than at Jorba, which is represented by the exponents of the rating curve equations (Table 8.3). Finally, although the scatter is high at both sites, it is greater at Jorba, in which discharge only explains 17% of the variance, while at Sant Sadurní it is 56% (Table 8.3 and Fig. 8.3).

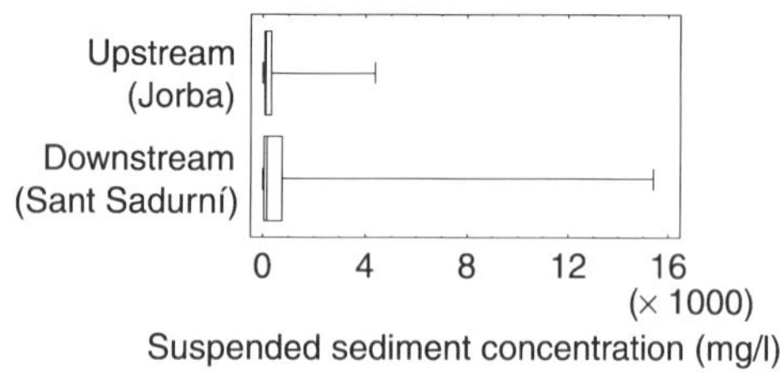

Fig. 8.4. Comparison of the general data sets between both sampling sites.

Seasonal Trends

The scatter on the concentration–discharge plots may be reduced by dividing the data set into seasons (Walling, 1977). As the rainfall period takes place during spring and autumn months, the values recorded from January to June were grouped as 'spring', while those recorded from August to December were grouped as 'autumn'.

Upstream site: Jorba

Table 8.2 shows the different seasonal concentrations measured at Jorba. All values are greater during autumn than during spring, including the dispersion parameters such as the standard deviation. In addition, a different trend between both rating curves is detected, showing that concentration increases differently with discharge depending upon the season. The scatter has been reduced compared to the general data set, and furthermore, the percentage of variance explained by discharge improves considerably, being 60% in autumn and 46% in spring

(Fig. 8.3c). However, the range of values recorded during both seasons is similar as the coefficients of variations are 137% for autumn concentrations and 132% during spring (Table 8.2).

Downstream site: Sant Sadurní

Table 8.3 shows that concentrations are greater during autumn than during spring; however, specific discharges are equal during both seasons. Both regression lines are nearly parallel, indicating that concentration increases with discharge at a similar rate during both seasons and exponents are alike, around two in both equations (Fig. 8.3d). Discharge explains 85% of the variance for autumn values and 61% during spring, which are greater percentages than the entire data set. The range of values during both seasons is large, as indicated by the coefficients of variation which are *c.*170% and 220% for autumn and spring, respectively.

Comparison between both sampling sites

Figure 8.5 shows concentrations measured at both sampling sites, which are, in turn, divided into seasons. The range of values is nearly four times greater at Sant Sadurní during both seasons. However, at both sites the range is larger during autumn than during spring. The median values are alike at both sites and during both seasons, which indicates that maximum values take place during small periods of time as the bulk of the data in Fig. 8.5 appears on the left side of the graph, where the lower values are located.

The seasonal relationship between suspended sediment and discharge shows that the slope of the rating curves is similar between seasons within each study site, with concentrations greater during autumn (Figs 8.3c, d). However, the steepness of the slopes is different between sites. The different rating curves plotted at Jorba show that the exponent of the equations are one or smaller, while the equations for Sant Sadurní show a value near two. Similar findings were documented by Walling (1974), who suggested that the different steepness of slopes could be related to the nature and calibre of the sediment load. A small basin composed entirely of clay- and silt-sized material had a value smaller than one, while a basin with sand-sized material showed a value of two. Thus, concentrations are greater downstream due to the greater rainfall amounts and the greater magnitude of events (Table 8.4). During late summer and early autumn severe convective cells, especially affecting the lower part of the basin, wash the material from the slopes loosened after the summer drought (Walling, 1974). During the remaining autumn months, the rain events become more generalized, affecting the whole basin and accounting for rainfall greater than 100 mm in 24 h (Table 8.4). These strong autumn events are widespread in all the Mediterranean coast of Spain, sometimes leading to catastrophic floods (Sala, 2003). During spring, rainfall amounts and intensities are lower. In addition, the magnitude of events is also smaller at both sites, but always greater downstream.

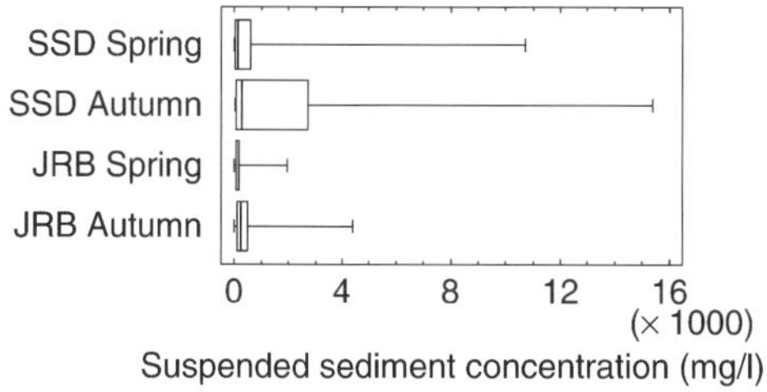

Fig. 8.5. Comparison of seasonal data sets between both sampling sites. SSD, Sant Sadurní; JRB, Jorba.

Conclusions

The relationships between suspended sediment concentration and discharge show that the general data sets are not strongly dependent on discharge, showing that other variables affect the concentration. However, the seasonal rating curves considerably improve the relationships at both sites, especially during autumn.

During autumn, greater concentrations at both sites have been measured. Nevertheless, these concentrations reach larger values at the

Table 8.4. Comparison between both sites and seasons during the study years.

		Number of events		Mean rainfall (mm/season)		Max. daily rainfall (mm/24 h)		Mean specific discharge (10^{-3} $m^3/s/km^2$)		Instantaneous peak discharge (m^3/s)	
		2001/02	2002/03	2001/02	2002/03	2001/02	2002/03	2001/02	2002/03	2001/02	2002/03
Jorba	Autumn	9	8	299	290	47.2	56.9	0.23	0.51	2.0	8.7
	Spring	8	4	183	192	30.8	34.2	0.54	0.54	2.1	4.0
Sant Sadurní	Autumn	7	7	416	404	60.1	137.8	0.9	0.9	72	44.1
	Spring	6	4	271	184	45.5	31.5	1.15	1.4	5	15

downstream site than the upstream one. Higher rainfall amounts and the occurrence of larger events of greater magnitude than during spring suggest greater suspended sediment concentrations during this season. In spring, the lower intensity events and the increased base flow level reduce such concentrations.

Variability of suspended sediment concentration is very high at both sites and during both seasons, indicating that it is strongly related to the occurrence of flow events.

Acknowledgements

The authors wish to thank the Water Authorities of Catalunya for allowing the installation of monitoring equipment and for the hydrological data supplied. We are indebted to Sara Bernia for her assistance in the field during the sampling period. Finally, the authors are very grateful to an anonymous reviewer and to Ramon Batalla, whose comments and suggestions greatly improved the manuscript.

References

Batalla, R.J. and Sala, M. (1994) Temporal variability of suspended sediment transport in a Mediterranean sandy gravel-bed river. In: Olive, L.J., Loughran, R.J. and Kesby, J. (eds) *Variability in Stream Erosion and Sediment Transport*. IAHS Publication 224. IAHS Press, Wallingford, UK, pp. 299–305.

Clotet-Perarnau, N., Gallart, F. and Balasch, C. (1986) Medium-term erosion rates in a small scarcely vegetated catchment in the Pyrenees. In: Harvey, A. and Sala, M. (eds) *Geomorphic Processes in Environments with Strong Seasonal Contrasts. Vol. II: Geomorphic Systems. Catena Supplement* 13, 37–47.

Gallart, F., Llorens, P., Latron, J. and Regüés, D. (2002) Hydrological processes and their seasonal controls in a small Mediterranean mountain catchment in the Pyrenees. *Hydrology and Earth Systems Sciences* 6, 527–537.

Llorens, P., Queralt, I., Plana, F. and Gallart, F. (1997) Studying solute and particulate sediment transfer in a small Mediterranean mountainous catchment subject to land abandonment. *Earth Surface Processes and Landforms* 22, 1027–1035.

Regüés, D., Balasch, C., Castelltort, X., Soler, M. and Gallart, F. (2000) Relación entre las tendencias temporales de producción y transporte de sedimentos y las condiciones climáticas en una pequeña cuenca de montaña mediterránea (Vallcebre, Pirineos Orientales). *Cuadernos de Investigación Geográfica* 26, 41–65.

Rovira, A., Batalla, R.J. and Sala, M. (2004) Balance sedimentario en el tramo bajo de un río Mediterráneo (La Tordera, Cordilleras Costeras Catalanas). *Revista de Cuaternario y Geomorfología* 18, 51–56.

Sala, M. (2003) Floods triggered by natural conditions and by human activities in a Mediterranean coastal environment. *Geografiska Annaler* 85A, 301–312.

Sala, M. and Farguell, J. (2002) Water and sediment yield in two representative Mediterranean catchments under different land uses in the Catalan Coastal Ranges. *Revista de Cuaternario y Geomorfología* 16, 97–109.

Walling, D.E. (1974) Suspended sediment and solute yields from a small catchment prior to urbanization. In: Gregory, K.J. and Walling, D.E. (eds) *Fluvial Processes in Instrumented Watersheds*. Institute of British Geographers Special Publication 6. IBG, London, pp. 169–192.

Walling, D.E. (1977) Assessing the accuracy of suspended sediment rating curves for a small basin. *Water Resources Research* 13, 531–538.

9 Suspended Sediment Transport during Rainfall and Snowmelt–Rainfall Floods in a Small Lowland Catchment, Central Poland

L. Hejduk, A. Hejduk and K. Banasik
Warsaw Agricultural University SGGW, Poland

Introduction

Sediment transport in lowland watersheds occurs mainly during floods and snowmelt periods. During rainfall events, a relation between suspended sediment concentration (SSC) and discharge may be found. Such an investigation has been carried out by different authors (e.g. Lenzi and Marchi, 2000; Jansson, 2002; Hudson, 2003) in basins of different scales. Usually, five different types of relations occur (Morris and Fan, 1998), and the clockwise and anticlockwise loops are the most common. The clockwise hysteresis type occurs when the peak of SSC occurs prior to the peak of discharge, whereas the opposite is true for the anticlockwise situation.

There are few publications about sediment concentrations and transport during snowmelt events (Chikita, 1996). This is because it is often difficult to observe these events, especially in lowland regions, and usually it is snowmelt–rainfall floods that are observed. In our study we present information on suspended sediment transport and loads during rainfall floods and snowmelt–rainfall events. Our analysis was based on continuous measurement of SSC. We also report information on the particle size composition of the material transported in the lowland river, based on a low-angle laser light scattering technique.

Zagożdżonka River Catchment

The Zagożdżonka watershed is located in central Poland, about 100 km south of Warsaw. The Department of Hydraulic Engineering and Environmental Recultivation (former Department of Hydraulic Structures) started its research in the watershed in 1962. The aims of the research were the estimation of river discharge, water budget analysis and sediment yield, and, since 1995, nutrient (phosphorus, nitrogen) transport. The total area of the watershed at the outlet at the Płachy Stare gauging station (Fig. 9.1) is 82.4 km^2. However, in this present study, a smaller (23.4 km^2) sub-watershed, with the outlet at the Czarna gauging station, was investigated.

The dominant soil type is sandy, ranging from almost pure to loamy sands. In depression areas like river beds, peaty soils can be found. Generally, sandy soils cover over 90% of the watershed to the Czarna gauging station. The Zagożdżonka watershed is located about 170 m above sea level, and absolute relative relief of the sub-catchment is 16.5 m, so it can be

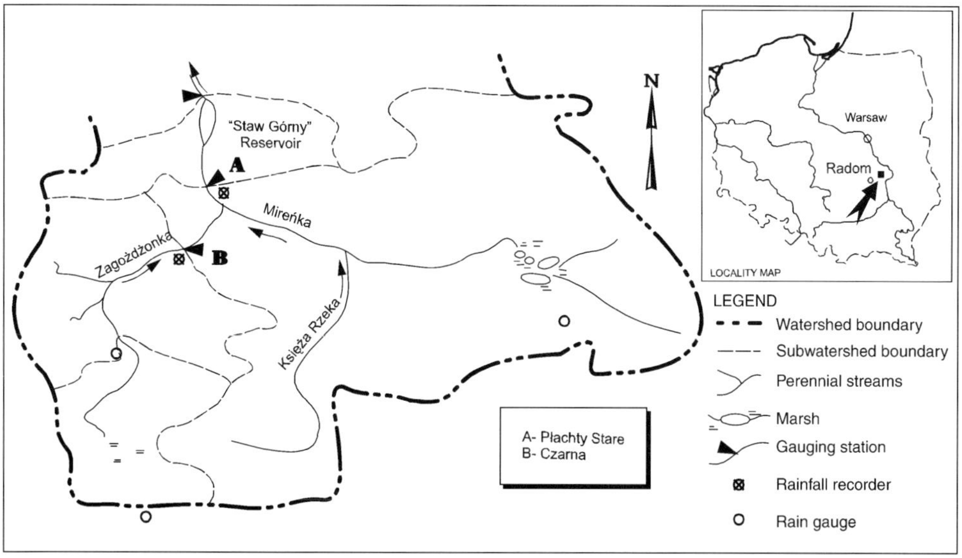

Fig. 9.1. Location of the Zagożdżonka catchment.

considered as lowland and typical for central Poland. The study area is mainly (64%) under cultivation, with 24.1% covered by forest and 10.7% under pasture. The remaining 1.2% is occupied by other forms of land use.

The Czarna gauging station is equipped with an electronic monitoring system consisting of the following: RC12 data logger, tipping bucket rainfall gauge, water level sensor, infrared Partech IR100C concentration sensor, and water, air temperature and humidity sensors. All data are collected in 10 min intervals. In addition, standard mechanical devices such as a limnigraph and pluviometer were also installed (Górski and Hejduk, 1998).

Precipitation, Runoff and Suspended Sediment Data

Mean annual precipitation for 1963–2000 is estimated as 614 mm for this area, and values are 581.1 and 628.7 mm for hydrological years 1999 and 2000, respectively. Annual precipitation during these years can be assumed to have been average because values were close to the long-term average. Usually the wettest months are June and July.

Mean annual outflow for the watershed at Płachty Stare was estimated as 112.3 mm. Since 1991 the yearly outflow has also been estimated for the Czarna station. In 1999 and 2000 it was 144.4 and 99.9 mm, respectively. Suspended sediment concentrations have been estimated periodically since 1991. During 1999–2000, different types of measurements were used. For the concentration data, the infrared sensor was used. For calibration purposes, water samples were taken using an automatic sampler (SIGMA) and SSC was estimated by filtering. Based on 38 paired data (from infrared sensor and filtering), the following equation was determined:

$$C = 2.43 \cdot (WR - 657)^{0.36} \quad r^2 = 0.90$$

where C is suspended sediment concentration (mg/l) and WR is data logger output (–).

This equation has been assumed as a calibration equation for the infrared sensor connected to the data logger system at the Czarna gauging station. Due to algal growth and deposition on the sensor optics, a temporary cleaning and calibration curve-checking approach was used. Except for the infrared sensor and automatic sampler, the water samples were taken by an observer once a week at a similar time and suspended sediment was determined by filtering.

For suspended sediment grain size determination, a special sediment tank was built. A detailed description of the station and devices is contained in Górski and Hejduk (1998). Grain size distribution was estimated using a low-angle laser light scattering technique using a Mastersizer MicroPlus (Malvern Instruments Ltd) (Hejduk and Banasik, 1999).

Results and Discussion

Relation between suspended sediment concentration and discharge

During the study period, eight flood events were measured with complete discharge and SSC data. Six of these were typical rainfall events. Events 3 and 4 can be considered as snowmelt or snowmelt–rainfall floods, because they were caused by temperature increases and had slight rainfall. Table 9.1 shows the basic characteristics of these events. In the most cases (i.e. in seven out of eight events, Fig. 9.2), clockwise hysteresis was found when discharge was plotted against SSC. This common hysteresis means that the SSC peak occurred prior to the peak in discharge. Often, this type of relation indicates that surface runoff, at the start of the event, causes the smallest and the weakest-bound particles to be washed away.

Event 1 shows an anticlockwise hysteresis loop as the SSC peak followed the discharge peak. There are several possible explanations of such a relation. In this case, the pattern of rain distribution had a strong impact as the rain that caused this flood had two peaks in intensity.

The concentration and pattern of suspended sediment varied much between flood events. The highest concentration was observed during event 5. This event was the first spring flood during 2000 and probably this is the main reason for high values of flow and SSC. As the rainfall was rather low (19.2 mm) and the mean intensity was 0.022 mm/min, this rain event can be classified as ordinary. The highest concentration was estimated as 53.6 mg/l.

For events 1 and 2 it is difficult to determine the peak concentration. The first flood was caused by ordinary rain and after a big flood on 20 April (peak discharge of about 1.9 m^3/s), which probably flushed particles from the surface. However, unfortunately there are no concentration data because of equipment failure. This big flood and a good vegetation cover probably reduced sediment concentration in events 1 and 2. A similar situation happened the following year. Flood event 5 had high SSC values, whereas the SSC during the next flood (event 6) was quite low.

The soil moisture condition during July and August (the most intensive vegetation period) varies considerably from year to year in this area. July 2000 was wetter than the long-term average (37 year average: 77.2 mm; June 2000: 138.6 mm). However, there was not a big storm during this month. Flood event 7 was caused by the highest rain recorded (48.5 mm), but the duration was almost 23 h. Consequently, this and the fact that it occurred in the middle of the vegetation cover season is probably the reason for low SSC values during this flood. Flood event 8 occurred in the middle of September. SSC was at the same level as during event 7, but the peak flow was considerably lower. The main period of vegetation cover

Table 9.1. Basic characteristics of the sampled events.

Event no.	Date of event	Rainfall (mm)	Runoff (mm)	Peak discharge (m^3/s)	Maximum SS concentration (mg/l)
1	May 1999	15.8	0.45	0.27	31.5
2	June 1999	23.0	0.64	0.34	32.3
3	December 1999	26.8	1.14	0.26	16.3
4	January 2000	10.3	0.58	0.24	23.4
5	March 2000	19.2	3.25	1.50	53.7
6	March 2000	11.1	0.56	0.34	23.8
7	July 2000	48.5	1.72	0.69	37.4
8	August 2000	36.9	1.19	0.35	33.5

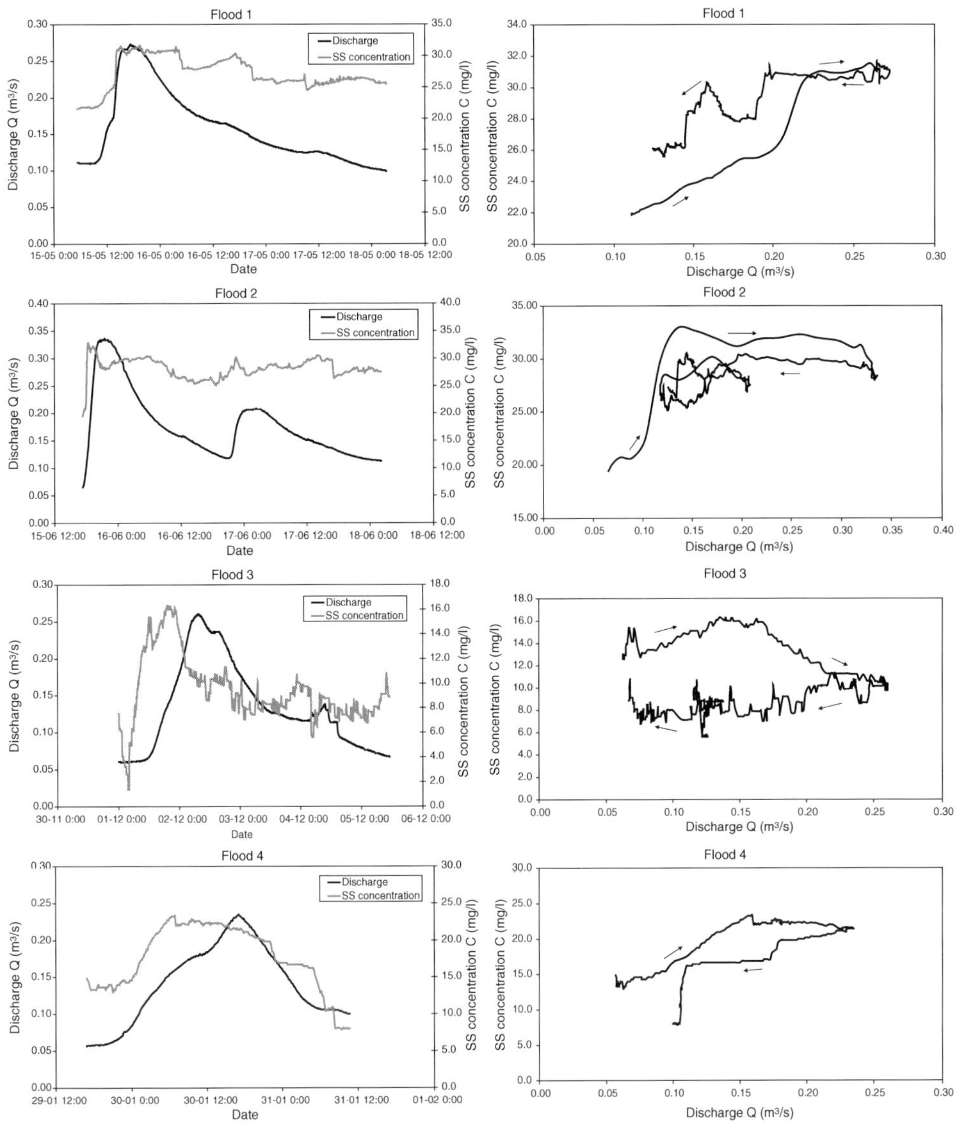

Fig. 9.2. Plots of discharge and suspended sediment concentration, and hysteresis loops for flood events 1–8.

ends during September, after which the preparation of arable land starts for the next season. It has an influence on erosion processes and SSC.

The lowest SSC was observed during the snowmelt–rainfall floods, events 3 and 4. Event 3 was caused by the following situation. From 16 to 26 November the air temperature was below zero with a minimum of –8°C and a water temperature of + 2°C. The air temperature rapidly (during 16 h) increased from –4.4 to + 2°C on 26 November and then it started to rain. After sunset the temperature dropped to –4°C and almost all rain became frozen. From 27 to 30 November the temperature oscillated from below zero during nights to above zero during days. The oscillations were not very high (from –3 to + 3°C). During these days, the discharge increased slightly, but there were no rises of suspended sediment transport. On 30 November, during the night temperatures were above zero

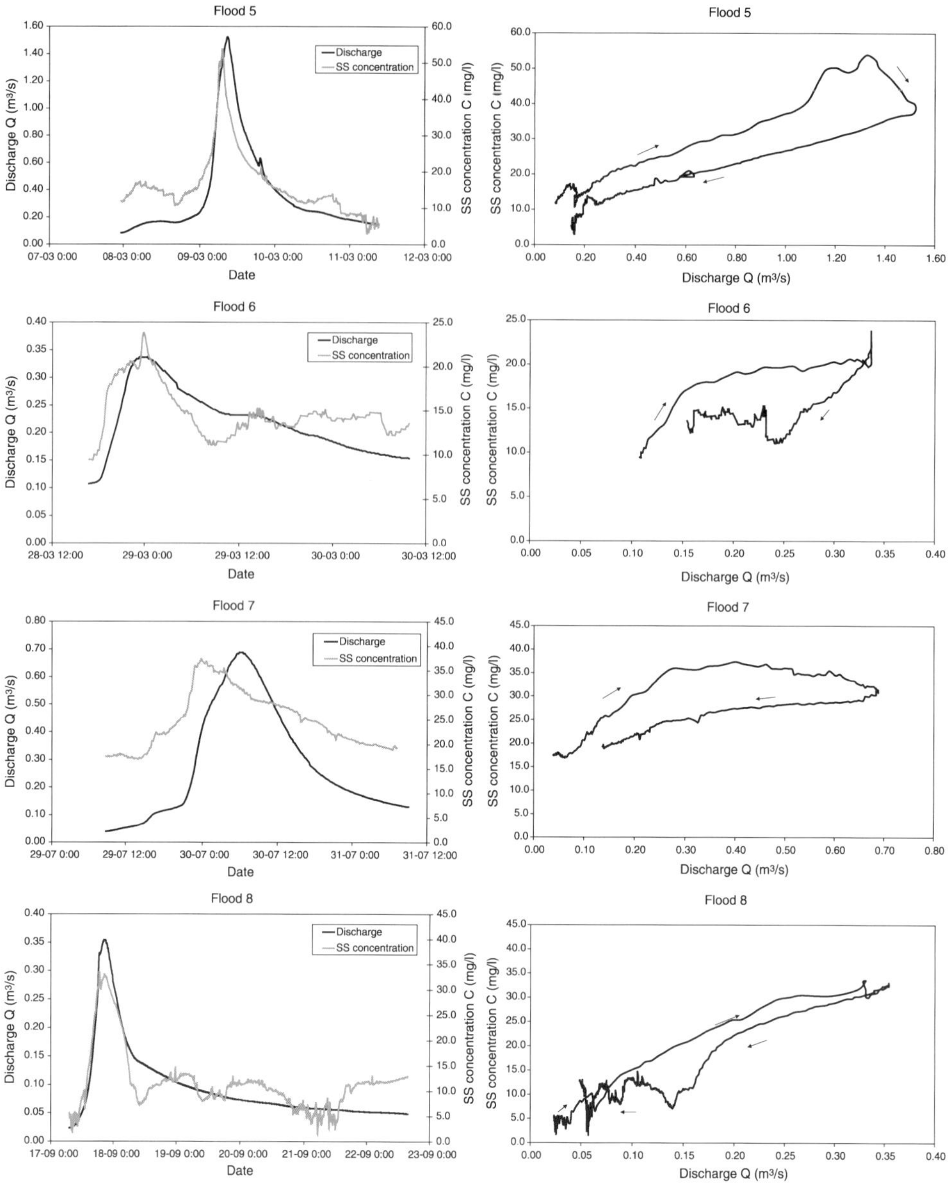

Fig. 9.2. *Continued.*

then increased to 5°C on 1 December and this marked the beginning of the flood. The flow peak was on 2 December in the morning. The SSC peak occurred on 1 December in the evening (time between peaks: 12 h). The soil was probably frozen and explains the very low SSC values.

A similar situation was observed before flood event 4. There were 6 days of low temperatures (with the lowest –10°C), then warming from –10°C to +1°C over a period of 48 h, followed by a small amount of rain. The time between the peak of SSC and the peak of discharge was 10 h.

Suspended sediment loads during floods and yearly total

Total suspended sediment loads were estimated based on weekly measurements. There was no correlation between discharge and SSC (Hejduk, 2001), so the annual loads were calculated as a product of concentration, discharge and time between each measurement. The total SS loads for hydrological years 1999 and 2000 were estimated as 68.6 and 57.4 t, respectively. The investigated floods provided 3% and 10.5% of the total load for 1999 and 2000, respectively, although the flood events presented here were not the only ones during those years.

Grain size distribution of suspended sediment

The suspended sediment grain size distribution varied during floods. Lenzi and Marchi (2000) reported the coarsening of transported material and an increase of sediment concentration during a flood. In our study the samples of suspended sediment for grain size distribution were taken 16 times during the considered period. Because of problems with securing enough sediment for grain size analysis, a special settling tank (collector) was built (Górski and Hejduk, 1998). Usually, the samples were collected over a month and a half, after which the sediment was removed and taken to the laboratory for grain size distribution analysis. The data represent average grain size composition during the collection period (Table 9.2).

An analysis of the relation between values of d_{50} and average discharge during sample collection (Fig. 9.3) shows a slight increase in d_{50} with discharge. However, this stops near a value of about 0.1 m^3/s. When the discharge was larger, there was significant scatter in the values of d_{50}. The reason for such a relation could be

Table 9.2. Characteristic particle size composition of suspended sediment samples.

Suspended sediment distribution curve	Characteristic diameters (mm) d_5	d_{10}	d_{50}	d_{90}
The finest sample	0.0053	0.0099	0.0455	0.1396
Average for all samples	0.0104	0.0188	0.0797	0.2402
The coarsest sample	0.0149	0.0265	0.1021	0.3054

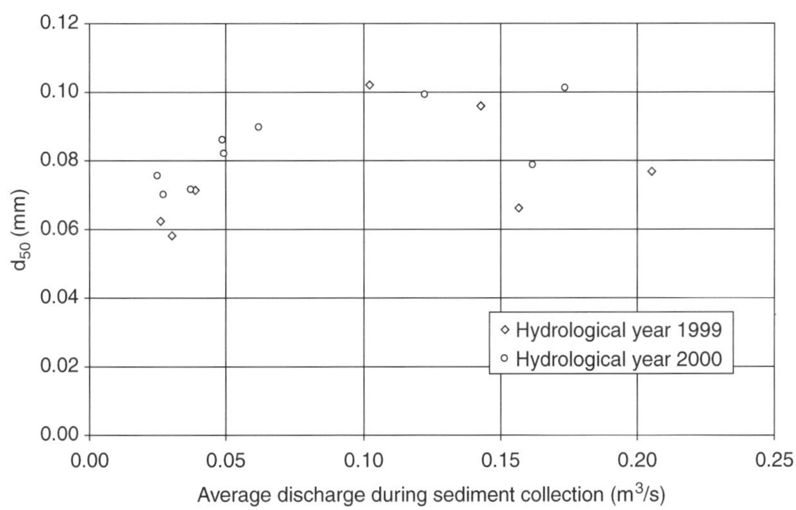

Fig. 9.3. Relationship between d_{50} and average discharge for sediment samples collected during the study period.

the washing of some of the particles from the settling tank during high flows. Thus there was not a strong relation between d_{50} and discharge.

Conclusion

Data gathered during 2 years of investigation on a lowland river of the Zagożdżonka watershed show variability of suspended sediment transport during floods. It seems that the time of flood occurrence is quite important. The first spring floods after winter provide much more material and have an influence on the amount of sediment transported in the subsequent floods. Almost all relations between discharge and suspended sediment concentration (including snowmelt floods) have a clockwise hysteresis pattern. This type of relation is caused by early suspended sediment depletion (exhaustion effect) (Morris and Fan, 1998). The suspended sediment loads were similar in 1999 and 2000 (68.6 t and 57.4 t). The grain size distribution did not vary much and the d_{50} ranged from 0.0455 mm to 0.1021 mm with an average of 0.0797 mm. An increase in d_{50} with increasing discharge was noticed, but only for discharges < 0.1 m^3/s. Above this value, high variability was observed, although this may have been caused by the settling tank not working properly. Future studies are needed to investigate suspended sediment transport during both low flow and floods, with special attention to snowmelt and snowmelt–rainfall events.

Acknowledgement

This research was partially funded by Polish Ministry of Research and Information Technology, State Committee for Scientific Research, Grant No. 3 P06S00123.

References

Chikita, K.A. (1996) Suspended sediment discharge from snowmelt: Ikushunbetsu River, Hokkaido, Japan. *Journal of Hydrology* 186, 295–313.

Górski, D. and Hejduk, L. (1998) Gauge station for runoff and water quality investigation from small agricultural watershed. *Przegląd Naukowy Wydziału Melioracji i Inżynierii Środowiska SGGW*. Zeszyt 15, sesja 1.

Hejduk, L. (2001) Suspended sediment and phosphorus transport during floods from small lowland catchment. Unpublished PhD thesis. The Warsaw Agricultural University SGGW, Warsaw, Poland.

Hejduk, L. and Banasik, K. (1999) Application of laser particle size analyzer for estimating the particle size distribution of suspended sediment. *Roczniki Akademii Rolniczej w Poznaniu* CCCX, part I.

Hudson, P.F. (2003) Event sequence and sediment exhaustion in the lower Panuco Basin. Mexico. *Catena* 52, 57–76.

Jansson, M.B. (2002) Determining sediment source areas in tropical river basin, Costa Rica. *Catena* 47, 63–84.

Lenzi, A.M. and Marchi, L. (2000) Suspended sediment load during floods in a small stream of the Dolomites (northeastern Italy). *Catena* 39, 267–282.

Morris, G.L. and Fan, J. (1998) *Reservoir Sedimentation Handbook*. McGraw-Hill, New York.

10 Sediment in the River Bush, Northern Ireland: Transport, Sources and Management Implications

D.J. Evans[1] and C.E. Gibson[2]
[1]*Queen's University Belfast, UK.* Present address *Atkins Limited, UK;* [2]*Department of Agriculture and Rural Development, UK*

Background

Many studies have demonstrated a link between land drainage and an increase in both the frequency and magnitude of peak flows and a decrease in baseflow due to reduced storage capacity within catchments (e.g. Bree and Cunnane, 1980; Novikov and Pokumeiko, 1980; Newson and Robinson, 1983; Guertin *et al.*, 1987). These hydrological changes have important implications for sediment dynamics. An increased potential for in-channel bed/bank erosion during storm flows and an increased potential for deposition of fines during baseflow has been observed in previous studies (e.g. Swales, 1982; Carling, 1988; Wilcock and Essery, 1991). In addition, conveyance loss of fine material on to floodplains during inundation in storm events (as reported in Walling *et al.*, 1998) would be limited.

The Bush catchment lies in the northern area of County Antrim, Northern Ireland, and drains an area of 340 km^2 (Fig. 10.1). The lower and upper zones of the catchment were extensively drained from the 1950s onwards in order to increase agricultural productivity. The extensive drainage work has resulted in entrenched river channels that are disconnected from their floodplain. This has implications for both the hydrological and sedimentological cycles in the Bush catchment. Following these changes, an increase in the percentage of fine material (< 2 mm) was observed in the River Bush channel (O'Connor and Andrew, 1998). Various studies have demonstrated that sedimentation of salmon spawning beds is detrimental to spawning success because of infiltration of fine material in the gravel interstices of redds leading to lower oxygen diffusion and impeded alevin emergence (Ottaway *et al.*, 1981; Pauwels and Haines, 1994; MacKenzie and Moring, 1998). Indeed, there has been a decrease reported in salmon spawning success in the River Bush (Heaney *et al.*, 2001). This trend has been noted in many other systems, such as the Erne catchment, Ireland (Mathers and Crowley, 2001) and the River Conon, Scotland (Gilvear *et al.*, 2002).

The aim of the Bush Integrated Monitoring Project was to examine hydrological response, quantify instream fine sediment loads and trace the sources of this sediment in one of Northern Ireland's principal salmon rivers. An integrated monitoring programme was conducted at four sampling sites (Fig. 10.1) between July 2002 and July 2003. Table 10.1 summarizes the methodology used. The ultimate goal was to use the collated scientific data to recommend management strategies aimed at reducing fine sediment transport in the Bush catchment.

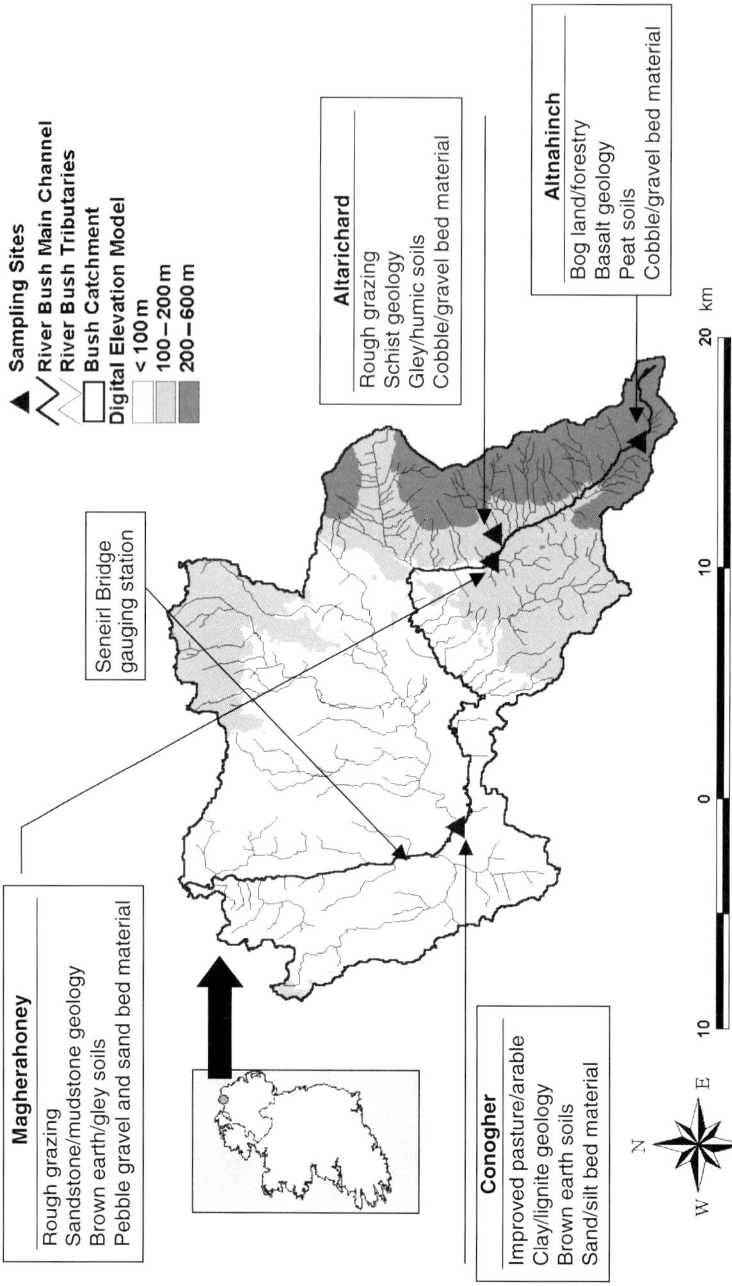

Fig. 10.1. Location of the River Bush catchment, Northern Ireland (54° 35′ N, 0.05° 56′ W).

Table 10.1. Methodology used in the Bush Integrated Monitoring Project.

Parameter	Method	Reference
Rainfall	Met office stations at Skerry Hill North (329 m AOD), Altnahinch Filters (213 m AOD) and Dunloy (110 m AOD)	–
	Rainfall monitored daily	
River flow	Rivers Agency velocity-area flow gauging station at Seneirl Bridge (monitored every 15 min)	–
	Regression to sampling sites (based on velocity-area manual monitoring)	
	Flow is summed to derive accumulated flow over week-long period (to compare to sediment load data)	
Suspended sediment transport	Storm integrating collection tubes (across channel cross-section). Sampled weekly	Phillips *et al.* (2000)
Fine bedload transport	Pit-type traps (across channel cross-section). Sampled weekly. Material wet and dry sieved	Evans *et al.* (2004)
Coarse bedload transport	Foil covered tracer pebbles (only at Magherahoney sampling site). Deployed 09/07/03, recovered 29/08/03 and 19/12/03	Sear *et al.* (2002)
Fine sediment sources	Erosion potential mapping (based on combination of rainfall, slope, soil texture and land cover)	–
	Visual observations (frequency ranked erosion)	–
	Bank erosion monitoring (erosion pin networks on 100 m reaches monitored after storm events, annual sediment yield calculated)	Lawler *et al.* (1999)
	Scanning electron microscopy	–
	Sediment fingerprinting of size fractionated soil and sediment (particle size analysis, mineralogy and mineral magnetics)	Collins *et al.* (1997)

Results

Hydrological response

The rainfall data show an orographic effect in the Bush catchment with decreasing rainfall with decreasing altitude (from a mean of 4.6 to 3.1 mm/day). Flow increases with increasing catchment size. Mean, maximum and minimum values of 6.86, 57.4 and 1.45 m^3/s, respectively, were recorded at the Seneirl Bridge gauging station for the period July 2002 to July 2003.

During the study period there were distinct periods of wetter weather in the winter of 2002 and the summer of 2003. Although the River Bush displayed marked variation in flow, no clear seasonal trend was evident. Periods of sustained high flow coincided with high rainfall in the winter period, particularly during October/ November 2002 and January/February 2003. However, high intensity storms were also observed throughout the spring and summer.

In comparison with the long-term mean (1440 mm/year), rainfall during the study period was slightly above average (1475 mm/year). However, although the mean flow for the study period was above the long-term average (6.81 m^3/s), the extremes in flow were more marked in the long-term record. The Q_{10} and Q_{95} values of 10.9 and 2.04 m^3/s, respectively, for the study period were lower than the Q_{10} and Q_{95} values of 14.9 and 1.10 m^3/s, recorded for the period 1972–2002.

The Bush system had limited flood storage capacity, with high rainfall events resulting in sharp rising and falling limbs on the hydrograph. This was confirmed by modelling the hydrological response to rainfall using a variant of CATCHMOD (Wilby *et al.*, 1994). Adequate simulation of flows was achieved ($r^2 = 0.62$), but the model grossly underestimated flow after heavy rainfall on to a dry antecedent catchment.

Suspended sediment transport

There was large spatial and temporal variation in suspended sediment transport in the River Bush (Fig. 10.2a). Median loads were largest at the sampling site furthest downstream (mean of 1.92 t/week at Conogher), whereas lowest loads were observed at the site furthest upstream (mean of 0.0320 t/week at Altnahinch). Peaks in suspended load generally coincided with large increases in flow. Rating curves revealed a strong log-log relationship between suspended sediment load and accumulated flow with no hysteresis at Conogher and Altarichard sampling sites ($r^2 = 0.61$ and 0.68, respectively). This indicated that transport capacity was the key control and could imply that sources of fine sediment were readily available for transport in suspension. At upland sites, there was no clear relationship between flow and suspended sediment load, suggesting that supply rather than transport controlled suspended sediment dynamics. Another factor responsible for large loads was the removal of bank side vegetation and disruption of the surface armouring of the channel bed by drainage maintenance work on the river. For instance, suspended sediment concentrations downstream from maintenance work were over 200 times greater than upstream concentrations between 9 and 12 June 2003 at Conogher. Other peaks in suspended sediment concentration were attributable to the collapse of banks.

The composition of the suspended sediment was investigated using scanning electron microscopy and varied on a temporal and spatial scale. Samples at the furthest site downstream (Conogher) were composed of silt/clay particles with some fine sand during high flow events. 'Aggregates' (particles that enter the river in an aggregated form and retain that structure) composed of particles derived from catchment sources contained both weathered and unweathered quartz grains, suggesting at least two sediment sources. 'Flocs' (aggregated particles formed within the channel) had a characteristic open structure. However, during the summer period, large increases in diatoms were observed in suspended sediment, explaining scatter during low flow periods on rating curves. Samples at Altnahinch, Altarichard and Magherahoney were normally composed of humic material and phytoplankton, except during high flow when fine sand was also observed.

Fine bedload sediment transport

Both the availability of sediment and the transport capacity of river flows were important in controlling fine bedload (defined arbitrarily here as < 4 mm) transport. High bedload transport was observed during storm events in the summer and winter of 2002, January to March 2003 and May 2003 (Fig. 10.2b). Conversely, bedload transport was extremely low during baseflow conditions (especially between March and May 2003) and anticlockwise hysteresis was noted. This suggests that large amounts of sediment were stored within the River Bush during low flow periods before being mobilized by subsequent rising flows and that exhaustion of channel storage limited sediment loads after storm events (as observed in a previous study, Evans *et al.*, 2003). No obvious bedform structures were observed at any of the sampling sites.

Highest loads were recorded at downstream sites (e.g. 0.0940 t/week at Conogher). Small differences between suspended and bed sediment loads here demonstrated the prevalence of fine sediment in this stretch. Indeed, although the bedload was composed mostly of very coarse, coarse and medium sand (65% summed annual total), the bedload also contained a significant proportion of silt-/clay-sized particles (6% annually). The percentage of fine and very fine sand, silt/clay and coarse particulate organic matter increased further (7, 4, 9 and 1%, respectively) during the winter period due to the dieback of macrophytes (releasing previously trapped material) coinciding with a number of high flow events (> 25 m^3/s). In contrast, further upstream at Magherahoney and Altnahinch, the bedload to suspended load ratio was far larger (> 60 times), reflecting an absence of fines in the water column and prevalence of gravel-sized particles in transport. Pebble and

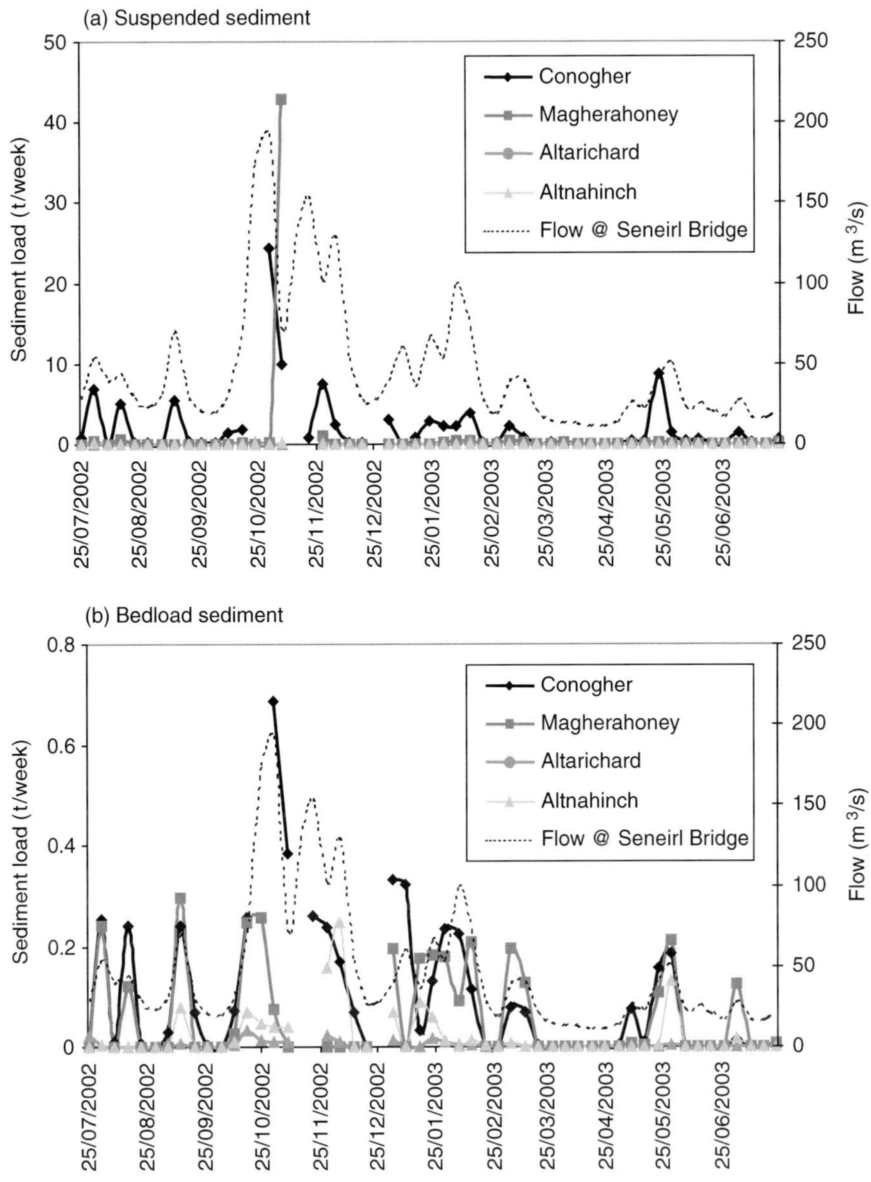

Fig. 10.2. Sediment transport in the River Bush (July 2002 to July 2003).

granule gravels, very coarse and coarse sand sized fractions accounted for over 80% of the bedload transported through these channels.

A strong, positive correlation between bedload transport and bed shear stress was noted at Conogher ($r^2 = 0.62$), Magherahoney ($r^2 = 0.69$) and Altarichard ($r^2 = 0.76$). This implies that flow competence exerted a key control on bedload transport and that, for a given shear stress, the variation in bedload transport was restricted (approximately 5, 9 and 3 N/m^2, respectively, for Conogher, Magherahoney and Altarichard). A decrease in critical entrainment threshold with decreasing grain size, as described in Evans *et al.* (2003), was also observed suggesting that higher flows were required to initiate the transport of larger particles. For instance, critical entrainment threshold values for pebble gravel and silt-/clay-sized fractions at Magherahoney were 12 and 2.8 N/m^2, respectively. Higher critical entrainment thresholds than predicted from this trend for very fine sand and silt/clay particles reflected both the lack of fine material present in the bedload due to resuspension into the water column and the cohesive nature of finer sediment on the bed. These inferences were supported by suspended sediment data.

Coarse bedload sediment transport

Tracer pebbles ($n = 250$) deployed at the Magherahoney sampling site were used to assess coarse bedload (4–160 mm) transport distance and burial processes. The study focused on this site because of its historic importance for salmon spawning and to appraise the benefit for seeding this particular stretch of the river with stones suitable for salmon spawning. These tracers were retrieved two months later following a peak flow of 1.78 m^3/s through the channel (81% recovery). Limited entrainment of intermediate sized local bed material was noted (maximum distance 7 m). The majority of the larger sized pebbles remained stationary. There was a tendency for the smallest tracers to be buried *in situ* by sand and gravel or trapped in the interstices between larger pebbles, suggesting that a depositional feature was developed at low flows in this channel.

Higher flows (maximum 6.8 m^3/s) occurred prior to a second retrieval after 5 months of deployment (only 14% recovery of tracers). Buried pebbles were exposed to the flow by scour and were generally deposited in the channel thalweg at distances of up to 75 m. However, some tracers were transported further on to the outside bend of the meander up to 160 m away (reflecting temporary storage) prior to semi-permanent deposition on gravel bars at distances of up to 200 m from the initial deployment location. Weak negative log relationships between transport distance and tracer pebble mass ($r^2 = 0.28$ and 0.44) and tracer pebble volume ($r^2 = 0.21$ and 0.47) were observed (at 95% significance level).

Fine sediment sources

Elucidation of the link between erosion and downstream fine sediment (< 1 mm) delivery was performed using a combination of desk based, observational and monitoring studies. A map based on a geographical information system was constructed to rank the potential soil erosion risk of grid areas (resolution 50 m^2) by combining four physical catchment derivatives (soil texture, slope, rainfall, land cover). A raster map was produced which aided the identification of areas prone to erosion. Cells with high erosion risk (4% of total land area) occurred either on high gradient land underlain by surface water humic gleys and schist in the east of the catchment or on tilled land on clay/lignite overlain by brown earths in the middle of the catchment (Fig. 10.1).

Visual observations of activities contributing to the existence of areas with bare ground were ranked by frequency of occurrence. It became clear that the most frequent cases of erosion were attributable to a combination of flow damage to riverbanks, livestock poaching and compaction of the topsoil in saturated fields by farm machinery (combined total of 74% of cases). However, a scale effect is evident here, because these activities did not cause damage to large tracts of land. In contrast, forestry clearfell and tillage did not account for a large number of observations (14% of total observations), but the extent of the areas of bare ground generated by these activities was large. The hydrological connectivity of these sources of fine sediment to the river channel must also be considered. Although observations on the frequency of drainage maintenance schemes within the catchment were small (5% of total observations), suspended sediment data highlighted the dramatic transfer of material caused by such work.

The annual sediment yield contributed by bank erosion was quantified by monitoring bank retreat at the four sampling sites. Highest yields were recorded at Altarichard (40.9 m^3) because of a combination of high erosion rate (mean 42.6 mm/storm) and steep banks composed of uncohesive material. Lower yields were calculated at Conogher despite high banks because material eroded gradually (9.4 mm/storm) with low spatial variation. At Altnahinch, extremely low yields were observed (0.6 m^3) because of low retreat rates (3.3 mm/storm) and shallow, stable peat banks.

Sediment tracing work involved measurement of a range of physical and chemical properties in size fractionated soil samples ($n = 75$) collected from pre-selected grid coordinates and areas of bare soil. These were compared statistically to the properties of size fractionated suspended sediments and bedload using principal components analysis in order to discern similar 'fingerprints' (Fig. 10.3). Sources identified were only localized due to complex mixing and storage processes within the Bush channel. At Altnahinch, the principal source of material was peat, but storm events liberated quantities of gravel lying under the peat layer. At Altarichard, both the suspended sediment and bedload were composed of gley, humic gley and peat soils from upstream areas of the catchment under pasture. The suspended load at Magherahoney was composed of brown earths and gleys from adjacent ploughed fields, but bedload almost entirely originated from riverbank material. However, storm events also transported peat material from long-distance sources downstream as bedload. Finally, at Conogher the suspended load was traced to channel sources. Subsurface/bed material liberated by drainage maintenance work and livestock poached banks provided the main components of bedload during baseflow conditions. However, during small storm events brown earths from adjacent fields were transferred into the river channel. During larger storms, peat from long-distance sources also contributed to the sediment load.

Management Recommendations

There are several examples of successful integrated catchment management plans addressing the sediment issue in other UK catchments. The Tarland Catchment Initiative, Scotland (Macaulay, 2003), aimed to assess and improve water quality through developing sustainable land management practices. The implementation of simple pragmatic measures (e.g. buffer zone creation, bank stabilization, soft engineering to increase channel habitat diversity, livestock fencing) have led to demonstrable habitat improvement. Perhaps as crucial in its success, the initiative encouraged participation from catchment stakeholders. Two similar projects (WHIP and pHish) are ongoing in the upper stretches of the River Wye, Mid-Wales (Wye and Usk Foundation, 2003). Significant amounts of river corridor work (e.g. removal of shading vegetation, fencing and bank repair) allowed habitats to recover and salmonid fish numbers to increase. Successful schemes such as these were used as a guide to habitat improvement in the Bush as the problems encountered in these catchments can be largely redressed by common strategies.

Instream fine sediment load and sediment source data, in addition to existing catchment management strategies, were used to guide a management plan for the Bush catchment. Table 10.2 contains a list of generic (after Hilton, 2002; RRC, 2002; Verstraeten *et al.*, 2002) and specific plans recommended to tackle high sediment loads in the Bush catchment.

Conclusions and the Future

Hydrological response and sediment source characterization data reported in this paper are important when assessing salmonid embryo survival and spawning in relation to the dynamic nature of sediment transport processes in natural channels. Sampling sites at Conogher and Magherahoney historically were known as areas of high quality salmon spawning. The processes reported in other papers (e.g. Heaney *et al.*, 2001; Mathers and Crowley, 2001; Gilvear *et al.*, 2002) obviously occur over a long time period, so the application of a 1-year monitoring programme to solve these issues is obviously limited. However, the delivery of large loads of fine sediment that infiltrate the coarse framework and accumulate in spawning gravels was noted and is consistent with other studies on the Bush system (e.g. O'Connor and Andrew, 1998). Data presented in this paper do not support the practice of fishery habitat improvement schemes that have 'seeded' the channel with material of an appropriate size for salmon spawning. Critical entrainment threshold for 30–80 mm pebbles was exceeded too frequently due to high bed

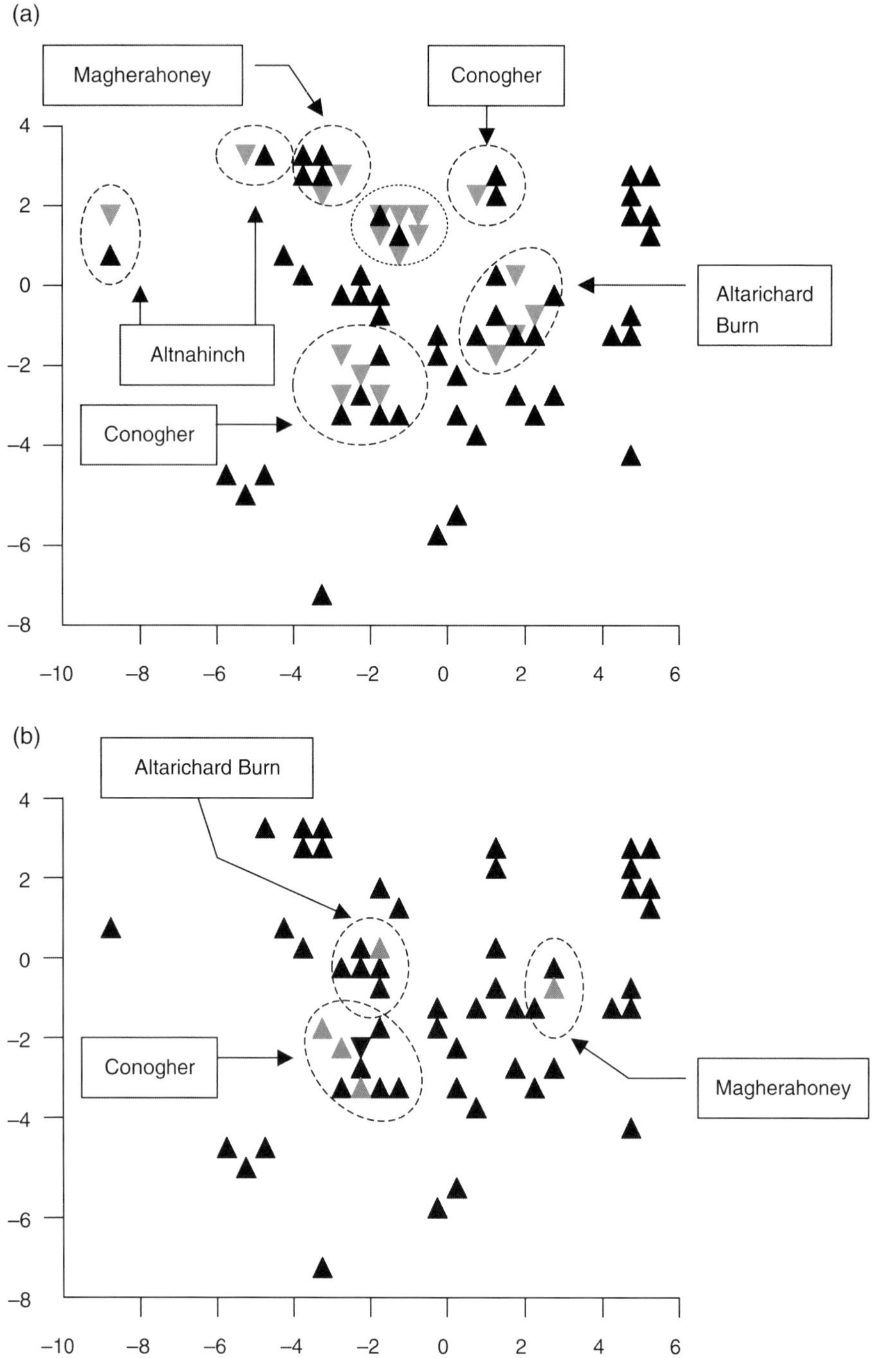

Fig. 10.3. PCA ordination for: (a) bedload sediments (▲) and source soil samples (▲); and (b) suspended sediments (▲) and source soil and bedload samples (▲).

Table 10.2. Best management practices recommended for the River Bush catchment.

Key action	Description	Region
1 – Stop drainage maintenance	(a) Stop widespread use of JCB long-arm digger to lower river bed level (b) Where work, critical cover disrupted soil with seed/synthetic turf reinforcement matting	Lower region
2 – Conifer plantations felling strategy plan	(a) Not working too close to the stream (b) Minimize use of temporary access roads. Avoid using natural drainage channels and gullies. Build on firm land (c) Loading areas kept as small as possible (d) Locate winches for dragging logs to the loading area uphill (e) When operations are complete roads and loading areas should be stabilized with water diversion devices and vegetation cover	Head/East region
3 – Livestock practices	(a) Fencing riparian/riverbank edges (b) Relocating feed/water troughs to hardened areas (c) Moving field access points away from river course	All regions
4 – Reduce bare ground	(a) Critical area planting (b) Cover crops (c) Planting time (d) Compaction management (e) Vehicle movement (f) Manure management (g) Riparian buffer zones	Lower and Mid regions
5 – Wetland restoration	(a) Wetland restoration of historically wet areas throughout the year adjacent to the River Bush (b) Slow down runoff rate through temporary storage plus trap fine sediment	All regions
6 – Macrophyte clearance	(a) Stop macrophyte clearance using JCB digger (b) Selective hand cutting of plants at the stem in heavily colonized stretches in the spring/summer period (c) Actively manage 'problem stretches' annually	Lower region
7 – BMPs for construction sites/housing developments	(a) Infiltration systems bordering such plots to reduce the volume of runoff and concentration of solids transported (b) Street sweeping to remove solid debris from vehicles accessing the site	All regions
8 – River warden	Employment and training of a catchment custodian	All regions
9 – Disseminate the project recommendations to the public	(a) Press release followed up by a series of information leaflets (b) Reconnection of the public with the river channel by providing access to the river to instil local pride	All regions

shear stress resulting in a highly mobile and unstable bed. In addition these gravels were covered rapidly by incoming fine sediment.

Feasibility studies to evaluate sites where best management practices could be applied to give the maximum benefit to fine sediment reduction in the River Bush were presented to the Department of the Environment. The sediment load and source data were used as a scientific justification for the cost of some of these practices.

This chapter has reported the first step towards combating further habitat degradation in the River Bush. However, the next stage will be to implement recommendations justified by this monitoring data into an integrated catchment management plan. The success of this will depend upon providing a framework for funding and legislation to encourage uptake of the plan – a slower and far more challenging process.

Acknowledgements

This work was part of a research fellowship awarded to Queen's University Belfast and financially supported by the Environment and Heritage Service, Department of the Environment, Northern Ireland.

References

Bree, T. and Cunnane, C. (1980) The effect of arterial drainage on flood magnitude. *Casebook of Methods of Computation of Quantitative Changes in the Hydrological Regime of River Basins Due to Human Activities.* Project 5.1 of the International Hydrological Programme, Studies and Reports in Hydrology, 28. UNESCO, Paris, pp. 116–126.

Carling, P.A. (1988) Channel change and sediment transport. *Regulated Rivers: Research and Management* 2, 369–388.

Collins, A.L., Walling, D.E. and Leeks, G.J.L. (1997) Source type ascription for fluvial suspended sediment based on a quantitative composite fingerprinting technique. *Catena* 29, 1–27.

Evans, D.J., Johnes, P.J. and Lawrence, D.S. (2003) Suspended and bedload sediment transport dynamics in two lowland UK streams – storm integrated monitoring. In: Bogen, J., Fergus, T. and Walling, D.E. (eds) *Erosion and Sediment Transport Measurement: Technological and Methodological Advances.* IAHS Publication 283. IAHS Press, Wallingford, UK, pp. 103–110.

Evans, D.J., Johnes, P.J. and Lawrence, D.S. (2004) Physico-chemical controls on phosphorus cycling in two lowland streams. Part 2 – The sediment phase. *The Science of the Total Environment* 329, 165–182.

Gilvear, D.J., Heal, K.V. and Stephen, A. (2002) Hydrology and the ecological quality of Scottish river ecosystems. *The Science of the Total Environment* 294, 131–159.

Guertin, D.P., Barten, P.K. and Brooks, K.N. (1987) The peatland hydrologic impact model – development and testing. *Nordic Hydrology* 18, 79–100.

Heaney, S.I., Foy, R.H., Kennedy, G.J., Crozier, W.W. and O'Connor, W.C. (2001) Impacts of agriculture on aquatic systems: lessons learnt and new unknowns in Northern Ireland. *Marine and Freshwater Research* 52, 151–163.

Hilton, J. (2002) *Dictionary of BMPs for the Reduction of Silt, Phosphorus and Other Pollutants Emanating From Diffuse Sources into Surface Waters.* Natural Environmental Research Council, Swindon, UK.

Lawler, D.M., Grove, J.R., Couperthwaite, J.S. and Leeks, G.J.L. (1999) Downstream change in river bank erosion rates in the Swale–Ouse system, Northern England. *Hydrological Processes* 13, 977–992.

Macaulay (2003) The Tarland Catchment Initiative. http://www.macaulay.ac.uk/tarland

MacKenzie, C. and Moring, J.R. (1988) Estimating survival of Atlantic salmon during the intergravel period. *North American Journal of Fisheries Management* 8, 45–49.

Mathers, R.G. and Crowley, K. (2001) *Erne Salmon Management Programme.* A report published by the Northern Regional Fisheries Board, Ballyshannon, Co. Donegal.

Newson, M.D. and Robinson, M. (1983) Effects of agricultural drainage on upland streamflow: case studies in Mid-Wales. *Journal of Environmental Management* 17, 333–348.

Novikov, S.M. and Pokumeiko, Y.N. (1980) Impact of drainage on stream flow. *Casebook of Methods of Computation of Quantitative Changes in the Hydrological Regime of River Basins Due to Human Activities.* Project 5.1 of the International Hydrological Programme, Studies and Reports in Hydrology, 28. UNESCO, Paris, pp. 103–107.

O'Connor, W.C. and Andrew, T.E. (1998) The effects of siltation on Atlantic salmon, *Salmo salar* L., embryos in the River Bush. *Fisheries Management and Ecology* 5, 393–401.

Ottaway, E.M., Carling, P.A., Clarke, A. and Reader, N.A. (1981) Observations on the structure of brown trout, *Salmo trutta* Linnaeus, redds. *Journal of Fish Biology* 19, 593–607.

Pauwels, S.J. and Haines, T.A. (1994) Survival, hatching and emergence success of Atlantic salmon eggs planted in three Maine streams. *North American Journal of Fisheries Management* 14, 125–130.

Phillips, J.M., Russell, M.A. and Walling, D.E. (2000). Time-integrated sampling of fluvial sediment: a simple methodology for small catchments. *Hydrological Processes* 14, 2589–2602.

RRC (2002) *Manual of River Restoration Techniques*. The River Restoration Centre, Bedfordshire, UK.

Sear, D.A., Lee, M.W.E., Carling, P.A., Oakey, R.J. and Collins, M.B. (2002) An assessment of the accuracy of the Spatial Integration Method (S.I.M.) for estimating coarse bed load transport in gravel-bedded rivers using tracers. In: Bogen, J., Fergus, T. and Walling, D.E. (eds) *Erosion and Sediment Transport Measurement in Rivers: Technological and Methodological Advances*. IAHS Publication 283. IAHS Press, Wallingford, UK, pp. 164–171.

Swales, S. (1982) Environmental effects of river channel works used in land drainage improvements. *Journal of Environmental Management* 14, 103–126.

Verstraeten, G., Van Oost, K., Van Rompaey, A., Poesen, J. and Govers, G. (2002) Evaluating an integrated approach to catchment management to reduce soil loss and sediment pollution through modelling. *Soil Use and Management* 19, 386–394.

Walling, D.E., Owens, P.N. and Leeks G.J.L. (1998) The role of channel and floodplain storage in the suspended sediment budget of the River Ouse, Yorkshire, UK. *Geomorphology* 22, 225–242.

Wilby, R., Greenfield, B. and Gleeny, C. (1994) A coupled synoptic hydrological model for climate-change impact assessment. *Journal of Hydrology* 153, 265–290.

Wilcock, D.N. and Essery, C.I. (1991) Environmental impacts of channelization on the River Main, County Antrim, Northern Ireland. *Journal of Environmental Management* 32, 127–143.

Wye and Usk Foundation (2003) Foundation Projects. http://www.wyeuskfoundation.org/

11 The Physical and Biological Influence of Spawning Fish on Fine Sediment Transport and Storage

E.L. Petticrew
University of Northern British Columbia, Canada. Present address *University of Plymouth, UK*

Introduction

The successful migration of Pacific salmon stocks back to their natal streams for spawning has long been known to be of economic importance, but only more recently has it been documented as being ecologically significant due to the contributions of organic nutrients to the aquatic ecosystem (Wipfli *et al.*, 1998; Gende *et al.*, 2002; Johnston *et al.*, 2004). The return of adult Pacific salmon to their spawning streams results in a transfer of oceanic biomass and nutrients to these freshwater systems. This semelparous species invests all their reproductive energy in this one season to dig a nest (redd) to deposit and fertilize their eggs after which they die, in the vicinity of their redd. While there is a nutrient contribution of excretion products from live fish and the decaying of dead eggs and sperm, the most significant contribution is from the decomposition of the fish carcasses (Johnston *et al.*, 2004).

In building a redd the fish rework the gravel bed streams to a depth of approximately 30 cm. In this process the finer sediments stored in the gravel matrices (e.g. Soulsby *et al.*, 2001) are resuspended into the flowing water and moved downstream (Chapman, 1988). The gravel stored sediment which is available to be released to the water column during redd building has been observed to be a combination of both sand and sand-sized aggregates comprised of silts, clays and organic matter (Petticrew, 1996). The settling rate of some of these larger aggregates is similar to that of fine sands (Petticrew and Droppo, 2000), indicating that the fine sediment (silts and clays) and the organic matter constituting the aggregates are not directly advected out of the system as hydrodynamic models would predict for these constituent grain sizes.

Terrestrial nutrients delivered from the watershed spiral down through the aquatic system (Vannote *et al.*, 1980; Webster and Meyers, 1997), but it is not clear what portion of the nutrients delivered from salmon carcass decay are retained in the system and for what length of time. Algae, periphyton and benthic insects as well as sediment-associated nutrients are potential retention pools within the stream system. While natural stream and mesocosm investigations of the effect of salmon nutrients on primary productivity and insects have been undertaken (Ritchie *et al.*, 1975; Schuldt and Hershey, 1995; Wipfli *et al.*, 1998), the retention by sediment has not been assessed directly.

Salmon die-off in productive spawning streams has been documented as contributing in excess of 250 g C/m^2 (Johnston *et al.*, 2004). These salmon carcass nutrients are readily bioavailable, as much of the decay process occurs within the stream. These nutrients drive instream bacterial activity, which has been linked to the development of sediment flocs and aggregates (Droppo, 2001; Leppard and Droppo, 2005). Therefore, as a means of clarifying if these nutrients have the potential to be retained in the stream, it is of interest to determine the magnitude of the suspended and gravel stored sediment compartments and if they contain salmon nutrients.

Stable isotope analysis (SIA) of nutrients (C and N) was adopted by ecologists as a method of tracing the flux of materials through food webs (Peterson and Fry, 1987). More recently it has been valuably employed to track fluxes through aquatic systems (Bilby *et al.*, 1996; France, 1997; Bouillon *et al.*, 2000). SIA has been used to characterize estuarine seston (Cifuentes *et al.*, 1998; Bouillon *et al.*, 2000), but its ability to distinguish organic source materials in freshwater stream suspended sediment has only recently been utilized (McConnachie, 2003; McConnachie and Petticrew, in press). The ratios of carbon isotopes ($^{13}C/^{12}C$) are usually very distinct between terrestrial sources and adult salmon, allowing the differentiation of source material of two potentially important pools of organic matter supply to stream sediments (McConnachie, 2003).

Some of the world's largest salmon stocks return to streams of the Pacific northwest, where the ecological effects of high numbers of semelparous fish on the flux of nutrients to the terrestrial and freshwater ecosystems have been investigated (Naiman *et al.*, 2002). However, the impact of the physical act of spawning and the die-back of the carcasses on the flux of inorganic sediment from watersheds has not been addressed in an ecological or geomorphic context. In order to assess the physical and biological effect of spawning salmon on fine sediment transport and storage, two objectives were identified, to determine: (i) if the digging of salmon redds modifies the transport and storage of fine sediments; and (ii) if salmonid nutrients are associated with the transported and stored fine sediment.

Study Area and Methods

The three study creeks, O'Ne-eil, Gluskie and Forfar, have small watersheds (36–75 km^2) located in the Stuart-Takla experimental forest in the central interior of northern British Columbia. These mountain tributaries of the larger Fraser River basin have highly productive sockeye salmon (*Oncorhynchus nerka*) returns which have been enumerated annually by the Canadian Department of Fisheries and Oceans since the 1950s. The streams are located in the Hogem Range of the Omenica Mountains, and have mouths at 700 m above sea level (m asl) and drainage divides at approximately 1980 m asl. The channels are all approximately 20 km in length and are straight with coarse bed and bank textures (Macdonald *et al.*, 1992). Each system has a steep upper reach which drains well developed cirques, a steeper middle reach that passes through a rock-walled canyon and a gentle, low gradient depositional reach (Ryder, 1995). In the lower 2 km of the streams the channel bed is comprised of clean gravels suitable for salmon redd construction. These spawning reaches are underlain by glaciolacustrine sediments and the only anthropogenic disturbance at the time of sampling consisted of a gravel road, constructed in 1980, that cuts through this fine-grained material. This work is part of the larger Stuart-Takla Fish–Forestry Interaction Project (MacIsaac, 2003).

Sampling took place in the creeks during 1995, 1996, 1997, 1998 and 2001. Active spawn, salmon die-off, spring melt and summer storms were sampled in the first 4 years respectively, while the 2001 sample period incorporated all of these hydrological and biological events of interest.

Suspended sediment analyses

Suspended sediment was collected and filtered for a range of analyses as well as photographed directly in the water column or in a settling column. Filtered sediment was used for estimates of suspended particulate matter (SPM), absolute particle size distribution (APSD) and isotopic carbon analysis. Photographic images were used to determine the effective particle size distribution (EPSD).

For determination of SPM, water samples were collected in the thalweg, just below the water surface, in large-mouthed 1-l bottles. The water was filtered through pre-combusted and pre-weighed 47 mm diameter, 0.7 μm pore size glass-fibre filters for gravimetric determination of suspended particulate matter. Water was collected in the same manner for APSD, but was returned to the laboratory and filtered through pre-weighed 8 μm SCWP Millipore cellulose-acetate filters. The weighed, dried filters were burned in a low-temperature asher (< 60°C) and wet digested with an excess of 35% H_2O_2 before analysis on a Coulter Counter (Milligan and Kranck, 1991). A Coulter Multisizer IIE was used to determine the inorganic, disaggregated or absolute particle size distribution (APSD). Results are expressed as a volume/volume concentration in ppm and are plotted as smoothed histograms of log concentration versus log diameter (Milligan and Kranck, 1991). Surface grab samples of water and suspended sediment were filtered through pre-ashed glass-fibre filters, freeze-dried and analysed for ^{12}C and ^{13}C by stable isotope mass spectrometry at the University of British Columbia, Oceanographic Stable Isotope Laboratory (see below).

Suspended sediment sizing

During the active spawn sampling in 1995 an underwater silhouette camera was used to photograph the particles as they moved in the water column. The camera was moved to various locations in the creek over the sampling period to collect images representing the background, or ambient, suspended sediment as well as the direct effect of fish digging their redds. Photographs were taken every 5 s of the volume of water (7.4 cm diameter by 4 cm thickness) passing through the camera aperture and grab samples of the water photographed were collected behind the camera aperture for APSD analyses. *In situ* or effective particle size distributions (EPSD), from the photographic negatives of the silhouette camera, were obtained by image analysis using Jandel Scientific's MOCHA program. The equivalent spherical diameters of the detected particles were counted and grouped into size classes which correspond to the same intervals as the Coulter Counter. The Multisizer has a lower detection limit of 0.63 μm and an upper detection limit of 1200 μm, while the silhouette camera has a lower detection limit of approximately 100 μm.

Effective particle size distributions of larger populations of suspended sediment were obtained using a rectangular plexiglass settling box (1.5 × 0.14 × 0.06 m) with two removable end caps, which held approximately 13 l of water. A scale, mounted on the outside back wall of the settling chamber using white adhesive paper, aided in photographing and sizing particles. The settling chamber was aligned into the stream flow such that water and suspended sediment passed through it. When a sample was required the ends were capped and the box carried in a horizontal position to the side of the creek, where it was placed vertically on to a stable platform 20–30 cm in front of a 35 mm single lens reflex (SLR) camera mounted on a tripod. After a period of several minutes, during which fluid turbulence decayed, a series of timed photographs were taken. Pairs of sequential images were then projected on to a large surface and examined to identify individual flocs. The particle size, shape and position in the two images were determined using image analysis packages (Mocha and/or Bioquant). Population means and other size statistics were derived from these data.

In the spring of 1997 the same settling chamber was used to collect suspended sediment samples from the snowmelt flood events in O'Ne-eil Creek. Due to the high overbank flows at this time, the box was lowered and returned to the bridge platform using a winch system. The box was filled and capped by persons standing in the stream. The photographic system employed in the field at this time was a video capture system. A black and white digital camera (a charged-coupled device – CCD), with a resolution of 512 × 512 pixels, was connected to a personal computer running Empix Imaging's NORTHERN EXPOSURE software. This field setup allowed an automated image grabbing system, which recorded the current time (accurate to 10^{-2} s) on each image. The resultant images had individual pixel resolution of 55 μm ± 10 μm. The images were then analysed using a custom-developed (Biickert, 1999) settling rate measurement program.

In 2001 a procedural change occurred mid-June, where the analog CCD was replaced by a Retiga 1300 digital CCD (resolution 1280 × 1024 pixels). At the same time the software was upgraded to Empix's NORTHERN ECLIPSE. These two changes were not found to bias the particle image collection. Dimensions (e.g. diameter, area, perimeter and shape) of 500–1500 particles for each sample date were measured and recorded using the NORTHERN ECLIPSE package with a detection minimum of 42 μm. D_{50} particle sizes were determined from linear interpolation of cumulative per cent volume data for both EPSD and APSD. While the lower resolution of particle diameters using these techniques was regulated by the pixilation, the upper limit was defined by the field of view of the cameras, which given the distance from the settling chamber allowed a photographic image of a particle with a long axis in excess of 10,000 μm.

Gravel stored fine sediment

As a means of characterizing the size and composition of the gravel stored sediments during the salmon die-off period in 1996, a resuspension technique was used that aimed to rework the surface gravels with approximately the same energy expended by spawning salmon. Photographs, water grab samples and settling column images were obtained following physical mixing of the top layer (0.04–0.06 m) of gravels by a field assistant, positioned 4–5 m upstream of the collection site. This distance provided sufficient travel time for the resettling of heavier sand particles, thereby allowing the collected material to comprise the aggregated and non-aggregated fine sediment stored within the surface gravel matrix. This material is referred to here as resuspended gravel stored fines.

Infiltration bags were used to collect gravel stored fine sediments over the summer season of 2001. On 13 July, prior to the return of the spawning fish, 12 infiltration gravel bags were installed in two riffles approximately 1500 m upstream of the mouth of O'Ne-eil Creek. Gravels were removed to dig a hole approximately 25 cm in depth and cleaned through a 2 mm sieve. Infiltration bags, modified from the design of Lisle and Eads (1991), consisting of a water-tight sack with a maximum volume of 10,000 cm^3 clamped on to a 20 cm diameter iron ring, were used to collect gravel stored sediment. The bag was folded down on itself at the bottom of the hole, while straps attached to the ring were placed along the sides of the hole and left at the gravel–water interface. The cleaned gravel, all > 2 mm, was placed on top of the folded bag, filling the hole, and left for a known period of time to accumulate fine sediments in the intergravel spaces. The bag traps were retrieved over a 71-day period following installation. The six retrieval dates represent: (i) the period before the fish return to the river to spawn (17 July); (ii) the early spawn (28 July); (iii) mid spawn (3 August); (iv) two dates during the major fish die-off (12 August and 16 August); and (v) a sample when there was no visual evidence of live or dead carcasses in the stream, termed post fish (22 September). Upon retrieval a lid was placed over the surface gravels between the emergent straps and pulled up, moving the iron ring and the bag up through the gravels ensuring a minimal loss of fine sediment. The gravels and water collected in the bags were washed through a 2 mm sieve such that all of the infiltrated sediment was collected in a calibrated bucket. This material was returned to the laboratory, dried, disaggregated in a mortar and sized using sieves of 1180, 500, 150 and 63 μm.

Stable isotopes

Suspended sediment collected in the three streams during: (i) the post spawning period of 1996; (ii) the spring melt discharges of 1997; and (iii) five summer storms in 1998 were used for analyses of isotopic carbon. The carbon isotope ratio of suspended sediment filters and potential source materials were measured and expressed relative to conventional standards as δ values defined as:

$$\delta X(‰) = (R_{sa} / R_{std} - 1) \times 1000$$

where X is ^{13}C determined as parts per thousand (‰), R_{sa} is the isotopic ratio of the sample ($^{13}C/^{12}C$), and R_{std} is the isotopic ratio of the standard (PeeDee Belemnite for carbon). The source materials represented terrestrial vegetation, salmon and free-floating algae. Multiple tissue samples from riparian vegetation comprising spruce needles, willow, alder and birch leaves, algae and salmon tissue were collected and stored in 1.2 ml centrifuge tubes and subsequently freeze-dried. Isotopes of carbon as well as the δ values of each were determined and reported in McConnachie (2003). This technique enables assessment of organic matter sources in

Table 11.1. Stuart-Takla Creek sampling schedule. Conditions and variables for 5 sample years.

Year	Stream	Date	Event type	Conditions sampled	Cumulative fish return	SPM (mg/l)	Figure number
1995	O'Ne-eil	1–8 Aug	Active spawn	Ambient average	20,648–26,456	14.12	na
1995	O'Ne-eil	8 Aug	Active spawn	Fish digging	26,456	241.07	Fig. 11.1
1995	O'Ne-eil	8 Aug	Active spawn	Post digging	26,456	24.81	Fig. 11.1
1996	O'Ne-eil	26 Aug	Die-off	Ambient	10,772	0.93	Fig. 11.4
1996	O'Ne-eil	26 Aug	Die-off	Resuspended gravel stored fines	10,772	7.22	Fig. 11.4
1996	Forfar	26 Aug	Die-off	Ambient	9,076	0.41	Fig. 11.4
1996	Forfar	26 Aug	Die-off	Resuspended gravel stored fines	9,076	15.45	Fig. 11.4
1997	O'Ne-eil	28 May	Spring melt rising limb	Ambient	0	8.38	Fig. 11.4
1997	O'Ne-eil	30 May	Spring melt rising limb	Ambient	0	6.79	Fig. 11.4
1997	O'Ne-eil	1 Jun	Spring melt rising limb	Ambient	0	8.70	Fig. 11.4
1997	Forfar	29 May	Spring melt rising limb	Ambient	0	3.92	Figs 11.4 and 11.6
1997	Forfar	1 Jun	Spring melt rising limb	Ambient	0	1.58	Figs 11.4 and 11.6
1997	Gluskie	29 May	Spring melt rising limb	Ambient	0	3.57	Fig. 11.4
1997	Gluskie	1 Jun	Spring melt rising limb	Ambient	0	1.57	Fig. 11.4
1998	Forfar	14 Jun	Summer rain storm	Ambient	0	3.08	Figs 11.4 and 11.5
1998	Forfar	18 Jun	Summer rain storm	Ambient	0	3.85	Figs 11.4 and 11.5
1998	Forfar	21 Jul	Summer rain storm	Ambient	0	0.32	Figs 11.4 and 11.5
1998	Forfar	8 Aug	Summer rain storm	Ambient	770	0.47	Figs 11.4 and 11.5

1998	Forfar	16 Aug	Summer rain storm	Ambient	956	0.29	Figs 11.4 and 11.5
1998	O'Ne-eil	14 Jun	Summer rain storm	Ambient	0	1.97	Fig. 11.4
1998	O'Ne-eil	18 Jun	Summer rain storm	Ambient	0	1.17	Fig. 11.4
1998	O'Ne-eil	21 Jul	Summer rain storm	Ambient	2	0.80	Fig. 11.4
1998	O'Ne-eil	8 Aug	Summer rain storm	Ambient	2,020	3.38	Fig. 11.4
1998	O'Ne-eil	16 Aug	Summer rain storm	Ambient	2,268	0.46	Fig. 11.4
1998	Gluskie	14 Jun	Summer rain storm	Ambient	0	0.51	Fig. 11.4
1998	Gluskie	18 Jun	Summer rain storm	Ambient	0	2.47	Fig. 11.4
1998	Gluskie	21 Jul	Summer rain storm	Ambient	0	0.17	Fig. 11.4
1998	Gluskie	8 Aug	Summer rain storm	Ambient	749	1.35	Fig. 11.4
1998	Gluskie	16 Aug	Summer rain storm	Ambient	812	4.26	Fig. 11.4
2001	O'Ne-eil	24 May–21 Aug	All event types	Ambient	0–13,893	0.87–18.09	Fig. 11.3A
2001	O'Ne-eil	17 Jul	Pre fish arrival	Gravel stored fines	0	na	Fig. 11.3B
2001	O'Ne-eil	28 Jul	Early spawn	Gravel stored fines	8,211	na	Fig. 11.3B
2001	O'Ne-eil	3 Aug	Mid spawn	Gravel stored fines	10,931	na	Fig. 11.3B
2001	O'Ne-eil	12 Aug	Die-off	Gravel stored fines	13,757	na	Fig. 11.3B
2001	O'Ne-eil	16 Aug	Die-off	Gravel stored fines	13,892	na	Fig. 11.3B
2001	O'Ne-eil	22 Sept	Post fish	Gravel stored fines	13,893	na	Fig. 11.3B

the suspended sediment at different times of the year by comparing isotopic ratios from source material contributed to the stream to those in suspended sediment samples.

Physical measurements

Stream flows were determined using a Swoffer current meter at the time of sample collection, while discharge on the three streams was monitored and calibrated over the open water season for each year by the Canadian Department of Fisheries and Oceans.

Results

Suspended sediment

Four hydrologically or biologically important periods, which comprised spring melt, summer rain storms, active spawning and fish die-off, were sampled for suspended sediment in the Stuart-Takla streams. Table 11.1 displays surface water concentrations of suspended sediment sampled during each of these event types over the 5 sample years. Measurements taken during the 1995 active spawn in the immediate vicinity of digging fish exhibit the highest suspended sediment concentrations (241 and 25 mg/l). The absolute and effective particle size distributions of these two fish-suspended sediment samples are shown in Fig. 11.1. The EPSD spectra were determined from analysis of the Benthos underwater camera images (triangles), while the APSD was measured using a Coulter Counter (circles). The solid symbols represent samples taken 0.5 m downstream of a fish digging its redd, while the open symbols represent samples collected approximately 4 m further downstream. The mode and maximum of the APSD in the immediate vicinity of fish digging are 294 and 512 μm, representing medium sands, while the EPSD of the same volume of suspended sediment has a mode and maximum of 588 and 1024 μm. Given that the EPSD comprises a large number of particles greater than the maximum size of the constituent particles (APSD), it is clear that the physical action of digging fish resuspends aggregated fines as well as sands. Images from the Benthos camera just preceding, during and following the sampled fish resuspension are shown in Fig. 11.2. Comparing the image on the right to the central one indicates an abrupt change in sediment concentration due to the fish digging, while the image on the left demonstrates the rate at which the water clears. The spectra for fish-suspended sediment sampled 4 m downstream of redd construction has modes of 169 μm for the EPSD and 16 μm for the APSD. Note that all of the aggregates moving in the water column at this time are smaller than 400 μm and are comprised of inorganic sediment less than 85 μm, indicating the loss, by settling, of the sands and larger aggregates.

The average suspended sediment concentration observed in O'Ne-eil Creek during the 1995 active spawn was 14 mg/l. This value does not include samples such as those mentioned above that directly tracked the plumes of fish-resuspended sediment, but rather represents the ambient water conditions in reaches of this stream, where by 8 August up to 24,000 salmon had returned to spawn. This ambient average suspended sediment concentration is higher than 1997 spring melt concentrations of 7–9 mg/l (Table 11.1) but less than the maximum of 18 mg/l observed in spring melt of 2001 (Fig. 11.3A). Note in 2001 the significant ($P < 0.05$) elevation of suspended sediment concentrations during the active spawn when compared to the pre spawn lower flow concentrations. These ambient average values ($n = 3$) are taken over 5 days following the midpoint of the fish returns (Fig. 11.3C) reflecting large numbers of active spawners in this reach.

Size characteristics of the suspended sediment populations from spring melt, active spawn and fish die-off are shown in Table 11.2. These results were obtained from image analysis of particles in the settling column and in all but 1996 represent the ambient or background conditions at that time. Aggregates moving during spring melt are smaller than those of active spawn, while the largest particles are observed in the sediment resuspended from on and within the gravels during salmon die-off in 1996.

Gravel stored fine sediment

Figure 11.3B shows the amount of sediment infiltration into the gravel bed before, during

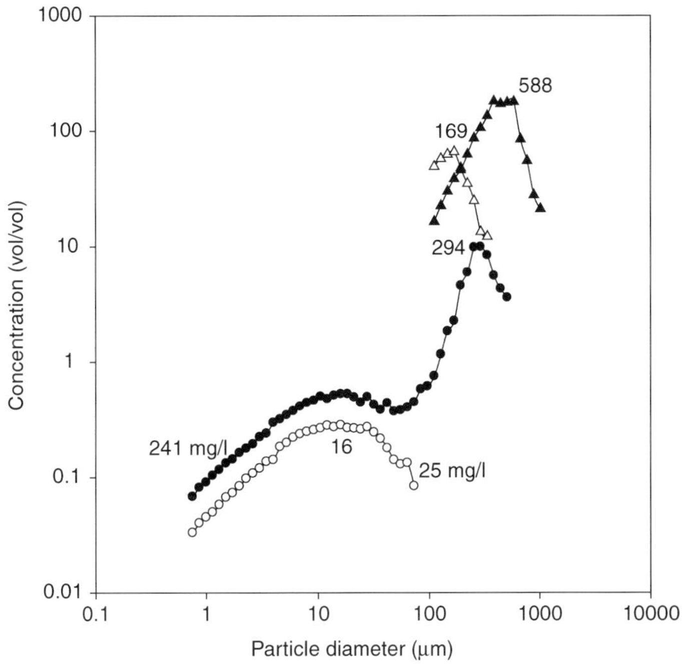

Fig. 11.1. Particle size spectra for sediment resuspended by a redd-digging fish (solid symbols) and 4 m downstream (open symbols). The circles represent the disaggregated inorganic component of the suspended sediment (APSD), while the triangles represent the *in situ* or effective particle size distribution (EPSD). The suspended sediment concentrations of the APSD samples are noted (mg/l) along with the mode size of each sediment spectra (μm).

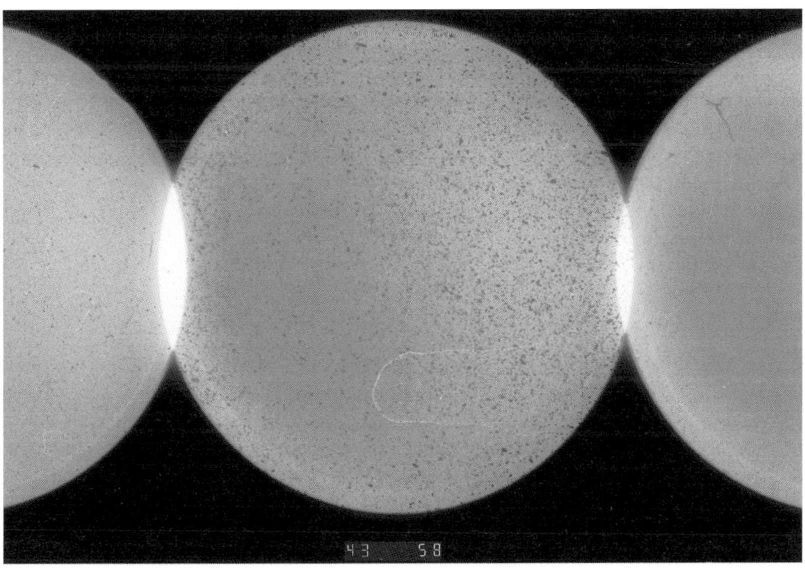

Fig. 11.2. Silhouette images of suspended sediment taken in sequence from right to left at intervals of 5 s. A fish digging its redd resuspended sands, aggregates and fine sediment 0.5 m in front of the Benthos camera in the central image. Ambient water column conditions are represented in the right image, while clearing rates of the water column can be assessed in the left image. The diameter of each image is 7.4 cm.

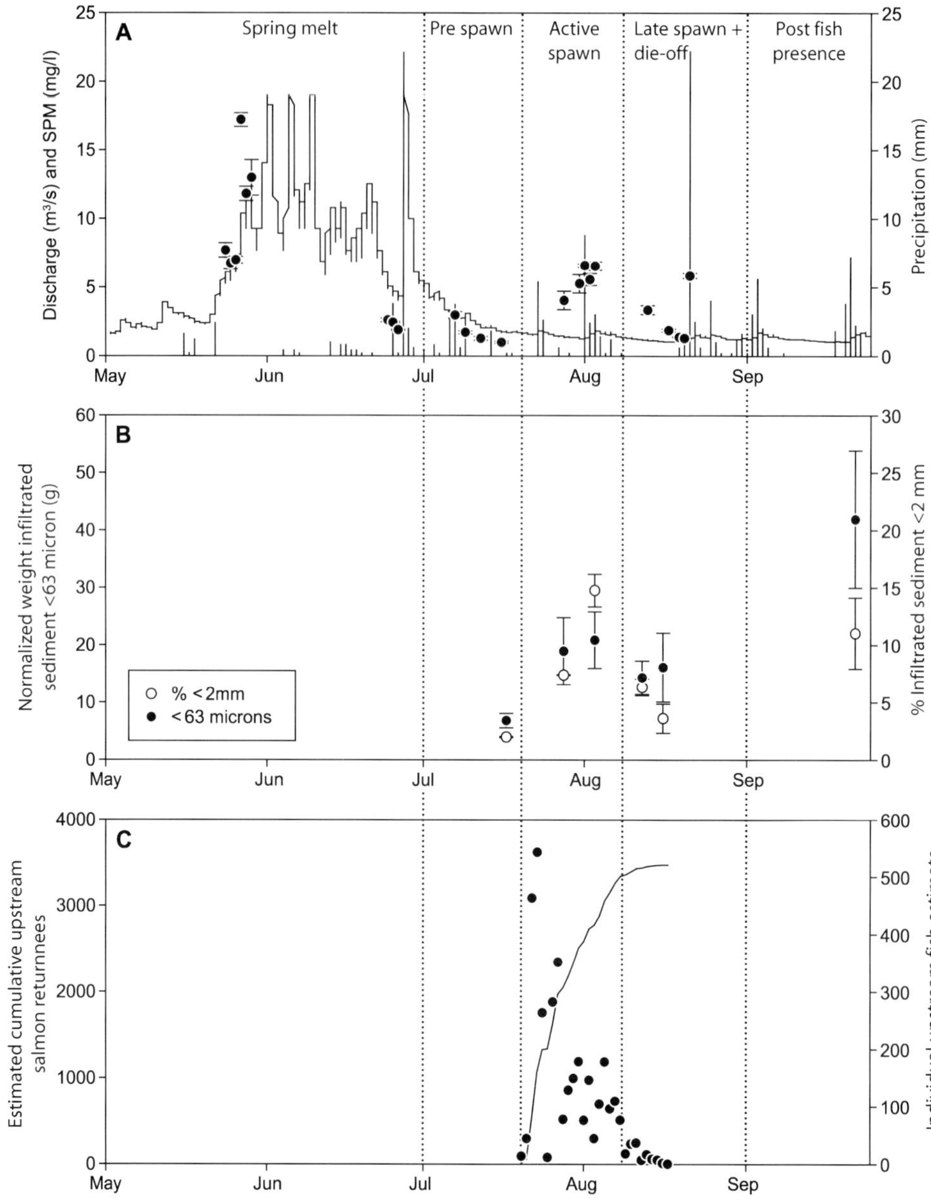

Fig. 11.3. (A) Discharge, precipitation and suspended sediment in O'Ne-eil Creek for periods of spring melt, pre spawning, active spawn and salmon die-off. (B) Weight of fine sediment infiltration into gravels in an active spawning reach and the proportion of material < 2 mm accumulated in the gravels before and following fish presence in the stream. (C) The estimated number of sockeye salmon entering the upstream spawning reach on a daily basis (circles) and the cumulative numbers (line) for the full 2001 spawning season. All error bars represent ±1 standard error (SE).

and following the salmon spawn of 2001. The temporal pattern is similar for both the per cent less than 2 mm and the normalized weight of sediment less than 63 μm. Increases are associated with the period of active spawning, when more than 50% of the returnees to the section of the river upstream of the sample reach are present (Fig. 11.3C). These gravels which had been

Table 11.2. Population particle sizes associated with event type.

Year*	Event	Sample type	Mean diameter and SE (μm)	D_{50} and SE (μm)	Maximum diameter (μm)	Maximum D_{99} (μm)
1997	Spring melt rising limb	Ambient	276 (6.5)		712	
1995	Active spawn	Ambient	514 (58.2)		1,162	
1996	Die-off	Resuspended	897 (107.8)		1,828	
2001	Spring melt rising limb	Ambient		369 (49.4)		894
2001	Active spawn	Ambient		697 (208.5)		2,033
2001	Die-off	Ambient		294 (42.5)		881

*Events are shown in seasonal order not chronological order.

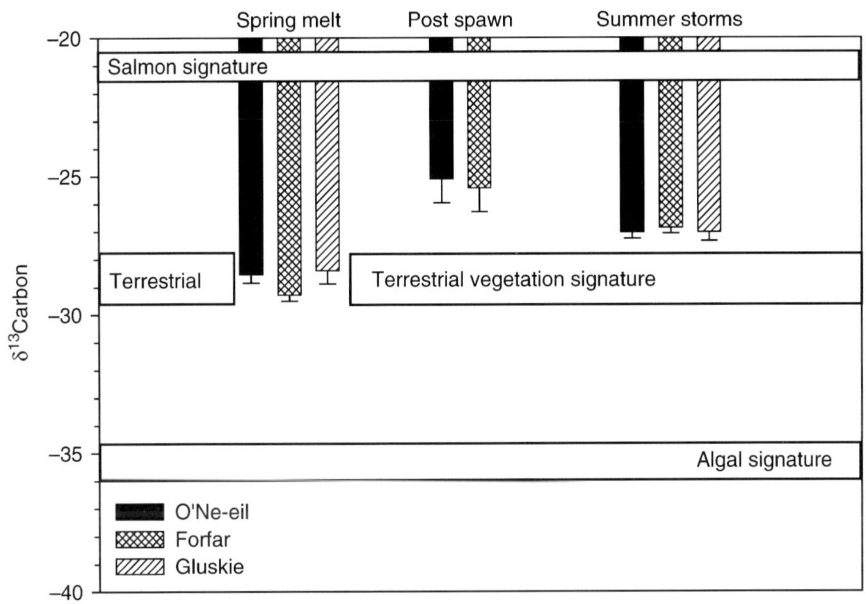

Fig. 11.4. Isotopic carbon results (± 1 SE) for the Takla streams in three periods of hydrologic or biologic interest. The isotopic signatures for the expected organic source material are also shown.

cleaned of sediment < 2 mm, before installation, collected 15% by weight of sediment smaller than this size after a period of 21 days, while up to 20 g of this was < 63 μm. Removal after 71 days indicated the final bags had on average 40 g of silt- and clay-sized particles (< 63 μm).

Stable isotopes

The isotopic carbon content of the suspended sediment was compared to the isotopic signatures for the dominant organic source materials to the creeks (Fig. 11.4). Samples from the 1997 spring melt exhibit values which fall directly within the terrestrial vegetation signature, while post spawn samples of 1996 are significantly different ($P < 0.05$) and are closer to the salmon tissue signature, indicating a dilution of terrestrial organic matter in the samples. Averages of the five summer storms sampled in the three creeks in 1998 reflect a mixture of materials which is due in part to the temporal spacing of the storms. When individual storms are viewed, for example on Forfar Creek (Fig. 11.5), a more interesting pattern emerges, although only single samples are taken for each storm event,

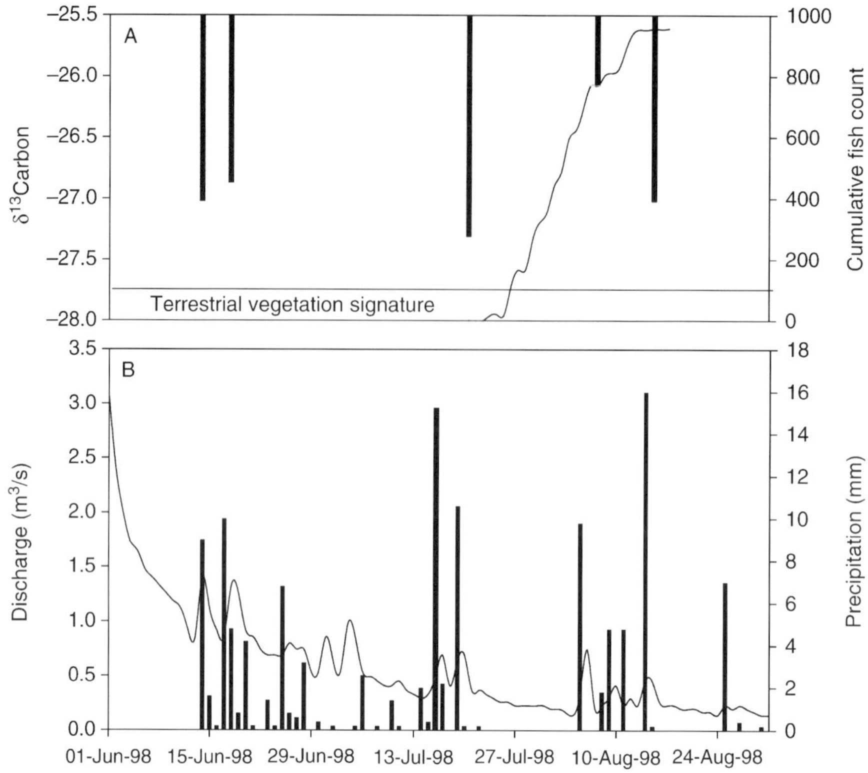

Fig. 11.5. (A) Carbon isotope results of suspended sediment collected for five storms sampled before and during the spawning season in Forfar Creek. Cumulative counts for the 1998 fish return indicate the timing and magnitude of the salmon biomass in the stream. (B) The hydrograph and daily precipitation for 1998 indicate the timing and the stream response of the summer rain events.

thereby restricting statistical comparison. The three storms preceding the return of the fish have values closer to the terrestrial signature, while a sample following a rain storm in the midst of the salmon spawn has a value closer to the salmon tissue signature, indicating the inclusion of fish-based organic matter in the suspended sediment. At the end of spawn the largest summer storm, with 16 mm of rain, has a suspended sediment carbon value that reflects an increased contribution of terrestrial materials.

Both the suspended sediment and the resuspended gravel stored fines were analysed for isotopic carbon at three locations in Forfar Creek in the die-off period of 1996. The upper reaches of stream (1700 and 1500 m from the mouth) exhibit values further away from the salmon tissue signature than the samples taken near the mouth (100 m). It is interesting to note that the isotopic values for the suspended sediment and gravel stored sediment at each site are not appreciably different. In this example, again only single filters were analysed for each $\delta^{13}C$ datum, so statistical tests for significance are not possible.

Discussion

Physical influences of salmon

The return of sockeye salmon to their natal streams for spawning results in the digging of a single redd for each salmon pair. The redds are dug to a depth of 25–30 cm, but tend to be much wider due to the subsidence of the gravels, therefore the volume of gravels moved to prepare a single redd can be in the order of 0.03–0.05 m^3. Given the numbers of salmon returning to these streams (approximately 13,000 pairs in 1995), in some years all of the viable gravel bed sites in the lower 2 km of the stream are reworked once, and sometimes more often, over the 3–4 week spawning period. The preparation of the redd results in the short term resuspension of sands and aggregated fines and presumably a longer term resuspension of less dense aggregates and disaggregated silts and clays. Ambient suspended sediment concentrations in these streams were elevated during the active spawning period in both 1995 and 2001, years that had comparatively high fish returns. It is useful to note that the concentration of suspended sediments reported here for spring melt and summer storms are relatively low compared to other streams, reflecting the fact that these mountain creeks are very clean with few anthropogenic disturbances, but also that the samples were surface grabs and not depth integrated samples.

The direct effect of digging fish was documented in 1995 with the images from the Benthos camera. It is clear that sands up to 500 μm in diameter as well as large aggregated particles exceeding 1000 μm are removed from the gravels and introduced into the water column by fish (Fig. 11.1). Comparison of the sample taken 4 m downstream of the digging fish to the sample in the vicinity of the redd indicates that the sands and larger aggregates settle out of the water column in this short distance but that smaller, less dense aggregates are still maintained in suspension.

In papers by Petticrew and Droppo (2000) and Petticrew (2005), the size, settling velocities and densities of the particle populations from these same 1996 artificially resuspended gravel stored fines from Forfar and O'Ne-eil Creeks are presented. The sediment collected 4–5 m downstream of these disturbances comprises two distinct populations of particles: (i) larger, less dense flocs; and (ii) smaller, denser compact aggregates. The settling velocities of a 400 μm floc and compact particle were determined to be 2.4 and 5.5 mm/s respectively. In low flows of 0.2–0.3 m/s, observed during post spawning periods, these particles could be maintained in suspension and advected downstream to the lake for settling. However, in shallow (< 0.25 m) turbulent flows the probability that the particles will make contact with the gravel surface or penetrate into the gravel matrix is high, increasing the likelihood of their instream storage.

The sediment stored on or in the gravels was observed to exhibit the largest particle sizes (Table 11.2), although the ambient active spawn populations in both 1995 and 2001 also had means and maximum sized particles appreciably larger than spring melt and ambient die-off suspended sediments. Given the extensive, continuous digging of the gravels in the spawning period, these larger sizes likely reflect the recently released gravel stored flocs and aggregates which take longer to settle.

Petticrew and Arocena (2003) reported on the size and density of the population of gravel stored aggregate particles (sands had been removed) collected in the infiltration bags discussed in this paper. The median size of the stored aggregates was smallest (244 ± 89 μm) at mid-spawn and exhibited a mean settling velocity of 2.5 mm/s. These particles would be available for resuspension in the water column by fish digging, and due to their density would settle more slowly, thereby increasing the ambient concentration of suspended particles. The larger aggregates from this population of gravel stored sediment would presumably settle back to the gravels in near-field distances (4–5 m) along with the sands. This assumption is supported by the results shown in Fig. 11.1 and in the artificial resuspension exercises of 1996 (reported here and in Petticrew and Droppo, 2000). The data showing that the proportion of sediment < 2 mm, and the mass of fine sediment (< 63 μm) collected in the infiltration bags (Fig. 11.3B) both increase during the active spawning period further corroborates this assumption. The final infiltration bags removed in late September had the highest mass of silt- and clay-sized particles, indicating a continued accumulation of fine sediments in the gravels during the period of salmon egg incubation.

While the observation that spawning salmon temporarily alter the suspended sediment concentrations in their natal streams is not surprising, what is of consequence is that much of the material they clean from their redds is being redeposited in short distances downstream. Silts and clays combined with organic matter as aggregated compact particles are settling out of the water column in the near-field, along with the sands resuspended by digging fish. This transfer of particles back to the gravels implies that both the inorganic sediment and the nutrients associated with the organic matter component of aggregates are being retained in the channel bed at least during the low flow period associated with salmon egg incubation. Presumably these sediments will be flushed out of the system during the high flows of spring melt floods.

Biological influences

Stable isotope analysis of carbon in these stream systems was used as a means of differentiating the source of organic material that contributes to the suspended and gravel stored sediment. The carbon signatures for the source materials of salmon tissue, terrestrial vegetation and instream algae for this region have been reported in McConnachie (2003) and Figs 11.4 and 11.6. As their $\delta^{13}C$ signatures are significantly different, they can be used to evaluate the contribution of the different sources to the organic component of the sediment. The results of this preliminary $\delta^{13}C$ work on suspended and gravel stored sediment indicate that stable isotope analysis is a viable technique to use for differentiating the presence of these organic matter sources in stream sediments. By sampling suspended sediment at several times (Figs 11.4 and 11.5) and locations within the streams (Fig. 11.6), the temporal and spatial influence of the salmon die-off on stream sediment organic matter composition has been determined.

The comparison of suspended sediment from three event types (Fig. 11.4) indicates a shift away from dominantly terrestrial sources in spring melt rising limb samples to a mixture of salmonid and terrestrial sources in a post spawn period. Intuitively this makes sense and is corroborated by Johnston *et al.*'s (2004) modelling of nutrient budgets in Gluskie and Forfar Creeks in 1996, 1997 and 1998. They determined that salmon were the dominant particulate organic carbon source to the streams from late July through October, in all years except 1998 when the fish returns were very low.

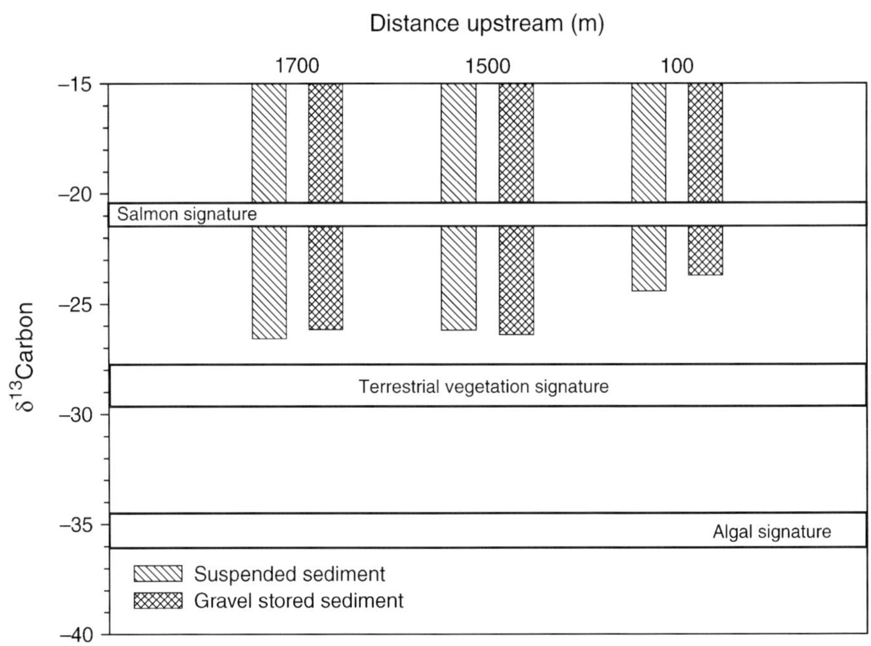

Fig. 11.6. Carbon isotope values for suspended and resuspended gravel stored sediment in three locations in Forfar Creek. Sampling occurred in the die-off period of 1996 when 9076 fish carcasses were contributing nutrients to the stream.

The pattern which emerges from the Forfar Creek rain storm samples also indicates that stable isotope analysis is able to detect temporal differences of organic matter sources in suspended sediments. The suspended sediment sampled from a storm occurring 2 weeks after the return of the fish exhibits a $\delta^{13}C$ closer to that of salmon tissue than the three pre-fish return storms. A storm sampled 8 days later, close to the end of the spawning period, exhibited the highest precipitation recorded that season, but its $\delta^{13}C$ signal indicated a return towards the terrestrial carbon signature. An analysis of two soil samples from the O'Ne-eil Creek streambank indicated that its carbon signature was –27.04 (SE = 0.16), which is very close to the terrestrial values, indicating the dominant organic source material in the soils is from the riparian vegetation. Therefore this directional change in carbon source dominance associated with high precipitation rates at the end of the fish spawn of 1998 likely reflects the contributions of bank sediments and/or the increased throughflows from the surrounding floodplains and soils. Interestingly as well, the salmon returns for 1998 were the smallest reported here, with only 956 salmon having passed the counting fence at the time of this storm event. Therefore the amount of salmon contributing to the carbon pool was much lower that year, suggesting that the $\delta^{13}C$ signals would have been weaker than in other years. This same pattern of results occurred in both Gluskie and O'Ne-eil Creek $\delta^{13}C$ data for these five rain storm dates of 1998, supporting the validity of the pattern obtained with only single data points.

Results from a downstream transect of both suspended and gravel stored sediment in Forfar Creek in the post spawn period of 1996 (Fig. 11.6) also reflects the association of fish nutrients with sediments. Samples of both types of sediment taken 100 m upstream of the mouth, where the influence of the 9076 decaying fish would be realized, have $\delta^{13}C$ values closer to that of the salmon tissue. Results, not presented here, but again based on single $\delta^{13}C$ samples from an O'Ne-eil Creek downstream transect, corroborate these results, exhibiting this same trend in both types of sediment.

The patterns observed in these three creeks over the 3 years where the preliminary isotopic carbon sampling was undertaken indicated that this technique was useful for differentiating source materials of organics in stream sediments and that fish nutrients were associated with both suspended and gravel stored sediment. Changing temporal and spatial contributions of terrestrial vegetation and fish nutrients were detected in the suspended and gravel stored sediment. McConnachie (2003) followed up this approach and sampled O'Ne-eil Creek suspended sediments in replicate, over all event types in 2001. She reported on the ability of stable isotopes of both C and N to differentiate the proportions of salmon nutrients in suspended sediments over the season. Mixing models (Phillips and Gregg, 2001) combined with isotopic results indicated that salmon nutrients comprised 33% of the organic C and N in the active spawn suspended sediment, which increased to 46% during the 2001 post spawn period (McConnachie and Petticrew, in press).

While Johnston *et al.* (2004) identified the importance of salmon carcass decay in the P and N budgets of these streams, they stated that the majority of the nutrients were exported from the study reaches. This was determined from their analyses of reach loadings calculated from water samples and discharge measurements. This approach assumes that all of the material collected in their grab samples of water and suspended sediment for total nutrient analysis remained in suspension until reaching the downstream receiving water body. Given that we have noted the propensity of aggregated fine sediment to settle on to and into the gravels, and that we have detected the influence of salmonid carbon on the gravel stored sediment, it would be relevant for these models to reconsider the importance of storage of aggregated fine sediments as a temporary sink for nutrients.

Conclusion

The physical action of large numbers of spawning salmon digging redds increases the contribution of fine sediment to the water column, enabling the transport and advection of the material that remains in suspension out of the riverine system. However, silt and clay sized particles, combined with organic matter as aggregates, exhibited increased settling velocities and when delivered to the water column, by fish digging redds, settled out of suspension over short distances. The implication of this process is that both aggregated fine sediments and their sediment-associated nutrients readily collect on and in the downstream gravels. Isotopic analyses ($\delta^{13}C$) of the stream sediment indicated that different sources of C were sequestered into the suspended and gravel stored fines of the spring melt flood flows versus post spawning low flows. Active and post spawn sediment exhibited the largest aggregate sizes and incorporated high quality nutrients from instream decaying salmon. This suggests the importance of the biological role that the die-off has on the structure of fine sediment, and therefore the transfers and storage of this material in these fish-bearing streams. While excessive storage of fine sediment and organic matter in the gravels could be deleterious to egg growth, the retention of some nutrients may not necessarily be problematic, as it could enhance stream productivity at both primary and secondary levels.

Acknowledgements

Many people helped with the collection of field samples, including A. Buller, N. Alexander, C. Spicer, S. Biickert, J. Yannick, A. Ullrich, J. McConnachie, R. McConnachie, M. Riley and H. Herunter. L. Chamberlist and T. Waqar assisted in the lab at UNBC, while T. Milligan of Bedford Institute of Oceanography loaned field equipment and determined APSDs. Data and field support by both the Canadian Department of Fisheries and Ocean and Canadian Forest Products was greatly appreciated. Funding for this work was provided by Natural Science and Engineering Research Council, Northern Land Use Institute and Fisheries Renewal British Columbia.

References

Biickert, S. (1999) The effect of pulp mill effluent on fine grained sediment morphology and storage in the Fraser River at Prince George, British Columbia. Unpublished MSc thesis. University of Northern British Columbia, Prince George, Canada.

Bilby, R.E., Fransen, B.R. and Bisson, P.A. (1996) Incorporation of nitrogen and carbon from spawning coho salmon into the trophic system of small streams: evidence from stable isotopes. *Canadian Journal of Fisheries and Aquatic Sciences* 53, 164–173.

Bouillon, S., Mohan, P.C., Sreenivas, N. and Dehairs, F. (2000) Sources of suspended organic matter and selective feeding by zooplankton in an estuarine mangrove ecosystem as traced by stable isotopes. *Marine Ecology Progress Series* 208, 79–92.

Chapman, D.W. (1998) Critical review of variables used to define effects of fines in redds of large salmonids. *Transactions of the American Fisheries Society* 117, 1–21.

Cifuentes, L.A., Sharp, J.H. and Fogel, M.L. (1988) Stable carbon and nitrogen isotope biogeochemistry in the Delaware estuary. *Limnology and Oceanography* 33, 1102–1115.

Droppo, I.G. (2001) Rethinking what constitutes suspended sediment. *Hydrological Processes* 15, 1551–1564.

France, R.L. (1997) Stable carbon and nitrogen isotopic evidence for ecotonal coupling between boreal forests and fishes. *Ecology of Freshwater Fish* 6, 78–83.

Gende, S.M., Edwards, R.T., Willson, M.F. and Wipfli, M.S. (2002) Pacific salmon in aquatic and terrestrial ecosystems. *BioScience* 52, 917–928.

Johnston, N.T., MacIsaac, E.A., Tschaplinksi, P.J. and Hall, K.J. (2004) Effects of the abundance of spawning sockeye salmon (*Oncorhynchus nerka*) on nutrients and algal biomass in forested streams. *Canadian Journal of Fisheries and Aquatic Science* 61, 384–403.

Leppard, G.G. and Droppo, I.G. (2005) Overview of flocculation processes in freshwater ecosystems. In: Droppo, I.G., Leppard, G.G., Liss, S.N. and Milligan, T.M. (eds) *Flocculation in Natural and Engineered Systems*. CRC Press, Boca Raton, Florida, pp. 25–46.

Lisle, T.E. and Eads, R.E. (1991) *Methods to Measure Sedimentation of Spawning Gravels*. USDA Forest Service Research Note PSW 411.

Macdonald, J.S., Scrivner, J.C. and Smith, G. (1992) *The Stuart-Takla Fisheries/Forestry Interaction Project*. Canadian Technical Report of Fisheries and Aquatic Science No. 1899. Ottawa, Canada.

MacIsaac, E.A. (2003) *Compendium of Research from the Stuart-Takla Fisheries/Forestry Interaction Project.* Canadian Technical Report of Fisheries and Aquatic Science No. 2509. Ottawa, Canada.
McConnachie, J.L. (2003) Seasonal variability of fine grained sediment morphology in a salmon bearing stream. Unpublished MSc thesis. University of Northern British Columbia, Prince George, Canada.
McConnachie, J.L. and Petticrew, E.L. (2005) Tracing organic matter sources in riverine suspended sediment: implications for fine sediment transfers. *Geomorphology* (in press).
Milligan, T.G. and Kranck, K. (1991) Electroresistance particle size analysers. In: Syvitski, J.P.M. (ed.) *Principles, Methods and Applications of Particle Size Analysis.* Cambridge University Press, Cambridge, UK, pp. 109–118.
Naiman, R.J., Bilby, R.E., Schindler, D.E. and Helfield, J.M. (2002) Pacific salmon, nutrients, and the dynamics of freshwater and riparian ecosystems. *Ecosystems* 5, 399–417.
Peterson, B.J. and Fry, B. (1987) Stable isotopes in ecosystem studies. *Annual Review of Ecology and Systematics* 18, 293–320.
Petticrew, E.L. (1996) Sediment aggregation and transport in northern interior British Columbia streams. In: Walling, D.E. and Webb, B.W. (eds) *Erosion and Sediment Yield: Global and Regional Perspectives.* IAHS Publication 236. IAHS Press, Wallingford, UK, pp. 313–319.
Petticrew, E.L. (2005) Flocculation and aggregation of suspended and gravel stored fine sediment in productive headwater streams – a case study of O'Ne-eil Creek, British Columbia, Canada. In: Droppo, I.G., Leppard, G.G., Liss, S.N. and Milligan, T.M. (eds) *Flocculation in Natural and Engineered Systems.* CRC Press, Boca Raton, Florida, pp. 71–93.
Petticrew, E.L. and Arocena, J.M. (2003) Organic matter composition of gravel stored sediments from salmon bearing streams. *Hydrobiologia* 494, 17–24.
Petticrew, E.L. and Droppo, I.G. (2000) The morphology and settling characteristics of fine-grained sediment from a selection of Canadian rivers. In: *Contributions to the International Hydrological Programme – V by Canadian Experts.* Technical Documents in Hydrology No. 33. UNESCO, Paris, pp. 111–126.
Phillips, D.L. and Gregg, J.W. (2001) Uncertainty in source partitioning using stable isotopes. *Oecologia* 127, 166–170.
Ritchie, J.E., Perkins, M.A. and Goldman, C.R. (1975) Effects of kokannee salmon (*Oncorhynchus nerka*) decomposition on the ecology of a subalpine stream. *Journal of Fisheries Research Board of Canada* 32, 817–820.
Ryder, J.M. (1995) *Stuart-Takla Watersheds: Terrain and Sediment Sources.* Working Paper 03/1995. British Columbia Ministry of Forests, Victoria, Canada.
Schuldt, J.A. and Hershey, A.E. (1995) Effect of salmon carcass decomposition on Lake Superior tributary streams. *Journal of North American Benthological Society* 14, 259–268.
Soulsby, C., Youngson, A.F., Moir, H.J. and Malcolm, I.A. (2001) Fine sediment influence on salmonid spawning habitat in a lowland agricultural stream: a preliminary assessment. *The Science of the Total Environment* 265, 295–307.
Vannote, R.L., Minshall, G.W., Cummins, K.W., Sedell, J.R. and Cushing, C.E. (1980) The river continuum concept. *Canadian Journal of Fisheries and Aquatic Sciences* 37, 130–137.
Webster, J.R. and Meyers, J.L. (1997) Stream organic matter budgets – introduction. *Journal of North American Benthological Society* 16, 5–13.
Wipfli, M.S., Hudson, J. and Caouette, J. (1998) Influence of salmon carcasses on stream productivity: response of biofilm and benthic macroinvertebrates in southeastern Alaska, USA. *Canadian Journal of Fisheries and Aquatic Science* 55, 1503–1511.

12 Lakes and Reservoirs in the Sediment Delivery System: Reconstructing Sediment Yields

I.D.L. Foster

Coventry University, UK and Rhodes University, South Africa

Introduction

The purpose of this chapter is threefold. First, to provide a brief history of lakes and reservoirs with particular reference to UK river catchments, secondly to review the palaeohydrological benefits and pitfalls of using lake sediment sequences to reconstruct the impacts of catchment disturbance and climate change on sediment yields and sources and, thirdly, to synthesize a body of research conducted in the UK on the change in sediment yields, sediment sources and sediment-associated phosphorus transport over the last 100–150 years. The emphasis here is upon catchments that have been impacted by changes in rural land use.

A Brief History of Lakes and Reservoirs

Freshwater lakes and reservoirs exist in almost all parts of the world, but they are transitory features on geological timescales (Hutchinson, 1975). In middle to high latitudes most lake basins were created during the Holocene after the wasting of ice sheets and glaciers at the end of Oxygen Isotope Stage 2.

Human intervention in the hydrological cycle has often been prompted by a need to store water, by the desire to regulate downstream river flooding and for hydro-electric power (HEP). Smith (1971), for example, dates the earliest known reservoir construction for river regulation at *c.* 3000 BC in Egypt and noted that dams had become widespread in the Mediterranean area by Roman times. By the late Middle Ages small dams were being constructed throughout Europe (Beckinsale, 1972) and many other notable periods of reservoir construction have been documented since that time. The Late Medieval period in England, for example, saw the construction of mill and fishponds (Sheail, 1988), while during the 18th century many formal gardens included the construction of ornamental lakes and engineers built small reservoirs to maintain a reliable water supply for the expanding canal systems. The 19th and 20th centuries witnessed the most rapid expansion of dam construction, and most of the world's major rivers are now regulated through reservoirs (Petts, 1984; McCully, 1996; Vörösmarty *et al.*, 1997).

While the exact number of lakes and reservoirs in the UK is not known, Smith and Lyle (1979), based on an analysis of 1 : 63,360 scale Ordnance Survey maps, estimated a figure in excess of 81,000. However, only a relatively small proportion (< 0.5%) of UK river catchments drain through lakes and reservoirs (Ward, 1981) and no known published studies have quantified their impact on the sediment

 Soil Erosion and Sediment Redistribution in River Catchments (eds P.N. Owens and A.J. Collins)

delivery of UK river catchments. The majority of UK lakes and reservoirs are in low order headwater tributaries that are located in the catchment zone identified by Schumm (1977) as the dominant sediment source in river catchments. Many storage reservoirs constructed in regions of high sediment yield over the last 50 years have lost a considerable proportion of their initial storage capacity as a result of sedimentation with, in extreme cases, deposition rates exceeding 0.9 m/year (e.g. Lahlou, 1996; Woodward and Foster, 1997). Such rapid rates of sedimentation suggest that storage reservoirs may have a short-lived impact on sediment export from drainage basins in areas of high sediment yield.

Not only do lakes and reservoirs exist in almost all climatic zones, but they also exist in a variety of locations within a drainage basin. From the uplands to the lowlands, a typical mid-latitude catchment could include tarns and large glacial lake basins, upland water supply, river regulating and HEP generating reservoirs, canal feeder reservoirs, ornamental lakes, urban retention basins, fish and mill ponds, farm ponds and coastal freshwater lagoons created behind barrier beaches or blown sand deposits. On the river floodplain, small meander cut-off lakes often exist and are defined here as *off-line* if they are only connected to the river at periods of high flow (above bankfull) or *on-line* if they are in direct contact with the river at flow stages below bankfull. In both cases, the reduction in flow velocity associated with an increase in effective channel cross-section, will induce sediment to settle from suspension to the lake bed and will preserve at least a partial history of sediment and contaminant transport (e.g. Foster and Charlesworth, 1996; Winter *et al.*, 2001; Paine *et al.*, 2002). In all situations, the episodic movement of fine and/or coarse sediments from uplands to lowlands is interrupted and the history of sediment and sediment-associated nutrient and contaminant transport could be reconstructed in most of these locations.

Palaeohydrological Benefits and Pitfalls

The development of a unifying methodology to reconstruct sediment yields and sediment-associated nutrient and contaminant fluxes has enabled the magnitude of the impacts of climate change and human disturbance over the last century or more to be reconstructed over decadal to millennial timescales (e.g. Likens and Davis, 1975; Oldfield *et al.*, 1985; Dearing, 1991; O'Hara *et al.*, 1993; Foster, 1995; Page and Trustram, 1997; Zolitschka, 1998; Foster and Lees, 1999a, b; Foster *et al.*, 2002; Smol, 2002).

It is important to recognize that the dynamics of natural lakes and reservoirs differ significantly, largely as a result of their physical location and morphology. Reservoirs are often located in narrow river valleys. In consequence the deepest point of the reservoir is usually close to the dam wall rather than at the centre of the water body. Hydraulically and ecologically reservoirs comprise three major zones (McCully, 1996). Close to the major inflows most coarse sediment is deposited and the hydraulic characteristics are similar to rivers. Between this zone and the fully limnic deepwater zone is a transition zone. However, the location of these zones may move towards the dam at times of low water level and towards the inflows at periods of high water level, often giving rise to complex sedimentation patterns across the reservoir bed. Furthermore, drawdown increases the opportunity for shoreline erosion as unconsolidated marginal sediments are exposed to wave action.

Considerable care has to be exercised in selecting appropriate reservoirs for reconstruction, especially in relation to the existence of by-pass channels and/or scour valves. The former can divert inflowing river water during high runoff periods because of poor quality or high sediment transport rates. The latter may be periodically opened to rapidly draw down the reservoir and remove some of the sediment from the basin. In both cases, where documentary records of these activities are not available, these problems provide uncertainties in reconstructing sediment yield.

Reconstructing sediment yield

Average sediment yields for reservoirs can be obtained by one of two methods. First, by estimating the water storage loss between the date of construction and the present-day capacity (estimated by bathymetric survey) and measuring or estimating sediment density in order to calculate sediment mass (e.g. Lahlou, 1996; White *et al.*, 1996). Secondly, yields may be estimated by ground surveys of sediment volume and density during drawdown (e.g. Duck and McManus, 1987). These approaches provide a rapid reconnaissance method to estimate sediment yield, but they suffer three major disadvantages. First, the reliability of the original storage capacity estimates may be questionable or unknown and will depend on the resolution of the Digital Terrain Models (DTMs) used for volume calculations. Secondly, sediment yields are averaged over variable time periods depending on the date of reservoir construction and, thirdly, no temporal patterns in sediment yield can be obtained. Of major importance in reconstructing temporal changes in the sediment yield record, therefore, has been the development of reliable methods for dating lake and reservoir sediments.

Laminated sediments can provide a precise chronological background, whereas most other dating methods, which usually rely on radionuclide activity measurements, have a wide margin of error. In environments where such annually laminated sediments exist, they provide the most precise estimates of sediment age and sediment yield (cf. Håkanson and Jansson, 1983; Saarnisto, 1986; Gilbert and Desloges, 1987; Desloges, 1994; Zolitschka, 1998).

In environments where laminations are not annual, or where no laminations are preserved, reliance has to be placed on known marker horizons or on one or more radiometric dating methods (e.g. ^{14}C). Tephra layers, exotic pollen and atmospheric pollutants have all been extensively used for dating sedimentary sequences (see Smol, 2002). However, the main focus of this chapter is on the last 100–200 years of environmental change and discussion will focus on the use of ^{137}Cs and ^{210}Pb dating methods.

In 1973, two independent studies demonstrated that ^{137}Cs could be detected in lake sediments (Pennington *et al.*, 1973; Ritchie *et al.*, 1973). Caesium-137 was first released into the environment as a result of atmospheric thermonuclear weapons testing in the early 1950s. Its presence in sedimentary deposits is a clear indication of the earliest possible date when these sediments could have been deposited. The pattern of atmospheric fallout, for which a long-term record is available in the UK from Chilton (Fig. 12.1), peaked in activity in 1963 (prior to the atmospheric test-ban treaty), and in 1986 (Chernobyl fallout). While the Chernobyl fallout record is spatially variable (Smith and Clark, 1989), these two events often provide clear dateable horizons in freshwater sedimentary sequences.

Lead-210 (^{210}Pb), a radionuclide produced through the natural decay of radioactive elements in the Earth's crust, provides a second means of dating sediments (Goldberg, 1963). Lead-210 is derived from the escape of radon gas from the Earth's crust and the subsequent decay of this radioactive gas via a series of short-lived radionuclides to ^{210}Pb. Since ^{210}Pb has a half-life of 22.26 years, it provides an opportunity for dating over the last 150 years (6–7 half lives) depending on activities within the sediment.

The calculation of sediment age using this dating method requires the separation of the supported $^{210}Pb_{(sup)}$ (in equilibrium with the parent radium-226, ^{226}Ra) and unsupported $^{210}Pb_{(uns)}$ in the sediments. Total $^{210}Pb_{(tot)}$ activities can be measured directly using low energy gamma spectrometry, but ^{226}Ra does not decay through the release of gamma radiation. The ^{226}Ra daughter isotope ^{214}Pb can be measured by gamma spectrometry to derive ^{226}Ra activities and, by calculation, the activity of $^{210}Pb_{(sup)}$ and $^{210}Pb_{(uns)}$ in the sample (Gilmore and Hemingway, 1995).

Several different procedures have been used to convert the unsupported ^{210}Pb profile into an age–depth relationship, but most of these were criticized by Robbins and Herche (1993), because of the lack of a strong theoretical model of ^{210}Pb transport. Most commonly, one of three point transformation methods (constant rate of supply, 'crs'; constant initial concentration, 'cic'; constant rate of accumulation, 'cra') are used depending on the assumptions made in applying the transformation (Appleby and Oldfield, 1978; Oldfield and Appleby, 1984; Appleby *et al.*, 1986; Olsson, 1986; Robbins and Herche, 1993). In studies where changes in sedimentation rate are strongly suspected, the assumptions that underpin the crs point transformation (that $^{210}Pb_{(uns)}$ supply remains constant to the point within the lake basin at which the core is taken but that

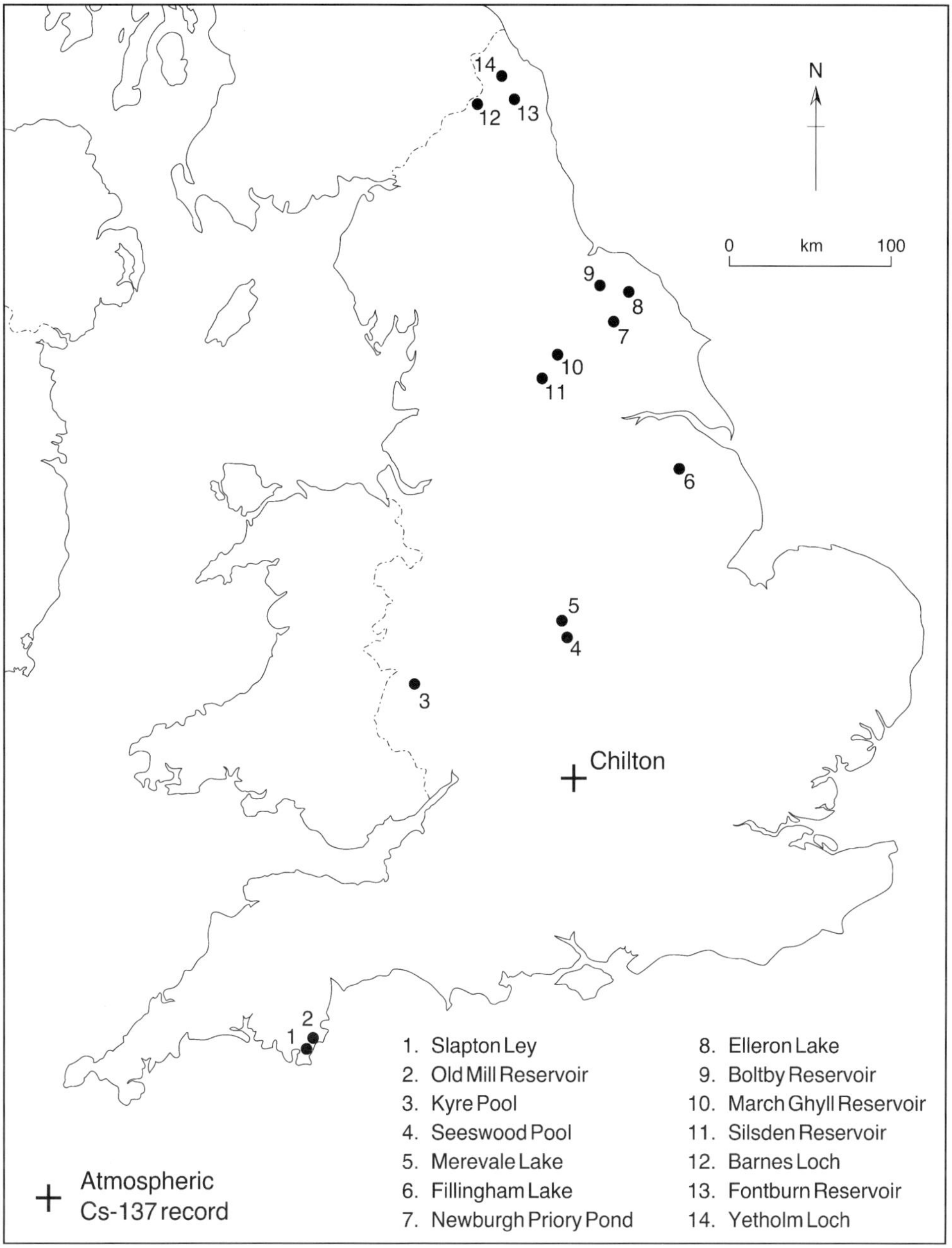

Fig. 12.1. Location of sites referred to in the text.

sedimentation rates may vary) often seems most appropriate. However, it must be remembered that this method assumes that there has been no episodic sedimentation or resuspension and that there has been no post-depositional remobilization of sediment and/or $^{210}Pb_{(uns)}$. Alternative procedures for estimating depth–age curves have attempted to overcome potential problems of post-depositional $^{210}Pb_{(uns)}$ diffusion within deposited sediments (e.g. Joshi and

Shukla, 1991). More recently, Appleby (2001) and Appleby *et al.* (2003) have discussed the limitations in using ^{137}Cs and the ^{210}Pb crs and cic transformation methods in a range of depositional environments.

On the basis of dated lake sediment sequences, the total mass of sediment deposited in a water body can be divided into time zones for estimating historical sediment yields. However, estimates cannot be based on a single sediment core because sedimentation rates are often non-uniform across the lake bed with respect to time (Dearing, 1986). Sediment yield estimation therefore requires the collection of representative sediment cores in which time-synchronous layers can be correlated across the whole deposit. Further adjustments to the reconstructed yields are required to account for dry bulk density, the area of sedimentation, contributions from biological productivity and losses resulting from variable lake trap efficiencies (see Foster *et al.*, 1985, 1986, 1990; Dearing, 1986; Dearing and Foster, 1993; Foster and Walling, 1994).

Additional information can be obtained from physical, mineral magnetic and/or chemical analysis of the deposited sediments. While the precise analysis largely reflects the purpose of the investigation, such as the reconstruction of pollution histories (Foster and Charlesworth, 1996), additional information on the characteristics of the deposited sediments may also be used to identify the most likely origins of the deposited sediments (e.g. Charlesworth and Foster, 1993; Foster and Walling, 1994; Foster and Lees, 1999b; Foster *et al.*, 2002).

Palaeolimnological Reconstructions in the UK

Thirteen sediment yield reconstructions from a range of UK locations (Fig. 12.1, Table 12.1) are plotted in Fig. 12.2, where catchment locations are arranged by land use. Sites were selected for a variety of purposes and included natural and ornamental lakes, canal feeder and water supply reservoirs, the oldest of which dates from the 16th century. The first three reconstructions of Fig. 12.2 are for catchments dominated by coniferous or deciduous woodland (Boltby, Fontburn and Merevale), while March Ghyll and Barnes Loch receive inputs from mixed moorland and rough grazing systems. The remaining eight sites are dominated by a variety of agricultural catchments ranging from 75% pasture to 92% arable at the time of survey. These catchments have a complex history of land use and management, and sediment yields over the last 100 years or so have ranged from a minimum of *c.* 5 t/km²/year or less (Fontburn and Elleron) to a maximum of *c.* 120 t/km²/year (Kyre Pool).

Since it was possible at all sites to identify the year 1963 from the ^{137}Cs and/or ^{210}Pb chronologies, the ratio of pre- to post-1963 sediment yields have been taken as an index of the magnitude of change from the earlier to the later part of the 20th century (Fig. 12.3a). This is also a time after which significant changes have been identified in the UK rainfall record, with an increase in winter rainfall, an increase in the number of daily rainfalls exceeding 25 mm and an increase in the amount of rainfall delivered by fewer events (Foster, 1995; Wilby *et al.*, 1997; Walling *et al.*, 2003; Foster and Lees, in press). The most dramatic increase in sediment yield (*c.* 60 t/km²/year) was reconstructed at Kyre Pool and was associated with the installation of land drainage (tile drains) that gave rise to a change in sediment delivery pathways (Foster *et al.*, 2002, 2003; Table 12.2). Site-specific explanations for increasing or decreasing sediment yields are complex and the summary of Table 12.2 identifies nine key factors that have been used to explain observed trends. These factors fall into three groups; those associated with catchment management; those associated with climate change and those associated with changes to the sediment delivery ratio. The latter is particularly evident in the Yetholm Loch sediment yield record, where the decrease in the most recent time period is associated with a major drainage diversion within the catchment (Foster and Lees, 1999a).

At all sites, total particulate phosphorus concentrations (PP) were measured in the dated lake sediments (Fig. 12.4). An additional site (Slapton Ley, see Fig. 12.1) has also been included here since the PP concentration was compared with reconstructions based on hindcasted export coefficient modelling and catchment records of total P influx (Johnes and

Table 12.1. Site characteristics for lake catchments where historical sediment yields have been reconstructed.

Site	Lake/reservoir type (year of construction)	Catchment area (km^2)	Lake area (km^2)	Catchment: lake ratio	Catchment maximum altitude (m)	Relative relief (m)	Approx. average annual rainfall (mm)
Boltby	Water supply (1882)	3.25	0.022	145	366	146	800
Fontburn	Water supply (1905)	27.74	0.323	86	440	250	1200
Merevale	Ornamental lake (1861)	1.95	0.065	30	175	57	675
March Ghyll	Water supply (1907)	4.04	0.057	71	380	170	1000
Barnes Loch	Water supply (1915)	1.78	0.059	30	400	150	1200
Silsden	Water supply (1867)	8.15	0.104	79	373	203	1000
Elleron	Ornamental lake (1919)	2.56	0.030	86	229	89	800
Seeswood	Canal feeder (1765)	2.21	0.067	33	160	35	675
Old Mill	Water supply (1942)	1.60	0.019	86	194	50	1228
Kyre Pool	Water supply (1584)	2.73	0.053	51	262	172	762
Newburgh	Ornamental lake (1760)	5.88	0.040	148	122	46	650
Fillingham	Ornamental lake (1790)	2.9	0.070	41	61	31	597
Yetholm	Natural	12.21	0.144	84	160	50	1000

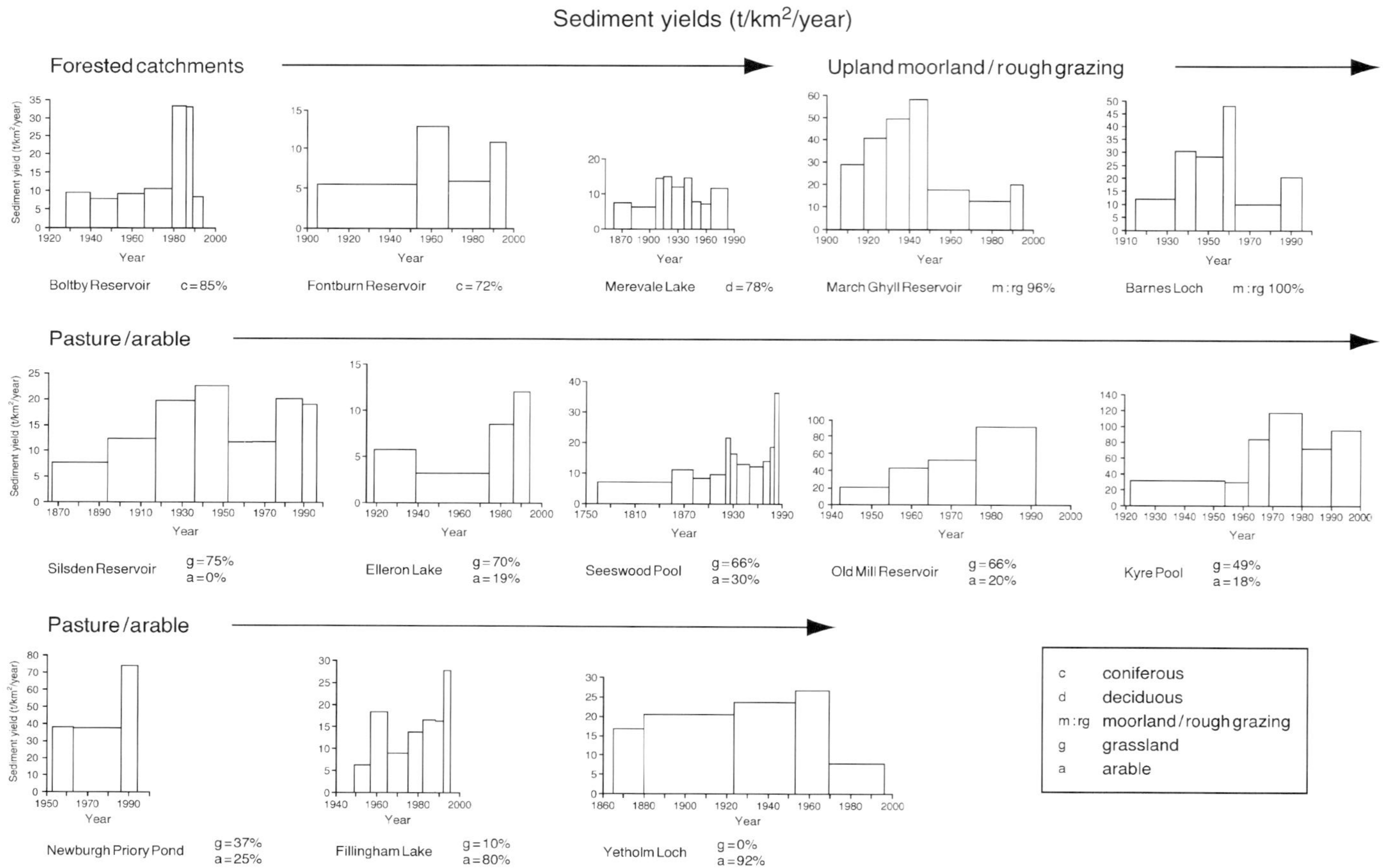

Fig. 12.2. Reconstructed sediment yield histories for 13 UK catchments. (Data from Foster *et al.* (1985) (Merevale); Foster *et al.* (1986) (Seeswood Pool); Foster and Walling (1994) (Old Mill); Foster and Lees (1999a) (Boltby, Fontburn, March Ghyll, Barnes, Silsden, Elleron, Newburgh, Fillingham and Yetholm); Foster *et al.* (2003) (Kyre Pool)).

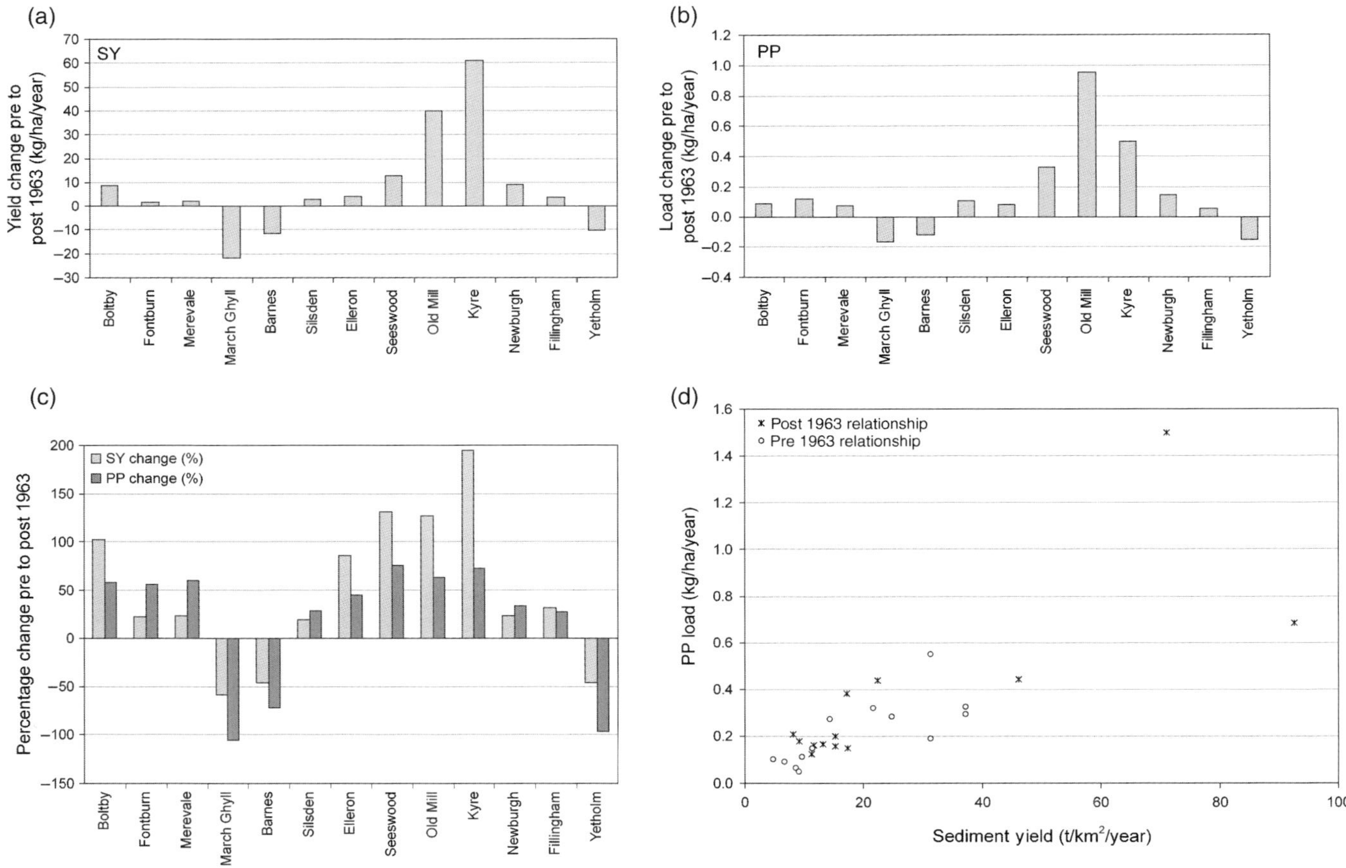

Fig. 12.3. Pre- to post-1963 changes in sediment yield (SY) and total sediment P (PP) loads on a mass specific (a and b) and percentage basis (c) and the pre- and post-1963 relationship between sediment yield and PP load (d).

Table 12.2. Site-specific explanations for changing sediment yields through time (X = major factor, x = minor factor).

Site	Forest management	Increased improved pasture area	Decreased stock density	Increased stock density	Increased arable area	Increased field size	Land drainage installed	Climate change	Decreased sediment delivery
Boltby	X								
Fontburn	X								
Merevale	x							x	x
March Ghyll			X						
Barnes Loch			X						
Silsden		X							
Elleron		X						x	
Seeswood				X		X		x	
Old Mill				X		X		x	
Kyre Pool							X	x	
Newburgh					X				
Fillingham					X			x	
Yetholm					X				X

Heathwaite, 1997). While concentrations in the lake sediments are consistently lower than those measured or modelled, there is a remarkable consistency in the temporal pattern suggesting that the sedimentary P record reflects the trends in lake PP concentrations through time. Of the 14 plots of Fig. 12.4, highest PP concentrations appear to be associated with forestry plantation (Fontburn) and with grazing systems (Silsden, Slapton, Elleron, Seeswood and Old Mill). Concentrations of PP in the lake sediments derived from most arable systems, with the exception of the most recent period in Yetholm Loch, appear to be much lower.

By combining the sedimentary P record with the sediment yield record, particulate phosphorus (PP) loads have also been reconstructed for the 13 sites (Fig. 12.5). PP loads range from < 0.5 kg/ha/year to almost 2.5 kg/ha/year (Old Mill). It is evident that both upland and lowland grazing systems are associated with the highest PP loadings to lakes and reservoirs. Absolute changes in the pre- to post-1963 PP loads are plotted in Fig. 12.3b. The most dramatic absolute increase is recorded in the sediments of Old Mill, Kyre and Seeswood. The absolute increase in load at Kyre is again associated with the major influx of sediment following land drainage, while at Seeswood and Old Mill the importance of increased stocking densities are apparent in the record.

Percentage increases in pre- to post-1963 sediment yield and PP yields are plotted in Fig. 12.3c. A generally consistent pattern of increased or decreased sediment yields and PP loads is apparent from these plots and the relationship between sediment yield and PP load is statistically significant for both the pre- and post-1963 periods ($r = 0.755$ and 0.784 and $r^2 = 57.0\%$ and 61.5% for pre- and post-1963 relationships respectively; $n = 13$ in both cases and both relationships are significant at $P < 0.01$; Fig. 12.3d). However, there is considerable scatter in the relationship and it is clear that the delivery of P to lakes and reservoirs at individual sites is controlled both by concentration and sediment yield.

Sediment Sources

While the focus of this discussion has centred largely upon sediment and particulate phosphorus loads in relation to a range of external controls, it must also be recognized that changes in sediment source through time may in part explain changes in the sediment yield and PP loadings. While space does not allow a detailed

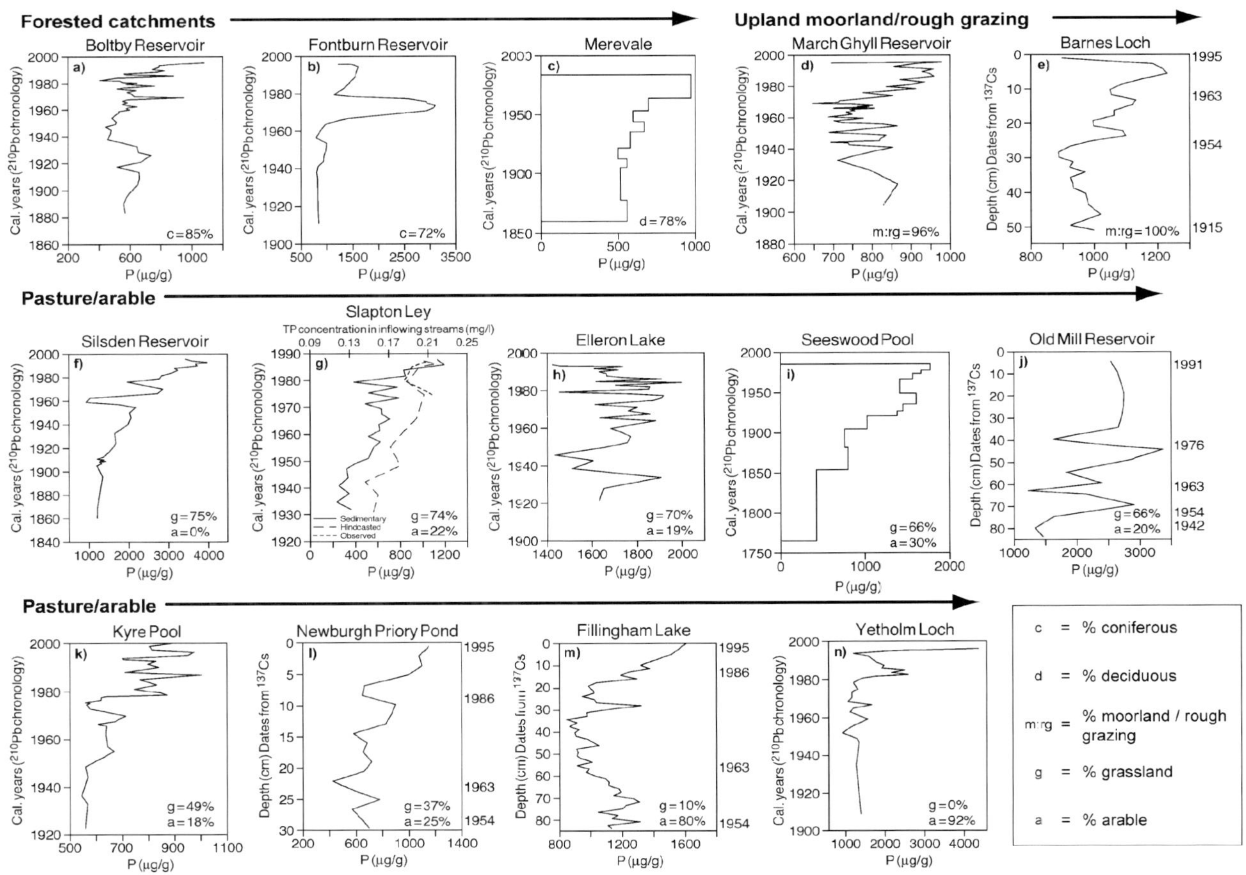

Fig. 12.4. Sediment total P concentrations from 14 UK sites. Sources of data are those given in the caption to Fig. 12.2. Additional data for Slapton Ley are from Foster *et al.* (1996) (sediment total P) and Johnes and Wilson (1996) (hindcasted export coefficient model and measured total P concentrations).

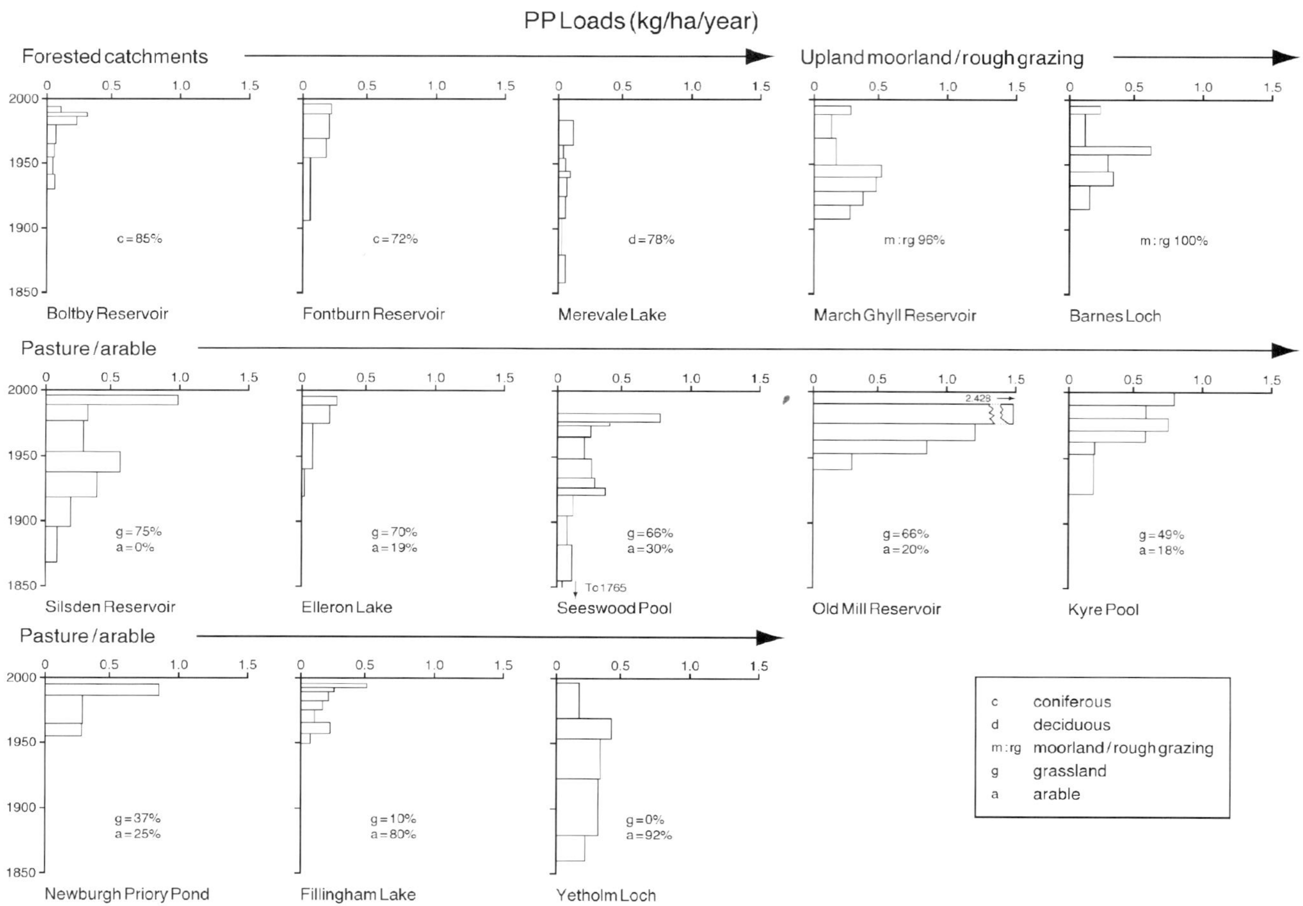

Fig. 12.5. Total sediment P (PP) loads for 13 lake catchments calculated from the data of Figs 12.3 and 12.4. (Data sources are given in caption to Fig. 12.2.)

discussion of sediment source tracing, an example based on the application of un-mixing models to the Kyre Pool sediments will be used to illustrate the point. Foster *et al.* (2002, 2003) reconstructed sediment yields for Kyre Pool and suggested that a combination of mineral magnetic and radionuclide signatures might be used to discriminate five potential sediment sources within the catchment. These sources included arable topsoil, pasture topsoil, woodland topsoil, channel bank, and pasture and arable subsoil. Discrimination between these sources was achieved using three mineral magnetic parameters (Xlf, IRM0.8T and HIRM) and three radionuclides measured by gamma spectrometry (^{40}K, ^{228}Ac and ^{226}Ra). Although ^{137}Cs provided good discrimination between contemporary sources, its complex fallout history precluded its used for reconstructing sediment sources through time.

Details of the procedure used to select parameters, correct for particle size effects and model the mixtures are beyond the scope of this paper. The procedures are explained in Slattery *et al.* (2000) and Gruszowski *et al.* (2003). Results of the application of these un-mixing models to the most recent (*c.* 80 years) of the sedimentary record in Kyre Pool are given in Fig. 12.6, where source contributions are plotted in proportion to total sediment yields reconstructed for six time zones. All sources were modelled as making a contribution to the sediment yields. However, in the pre-1963 period, channel banks made a significant contribution to the yield in both time zones (13% and 48%, respectively), but did not contribute to post-1963 yields. Subsoil is modelled to contribute in two zones after 1963 and probably reflects disturbance associated with land drain installation. Only one zone in the record is modelled to contain a significant proportion of arable topsoil and this period also appears to have no contribution from woodland topsoil. The presence of a woodland topsoil signal in five of the six time zones is unsurprising, since managed coniferous and deciduous woodland surrounds most of the lake. Despite the change in transport pathways associated with land drainage, pasture topsoil has dominated sediment sources in the catchment since 1922. The effect of drain installation, therefore, appears to have been to increase the total amount delivered from this source to the lake.

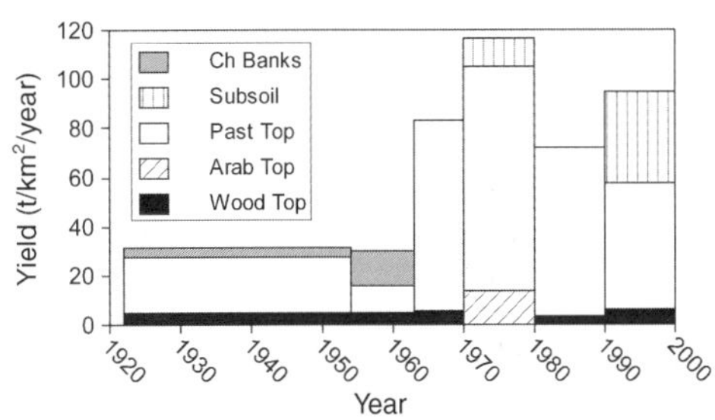

Fig. 12.6. Un-mixing models based on mineral magnetic and radionuclide signatures of five potential contributing sources (channel banks, subsoil, pasture topsoil, arable topsoil and woodland topsoil) to the sediment yields of Kyre Pool during the 20th century.

Conclusions

The methods of palaeoenvironmental reconstruction reported in this contribution offer tremendous potential for understanding and exploring the impact of a range of disturbances on river catchments. Explanations of the impact of catchment disturbance are, by definition, complex and often site-specific, yet it is only through historical reconstruction that some of the impacts of recent environmental change can be studied, since the UK has a paucity of fluvial sediment and sediment-associated contaminant transport data. The reconstruction of the impact of land drainage on sediment yields at Kyre Pool, for example, was the only practical method for quantifying the impacts of land drainage given the fact that most areas of Herefordshire utilized available UK Government grant-aid in order to install land drainage (Foster *et al.*, 2002, 2003). In consequence, there was no opportunity to undertake a paired catchment experiment in order to quantify the impact. While the palaeoenvironmental approach may be viewed as 'less than satisfactory' to geomorphologists and hydrologists, palaeolimnologists have used these methods in order to answer pertinent questions about the impacts of recent environmental change on catchment systems (see Smol, 2002).

As Deevey (1969) so eloquently suggested: 'When time is needed to conduct an experiment, there is no substitute for history' (Deevey, 1969, p. 43).

Students of geomorphology and hydrology can, and should, benefit from such an approach. Failure to recognize the contribution that this approach might make to our understanding of sediment dynamics in river catchments over short to medium timescales would deny us a critical future opportunity for understanding the consequences of environmental change on sediment delivery and the sediment delivery system.

Acknowledgements

Thanks go to a large number of colleagues who have stimulated many of the ideas summarized in this contribution and whose names appear extensively in the references. Special thanks also go to Stuart Gill, Coventry University Cartographic Unit, for the artwork.

References

Appleby, P.G. (2001) Chronostratigraphic techniques in recent sediments. In: Last, W.M. and Smol, J.P. (eds) *Tracking Environmental Change using Lake Sediments: Volume 1, Basin Analysis, Coring and Chronological Techniques*. Kluwer, Dordrecht, The Netherlands, pp. 171–203.

Appleby, P.G. and Oldfield, F. (1978) The calculation of Pb-210 dates assuming a constant rate of supply of unsupported Pb-210 to the sediment. *Catena* 5, 1–8.

Appleby, P.G., Nolan, P., Gifford, D.W., Godfrey, M.J., Oldfield, F., Anderson, N.J. and Battarbee, R.W. (1986) Pb-210 dating by low background gamma counting. *Hydrobiologia* 141, 21–27.

Appleby, P.G., Hawarth, E.Y., Michel, H., Short, D.B., Lapte, G. and Piliposian, G.T. (2003) The transport and mass balance of fallout radionuclides in Blelham Tarn, Cumbria (UK). *Journal of Paleolimnology* 29, 459–473.

Beckinsale, R.P. (1972) Clearwater erosion: its geographical and international significance. *International Geography* 2, 1244–1246.

Charlesworth, S.M. and Foster, I.D.L. (1993) Effects of urbanization on lake sedimentation: the history of two lakes in Coventry, UK – preliminary results. In: McManus, J. and Duck, R.W. (eds) *Geomorphology and Sedimentology of Lakes and Reservoirs*. Wiley, Chichester, UK, pp. 15–29.

Dearing, J.A. (1986) Core correlation and total sediment influx. In: Berglund, B. (ed.) *Handbook of Holocene Palaeoecology and Palaeohydrology*. Wiley, Chichester, UK, pp. 247–270.

Dearing, J.A. (1991) Erosion and land use. In: Berglund, B. (ed.) The cultural landscape during 6000 years in Southern Sweden. *Ecological Bulletins* 41, 283–292.

Dearing, J.A. and Foster, I.D.L. (1993) Lake sediments and geomorphological processes: some thoughts. In: McManus, J. and Duck, R.W. (eds) *Geomorphology and Sedimentology of Lakes and Reservoirs*. Wiley, Chichester, UK, pp. 5–14.

Deevey, E.S. Jr (1969) Coaxing history to conduct experiments. *BioScience* 19, 40–43.

Desloges, J.R. (1994) Varve deposition and sediment yield record at three small lakes of the southern Canadian Cordillera. *Arctic and Alpine Research* 26, 130–140.

Duck, R.W. and McManus, J. (1987) Sediment yields in lowland Scotland derived from reservoir surveys. *Transactions of the Royal Society of Edinburgh, Earth Science* 78, 369–377.

Foster, I.D.L. (1995) Lake and reservoir bottom sediments as a source of soil erosion and sediment transport data in the UK. In: Foster, I.D.L., Gurnell, A.M. and Webb, B.W. (eds) *Sediment and Water Quality in River Catchments*. Wiley, Chichester, UK, pp. 265–283.

Foster, I.D.L. and Charlesworth, S.M. (1996) Heavy metals in the hydrological cycle: trends and explanation. *Hydrological Processes* 10, 227–261.

Foster, I.D.L. and Lees, J.A. (1999a) Changing headwater suspended sediment yields in the LOIS catchments over the last century: a palaeolimnological approach. *Hydrological Processes* 13, 1137–1153.

Foster, I.D.L. and Lees, J.A. (1999b) Physical and geochemical properties of suspended sediments delivered to the headwaters of LOIS River Basins over the last 100 years: an analysis of lake and reservoir bottom-sediments. *Hydrological Processes* 13, 1067–1086.

Foster, I.D.L. and Lees, J.A. (2000) Tracers in geomorphology: theory and applications in tracing fine particulate sediments. In: Foster, I.D.L. (ed.) *Tracers in Geomorphology*. Wiley, Chichester, UK, pp. 4–20.

Foster, I.D.L. and Lees, J.A. (in press) Evidence for past erosional events from lake sediments. In: Boardman, J. and Favis-Mortlock, D. (eds) *Climate Change and Soil Erosion*. Imperial College Press, London.

Foster, I.D.L. and Walling, D.E. (1994) Using reservoir deposits to reconstruct changing sediment yields and sources in the catchment of the Old Mill reservoir, South Devon, UK over the past 50 years. *Hydrological Sciences Journal* 39, 347–368.

Foster, I.D.L., Dearing, J.A., Simpson, A., Carter, A.D. and Appleby, P.G. (1985) Lake catchment based studies of erosion and denudation in the Merevale Catchment, Warwickshire, UK. *Earth Surface Processes and Landforms* 10, 45–68.

Foster, I.D.L., Dearing, J.A. and Appleby, P.G. (1986) Historical trends in catchment sediment yields: a case study in reconstruction from lake sediment records in Warwickshire, UK. *Hydrological Sciences Journal* 31, 427–443.

Foster, I.D.L., Dearing, J.A., Grew, R.G. and Orend, K. (1990) The sedimentary data base: an appraisal of lake and reservoir sediment based studies of sediment yield. In: Walling, D.E., Yair, A. and Berkowicz, S. (eds) *Erosion, Transport and Deposition Processes.* IAHS Publication 189. IAHS Press, Wallingford, UK, pp. 15–43.

Foster, I.D.L., Owens, P.N. and Walling, D.E. (1996) Sediment yields and sediment delivery in the catchments of Slapton Lower Ley, South Devon, UK. *Field Studies* 8, 629–661.

Foster, I.D.L., Lees, J.A., Owens, P.N. and Walling, D.E. (1998) Mineral magnetic characterisation of sediment sources from an analysis of lake and floodplain sediments in the catchments of the Old Mill reservoir and Slapton Ley, South Devon, UK. *Earth Surface Processes and Landforms* 23, 685–703.

Foster, I.D.L., Lees, J.A., Jones, A.R., Chapman, A.S., Turner, S.E. and Hodgkinson, R.A. (2002) The possible role of agricultural land drains in sediment delivery to a small reservoir, Worcestershire UK; a multiparameter tracing study. In: Dyer, F.J., Thoms, M.C. and Olley, J.M. (ed.) *The Structure, Function and Management Implications of Fluvial Sedimentary Systems*. IAHS Publication 276. IAHS Press, Wallingford, UK, pp. 433–442.

Foster, I.D.L., Chapman, A.S., Hodgkinson, R.A., Jones, A.R., Lees, J.A., Turner, S.E. and Scott, M. (2003) Changing suspended sediment and particulate loads and pathways in underdrained lowland agricultural catchments, Herefordshire and Worcestershire, UK. *Hydrobiologia* 494, 119–126.

Gilbert, R. and Desloges, J.R. (1987) Sediments of ice-dammed, self draining Ape Lake, British Columbia. *Canadian Journal of Earth Science* 24, 1735–1747.

Gilmore, G. and Hemingway, J. (1995) *Practical Gamma Spectrometry*. Wiley, Chichester, UK.

Goldberg, E.D. (1963) Geochronology with Pb^{210}. In: *Radioactive Dating (Proceedings of the Athens Symposium).* International Atomic Energy Authority, Vienna, pp. 121–131.

Gruszowski, K.E., Foster, I.D.L., Lees, J.A. and Charlesworth, S.M. (2003) Sediment sources and transport pathways in a rural catchment, Herefordshire, UK. *Hydrological Processes* 17, 2665–2681.

Håkanson, L. and Jansson, M. (1983) *Principles of Lake Sedimentology*. Springer-Verlag, Berlin, Germany.

Hutchinson, G.E. (1975) *A Treatise on Limnology. Part 1. Geography and Physics of Lakes*. Wiley, New York.

Johnes, P.J. and Heathwaite, A.L. (1997) Modelling the impact of land use change on water quality in agricultural catchments. *Hydrological Processes* 11, 269–286.

Johnes, P.J. and Wilson, H.M. (1996) The limnology of Slapton Ley. *Field Studies* 8, 585–612.

Joshi, S.R. and Shukla, B.S. (1991) *Ab Initio* derivation of formulations for Pb-210 dating of sediments. *Journal of Radioanalytical and Nuclear Chemistry, Articles* 148, 73–79.

Lahlou, A. (1996) Environmental and socio-economic impacts of erosion and sedimentation in North Africa. In Walling, D.E. and Webb, B.W. (eds) *Erosion and Sediment Yield: Global and Regional Perspectives*. IAHS Publication 236. IAHS Press, Wallingford, UK, pp. 491–500.

Likens, G.E. and Davis, M.B. (1975) Postglacial history of Mirror Lake, New Hampshire, USA: an initial report. *Verhandlungen Internationale Vereiningen Limnologie* 19, 982–993.

McCully, P. (1996) *Silenced Rivers. The Ecology and Politics of Large Dams*. Zed Books, London.

O'Hara, S., Street-Perrott, F.A. and Burt, T.P. (1993) Accelerated soil erosion around a Mexican highland lake caused by prehispanic agriculture. *Nature* 362, 48–51.

Oldfield, F. and Appleby, P.G. (1984) Empirical testing of ^{210}Pb dating models for lake sediments. In: Hawarth, E.Y. and Lund, J.W.G. (eds) *Lake Sediments and Environmental History*. Leicester University Press, Leicester, UK, pp. 93–124.

Oldfield, F., Worsley, A.T. and Appleby, P.G. (1985) Evidence from lake sediments for recent erosion rates in the highlands of Papua New Guinea. In: Douglas, I. and Spencer, T. (eds) *Environmental Change and Tropical Geomorphology*. Allen and Unwin, London, pp. 185–196.

Olsson, I.U. (1986) Radiometric dating. In: Berglund, B. (ed.) *Handbook of Holocene Palaeoecology and Palaeohydrology*. Wiley, Chichester, UK, pp. 273–312.
Page, M.J. and Trustrum, N.A. (1997) A late Holocene lake sediment record of the erosion response to land use change in a steepland catchment, New Zealand. *Zeitschrift fur Geomorphologie NF* 41, 369–392.
Paine, J.L., Rowan, J.S. and Werritty, A. (2002) Reconstructing historic floods from embanked floodplains: a case study of the River Tay in Scotland. In: Dyer, F.J., Thoms, M.C. and Olley, J.M. (ed.) *The Structure, Function and Management Implications of Fluvial Sedimentary Systems*. IAHS Publication 276. IAHS Press, Wallingford, UK, pp. 211–218.
Pennington, W., Cambray, R.S. and Fisher, E.M. (1973) Observations on lake sediments using fallout ^{137}Cs as a tracer. *Nature* 242, 324–326.
Petts, G.E. (1984) *Impounded Rivers: Perspectives for Ecological Management*. Wiley, Chichester, UK.
Ritchie, J.C., McHenry, J.R. and Gill, A.C. (1973) Dating recent reservoir sediments. *Limnology and Oceanography* 18, 254–263.
Robbins, J.A. and Herche, L.R. (1993) Models and uncertainty in ^{210}Pb dating of sediments. *Verhandlungen Internationale Vereiningen Limnologie* 25, 217–222.
Saarnisto, M. (1986) Annually laminated lake sediments. In: Berglund, B. (ed.) *Handbook of Holocene Palaeoecology and Palaeohydrology*. Wiley, Chichester, UK, pp. 343–370.
Schumm, S.A. (1977) *The Fluvial System*. Wiley, New York.
Sheail, J. (1988) River regulation in the United Kingdom: an historical perspective. *Regulated Rivers, Research and Management* 2, 221–232.
Slattery, M.C., Walden, J. and Burt, T.P. (2000) Use of mineral magnetic measurements to fingerprint suspended sediment sources: results from a linear mixing model. In: Foster, I.D.L. (ed.) *Tracers in Geomorphology*. Wiley, Chichester, UK, pp. 309–322.
Smith, F.B. and Clark, M.J. (1989) *The Transport and Deposition of Airborne Debris from the Chernobyl Nuclear Power Plant Accident with Special Emphasis on the Consequences to the United Kingdom*. Meteorological Office, Scientific Paper No. 42. HMSO, London.
Smith, I. and Lyle, A. (1979) *Distribution of Freshwaters in Great Britain*. Institute of Terrestrial Ecology, Edinburgh, UK.
Smith, N. (1971) *A History of Dams*. Peter Davies, London.
Smol, J.P. (2002) *Pollution of Lakes and Reservoirs, a Palaeolimnological Perspective*. Arnold, London.
Vörösmarty, C.J., Sharma, K.P., Balázs, M.F., Copeland, A.H., Holden, J., Marble J. and Lough, J.A. (1997) The storage and aging of continental runoff in large reservoir systems of the world. *Ambio* 26, 210–219.
Walling, D.E., Owens, P.N., Foster, I.D.L. and Lees, J.A. (2003) Changes in the sediment dynamics of the Ouse and Tweed basins in the UK, over the last 100–150 years. *Hydrological Processes* 17, 3245–3269.
Ward, R.C. (1981) River systems and river regimes. In: Lewin, J. (ed.) *British Rivers*. Allen and Unwin, London, pp. 1–33.
White, P., Labadz, J.C. and Butcher, D.P. (1996) Sediment yield estimates from reservoir studies: an appraisal of variability in the southern Pennines of the UK. In: Walling, D.E. and Webb, B.W. (eds) *Erosion and Sediment Yield: Global and Regional Perspectives*. IAHS Publication 236. IAHS Press, Wallingford, UK, pp. 491–500.
Wilby, R.L., Dalgleish, H.Y. and Foster, I.D.L. (1997) The impact of weather patterns on historic and contemporary catchment sediment yields. *Earth Surface Processes and Landforms* 22, 353–363.
Winter, L.T., Foster, I.D.L., Charlesworth, S.M. and Lees, J.A. (2001) Floodplain lakes as sinks for sediment-associated contaminants – a new source of proxy hydrological data? *Science of the Total Environment* 266, 187–194.
Woodward, J.C. and Foster, I.D.L. (1997) Erosion and suspended sediment transfer in river catchments. *Geography* 82, 353–376.
Zolitschka, B. (1998) A 14,000 year sediment yield record from Western Germany based on annually laminated lake sediments. *Geomorphology* 22, 1–17.

III

Modelling

Sediment flume, Chinese Academy of Sciences, Chengdu, China (photo: P.N. Owens).

13 Can Soil Erosion be Predicted?

M.A. Nearing
USDA-ARS Southwest Watershed Research Center, USA

Introduction

The purpose of this chapter is to present a realistic overview of soil erosion modelling capabilities and limitations. The data and model applications will focus on hillslope-scale processes, but have obvious implications for sediment generation and sediment yield across larger scales as well. There are three major points upon which the paper will focus. The first has to do with variability of erosion in nature and its implications for erosion prediction. There have been many studies of soil erosion model application and validation using measured erosion data, but it is difficult to get a general or broad perspective and quantification of variability until you have relatively large data sets to work with, which is rare. It is also difficult to address variability unless one can replicate experiments, which is a challenge for watersheds. Here we will focus on plot-scale erosion and variability associated with hillslope erosion. The basic message is that erosion in nature at the hillslope scale is quite variable, and that variability has major implications for models and prediction. On the other hand, patterns are evident, and we will discuss those patterns that are observable.

The second issue has to do with the importance of continuous simulation for erosion modelling. By continuous simulation, we refer to a model that calculates erosion through the year and over many years. Most importantly, it is a model that has the capability to update the parameters that define the state of the system as it influences erosion resistance or susceptibility, such as standing plant biomass that acts as soil cover, plant residues in contact with the surface, soil moisture, soil consolidation, etc. We will argue that one cannot effectively evaluate land use scenarios without a reliable form of continuous simulation. Lastly, we will argue that soil erosion models can be used as effective tools for many purposes, as long as they are used with the understanding of their capabilities and limitations.

Materials and Methods

In order to evaluate variability of measured soil erosion and expectations for model predictions we introduce the concept of the physical model of soil erosion. One can argue that the best possible model for erosion from a given plot will be the physical model, that is, a replicate of the plot on the same hillslope with the same slope, soil, land use and weather input. By 'the same', it is meant that we would characterize the plots the same for the purposes of modelling the erosion. Another way of looking at it is that the measured erosion values from the two plots would be samples from the same treatment distribution.

A large number of experimental natural rainfall–erosion plot data were used for the analyses presented. For each section of the information presented, data from some or all of the following locations in the USA were

 Soil Erosion and Sediment Redistribution in River Catchments (eds P.N. Owens and A.J. Collins)

used: Holly Springs, MS; Madison, SD; Morris, MN; Presque Isle, ME; Watkinsville, GA; Bethany, MO; Guthrie, OK; Castana, IA; Tifton, GA; Pendleton, OR; Geneva, NY; and Kingdom City, MO. The experimental erosion plots used here represent a wide range of cropping systems, including fallow, cotton, grass–maize–oats, lucerne, wheat–clover–cotton, bermuda grass, red clover, winter rye, autumn-tilled maize, conservation-tilled maize, no-till maize, oats, no-till maize and soybeans, no-till soybeans, conventional-tilled soybeans, and potato.

Results and Discussion

Data variability and modelling implications

Using the data from replicated erosion plots, we were able to first of all obtain an idea of the variance associated with the erosion data from the plots. The details of the methodology that was used to generate the graph (Fig. 13.1) of the coefficient of variation in the measured data, expressed as a fraction, versus the magnitude of soil loss measured on the plots is presented in Nearing *et al.* (1999). There are several key points to be made. One is that the level of variance in the measured data is high in general. At a measured soil loss level of 0.1 kg/m^2, which translates to 1 t/ha, the coefficient of variation is approximately 1, or in other words, 100%.

The second obvious point here is that the level of variance between plots was dependent on the magnitude of soil erosion that was measured (Fig. 13.1). At low erosion levels, variance was quite large. As erosion level increases, we see the coefficient of variation reduced to tens of per cent. Implicit in this, but not stated in the graph, is that other system parameters such as the geographic location, type of soil and crop type, did not enter into the picture for explaining the differences in variance found in the data. Variance was, as far as was discernable from the data, a function only of the magnitude of soil loss measured.

What is not clear in this graph (Fig. 13.1) is that the *x*-axis represents measured soil loss for the plots over three different time scales: events, individual year values and average annual erosion values. In other words, the variance depended on the magnitude of soil erosion measured, but it did not matter over what time period the erosion was measured. An overlay of the same graph for event data exactly overlaps the same graph for annual average erosion, though event data values on average were less than the annual data values.

These results have significant implications for modelling erosion. One immediate implication of data variability is that there is a limit in terms of accuracy for models. For example, using the data from replicated plots, Fig. 13.2

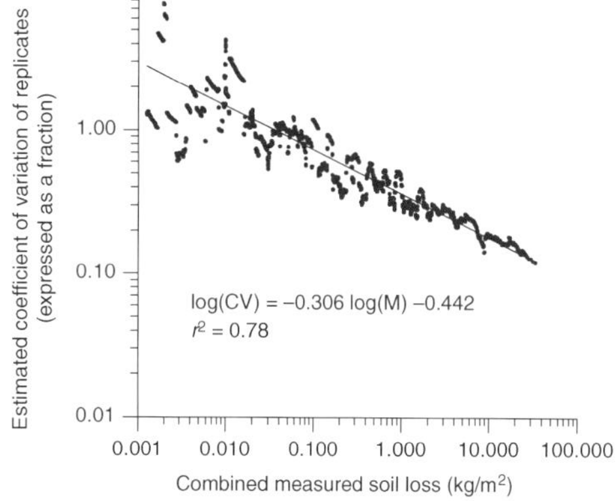

Fig. 13.1. Coefficients of variation between replicated plots as a function of magnitude of measured soil loss (from Nearing *et al.*, 1999).

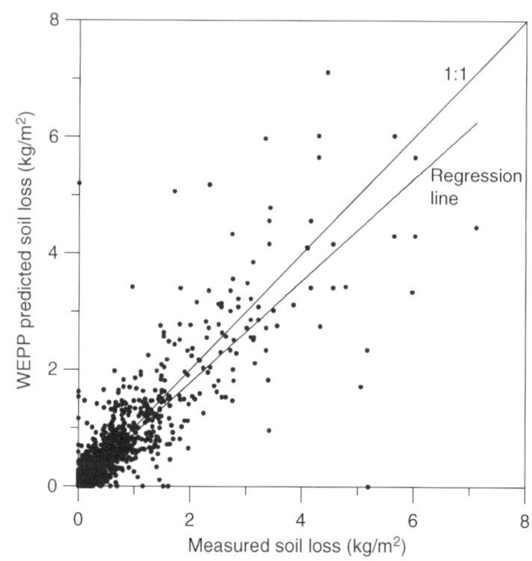

Fig. 13.2. Measured vs predicted soil loss for the physical model as represented by the replicate plot (from Nearing, 1998).

represents prediction accuracy for the best-case 'physical model' (Nearing, 1998). All of the plot data were paired by replication, and event soil loss was plotted with one plot assigned as the treatment, and the other plot as the physical model. The level of fit obtained, in this case an r^2 of 0.77, can be considered as a benchmark, or 'best-case' prediction scenario. One cannot reasonably expect a simulation model to fit better than this. Using the information from the previous graph, the coefficient of variation for measured data in the range of 1–5 kg/m^2 is of the order of 30–50%. If the measured data for the physical model in this graph were lesser in magnitude, the fit would reduce accordingly.

Another implication of variance in data is that it is much harder to predict low erosion rates than to predict high ones. So, for example, even though we know that the relationship between variance and soil erosion magnitude does not appear to depend on whether the erosion is from single storms or from long-term averages, erosion values on average will tend to be lower for individual storms, and erosion predictions will tend to be poorer.

Figure 13.3 shows single storm predictions using the Water Erosion Prediction Project (WEPP) model (Zhang *et al.*, 1996). WEPP is a process-based, continuous simulation model for soil erosion. It contains a model for predicting soil erosion for daily storm events, but also auxiliary models for plant growth and canopy cover; residue production, decay and burial; tillage; soil consolidation; soil moisture; infiltration; runoff; and many other system dynamics. Here Zhang *et al.* (1996) have used the model, uncalibrated, for 2119 storm events, and received a fit of approximately 0.4, which compares to a fit of 0.77 for the physical model results. One way of interpreting this is that the WEPP model is predicting the events approximately half as well as would our ideal physical model of the replicate plot. There obviously may be some room for improvement here in the WEPP predictions, but here at least we have a more realistic idea of where we stand and how much we could improve if our model was 'perfect'. It is important to stress that these predictions were using the model in the uncalibrated state. One finds very few published evaluations of uncalibrated erosion models, though this is how we most often need to apply them for solving problems.

When looking at the predictions of the annual average erosion rates using the same data, things look much better (Fig. 13.4) (data from Zhang *et al.*, 1996). The fit in this case was much higher, at an r^2 of 0.77. This is a result of the fact that the random variation inherent in the plot data tends to smooth out when we look at higher erosion rates, which also often happens to be correlated to measuring over longer time scales. But the key point is that it is not, apparently, the time of record itself that governs the level of accuracy that can be and is achieved, but rather the magnitude of erosion measured. The effect is correspondent to Fig. 13.1, wherein the variance in the measured data decreased as a function of measured soil erosion magnitude.

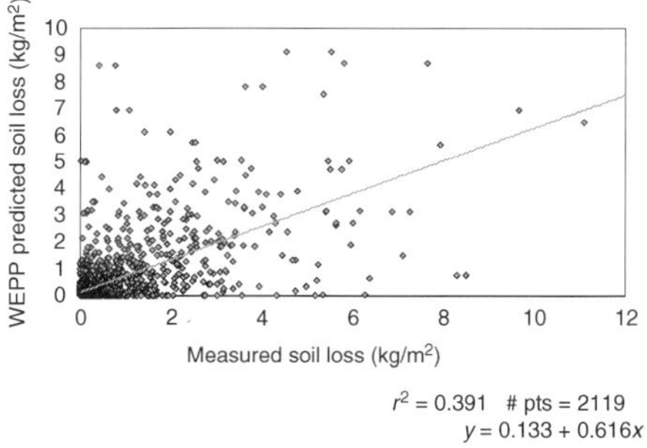

Fig. 13.3. Measured vs predicted soil loss for daily results of the WEPP model for 2119 storm events (data from Zhang *et al.*, 1996).

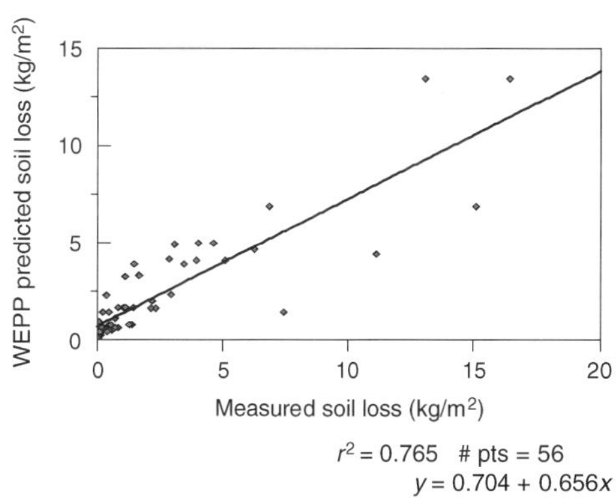

Fig. 13.4. Measured vs predicted soil loss for average annual results of WEPP for 56 natural runoff plots (data from Zhang *et al.*, 1996).

Figure 13.5 shows similar data for average annual erosion using the Universal Soil Loss Equation (USLE). The accuracy of the results is approximately similar as that for WEPP. According to its developers, the USLE was never intended to be used as an event model, but only for predicting annual averages. However, if one looks at the structure of the model one sees that the average annual erosivity factor, R, is simply an average annual summation of individual storm erosivities, or EI_{30}s. Hence there is no fundamental reason why the USLE could not be used as an event model. The probable reason that the USLE was designated to be used only for annual averages was that the developers had access to enough data to know that predicting erosion for individual events, particularly with an uncalibrated model on a routine basis, was simply not possible with any reasonable level of accuracy. The individual event predictions using the USLE probably would not differ in accuracy much from the WEPP predictions for the same data.

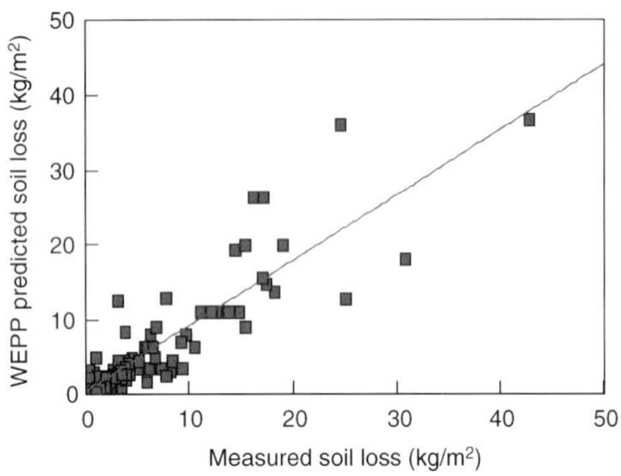

Fig. 13.5. Measured vs predicted soil loss (kg/m²) for 208 plot-years of average annual results of the USLE for natural runoff plots (data from Risse *et al.*, 1993).

Another approach that can be taken to the problem of validation, application and calibration of models is the use of the event soil loss frequency distributions. Here (Fig. 13.6) we have a frequency distribution of measured and predicted soil erosion plotted in terms of recurrence interval (Baffaut *et al.*, 1998). Baffaut *et al.* (1998) found that even though the fit for measured versus predicted events is usually relatively low, such as the r^2 of 0.39 shown in Fig. 13.3, the frequency distribution of soil loss may compare well with the measured data. They also showed that the frequency distribution of events can be used for calibration purposes. In general, the lower end of the frequency curve, or the small events, tends to be dominated by splash erosion. Thus the lower part of the curve can be used to calibrate the data for the splash or interrill parameters. The upper end of the curve is dominated by rill erosion, and correspondently the rill erodibility

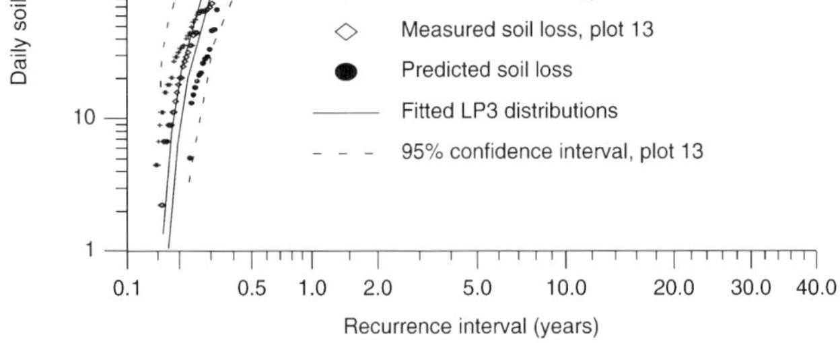

Fig. 13.6. Comparison of distributions of measured and predicted daily soil losses on the fallow plots in Morris, Minnesota, USA, from 1962 to 1971 (from Baffaut *et al.*, 1998).

parameters can be calibrated on that portion of the curve.

Continuous simulation models

Much of the discussion above, including the idea of using long-term modelling averages such as average annual and frequency distributions of individual events, relies implicitly on the idea of utilizing continuous simulation models for predicting erosion. A large number of the models that are being used are not continuous simulation, but rather single event models. By a continuous simulation model, we refer to a model that calculates erosion through the year and over many years. Most importantly, it is a model that has the capability to update the parameters that define the state of the system as it influences erosion resistance or susceptibility, such as standing plant biomass that acts as soil cover, plant residues in contact with the surface, soil moisture, soil consolidation, etc.

Why is this so important? The issue revolves around the temporal variability in the system characteristics that influence so dramatically the erosion rates for a given storm event. One can think of the erosional response as being a function of the overlap of two distributions, the driving force of rainfall (in this case) and the state of the system in terms of its resistance to the driving forces (Fig. 13.7).

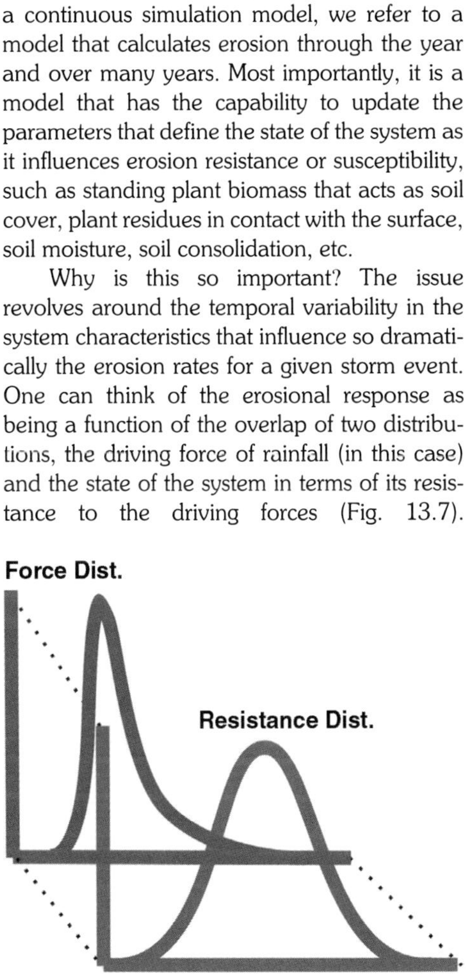

Fig. 13.7. Schematic diagram representing the overlap of two distributions, one representing the driving force of erosion (e.g. rainfall) and the second representing the system resistance to erosion.

Obviously, the reality is very complex, and Fig. 13.7 simplifies the reality to a conceptual level. For example, there is no guarantee that the resistance distribution itself is independent of the driving force distribution.

For a given set of force and resistance distributions we can expect a specific erosion response distribution. If we look, for example, at a 50-year return frequency storm occurring on a field that has recently been planted, the impact of that event will be a relatively large erosion response. For the same storm on the same crop rotation except offset by 1 year, thus occurring at a time during the rotation when the system resistance is very high, the erosion response may be very small. In this case the first field might be devastated by this storm, where in the second field there was no visible sign of erosion at all, even though both fields are under the same cropping system.

One possible way to deal with the questions that continuous simulation attempts to address, without actually doing long-term simulation, is to run a single storm model on a distribution of individual events. The limitation is that this does not in itself address the fact that the erosional response is a complex overlap of the two dynamic distributions, rather than just the storm distribution alone. As such, attempts by the author at using the WEPP model in this way have indicated that this method was not sufficient for characterizing erosional response differences among land-used treatments or for producing accurate long-term erosion estimates. The exception to this might be in a special circumstance where the erosivity is highly packed into a short (e.g. 2-month) window, and where the cover and soil during that time do not change much. Such may be the case, for example, in many semiarid environments.

With the WEPP model we also attempted to develop a technique to use 3 representative years of continuous simulation to obtain good estimates of long-term averages of erosion. The idea was to select representative wet, dry and average precipitation years that, when used in the WEPP model, would result in erosion estimates that mimicked long-term average trends. We found that one might be able to do this, as long as one was dealing with nearly identical systems, such as summer crops with similar planting and harvest dates. If one changed to winter crops, such as wheat for example, the process no longer worked. One had to choose a different representative 3 years. Relative erosion rates from the long-term averages compared to the 3-year averages were not consistent internally between management systems, which were what we were trying to differentiate.

The reason for our lack of success in the above-mentioned problem is evident in Fig. 13.8. Even with a continuous simulation model one has to deal with extreme variability in the erosion predictions. Baffaut *et al.* (1996) conducted a simulation study using the WEPP model to determine how many years of simulation were necessary in order to obtain a stable long-term erosion estimate. The predicted erosion rate was considered to be stable when its value was within 10% of the 200-year average soil loss, and remained within that interval for all subsequent years. The results (Fig. 13.8) indicated a need for periods of simulation approaching 150 years in order to reach stability. In general, the simulation time tended to decrease as a function of the magnitude of measured soil loss, but as one can see from the graph the relationship was a weak one. The simulations were primarily done using fallow conditions, and in that sense probably reflect conservative values since erosion magnitude for cropped conditions would tend to be less, and hence variability greater.

It is useful to mention again that WEPP model predictions of event distributions were shown to mimic distributions of measured erosion quite well (Baffaut *et al.*, 1998). We do not believe that the long simulation times reported here are a function of the instability of the model. If anything, the model may not reflect the variability that actually could exist over decadal time periods. These types of results of model simulations leave one feeling discouraged about the possibility of constructing a method of using a single storm model in conjunction with rainfall frequency distributions to obtain accurate long-term erosion estimates or for quantifying erosion differences between land uses.

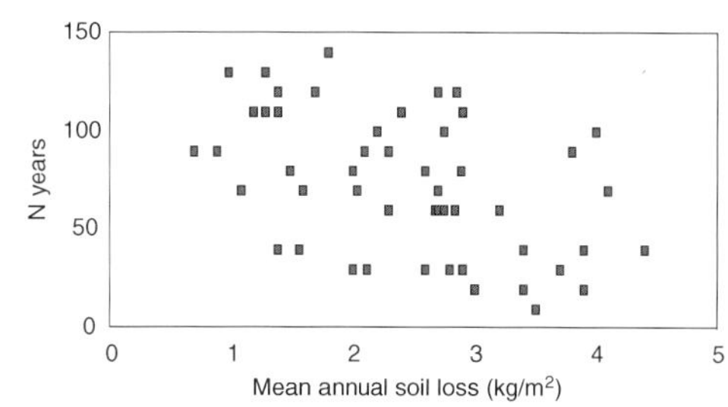

Fig. 13.8. Minimum simulation times required to reach stable long-term average predictions of soil loss using the WEPP model applied to fallow conditions (Baffaut *et al.*, 1996).

These results also leave one questioning the use of a 'design storm' for erosion prediction. Land use systems cannot be evaluated with the concept of the design storm. One could take the 'worst case' scenario of the designated return storm frequency modelled at the least resistant time of the year for designing erosion control structures, for example. However, the only way to determine the probability of occurrence of that magnitude of an erosion event occurring within a given return period would be to run the continuous simulation model and determine the predicted frequency distribution curves for erosion. In other words, using the worst-case-design-storm method one will predict an erosion rate for the storm. But what is the probability that that level of erosion might occur in any given year? A slightly larger storm at a slightly different time of year, and hence theoretically more resistant condition of the system, would produce the same level of erosion. On the other hand, with a continuous simulation model one can design structures and conservation practices for a design erosion event. If one runs the simulation model for 200 years, then it is possible to pick from the event distribution data the 10- or 20-year return frequency erosion event.

Another way of summarizing this is that with a single storm model one can plan for a certain return frequency storm precipitation for the system, but only in a single specific system state. With a continuous simulation model one can plan for a design, return frequency erosion event. Those are two very different things, and the former is not useful relative to erosion.

Useful applications of erosion models

To this point we have discussed the limitations and the problems of natural variability in erosion and long simulations required for obtaining stable erosion values. Erosion models can, none the less, act as very valuable tools for a variety of purposes.

1. They can help the land owner or manager choose suitable conservation practices, because they are able to assess relative effects of land use even in individual hillslope cases where the accuracy of any given prediction might be uncertain.
2. They can be used to make broad-scale erosion surveys in order to understand the scope of the problem over a region and to track changes in erosion over time, because if the model is predicting the mean erosion well for a given land use, then it will reflect the mean of the population of erosion values for land use treatments.
3. Models can be used to regulate activities on the land for purposes of conservation compliance, because they can provide a consistent and fair evaluation system to compare agricultural fields. The model might not give the exact quantification of erosion on every field for every year, but in the long term the predictions are fair and reasonably accurate.
4. They can be used to estimate long-term loadings to streams and other water bodies, because as the time period increases, the accuracy increases.
5. If used properly, they are useful as storm-response design tools. But in this case the storm design must be done within the context of a continuous simulation model run over a sufficiently long period of time to obtain a clear quantification of the size of the erosion event for a specified return frequency.

All of the above applications require explicitly the use of a continuous simulation model. One cannot accurately assess changes or differences in land use scenarios or conservation practices without continuous simulation. USLE and RUSLE are included in this class of models, since erosion is calculated based on time variations in the cropping and erosivity factors. However, one cannot use USLE or the Revised USLE (RUSLE) for the last two applications (4 and 5). They cannot be used for estimating long-term off-site loadings, because they do not include concentrated flow routing for off-site sediment yields, and they cannot be used for determining design storms, since they are not event models. All of the above applications also implicitly take into account the issue of variability that we talked about earlier. Simulations must be made for long time periods in order to make reliable quantitative assessment.

Where mistakes are most often made in the application of models is when we do not recognize the inability of models to accurately predict erosion at low levels, such as for events or even for a few years of erosion in cases where erosion rates are low and events infrequent. We really can't accurately measure erosion as a function of treatments when we try to do it over short time periods. Natural variation is huge, and our model variation is even greater. Secondly, it is not adequate to use single storm models to assess land use treatments or to define design erosion events. Single storm models do not have the capability to function in this capacity.

This chapter was limited to the discussion of hillslope-scale erosion in a simple context of sheet and rill erosion. But even at hillslope scales the situation can be much more complex. There is enormous complexity in the many processes that take place in a real landscape, as well as complexity of surface morphology and the interactions of the morphology with processes, which our models do not attempt to take into account. Along the same line, we are also finding that the basic concepts of our process-based models appear not to function in natural areas and rangelands. The entire concept of rill and interrill erosion breaks down in these areas.

Summary

Variability in erosion data from plots is high. In an analysis of several thousands of data points it was determined (Nearing *et al.*, 1999) that the coefficient of variation (CV) among replicates was a function of the level of erosion measured: CV was 100% at 1 t/ha and CV was 35% at 10 t/ha of measured erosion. The relationship between CV and erosion magnitude was independent of time period of collection, land use, soil, geographic location or slope. Because of the natural uncertainty in measured erosion data there is a limit in terms of accuracy for erosion models. For erosion data that ranges within the order of 0–75 t/ha, an r^2 of approximately

0.77 is as high as can be expected (Nearing, 1998). It is much harder to predict low erosion rates than to predict large ones, and for predicting data that is less than 75 t/ha, an r^2 of less than 0.77 is expected. It is easier to duplicate measured event distributions than to predict individual events on a one-to-one basis (Baffaut *et al.*, 1998).

Continuous simulation models are necessary to account for the complex overlap of temporally and spatially heterogeneous distributions of both the driving force of erosion (e.g. rainfall) and the system resistance (e.g. erodibility). Because of this complex interplay between these distributions, single storm models cannot be used to accurately predict erosion response as a function of land use, single storm models cannot be used to define design-level erosion events for specified return frequencies, and simulation times with continuous models of the order of 150 years are needed to obtain stable long-term erosion estimates (Baffaut *et al.*, 1996).

None the less, erosion models can serve many valuable purposes. They can be used for many applications related to conservation planning and engineering design if the natural data and model uncertainties are recognized, understood and accounted for, and if a continuous simulation model is used.

References

Baffaut, C., Nearing, M.A. and Nicks, A.D. (1996) Impact of climate parameters on soil erosion using CLIGEN and WEPP. *Transactions of the American Society of Agricultural Engineers* 39, 447–457.

Baffaut, C., Nearing, M.A. and Govers, G. (1998) Statistical distributions of daily soil loss from natural runoff plots and WEPP model simulations. *Soil Science Society of America Journal* 62, 756–763.

Nearing, M.A. (1998) Why soil erosion models over-predict small soil losses and under-predict large soil losses. *Catena* 32, 15–22.

Nearing, M.A., Govers, G. and Norton, L.D. (1999) Variability in soil erosion data from replicated plots. *Soil Science Society of America Journal* 63, 1829–1835.

Risse, L.M., Nearing, M.A., Nicks, A.D. and Laflen, J.M. (1993) Assessment of error in the Universal Soil Loss Equation. *Soil Science Society of America Journal* 57, 825–833.

Zhang, X.C., Nearing, M.A., Risse, L.M. and McGregor, K.C. (1996) Evaluation of runoff and soil loss predictions using natural runoff plot data. *Transactions of the American Society of Agricultural Engineers* 39, 855–863.

14 Erodibility Assessment in Dynamic Event-based Erosion Models

N.J. Kuhn
University of Exeter, UK

Erodibility Assessment and Global Environmental Change

Climate and land use change potentially accelerate soil erosion in the 21st century, causing land degradation and off-site sedimentation (IPCC, 2001). Intensification of farming will also increase off-site pollution by eroded soil enriched in organic contaminants, disease vectors, pesticides and fertilizers (Lal, 1990; De Walle and Sevenster, 1998; Van den Akker *et al.*, 2003). The significance of erosion for pollution and soil degradation is stressed in the European Union Water Framework Directive (EC, 2000) and the UK Soil Action Plan (DEFRA, 2003). The mitigation of future negative impacts of environmental change requires effective dynamic event-based erosion prediction (Valentin, 1998). Over the past two decades, several models, for example WEPP, LISEM, EUROSEM, have been developed to achieve process-based prediction of soil erosion (Jetten *et al.*, 1999). However, the interaction between rainfall, running water and soil surface leading to erosion is highly complex and the understanding of physical processes that determine short-term changes of soil resistance to erosion (erodibility) is still limited (Owoputi and Stolte, 1995; Bryan, 2000). Even sophisticated process-based erosion models often use only one erodibility factor for interrill and rill erosion, respectively, while the factors related to the forces that drive erosion, such as rainfall, runoff and topography, are clearly distinguished. Furthermore, no process-based model describing short-term changes of erodibility has been developed. With the exception of WEPP, most erosion models rely on calibration. Due to the limited understanding of soil–climate interaction the selection of calibration events is difficult and limits effective prediction (Quinton, 1997). WEPP contains a subroutine to simulate interrill and rill erodibility, based on days after tillage, generating daily erodibility values (Lane and Nearing, 1989). This empirical approach ignores the effect of varying seasonal patterns of rainfall and drying on soil resistance to erosion and does not provide information on within-storm changes of erodibility, seriously limiting the ability to predict erosion for individual storm events (Jetten *et al.*, 1999; Kuhn *et al.*, 2003; Kuhn and Bryan, 2004).

Improving dynamic process-based erosion prediction requires the integration of a wide range of processes that affect soil resistance to erosion, such as sealing, crusting, compaction, erosion itself, and drying-induced changes of relevant soil properties such as shear strength and aggregate stability. In the light of the complex interaction between rainfall, runoff and soil that determines erodibility, one critical question has to be addressed first: can erodibility be expressed in one factor, or does effective dynamic event-based erosion prediction require a process-based treatment of soil resistance to erosion? Using two examples, the potential for developing a process-based model for the

 Soil Erosion and Sediment Redistribution in River Catchments (eds P.N. Owens and A.J. Collins)

assessment of short-term changes of soil resistance to erosion, using a one-erodibility approach, will be discussed.

Interrill and Rill Erosion Studies

Analysis of the effects of rainfall, erosion and drying on soil properties relevant for erodibility requires detailed monitoring and a high degree of control over rainfall and soil moisture. Therefore, laboratory experiments simulating different rainfall and drying sequences were carried out in the Soil Erosion Laboratory, University of Toronto. The laboratory is equipped with a state-of-the-art rainfall simulator and provides relative control over temperature and humidity, enabling approximately similar rates of drying and initial soil moisture. Erosion tests involved simulated rainfall on tilted (5°) interrill flumes (1 m long × 0.2 m wide) and a 15-m rill erosion flume (15 m long × 0.8 m wide). In each flume, a 0.2 m deep soil layer was packed to a density of 1.2 g/cm^3, resting on a perforated floor covered by fine cloth. This arrangement permitted vertical drainage and inhibited the development of a perched water table. In the 15-m flume, a furrow was moulded along the middle of the flume, resembling an agricultural field after tillage, to enhance runoff concentration and flow competence. Artificial rainfall intensities ranged from 60 to 75 mm/h, with median drop diameter of 2 mm and a kinetic energy of 0.32 MJ/ha/mm. Sequences of storms, interrupted by drainage to field capacity or drying to air-dry conditions, were designed to observe the effect of repeated wetting and drying on soil resistance to interrill and rill erosion. During storms water and soil discharge were collected at terminal weirs and soil moisture and properties relevant for erosion were measured in regular intervals during and between storms. For a detailed description of the rainfall simulation facilities see Bryan (1990), Kuhn *et al.* (2003) and Kuhn and Bryan (2004).

The results observed on two smectite-rich soils from NE Mexico and a clay-loam from southern Ontario, Canada, are discussed in this study. The Mexican soils, a Kastanozem and a Vertisol, were sampled in Linares, Nuevo Leon, and are common in the piedmont of the Sierra Madre Oriental. They differ in texture, organic matter content, aggregate stability and land use (Table 14.1). The Vertisol is used for extensive grazing and the Kastanozem under intensive crop farming. Both soils are vulnerable to sealing, have a strong tendency for aggregate formation, and a high shrink–swell capacity. The Schomberg clay loam is common on the Oak Ridges Moraine in Southern Ontario. On sealing soils such as the Schomberg, the sloping terrain creates an environment prone to rill erosion when used for crop farming (Hoffman and Richards, 1955).

Interrill Erosion of Two Smectite-rich Soils from NE Mexico

Farmland Kastanozem

Samples of the Kastanozem (Table 14.2) were subjected to two different rainfall tests, each totalling 180 mm of rainfall. The first test consisted of a 3-h storm with an intensity of 60 mm/h. During the second test rainfall was interrupted after 60 and 120 min and the soil was dried to air-dry conditions before the next storm. Initial and final soil moisture, shear strength and percentage of water-stable aggregates after wet-sieving (WSA) were measured to assess the effect of rainfall, erosion and drying on soil condition. WSA, used as an index for interrill erodibility (Bryan, 1968), was measured by wet-sieving of aggregates or crust fragments. Shear strength, reflecting changes

Table 14.1. Properties of soils used for interrill and rill erodibility tests.

Soil	FAO type	% Sand	% Silt	% Clay	% C_{org}	% WSA > 0.25 mm
Linares cropland	Calcic Kastanozem	6.1	49.9	44	1.5	90.8
Linares grassland	Calcic Vertisol	11.6	61.1	28.2	3.0	81

Table 14.2. Properties of the Kastanozem during experimental storms.

Time (min)	RRR[a]	Shear strength (kPa)	% WSA > 0.5 mm
Initial	1	0.07	75.4
60	0.44	0.17/< 0.04[b]	33.2
120	0.32	0.14/< 0.04[b]	33.6
180	0.2/0.24[c]	0.14/0.15[c]	37/29.7[c]

[a]Relative Random Roughness, random roughness divided by initial random roughness.
[b]Washed-in/washed-out layer.
[c]Three 60-min storms/180-min storm.

Fig. 14.1. Kastanozem with washed-out layer after 60 mm (left), and without after 180 mm of rainfall (centre), note the clearing of flow pathways after 60 min, indicated by linear pattern of exposed washed-in layer; right: Vertisol after 180 mm of rainfall (size of interrill areas 20 by 20 cm).

in seal cohesion (Slattery and Bryan, 1992), was measured by a custom-designed shear vane, penetrating only 2 mm into the soil (Kuhn and Bryan, 2004). Surface soil condition was recorded by digital imaging before and after each rainfall to document changes in seal type relevant for erosion.

A double-layered seal which consisted of a washed-out layer of loose aggregates of up to 3 mm in diameter at the surface, and a cohesive washed-in layer of 1–2 mm in thickness underneath (West *et al.*, 1992), formed within 60 min during the simulated rainstorms (Kuhn *et al.*, 2003). The washed-out layer was removed between 100 and 180 min of the 3-h storm, exposing a cohesive washed-in layer (Fig. 14.1). The removal of the washed-out layer led to a marked increase of flow velocities from 0.8 to 1.2 cm/s and shear strength (from < 40 to 150 Pa). Interrill erodibility (Fig. 14.2), using the formula suggested by Zhang *et al.* (1998), declined during washed-out layer removal, showing that the increase of seal shear strength dominated soil resistance to erosion, rather than the decline of roughness that increased flow erosivity (Kuhn *et al.*, 2003). Interrupting the rainfall for drying produced a distinctly different pattern of erodibility during the latter part of storm 2 and the entire storm 3 (Fig. 14.2). More rainfall was required to reach peak erodibility after drying. Redistribution of aggregates by splash was observed before ponding at the beginning of storms 2 and 3. This change of surface conditions may be responsible for the initially lower, but higher erodibility later during both storms. Splash moved aggregates from areas with a washed-out layer to areas without washed-out layer. This initially increased surface roughness, reducing flow velocity and thus interrill erosion. Later, once pathways of continuous flow across the washed-in layer had formed again, the aggregates that had been moved by splash on to the washed-in layer were available for entrainment, increasing erodibility compared to the 3-h storm. Reduction of aggregate size due to low initial soil moisture after

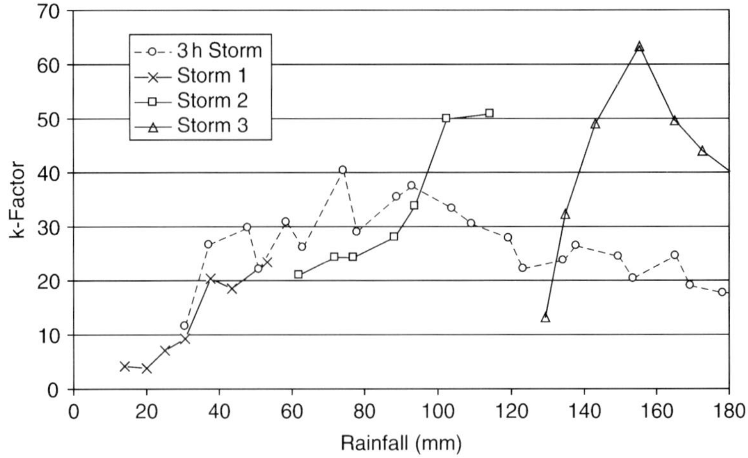

Fig. 14.2. Kastanozem interrill erodibility during 3-hour and three 1-hour storms (based on formula suggested by Zhang *et al.* (1998), $k = D/(Iq^aS^b)$, *k*: erodibility, *D*: soil loss (g/min/m^2), *I*: rainfall intensity (mm/min), *q*: unit discharge (mm/min), *S*: slope, a, b: coefficients).

drying may have also enhanced transportability of aggregates, thus also increasing erosion after drying. Towards the end of storm 3, the washed-out layer was removed almost completely, reducing erodibility similar to the 3-h storm. Total soil loss during the three-storm sequence was higher than during the 3-h storm (188 versus 144 g), indicating that drying did not only affect temporal patterns, but also led to an increase of erodibility.

Grassland Vertisol

On the Vertisol the formation of the washed-out layer required 120 mm of rainfall, and removal was much slower, leaving an almost continuous cover after 180 mm of rainfall (Fig. 14.1). The difference in seal formation is attributed to higher aggregate stability (68% WSA > 0.25 mm after 180 min of rainfall compared to 49% on the Kastanozem) (Table 14.1). No significant erodibility difference was observed during rainstorm sequences interrupted by drying, attributed to the slow rate of washed-out layer formation and removal caused by the high aggregate stability. However, under prolonged rainfall, e.g. during an entire rainy season in NE Mexico with 390–1850 mm of rainfall, seal formation and washed-out layer removal are likely to affect erodibility on the Vertisol as well.

Soil Resistance to Rill Erosion on a Clay Loam from Southern Ontario

Rill erosion occurs when shear stress of overland flow overcomes shear strength of the soil and non-selective transport is initiated (Savat and De Ploey, 1982). Controlling soil properties are texture, aggregate stability, soil cohesion and microtopography. Apart from texture, each of them changes as a result of rainfall and drying. The aim of the experiment described below was to examine the effects of varying degrees of drying during two contrasting rainstorm sequences on soil cohesion and flow hydraulic conditions relevant for incipient rill erosion.

Two rainstorm sequences were carried out to induce sealing and establish similar initial soil moisture conditions in the experimental flume. The first sequence consisted of two storms applied to the soil within a week. Both storms lasted 60 min. Drying was limited, so that the second storm was carried out at field capacity (wet storm). The second sequence consisted of three rainfalls. After the initial 60-min storm the soil was left to dry to air-dry conditions, requiring approximately 6 weeks. Then a second storm was applied until runoff rates were constant, indicating that seal condition and infiltration had readjusted to rainfall after drying. Within a week after the second storm, a third storm was applied to simulate rill erosion in 'dried' conditions (dried storm). In this way, soil moisture during the wet and dried storms was similar, providing similar runoff patterns during both storms. Soil resistance to rill incision was determined by measuring soil shear strength (Torri *et al.*, 1987), using the shear vane described above, and flow width, depth and velocity at the point of incipient incision. Shear velocity was calculated to describe flow hydraulic conditions critical for rill incision. Sediment concentration of the concentrated flow was measured to assess the effect of interrill sediment input on capacity of the concentrated flow to cause rill incision. For a detailed description of the techniques used during this test see Bryan (1990).

The preparatory storms produced a continuous depositional seal, embedded with single aggregates of various sizes and degrees of destruction. Drying increased aggregate size and shear strength. Subsequent rainfall caused a decline of both properties, however aggregate size remained higher (60% WSA > 0.25 mm) than before drying, while shear strength always reached the same minimum values (150 kPa) within 10–15 min of rainfall. In wet, non-dried conditions runoff started after 4 min and reached peak values within 20 min (Fig. 14.3). Rill incision started after 18 min, producing one joint rill reaching from 1.5 to 7 m along the centre of the flume by the end of the storm. Shear strength had reached minimum values after 10 min, before rill incision started, showing that increasing flow competence rather than lower soil cohesion controlled the timing of incision.

Runoff followed a similar pattern in dried conditions (Fig. 14.3). Soil shear strength also dropped to values similar to those of the wet storm within 10 min, but no rill incision occurred, and soil loss rates were an order of magnitude below those observed during the wet storm. Flow was also significantly slower, wider and shallower than during the wet storm (Table 14.3), so that shear velocity remained below the threshold critical for rill incision (Table 14.4). Cross-sectional profiles taken in the lower part of the flume confirm that the channel was wider (Fig. 14.4). Apparently, the morphology of the furrow bed developed differently during the two rainfall sequences. Since soil shear strength and runoff were similar in both storms, the lower competence of the concentrated flow in the wider furrow during the dried storm was responsible for lack of rill incision.

Assessing Short-term Changes of Erodibility

The two case studies demonstrate that soil erodibility varies as a result of several processes, controlled by soil properties, and the characteristics of the rainfall and drying sequence. On

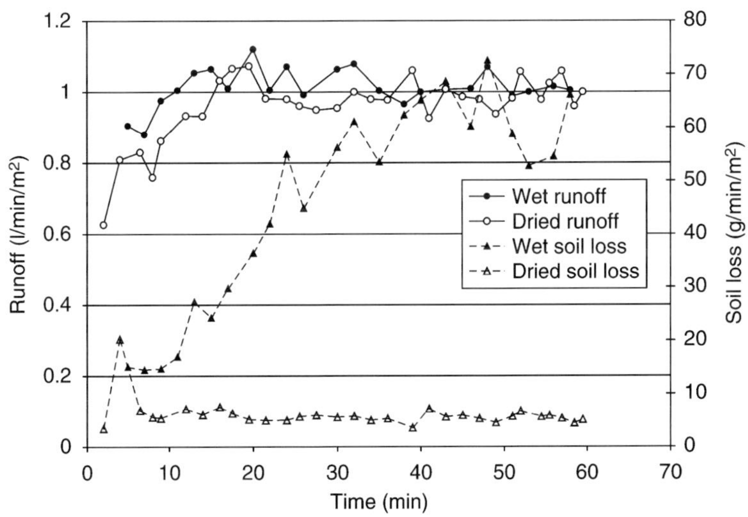

Fig. 14.3. Runoff and soil loss during rill erosion experiments on Schomberg clay loam.

Table 14.3. Schomberg clay loam: flow width, depth and velocity in wet and dried conditions.

	Wet storm			Dried storm		
	Width (cm)	Depth (mm)	Velocity (cm/s)	Width (cm)	Depth (mm)	Velocity (cm/s)
	14	8	45.45	15	3.5	37.31
	15	7	46.30	16	3	42.37
	13	8	37.31	16	4	43.48
	12	8	32.89	16	3	40.32
	14	10	28.74	16	4.5	43.48
	15	4	23.47	6	4.5	37.31
	6	10	38.17	16	2.5	34.48
	6.5	12	31.25	19	3	35.46
	6	8	32.47	19	3	37.31
	2.5	9	33.78	22	4.5	35.71
	3.5	8	26.60	21	4.5	35.71
				21	5	44.25
				19	5	36.76
				21	5	34.48
				20	5	30.49
Mean	9.77	8.36	34.22	17.36	3.93	38.46

Table 14.4. Shear velocity (V*) and sediment concentration (C_{sed}) observed in the furrows on the Schomberg clay loam during wet and dried conditions; values from the wet storm were collected immediately above developing headcuts, indicative of critical conditions for incipient rill incision.

Wet storm		Dried storm	
V* (cm/s)	C_{sed} (g/l)	V* (cm/s)	C_{sed} (g/l)
8.29	24.02	5.48	9.54
7.75	26.44	5.07	11.00
8.29	5.63	5.86	12.46
8.29	34.83	5.07	10.23
9.26	58.93	6.22	9.29
5.86	46.51	6.22	7.80
9.26	24.88	4.63	7.39
10.15	22.49	5.07	8.05
8.29	14.86	5.07	9.31
8.79	18.69	6.22	7.11
8.29	13.44	6.22	8.28
		6.55	8.96
		6.55	6.51
		6.55	8.99
		6.55	7.32

the Mexican soils, the formation and removal of the washed-out layer of small aggregates above a cohesive seal determined shear strength of soil exposed to raindrop impacted flow. The observed changes in soil properties relevant for erodibility are controlled by the kinetic energy of rainfall, determining seal formation and detachment, and runoff effective rainfall, controlling the removal of the washed-out layer. Drying between storms had a more complex effect on erodibility. Low initial soil moisture caused further destruction of aggregates, enhancing erodibility. Splash redistributed loose surface material, leading to a different availability of easily entrained material than during the 3-h storm. The effect of initial soil moisture on aggregate destruction has been described quantitatively (Torri *et al.*, 1998), and redistribution of aggregates can probably be integrated into sophisticated erosion models such as RillGrowth (Favis-Mortlock *et al.*, 1998). However, the complexity of these processes and the data requirements for successful modelling render any attempt of process-based prediction of the soil properties responsible for short-term changes of interrill erodibility most likely unfeasible for practical purposes.

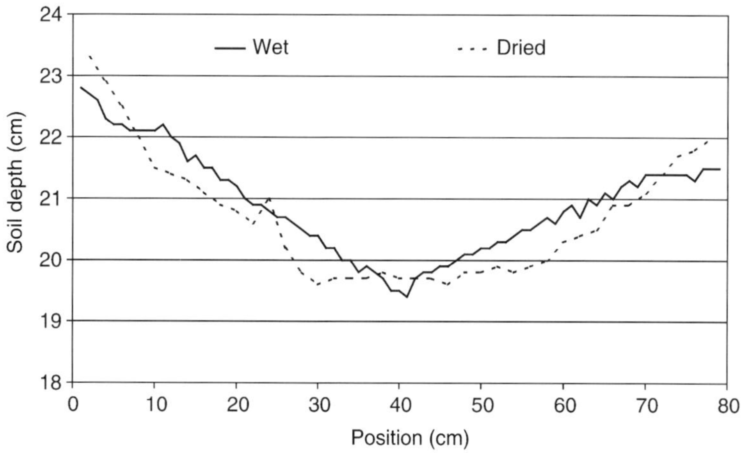

Fig. 14.4. Cross-sectional profiles on Schomberg clay loam before wet and dried storm at location of rill incision during wet storm.

The effect of interstorm drying on resistance of the Schomberg clay loam to rill erosion is more difficult to assess. Soil shear strength, often used as an index for soil resistance to rill erosion, increased in response to drying. However, the increase was reversed by rainfall during subsequent storms and did not affect erodibility at the time of incipient rill incision. The different development of furrow bed morphology during the two rainstorm sequences played a more critical role for rill development. This highlights the need for more realistic representation of rill channel morphology and its changes over time, rather than the commonly assumed square or hexagonal forms with flat beds. The exact reasons for the different development of furrow bed morphology during the wet and dry sequences remain unclear. The factors that determine the development of furrow bed morphology during a sequence of rainstorms, most notably runoff amount and flow hydraulics, sediment load and bed resistance to flow and erosion, are analogous to those in fluvial channels. However, the processes and resulting morphology are much more dynamic. Essentially, each rainfall event generates a different regime of rill flow, depending on rainfall duration and intensity, initial soil moisture, interrill sediment input and inherited form of the furrow. Due to the changing flow regimes, furrow bed morphology develops differently during each event. Process-based assessment of the effects of varying rainfall and drying patterns on soil resistance to rill incision would therefore require sophisticated models, linking temporal and spatial patterns of soil properties, interrill erosion and microtopography to entrainment and transport by concentrated overland flow.

Conclusions

On the Schomberg and the two Mexican soils, all soil properties relevant for erodibility changed in response to rainfall and drying, but not all of them were relevant for the actual soil resistance to erosion. On the Mexican soils aggregate stability and size of the washed-out layer initially determined resistance to entrainment and transport, while cohesion of the washed-in layer became dominant when the washed-out layer had been removed. Furrow morphology controlled soil resistance to rill incision on the Schomberg because soil shear strength had reached similar values before peak shear velocity values developed in both wet and dry conditions. Since several processes, driven by different forces, controlled soil properties responsible for resistance to erosion, one erodibility factor cannot fully represent soil resistance to erosion in dynamic process-based erosion models. There is clearly a need for further research on a wider range of soils and rainfall conditions to evaluate this conclusion.

Ultimately, the results would indicate that new erosion models with more sophisticated representation of the soil properties and processes that control soil resistance to erosion have to be developed. Alternatively, erodibility models have to integrate all the processes and properties that determine their short-term changes. Perceptually, such a process-based erodibility model can be described as follows:

$$E_t = f(T,H,C,S)$$

where E_t is soil resistance to interrill or rill erosion at time t, T microtopography determining distribution of rainfall and runoff, H hydraulic roughness of the surface where flow occurs, C cohesion of the surface determining resistance to detachment, and S sediment size controlling entrainment and transport capacity of the overland flow. In the light of the data requirements of such models they are also unlikely to be feasible for practical purposes. Therefore, careful selection of erosion events for calibration of erodibility parameters in dynamic process-based models may be the best available option. Calibration events should be selected by using similar preceding rainfall and drying history to those of the event to be predicted. Further research has to focus on improving our understanding of short-term soil–climate interaction as well as the transfer of the results into an expert system for calibration of erosion models.

References

Bryan, R.B. (1968) The development, use and efficiency of indices of soil erodibility. *Geoderma* 2, 5–26.

Bryan, R.B. (1990) Knickpoint evolution in rillwash. *Catena Supplement* 17, 111–132.

Bryan, R.B. (2000) Soil erodibility and processes of water erosion on hillslope. *Geomorphology* 32, 385–415.

De Walle, F.B. and Sevenster, J. (1998) *Agriculture and the Environment: Minerals, Manure and Measures.* Kluwer Academic, Dordrecht, The Netherlands.

DEFRA (2003) *The First Soil Action Plan for England: 2004–2006.* Department for Environment, Food and Rural Affairs, London.

EC (2000) *Directive 2000/60/EC of the European Parliament and of the Council of 23 October 2000 Establishing a Framework for Community Action in the Field of Water Policy.* Official Journal of the European Commission, Brussels.

Favis-Mortlock, D.T., Guerra, A.J.T. and Boardman, J. (1998) A self-organising dynamic systems approach to hillslope rill initiation and growth: model development and validation. In: Summer, W., Klaghofer, E. and Zhang, W. (eds) *Modelling Soil Erosion, Sediment Transport and Closely Related Hydrological Processes.* IAHS Publication 249. IAHS Press, Wallingford, UK, pp. 53–61.

Hoffman, D.W. and Richards, N.R. (1955) *Soil Survey of York County.* Report No. 19 of the Ontario Soil Survey. Guelph, Ontario.

IPCC (Intergovernmental Panel on Climate Change) (2001) *The Scientific Basis: Advancing our Understanding.* Cambridge University Press, Cambridge, UK.

Jetten, V., de Roo, A. and Favis-Mortlock, D. (1999) Evaluation of field-scale and catchment-scale soil erosion models. *Catena* 37, 521–541.

Kuhn, N.J. and Bryan, R.B. (2004) Drying, soil surface condition and interrill erosion on two Ontario soils. *Catena* 57, 113–133.

Kuhn, N.J., Bryan, R.B. and Navar, J. (2003) Seal formation and interrill erosion on a smectite-rich Kastanozem from NE-Mexico. *Catena* 52, 149–169.

Lal, R. (1990) *Soil Erosion in the Tropics.* McGraw Hill, New York.

Lane, L.J. and Nearing, M.A. (eds) (1989) *USDA – Water Erosion Prediction Project: Hillslope Profile Model Documentation. NSERL Report No. 2.* United States Department of Agriculture, West Lafayette, Indiana.

Owoputi, L.O. and Stolte, W.J. (1995) Soil detachment in the physically based soil erosion process: a review. *Transactions of the ASAE* 38, 1099–1110.

Quinton, J.N. (1997) Reducing predictive uncertainty in model simulations: a comparison of two methods using the European Soil Erosion Model (EUROSEM). *Catena* 30, 101–117.

Savat, J. and De Ploey, J. (1982) Sheetwash and rill development by surface flow. In: Bryan, R. and Yair, A. (eds) *Badland Geomorphology and Piping.* Geo Books, Norwich, UK, pp. 113–126.

Slattery, M.C. and Bryan, R.B. (1992) Laboratory experiments on surface seal development and its effects on interrill erosion processes. *Journal of Soil Science* 43, 517–529.

Torri, D., Sfalanga, M. and Chisci, G. (1987) Threshold conditions for incipient rilling. *Catena Supplement Band* 8, 97–105.

Torri, D., Ciampalini, R. and Accolti Gil, P. (1998) The role of soil aggregates in soil erosion processes. In: Boardman, J. and Favis-Mortlock, D.T. (eds) *Modelling Soil Erosion by Water*. NATO-ASI Series I-55. Springer, Berlin, Germany, pp. 247–258.

Valentin, C. (1998) Towards an improved predictive capability for soil erosion under global change. In: Boardman, J. and Favis-Mortlock, D.T. (eds) *Modelling Soil Erosion by Water*. NATO ASI Series I-55. Springer, Berlin, Germany, pp. 7–17.

Van den Akker, J.J.H., Arvidson, J. and Horn, R. (2003) Introduction to the special issue on experiences with the impact and prevention of subsoil compaction in the European Union. *Soil and Tillage Research* 73, 1–8.

West, L.T., Chiang, S.C. and Nelson, L.D. (1992) The morphology of surface crusts. In: Summer, M.E. and Stewart, B.A. (eds) *Soil Crusting: Chemical and Physical Processes*. Lewis Publishers, Boca Raton, Florida, pp. 301–338.

Zhang, X.C., Nearing, M.A., Miller, W.P., Norton, L.D. and West, L.T. (1998) Modelling interrill sediment delivery. *Soil Science Society of America Journal* 62, 438–444.

15 Double-averaging Methodology in Stochastic Modelling of Soil Erosion

A.Yu. Sidorchuk,[1] A. Smith[2] and V. Nikora[2]

[1]*Moscow State University, Russia.* Present address *Landcare Research, New Zealand;*
[2]*National Institute of Water and Atmospheric Research, New Zealand*

Introduction

The fundamentals of water erosion theory for cohesive soil are still largely undeveloped, in spite of their major significance for strategic estimates and predictions related to many aspects of human activity. For many years, efforts have focused mainly on the development of the empirical relationships based on data collected for different climatic and land-use conditions (Wischmeier and Smith, 1978). These empirical erosion models may be identified as the first-generation models. Although development of these first-generation models required significant resources and was time-consuming, progress was not satisfactory. As a result, a new generation of erosion models, known as physically based, appeared and provided some competition for purely empirical models. However, all available models are still semi-theoretical or semi-empirical. They are based on stream power Ω (or critical bed shear stress) relationships of the type $DER = K_{er}(\Omega - \Omega_0)^m$ for describing such complicated phenomenon as detachment rate *DER*, while the whole complexity of soil resistance to erosion is expressed by critical stream power Ω_0 and erodibility coefficient K_{er}. These relationships are derived from empirical measurements and are therefore restricted to a field of the measured characteristics. The main coefficients (like K_{er} and m) are not stable (see, for example, WEPP model database in Elliot *et al.*, 1989) and it is impossible to choose in advance the proper ones for a given soil type and flow characteristics without additional measurements and calibration of the relationships.

A new, third generation of erosion models is urgently needed, one that would account for the stochastic nature of soil erosion. The complicated interactions between flowing water and the soil surface cause stochastic detachment and deposition of soil particles (or aggregates), and this stochasticity seems to be the key factor in soil erosion mechanics. The original idea of stochastic modelling of erosion was formulated and used by H.A. Einstein (1937) for modelling erosion of non-cohesive sediments. There have been several attempts (Mirtskhulava, 1988; Nearing, 1991; Wilson, 1993a,b) that incorporate the probability of soil particle detachment being related to the excess of driving forces over stabilizing forces, and the rate of detachment being related to the frequency of flow velocity fluctuations. In the stochastic models the detachment rate can be estimated by two main approaches (Sidorchuk, 2005). The first is the 'velocity-concentration' (continuum) approach, where the detachment rate is calculated as the product of the concentration of unstable soil aggregates in the bed surface layer and the vertical velocity of soil aggregates. This method is discussed in Sidorchuk *et al.* (2004). The second is the 'double-averaging' (discrete particle) approach, where the mean detachment rate is calculated by space and time integration of individual discrete aggregate detachment rates.

The double-averaging approach in soil modelling has an important advantage of being directly linked to the double-averaged hydrodynamic variables, which appear in the double-averaged hydrodynamic equations (Nikora *et al.*, 2001). The double (in time/ensemble and space domains)-averaged equations relate to the time/ensemble averaged equations, as the latter relate to the Navier-Stokes equations for instantaneous hydrodynamic variables. The double-averaging procedure gives new momentum and continuity equations for fluid, which explicitly contain important additional terms such as form-induced stresses and, for the flow region below roughness tops, form and viscous drag terms. The time–space window for averaging hydrodynamic variables is the same as that for obtaining the double-averaged detachment rate introduced and discussed in this chapter. Analysis of the conceptual framework of the double-averaging approach in soil erosion modelling is the main goal of this paper.

The Double-averaging Approach

We describe an erosion process for a soil surface layer on the flow bed. The soil is composed of particles and water-stable aggregates of particles (also called particles). They are regarded as solid particles (without slaking or dispersion) with given boundaries described by equivalent diameter D_a, vertical projection area S_a, volume V_a (a total volume, including intra- and inter-aggregate pores) and other size and shape characteristics. Soil particles, joined to each other by resistance forces, are interacting with flow, which produces driving forces.

The principal scheme of soil detachment by flow is shown in Fig. 15.1. It is this scheme of detachment that will be expressed in double-averaged (over space and time) terms in this chapter. If total soil volume V is detached during

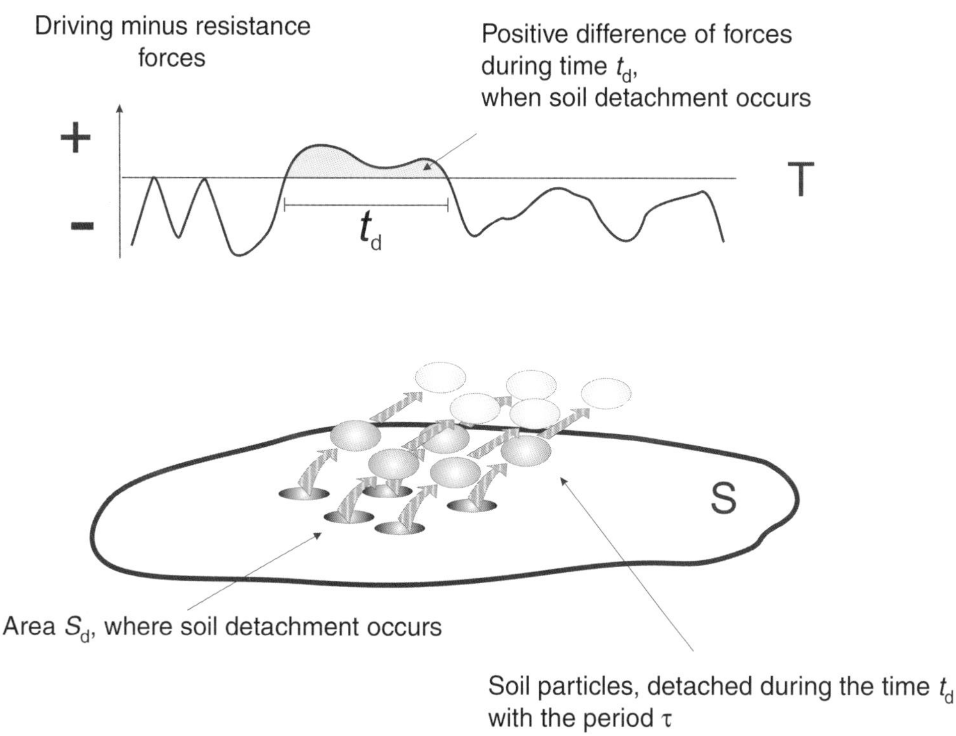

Fig. 15.1. General scheme of the detachment of soil particles in time and space.

some observation time T from some area S, the mean detachment rate DER will be:

$$DER = \frac{V}{ST} \tag{15.1}$$

The averaging time T and area S should be large enough to cover small-scale random variability of flow and soil particle properties. At the same time, they should be smaller than large-scale variability due to rainfall events and slope pattern. During this time T, and over this area S, the hydrodynamic driving forces and the resistant soil structural forces change randomly. A significant component of the resistant force is reactive; its magnitude and direction are determined by the sum of active forces, to ensure the total force balance remains at zero, for a particle in equilibrium. This reactive force is not infinitely flexible, however. It has a maximum magnitude, which will depend on the soil structure and direction of the active force. During some of time T, and in parts of area S, the magnitude of the active force will exceed this maximum, making soil particles unstable. We can denote this excess by introducing a binary filtering variable H, which equals 1 where and when it occurs, and 0 otherwise. Then, individual soil particles with volume V_a and area S_a are detached by the flow, requiring time τ. We can thus consider the detachment rate at the scale of individual discrete particles:

$$DER_{IND} = \frac{V_a H}{S_a \tau} \tag{15.2}$$

The overall mean (double-averaged) detachment rate is therefore:

$$DER = \frac{1}{ST} \iint_{ST} DER_{IND} \, dsdt = \frac{1}{ST} \iint_{ST} \frac{V_a H}{S_a \tau} dsdt \tag{15.3}$$

In order to solve this equation, we need: the physical conditions in which $H = 1$, an expression for τ, and statistics for V_a and S_a. These three aspects will be addressed individually in the remainder of the chapter.

The Force Balance

To predict when and where $H = 1$, we would ideally require a list of all forces involved in the flow-soil and within-soil interactions. It is extremely difficult to determine the full list of forces that occur in reality, but a good approximation will be provided by the following, which are well-known (see, for example, Wilson, 1993a) and whose effects have been observed to be significant.

We begin with submerged weight:

$$F_{wn} = V_a (\rho_s - \rho) g \tag{15.4}$$

Here ρ_s and ρ are soil and water density, and g is acceleration due to gravity.

Secondly, there is the static pressure of water with a mean depth d above a soil particle:

$$F_{sp} = g\rho S_a I_{sv} d \tag{15.5}$$

This force is controlled by soil integrity in vertical projection I_{sv}, which can be defined as the proportion of vertical projection area with strong contact between particles. I_{sv} is more specific for cohesive soil, but also can be applied to loose particles.

Form drag is proportional to the square of along-the-flow near-bed velocity U, frontal cross-section area of soil particle S_f and to particle exposure E (the proportion of S_f exposed to flow):

$$F_{fd} = C_R \rho E S_f \frac{U^2}{2} \tag{15.6}$$

Form drag coefficient C_R is a complicated function of Reynolds number and of the degree of particle submergence, but is close to constant for turbulent flows with submerged particles.

For small particles, viscous drag may be significant and can be expressed as:

$$F_{vd} = \rho \upsilon S_a \frac{dU}{dZ} \tag{15.7}$$

where υ is kinematic viscosity and Z is the spatial coordinate perpendicular to S_a.

Less is known about wave drag. The force is significant when flow depth is of the same order as particle size. By definition, this force is proportional to the difference Δz between water surface altitudes above the front and back side of a particle:

$$F_{sd} = g\rho E S_a \Delta z \tag{15.8}$$

Experiments by Lawrence (2000) show that:

$$\Delta z = C_{Rw} \frac{kU^2}{2gd} \quad (15.9)$$

Here k is the exposed height of the particle (which is close to $E^{1/2}D_a$), C_{Rw} is wave drag coefficient. The latter depends on particle inundation d/k. Drag force is applied to both cohesive and loose soil particles.

The lift force is controlled by the velocity difference around the 3D particle. It is equal to:

$$F_l = C_y \rho S_a \frac{U^2}{2} \quad (15.10)$$

where C_y is lift coefficient.

Finally, there is the reactive force of cohesion F_c. In cohesive soil, particles are attached to each other by a complex combination of friction, hydrostatic and electro-chemical forces. The maximum magnitude for this force is estimated, as an average over all directions, to be:

$$F_{c\_max} = C_0 S_a I_s \quad (15.11)$$

The force of cohesion (C_0 is the cohesive force per unit of contact area) works only on the area S_b of strong surface attachment between adjacent particles, when the distance between particles is less than the thickness of gravitational water intercalation (usually less than ~ 10^{-6} m). Therefore, the force of cohesion is controlled also with soil integrity I_s, which can be defined as the ratio of area of strong contacts between particles and vertical projection area S_b/S_a. In general, the force of cohesion can be applied also for loose particles, but the integrity of those particles is very low.

Now we can define the force balance θ as the magnitude of the sum of vectors of the individual forces:

$$\theta = |F_{wn} + F_{sp} + F_{fd} + F_{sd} + F_l + F_c| \quad (15.12)$$

The filtering variable H is equal to 1 when and where θ is greater than zero. In this case, the soil particle becomes unstable and is in the process of detachment.

Time Taken for Detachment

Even if θ operating on a given particle is large, that particle is not removed from its place immediately: it takes some time. We use the hypothesis that a particle is detached when it is removed from the initial position to the distance of ~1 diameter D_a and all cohesive contacts are torn away. Thus τ, for soil particles of a given size, can be derived from the second Newton law, written for particle acceleration (here Z is measured in the direction of the sum of forces, from the particle's initial position):

$$\frac{d^2 Z}{dt^2} = \frac{\theta}{\rho_s V_a} \quad (15.13)$$

If we assume that the inertial force θ is constant throughout the period of detachment, this gives:

$$\tau = \sqrt{\frac{2\rho_s V_a D_a}{\theta}} \quad (15.14)$$

For a more general case, we can consider that the force balance for a particle changes while it moves from its initial position in the soil. In particular, resistant soil forces decrease as the particle's contact with adjacent particles is reduced, and driving forces increase with the increase of particle exposure. We can incorporate this with a function F, so that:

$$\tau = \frac{1}{F(D_a, V_a, \theta)} \quad (15.15)$$

Detachment Rate Calculation

Using this expression for τ, we now have from Eqn 15.3:

$$DER = \frac{1}{ST} \iint_{ST} H \alpha \, ds \, dt \quad (15.16)$$

where

$$\alpha = \sqrt{\frac{V_a \theta}{2 S_a^2 \rho_s D_a}} \quad (15.17)$$

or

$$\alpha = \frac{V_a F}{S_a} \quad (15.18)$$

Note from the meaning of H and Eqns 15.17–15.18 that, when θ (and F) equals 0, both H and α equal 0. This means that H is redundant in Eqn 15.16; α is the discrete particle detachment rate. Secondly, Eqn 15.16 can be interpreted as the average value of α over the spatiotemporal field. Thus, if we consider $f(\alpha)$, which is the PDF (probability density function) of α, we can calculate this same average with the integral formula:

$$DER = \int_{\alpha} \alpha f(\alpha) d\alpha \qquad (15.19)$$

Equation 15.19 shows how the detachment rate of cohesive soil can be calculated within discrete DA approach, once the function α is known and its PDF is determined.

A Simplified Example

One of the main difficulties in the calculation of α is the complexity of local soil structure and the variability of the angle of detachment. The direction of the vector of active force balance is largely unknown. In the following simplified example we take into account only the forces that are normal to the local soil surface (having slope β), looking for a process of soil particle 'corking'. Considering only the most significant mechanisms, we have an expression for the inertial force:

$$\theta = C_y \rho S_a \frac{U^2}{2} - V_a g(\rho_s - \rho)\cos\beta - C_0 S_a I_s \qquad (15.20)$$

From Eqns 15.17 and 15.20, we have:

$$\alpha = \sqrt{\frac{V_a\left[C_y \rho S_a \frac{U^2}{2} - V_a g(\rho_s - \rho)\cos\beta - C_0 S_a I_s\right]}{2S_a^2 \rho_s D_a}} \qquad (15.21)$$

The probability density $f(\alpha)$ can be calculated once the probability densities of all random arguments of α are known. In this example, we can assume that the aspect ratio $\frac{V_a}{S_a D_a}$ is near constant for all particles. Then, we are left with three terms inside the square root: multiples of U^2, $D_a \cos\beta$ and I_s, respectively. These variables are statistically independent, so the following formulae can be used to calculate $f(\alpha)$ from their particular PDFs.

Let us consider independent random variables A_1 and A_2, described by the PDFs $p_1(a_1)$ and $p_2(a_2)$ respectively. Their product, $B = A_1A_2$, has a PDF $p_B(b)$ given by:

$$p_B(b) = -\int_{-\infty}^{0} \frac{1}{a_1} p_1(a_1) p_2\left(\frac{b}{a_1}\right) da_1 + \int_{0}^{\infty} \frac{1}{a_1} p_1(a_1) p_2\left(\frac{b}{a_1}\right) da_1 \qquad (15.22)$$

Secondly, their sum, $B = A_1 + A_2$, has a PDF $p_B(b)$ given by:

$$p_B(b) = \int_{-\infty}^{\infty} p_1(a_1) p_2(b - a_1) da_1 \qquad (15.23)$$

Thirdly, if B is the square root of A (with PDF $p_A(a)$), then its PDF $p_B(b)$ is given by:

$$p_B(b) = 2\sqrt{a}\, p_A(a) \qquad (15.24)$$

In the general case, with full forces balances and statistically dependent variables, calculating $f(\alpha)$ becomes more complicated. But the same principle applies; the method requires values and PDFs of the random arguments for the function α. In the remaining sections of this chapter, we discuss the ways to obtain these for flow velocity and depth, as well as for soil particle diameter.

Outlook

The hydrodynamic characteristics

To underpin and complement the probabilistic approach for modelling erosion and sedimentation processes, we require measurements of instantaneous values of velocity and pressure, at all near-bed points in the flow. Generating such data for flow over a soil surface is a difficult task, requiring a high-resolution mechanism-based hydrodynamic model.

We believe the Large Eddy Simulation (LES), which averages the Navier-Stokes equations in space, accounting for small-scale processes artificially, to be the most appropriate for our research considerations. We have employed the LES turbulence model in the commercially available FLUENT hydrodynamic software (www.fluent.com). The defining attribute of LES is the

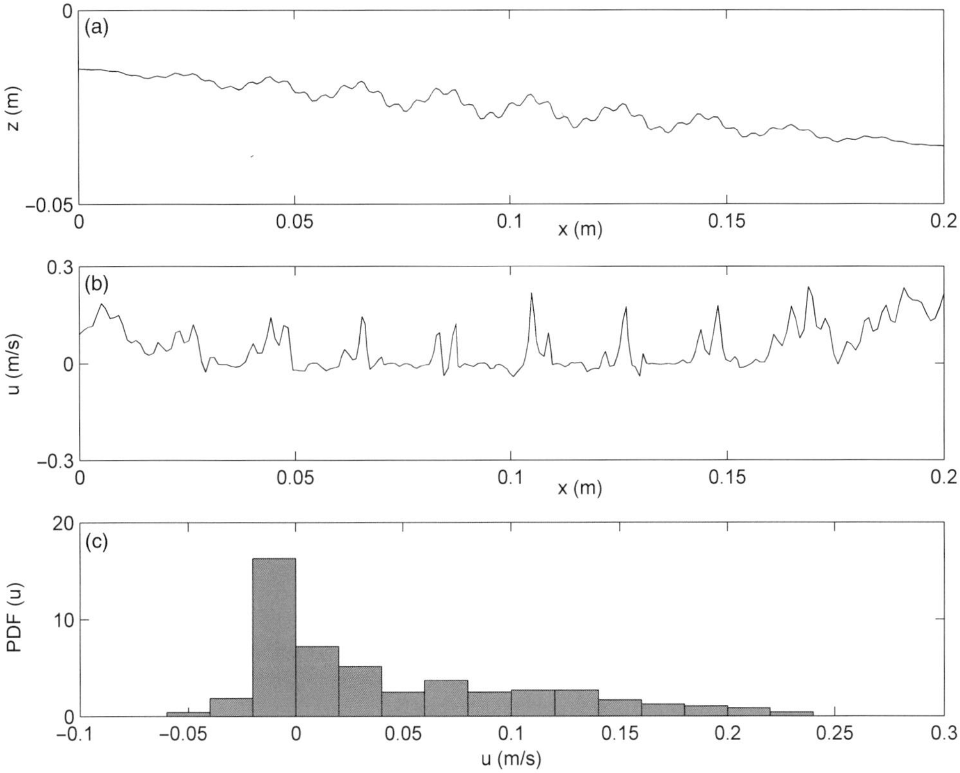

Fig. 15.2. Simulation of 2D free-surface gravity-driven flow over a rough soil surface. The domain length is 0.2 m; the spatial resolution (of both the soil surface and the LES grid) is 0.001 m. Flow is from left to right, at a rate of 0.0005 m^2/s, with gravity acting vertically downward. (a) Soil surface, z (m) with respect to x (m); (b, c) show the near-bed (tangential) flow velocity u (m/s), with respect to x (m) and its PDF respectively.

application of a spatial averaging that filters out all viscous, and some inertial, effects that operate beneath a certain cut-off scale. Thus, only the eddies that are larger than this scale are resolved explicitly. The concept behind this approach is that small-scale eddies are more isotropic, problem-independent, and less significant in terms of transporting momentum and energy; they are therefore more universal and better suited to approximation than are large-scale processes. For convenience, the cut-off scale is usually chosen to be the grid resolution scale.

As an example of the results obtainable from this modelling approach, we performed simulations of flow over a random soil surface cover with several scenarios involving flow, rain (as a significant source of energy to the system) and erosion. We present (Fig. 15.2) an initial case, which exemplifies the generation of pressure profiles from the LES modelling approach: 2D free surface flow over a rough bed.

Soil structure modelling

Soil structure is the spatial/temporal distribution of soil physical characteristics within the soil body. One of these characteristics is the size (linear, by the area, volumetric, by the weight) of soil particles and aggregates. The distribution of aggregates by size changes in time due to their fragmentation and aggregation. Nevertheless, there are quite a few main types of PDF, estimated empirically and associated with all variety of soils in different conditions (Perfect *et al.*, 1993). Only the logarithmically Normal distribution has a theoretical basis (Kolmogorov, 1941). This work described the process of random failure of soil particles, when the probability of fragmentation of a particle to some number of parts was scale-invariant, and the result was asymptotically logarithmically Normal.

The Kolmogorov-type algorithm of soil particles failure can be simulated numerically, and in numerical experiments the assumption of scale independence of fragmentation can be avoided. These experiments with different relationships between probability of failure and particle size show a great stability of result. The logarithmically Normal distribution of soil particles is valid in a broad range of scenarios of fragmentation, including those where probability of failure is different for particles of different size. This is important because of the great range of particle sizes within real soil. This distribution is asymptotic, but is developed within a first few steps of simulation. Each type of fragmentation process is characterized by specific rates of mean size decrease and particle size variability increase.

Conclusion

Stochastic approaches to soil erosion modelling already have about 60 years of history (from Hans Einstein's pioneering works), but are still at the very beginning of their development. We believe that the approach presented in this chapter is one of the main ways to change the current significantly empirical basis of soil erosion investigations to a theoretical one, based on fundamentals of flow and soil mechanics. It is far from any applicable results, but it shows our understanding of the further development of soil erosion theory.

We can also relate the double-averaging approach to the velocity-concentration approach (discussed in detail in Sidorchuk *et al.*, 2004): the active soil particle concentration in the surface soil layer C (or the probability of particle detachment) can be expressed as:

$$C = \int_{\alpha>0} f(\alpha)\,d\alpha \tag{15.25}$$

and the mean vertical velocity of active particles $U_{\uparrow}$ is:

$$U_{\uparrow} = \frac{\int_{\alpha} \alpha f(\alpha)\,d\alpha}{\int_{\alpha>0} f(\alpha)\,d\alpha} \tag{15.26}$$

The equivalence of these two approaches accentuates the physical meaning of each and makes possible cross-verification of the results of theoretical investigations.

Acknowledgement

This research is funded by the Marsden Fund administered by the Royal Society of New Zealand (grant LCR-203).

References

Einstein, H.A. (1937) Bedload transport as a probability problem. PhD thesis. ETH, Zurich (in German). Translation by Sayre, W.W. (1972) in Shen, H.-W. (ed.) *Sedimentation*. Appendix C. Colorado State University, Fort Collins, Colorado, C1–C105.

Elliot, W.J., Liebenov, A.M., Laflen, J.M. and Kohl, K.D. (1989) *A Compendium of Soil Erodibility Data from WEPP Cropland Soil Field Erodibility Experiments 1987&88*. NSERL Report No. 3.

Kolmogorov, A.N. (1941) About a logarithmically normal distribution of particles during the fragmentation. *Doclady Akademii Nauk SSSR* 31, 99–101 (in Russian).

Lawrence, D.S.L. (2000) Hydraulic resistance in overland flow during partial and marginal surface inundation: experimental observations and modelling. *Water Resources Research* 36, 2381–2393.

Mirtskhulava, T.E. (1988) *Principles of Physics and Mechanics of Channel Erosion*. Gidrometeoizdat, Leningrad (in Russian).

Nearing, M.A. (1991) A probabilistic model of soil detachment by shallow flow. *Transactions of the American Society of Agricultural Engineers* 34, 81–85.

Nikora, V.I., Goring, D.G., McEwan, I. and Griffiths, G. (2001) Spatially-averaged open-channel flow over a rough bed. *Journal of Hydraulic Engineering* 127, 123–133.

Perfect, E., Kay, B.D., Ferguson, J.A., da Silva, A.P. and Denholm, K.A. (1993) Comparison of functions for characterizing the dry aggregate size distribution of tilled soil. *Soil and Tillage Research* 28, 123–139.

Sidorchuk, A. (2005) Stochastic modelling of erosion and deposition in cohesive soils. *Hydrological Processes* 19, 1399–1417.

Sidorchuk, A., Smith, A. and Nikora, V. (2004) Probability distribution function approach in stochastic modelling of soil erosion. In: Golosov, V., Belyaev, V. and Walling, D.E. (eds) *Sediment Transfer Through the Fluvial System*. IAHS Publication 288. IAHS Press, Wallingford, UK, pp. 345–353.

Wilson, B.N. (1993a) Development of fundamentally based detachment model. *Transactions of the American Society of Agricultural Engineers* 36, 1105–1114.

Wilson, B.N. (1993b) Evaluation of fundamentally based detachment model. *Transactions of the American Society of Agricultural Engineers* 36, 1115–1122.

Wischmeier, W.H. and Smith, D.D. (1978) Predicting rainfall erosion losses – a guide to conservation planning. *USDA Agricultural Handbook 537*. US Government Printing Office, Washington, DC.

16 Runoff and Predicting Erosion on Hillslopes within Catchments

P.I.A. Kinnell
University of Canberra, Australia

Introduction

Erosion on hillslopes within catchments contributes to a decline in agricultural productivity and produces pollutants that adversely affect the quality of water in rivers, reservoirs and lakes. In many places, sheet, rill and interrill erosion dominate erosion on hillslopes within catchments. Rainfall erosion results from the detachment of particles from within the soil surface followed by the transport of detached particles away from the site of detachment. Four detachment and transport systems exist:

1. Raindrop Detachment with transport by Raindrop Splash (RD-ST).
2. Raindrop Detachment with transport by Raindrop-induced Flow Transport (RD-RIFT).
3. Raindrop Detachment with transport by Flow (RD-FT).
4. Flow Detachment with transport by Flow (FD-FT).

Raindrop Detachment with transport by Raindrop Splash (RD-ST) is the system that operates in what is commonly known as splash erosion. Raindrop-induced Flow Transport (RIFT) is a process where each drop impact causes soil particles to saltate underwater. Each drop impact causes soil material to be lifted into the flow and settle back to the bed some distance downstream. Flow transport (FT) occurs when loose particles travel with the flow without the aid of raindrop impact. Whether a particle detached by raindrop impact (RD) is transported by RIFT or FT depends on its size, density and the flow conditions. Rill erosion is dominated by Flow Detachment with transport by Flow (FD-FT). The RD-ST, RD-RIFT, RD-FT and FD-FT systems that operate on hillslopes within catchments result in sediment being discharged with flow. As a consequence, runoff is a factor in determining soil loss.

Erosion by Rain-impacted Flow

So-called process-based models like WEPP (Flanagan and Nearing, 1995) and EUROSEM (Morgan *et al.*, 1998) attempt to model the detachment and transport processes explicitly and recognize that runoff is a factor in determining soil loss. In sheet and interrill erosion, RD-RIFT tends to control the movement of silt- and sand-sized material, while RD-FT tends to control the movement of the finer material. Because erosion results from the discharge of this sediment, the equation:

$$q_s = q_w c \qquad (16.1)$$

where q_s is the sediment discharge (mass per unit width of flow), q_w is the water discharge (mass per unit width of flow) and c is the sediment concentration (mass of sediment per unit mass of water), is relevant to determining the erosion rate. RIFT is the major transport system that operates in sheet and interrill erosion areas and Kinnell (1993) showed that when RIFT dominated transport of sediment:

$$q_{sR}(p,d) = a_p I_d u\, f[h,d] \quad (16.2)$$

where $q_{sR}(p,d)$ is the mass of sediment of size p discharged per unit width of flow associated with the impacts of drops of size d, a_p is an empirical coefficient that is dependent on particle size and density, I_d is intensity of rain of drops of size d, u is flow velocity and $f[h,d]$ is a function that varies with flow depth (h) and drop size (d).

As noted above, in rain-impacted flows, RIFT tends to control the movement of silt- and sand-sized material, while FT tends to control the movement of the finer material. In RIFT, particles are lifted into the flow by drop impacts but then fall back to the bed under the force of gravity. Downstream movement during fall occurs because the flow exerts a horizontal force on the falling particle. With splash erosion, the tendency for raindrop splash to transport material radially from the point of impact means that on large level or near level surfaces, a layer of pre-detached material builds up on the surface over time. This is because the transport system is extremely inefficient. Any material splashed may come from this layer and from the soil surface beneath it. Also, because the pre-detached material sits on top of the soil surface, it provides a degree of protection (H) against detachment from that surface. Consequently, the erodibility of the surface (k_s) is given by:

$$k_s = (1 - H)\, k_{sm} + H\, k_{pdl} \quad (16.3)$$

where k_{sm} is the erodibility of the surface of the soil matrix (sm) when no pre-detached particles are present, k_{pdl} is the erodibility of the pre-detached layer (pdl) of particles and H has values of 0 to 1. Like splash, RIFT is a transport system that produces a layer of pre-detached material sitting on the soil surface and consequently the erodibility of the surface will vary depending on the depth and characteristics of this layer of pre-detached material. Thus, the erodibility of a surface eroding under a rain-impacted flow where raindrop-induced flow transport dominates ($k_{s.RIFT}$) is given by:

$$k_{s.RIFT} = (1 - H) k_{sm.RIFT} + k_{pdl.RIFT} \quad (16.4)$$

where $k_{sm.RIFT}$ is the erodibility of the surface of the soil matrix (sm) when no pre-detached particles are present, $k_{pdl.RIFT}$ is the erodibility of the pre-detached layer (pdl) of particles, and H has values of 0 to 1. Consequently, the erodibility of such a surface is not given by a single value but may range between $k_{sm.RIFT}$ and $k_{pdl.RIFT}$. Currently, so-called process-based models do not include any consideration of this and use a single experimentally derived erodibility factor which lies at some unknown point between the two extremes. This makes it difficult to relate these erodibility factors to measured soil physical and chemical factors because the physical and chemical properties of the two materials are quite different, and the dominance of one over the other is unknown.

Runoff as a Factor in Predicting Erosion on Hillslopes within Catchments

Process-based models like WEPP and EUROSEM require a considerable amount of data and it is common for erosion within catchments to be predicted using the Universal Soil Loss Equation (USLE, Wischmeier and Smith, 1978) or the revised version of it (RUSLE, Renard *et al.*, 1997) because they are less data and computationally intensive. While the USLE/RUSLE was not developed for predicting event erosion, it follows that:

$$A_e = R_e K_e L S C_e P_e \quad (16.5)$$

where A_e is the erosion that takes place during a rainfall event, $R_e = EI_{30}$ (where E is event rainfall kinetic energy and I_{30} is the maximum 30-minute rainfall intensity), L and S are the USLE topographic factors which vary in space but not time, C_e is the crop and crop management factor that is associated with the event and P_e is the soil conservation protection factor that applies during the event. Figure 16.1 shows how the USLE predicts event erosion on a bare fallow plot at Morris, MN in the USA. In this case, low soil losses were severely overpredicted.

As noted above, erosion on hillslopes results from sediment being discharged with flow. Equation 16.1 applies to all situations where sediment is discharged with flowing water. However, models like the USLE and the RUSLE do not consider runoff as a primary independent factor in

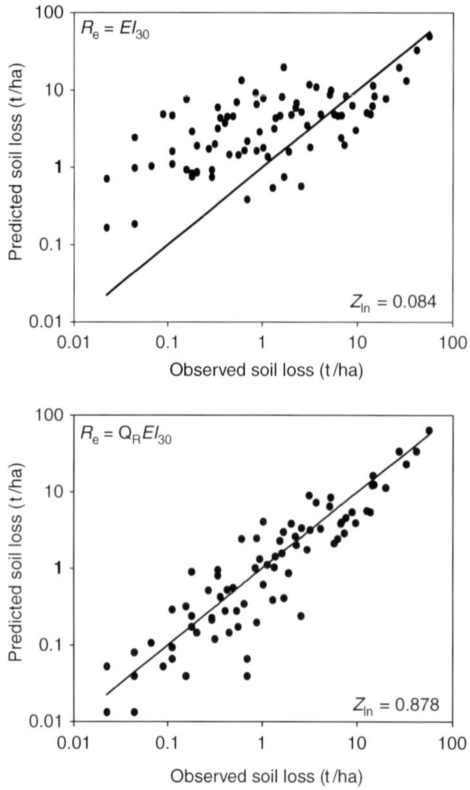

Fig. 16.1. Relationships between observed and predicted event soil loss for plot 10 (bare fallow) in experiment 1 at Morris, Minnesota, USA when predicted = bR_e where R_e is EI_{30} and $Q_R EI_{30}$ and b is a fitted parameter. Eff_{ln} is the Nash-Sutcliffe efficiency factor for the ln transforms of the data and reflects the amount of variation from the 1 : 1 lines shown in these figures. NB: This analysis takes no account of short-term variations in K. From Kinnell (2003).

the prediction of erosion from field-sized areas. It follows from Eqn 16.1 that if runoff is considered as a primary independent term in predicting erosion, then event sediment concentrations on a bare fallow area will vary between soils and with rainfall kinetic energy level of the rainfall and some measure of event rainfall intensity. The kinetic energy level of the rainfall is given by dividing E by the rainfall amount and I_{30} is a measure of event rainfall intensity. Thus:

$$A_e = k Q_e I30E \,/\, \text{rainfall amount} \quad (16.6)$$

where k is an empirical coefficient that is dependent in part on the soil, and Q_e is event runoff. Q_e divided by rainfall amount is the runoff coefficient (Q_{Re}) for the event. Consequently:

$$A_e = k Q_{Re} E I 30 \quad (16.7)$$

Figure 16.1 shows how Eqn 16.7 predicts event erosion for the bare fallow plot at Morris when event runoff is known. The variant of the USLE that uses Q_{Re} EI_{30} as its event erosivity index is known as the USLE-M (Kinnell and Risse, 1998). The total loss from the plot was 374 t/ha from 80 events over 10 years. The top five events produced 177 t/ha. The USLE (Fig. 16.1) predicted 123 t/ha (–31% error), while the USLE-M predicted 164 t/ha (–7% error). The 10 events producing the lowest soil loss contributed 0.83 t/ha. The USLE predicted 25 t/ha for these events, the USLE-M 1.12 t/ha.

The USLE-M is not the only USLE variant to include runoff as a parameter in the event erosivity factor. The MUSLE (Williams, 1975) uses the product of event runoff (Q_e) and peak runoff ($q_{p.e}$) in place of EI_{30}. However, it uses USLE factor values for K, L, S, C and P when these should only be used when $R_e = EI_{30}$. K, the soil erodibilty factor, has units of soil loss per unit erosivity index and must be re-evaluated if R_e is changed from EI_{30}. Also, even if this is done, C and P values cannot be applied if the values of Q_e and $q_{p.e}$ are determined for anything but bare fallow and cultivation up and down the slope. If they are determined for a vegetated area, then the effect of runoff is considered twice. In addition, the MUSLE event erosivity index does not account for erosion at the plot scale well. Figure 16.2 shows the relationship between that erosivity index and event soil losses from a cropped plot at Zanesville, Ohio, USA. The Nash-Sutcliffe efficiency factor for the index in this case is 0.283, assuming that C is constant with time. The efficiency factor value for the USLE-M was 0.619.

It is common to model erosion in catchments using grid cells. When a hillslope is uniform with respect to soil and vegetation, the effect of slope length for a cell with coordinates i, j can be determined using the approach proposed by Desmet and Govers (1996):

$$L_{i,j} = \frac{(A_{i,j\text{-in}} + D^2)^{m+1} - A_{i,j\text{-in}}^{\;m+1}}{D^{m+2}\, x_{i,j}^{\;m}\, (22.13)^m} \quad (16.8)$$

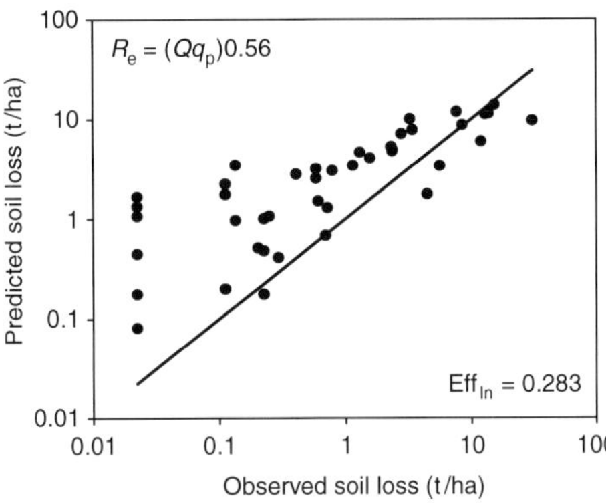

Fig. 16.2. Relationships between observed and predicted event soil loss for plot 1 in experiment 1 with maize at Zanesville, Ohio, USA when predicted = bR_e where $R_e = (Qq_p)^{0.56}$ and b is a fitted parameter. NB: This analysis takes no account of short-term variations in C. Eff$_{ln}$ for $R_e = Q_R EI_{30} = 0.619$. From Kinnell (2003).

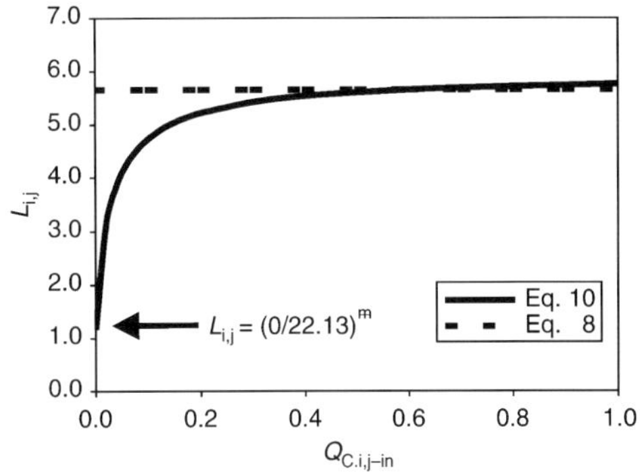

Fig. 16.3. The effect of $Q_{C.i,j-in}$ on $L_{i,j}$ for the outlet cell to a 1 ha area when $D = 30$ m and the runoff coefficient for the cell = 0.6.

where $A_{i,j-in}$ is the contributing area (m^2) upslope of the cell, D is cell size (m), m is the USLE slope length exponent (Renard *et al.*, 1997) and x is a factor that depends on the direction of flow with respect to grid orientation. Equation 16.8 is an adaptation equation for the L factor for a slope segment developed by Foster and Wischmeier (1974) to the grid cell situation. However, the L factor for a uniform rectangular slope is based on the distance from the onset of runoff (Wischmeier and Smith, 1978). Thus, if no runoff occurs from upslope then:

$$L_{i,j} = \frac{D^m}{(22.13)^m} \qquad (16.9)$$

i.e. equal to the USLE L factor for an area D metres long. Setting $A_{i,j-in}$ to zero results in Eqn 16.8 producing the correct result. However, it follows that if the upslope area has a runoff coefficient that lies somewhere between zero and that for the grid cell, $L_{i,j}$ should lie somewhere between that given by Eqn 16.8 when $A_{i,j-in}$ is zero and $A_{i,j-in}$ is equal to the physical area upslope of the grid cell. Similarly, if the runoff coefficient of the upslope area is greater than that of the cell, $L_{i,j}$ should be greater than that given when $A_{i,j-in}$ is equal to the physical area upslope of the grid cell. This can be achieved by replacing $A_{i,j-in}$ by an effective value of $A_{i,j-in}$ ($A_{i,j-in.eff}$) to give:

$$L_{i,j} = \frac{(A_{i,j-in.eff} + D^2)^{m+1} - A_{i,j-in.eff}^{\;m+1}}{D^{m+2}\, x_{i,j}^{\;m} (22.13)^m} \qquad (16.10)$$

where:

$$A_{i,j-in.eff} = A_{i,j-in} Q_{C.i,j-in} / Q_{C.i,j-all} \qquad (16.11)$$

and $Q_{C.i,j-in}$ is the runoff coefficient for the upslope area and $Q_{C.i,j-all}$ is the runoff coefficient for the area including the cell. Figure 16.3 shows how $L_{i,j}$ for the outlet cell to a 1 ha area varies with the upslope runoff coefficient when the cell size is 30 m.

The value of $L_{i,j}$ produced using Eqn 16.10 only differs significantly from that produced by the Desmet and Govers (1998) approach (Eqn 16.8) when the runoff coefficient of the upslope area ($Q_{C.i,j-in}$) is less than that of the cell. Basing the calculation of $A_{i,j-in.eff}$ on the runoff coefficient of the cell as an alternative using $Q_{C.i,j-all}$ results in greater departures from Eqn 16.8, but produces a value of infinity when the cell is pervious enough to absorb all the rain that falls on it when some runoff enters the cell from upslope. Such rainfall–runoff conditions can occur, but obviously a value of infinity for $L_{i,j}$ is inappropriate. Consequently, the combination of Eqns 16.10 and 16.11 has the appropriate characteristics to deal with this situation.

When the USLE-M is applied to grid cells:

$$A_{e.i,j} = [Q_{Re}EI_{30}]_{i,j}K_{UMe.i,j}L_{UMe.i,j} \times SC_{UMe.i,j}P_{UMe.i.j} \quad (16.12)$$

where the subscript UM indicates factors whose values differ from those of the USLE and the subscript *e* indicates parameters that vary between rainfall events. When the whole area in which the grid cell occurs is uniform, the USLE-M L factor for the cell with coordinates i, j is given by (Kinnell, 2001):

$$L_{UMe\ i,j} = \frac{Q_{Ce.i,j}(A_{i,j-in} + D^2)^{m+1} - Q_{Ce.i,j-in}A_{i,j-in}^{\ m+1}}{Q_{Re.i,j-cell}D^{m+2}x_{i,j}^{\ m}(22.13)^m} \quad (16.13)$$

For both uniform and non-uniform areas, it follows from Eqns 16.10 and 16.11 that:

$$L_{UMe\ i,j} = \frac{Q_{Ce.i,j-eff}(A_{i,j-in.eff} + D^2)^{m+1} - Q_{Ce.i,j-eff}A_{i,j-in.eff}^{\ m+1}}{Q_{Re.i,j-cell}D^{m+2}x_{i,j}^{\ m}(22.13)^m} \quad (16.14)$$

where:

$$Q_{Ce.i,j-eff} = [(Q_{Ce.i,j-cell}D^2) + (Q_{Ce.i,j-in}A_{i,j-in.})] / (A_{i,j-in.eff} + D^2) \quad (16.15)$$

Thus, when applied to grid cells or segments in a hillslope, the direct inclusion of runoff as a factor in the USLE-M results in an approach that takes account of temporal and spatial variations in runoff on erosion which is not possible using the USLE or the RUSLE.

Discussion

Although, in theory, the USLE-M has the capacity to predict event erosion better than the USLE, that capacity can only be realized if appropriate procedures exist for predicting runoff and values of $K_{e.UM}$, $C_{e.UM}$ and $P_{e.UM}$. There are a number of methods for predicting event runoff and it is up to the user to select an appropriate method. The USDA Curve Number approach (USDA, 1972) is commonly used to predict event runoff in water quality models and can be used with the USLE-M. At this time, procedures for determining $K_{e.UM}$, $C_{e.UM}$ and $P_{e.UM}$ have yet to be determined to the same degree as for the USLE. Alternatively, because they are ratios with respect to the soil loss from the bare fallow cultivation up and down the slope condition, it can be suggested that USLE parameter values for *L*, *S*, *C* and *P* can be used with an event erosivity factor other than EI_{30} if the event erosivity factor is applied to predicting erosion from bare fallow with cultivation up and down the slope. Thus:

$$A_e = [Q_{R1}EI_{30}]_e K_{UMe} LSC_e P_e \quad (16.16)$$

where Q_{R1} is the runoff coefficient for the bare fallow cultivation up and down the slope condition and C_e and P_e are event values for the USLE *C* and *P* factors respectively, may initially appear to be valid. However, although, as shown by Fig. 16.1, the approach takes advantage of predicting erosion better than the USLE on the bare fallow condition, it assumes that an erosion event will occur on a vegetated area whenever there is an erosion event on the bare fallow area and that assumption is not always correct. This can lead to erosion being predicted on vegetated areas for events when there is none (Fig. 16.4). Thus event

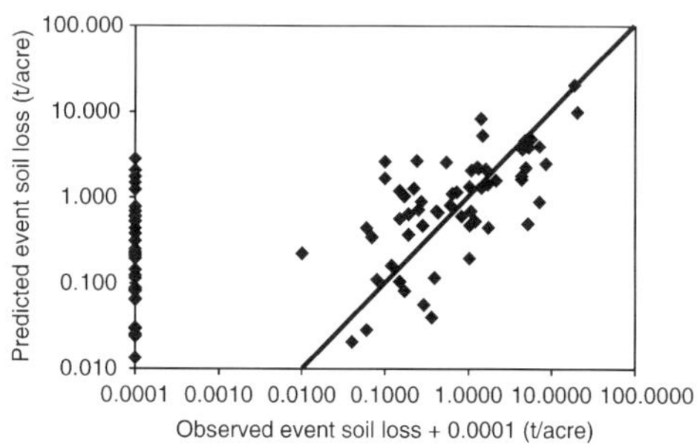

Fig. 16.4. Relationship between event soil losses predicted by multiplying event soil losses from a nearby bare fallow plot by RUSLE period Soil Loss Ratios (fortnightly *C* factor values) and event soil losses observed for conventional maize at Clarinda, Iowa, USA plus 0.0001 t/acre to enable predicted losses to be displayed when observed losses are zero.

erosion can only be predicted appropriately if runoff for the area being eroded is determined and used to calculate the USLE-M event erosivity index in conjunction with the appropriate values of $K_{e.UM}$, $C_{e.UM}$ and $P_{e.UM}$. However, this is not the case when erosion is being predicted on an annual basis. Under these circumstances:

$$A_A = R_{UM.A} K_{UM.A} LSC_A P_A \quad (16.17)$$

applies where $R_{UM.A}$ is total value of $[Q_{R1}EI_{30}]_e$ over the year, $K_{UM.A}$ is the associate erodibility factor and C_A and P_A are annual values for the USLE C and P factors respectively. Equation 16.17 takes advantage of the ability of the product of $R_{UM.A}$ and $K_{UM.A}$ to predict variation in annual soil better that can be achieved using the EI_{30} index. Procedures exist for determining K_{UM} values from USLE K_s (Kinnell and Risse, 1998).

Conclusions

Erosion resulting from sediment moving with runoff is directly related to the product of runoff and sediment concentration. At the small scale, variations in flow depth in rain-impacted flows influence sediment concentrations because the surface water absorbs raindrop energy. However, variations in flow velocity in rain-impacted flows do not cause variations in sediment concentration when RIFT is dominant. This is because particles travel limited distances in the flow following each drop impact, and those distances vary directly with flow velocity. The deposition of detached particles between drop impacts results in a layer of pre-detached material sitting on the surface of the soil matrix. Raindrop impact lifts soil material into the flow from this layer and from the underlying surface if the protective effect of the layer of pre-detached material is not too great. The erodibility of the pre-detached material differs from that of the surface of the soil matrix with the consequence that the erodibility of the eroding area lies somewhere between the two erodibilites. The physico-chemical differences between the two materials, and the lack of knowledge about where between the two erodibilities the actual erodibility of an eroding area lies, makes for difficulties when attempting to relate soil erodibility to measurable soil properties.

At the larger scale, erosion is often modelled using the USLE, a model that contains no direct consideration of runoff. There are variants of the USLE that do consider erosion to be directly dependent on runoff. One variant is the MUSLE, another the USLE-M. Both models use event erosivity indices that differ from that used by the USLE, but the MUSLE uses the USLE factors for K, L, S, C and P inappropriately. The USLE-M does not, and has been observed to account for event soil loss better than both the USLE and the MUSLE at the plot scale. The need to use factor values other than the ones for the USLE when the event erosivity index is changed from EI_{30}, the product of event kinetic energy and the maximum 30-min intensity, to the $Q_R EI_{30}$ index (where Q_R is the runoff ratio) used in the USLE-M can be reduced if the runoff ratio for bare fallow with cultivation up and down the slope is used in the calculation of the index. In this case only the soil erodibility factor has to be changed from that used with the USLE and procedures exist for determining annual values of K_{UM}. However, the approach where USLE C and P factors can be used because soil loss for the bare fallow up and down condition is being predicted using the $Q_R EI_{30}$ index only applies when erosion on the vegetated area occurs whenever erosion on the bare fallow also occurs. This condition does not apply on an event basis, but does at the annual time scale.

References

Arnold, J.G., Srinivasan, R., Muttiah, R.S. and Williams, J.R. (1998) Large area hydrologic modelling and assessment – Part I. *Journal of the American Water Resources Association* 34, 73–89.

Desmet, P.J.J. and Govers, G. (1996) A GIS procedure for automatically calculating the USLE LS factor on topographically complex landscape units. *Journal of Soil and Water Conservation* 51, 427–433.

Flanagan, D.C. and Nearing, M.A. (eds) (1995) *USDA-Water Erosion Prediction Project: Technical Documentation*. NSERL Report No. 10. National Soil Erosion Research Laboratory, West Lafayette, Indiana.
Foster, G.R. and Wischmeier, W.H. (1974) Evaluating irregular slopes for soil loss prediction. *Transactions of the American Society of Agricultural Engineers* 17, 305–309.
Kinnell, P.I.A. (1993) Sediment concentrations resulting from flow depth/drop size interactions in shallow overland flow. *Transactions of the American Society of Agricultural Engineers* 36, 1099–1103.
Kinnell, P.I.A. (2001) Slope length factor for applying the USLE-M to erosion in grid cells. *Soil and Tillage Research* 58, 11–17.
Kinnell, P.I.A. (2003) Event erosivity factor and errors in erosion predictions by some empirical modes. *Australian Journal of Soil Research* 41, 991–1003.
Kinnell, P.I.A. and Risse, L.M. (1998) USLE-M: empirical modeling rainfall erosion through runoff and sediment concentration. *Soil Science Society of America Journal* 62, 1667–1672.
Morgan, R.P.C., Quinton, J.N., Smith, R.E., Govers, G., Poesen, J.W.A., Auerswald, K., Chisci, G., Torri, D. and Styczen, M.E. (1998) The European Soil Erosion Model (EUROSEM): a dynamic approach for predicting sediment transport from fields and small catchments. *Earth Surface Processes and Landforms* 23, 527–544.
Renard, K.G., Foster, G.R., Weesies, G.A., McCool, D.K. and Yoder, D.C. (1997) *Predicting Soil Erosion by Water: A Guide to Conservation Planning with the Revised Universal Soil Loss Equation (RUSLE)*. Agricultural Handbook No. 703. US Department of Agriculture, Washington, DC.
USDA (1972) *National Engineering Handbook*. Soil Conservation Service, US Department of Agriculture, Washington, DC.
Williams, J.R. (1975) Sediment-yield prediction with universal equation using runoff energy factor. In: *Present and Prospective Technology for Predicting Sediment Yield and Sources*. ARS.S-40. US Government Print Office, Washington, DC, pp. 244–252.
Wischmeier, W.C. and Smith, D.D. (1978) *Predicting Rainfall Erosion Losses – A Guide to Conservation Planning*. Agricultural Handbook No. 537. US Department of Agriculture, Washington, DC.

17 The Roles of Natural and Human Disturbances in Forest Soil Erosion[1]

W.J. Elliot
USDA Forest Service Rocky Mountain Research Station, USA

Introduction

Forests provide numerous benefits for society, including fibre, wildlife and recreation. Forest managers are challenged to balance ecosystem health with maintaining public forest lands for multiple uses. During the first half of the last century, public forest management emphasized the harvesting of forest resources. In recent years, public forest management goals have shifted to long-term sustainability.

During most of the last century, fire suppression and timber harvest were the main fuel management practices. Timber harvest generally implies the removal of logs that can be processed into lumber. Most timber harvest activities removed almost all of the standing timber, leaving behind smaller trees, diseased trees or undesirable species. Selective harvesting and fire suppression activities have resulted in a surplus of fuels in many forests (Schmidt *et al.*, 2002). These fuels are causing an increase in high-severity wildfires. Fuel management in forests is a new challenge for many agencies to address in the USA. Forest managers are now carrying out practices to reduce this excess fuel, including thinning and prescribed fire. Thinning removes small diameter material, much of which has limited market value. Larger trees are frequently left behind during thinning operations.

Soil erosion is another major concern in forest management. In forested watersheds, erosion includes upland surface processes such as rill and interrill erosion, gullying and channel processes, and mass wasting. This paper will focus on surface erosion, and the delivery of that source of sediment to and through stream systems. Fires, timber harvest and roads increase soil erosion and sediment delivery from forest watersheds. Soil erosion reduces forest productivity and eroded sediment may adversely affect water quality in forest streams. Managers are seeking to minimize erosion by applying improved management practices for forestry operations and roads. One of the considerations when seeking to minimize erosion is whether fuel management operations like prescribed fire or thinning (and the associated road network) cause more or less erosion than wildfires. The purpose of this paper is to compare predicted upland erosion rates following forest disturbances associated with fuel management, such as harvesting, thinning, prescribed fire and road networks, to erosion rates following wildfires.

Forest Erosion Processes

Forest disturbances such as forestry operations and wildfire have major effects on both the vegetation and the soil properties. Soil erodibility depends on both the surface cover and the soil texture (Robichaud *et al*., 1993; Elliot and Hall, 1997). Erodibility following a wildfire is much greater than that in an undisturbed forest (Robichaud *et al*., 1993). Forests are highly susceptible to erosion in the year following a fire or a forestry operation. Forests do, however, recover quickly as vegetation regrowth is rapid when smaller plants do not have to compete with trees for sunlight, nutrients and water. For example, erosion rates following a wildfire in Eastern Oregon dropped more than 90% in the first year, with no erosion observed in year 4 (Fig. 17.1).

Erosion in forests is highly variable. In many ecosystems it is driven by a few extreme events each decade, and highly influenced by recent disturbances. Sediment production during years without large runoff events or disturbances are likely to be well below 'average' erosion rates, and the production during a year with a major runoff event or following a major disturbance is likely to be well above the 'average' rate (Kirchner *et al*., 2001). For example, road erosion is influenced by weather and presence of traffic (Fig. 17.2). In the 4-year study in Fig. 17.2 (Foltz, 1998), there was more than

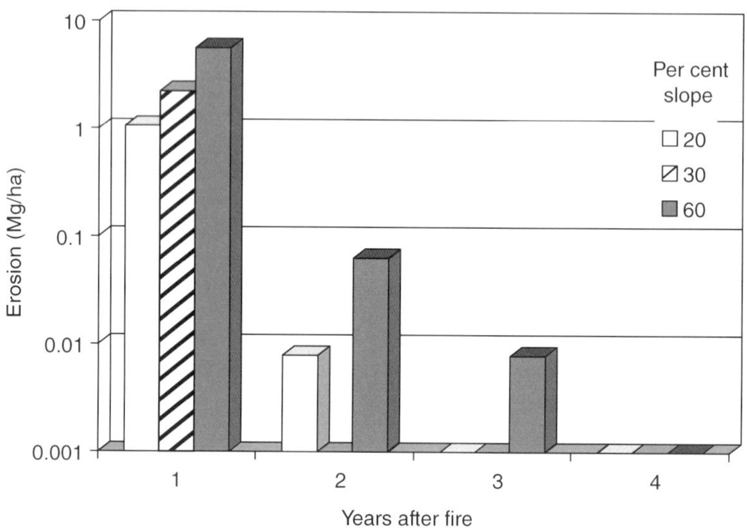

Fig. 17.1. Annual erosion rates measured following a wildfire in Eastern Oregon, for plots 25–40 m long with three different slopes for the first 4 years following a wildfire (adapted from Robichaud and Brown, 1999).

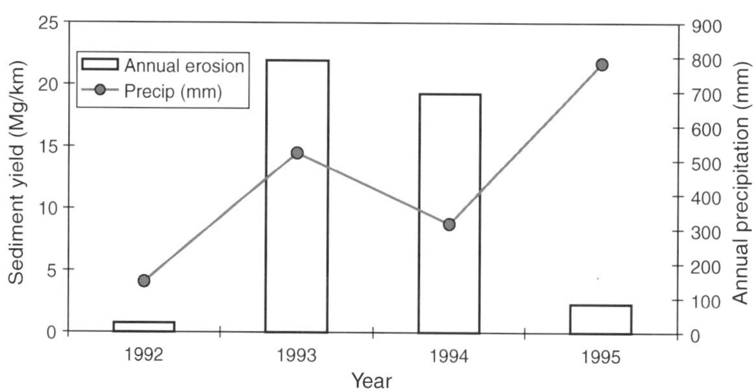

Fig. 17.2. Effect of weather and traffic on road sediment yields for a forest road in the Cascade Range in Oregon. Years 1992–1994 had traffic and 1995 had no traffic (based on Foltz, 1998).

an order of a magnitude difference in annual erosion rates due to differences in precipitation between years 2 and 3. In the final year of the study (1995), erosion rates were much less without traffic than in the previous 2 years, even though it was the wettest year of the study.

Eroded sediments are frequently deposited in stream channels where they may remain for years to decades, slowly moving through the stream system in response to high runoff events (Trimble, 1999). The attenuation of sediment in stream channels and its role in watershed processes increases the importance of the scale at which sedimentation is measured. Hillslope scales will show large variations in erosion rates as disturbed sites recover (Fig. 17.1). Watershed scale observations will tend to reflect decade to century trends in erosion rates, with large sedimentation events associated with infrequent watershed disturbances or flood events (e.g. McClelland *et al.*, 1997; Kirchner *et al.*, 2001). Both managers and the public tend to focus on erosion and sediment delivery occurring in the first year or two following a disturbance. They generally fail to consider the impacts of that sediment as it is transported through the watershed stream systems in the decades that follow.

Erosion Prediction

Prediction of soil erosion by water is a common practice for natural resource managers evaluating impacts of management activities on upland erosion and downstream water quality. Erosion prediction tools are used to evaluate different management practices and erosion control techniques. One of the prediction tools recently developed is the Water Erosion Prediction Project (WEPP) model (Flanagan and Livingston, 1995). The WEPP model is physically based, and is particularly suited to modelling common forest conditions. A set of input files describing forest conditions was developed for the model (Elliot and Hall, 1997) and later a user-friendly suite of Internet interfaces was developed (Elliot, 2004). Included with these interfaces is a database of typical forest soil and vegetation conditions. These databases are comprised of soil erodibility values determined from rainfall simulation and from natural rainfall events by scientists within the Rocky Mountain Research Station and elsewhere. Validation of the WEPP model to estimate erosion rates due to forest disturbances has been encouraging (Elliot and Foltz, 2001).

Modelling Typical Management Scenarios

The WEPP model Internet interfaces for forests (Elliot, 2004) were used to compare the estimated sediment yields from fuel management for wildfire for two different sites (Table 17.1): the Bitterroot Mountains in Western Montana, and the western slopes of the Cascade Mountains in Western Oregon. The assumptions for the

Table 17.1. Model inputs for two example erosion analyses using the WEPP model. Both total slope lengths were assumed to be 200 m.

Site	Bitterroot Range, Montana	Cascade Range, Oregon
Annual precipitation (mm)	1021	2640
Wildfire cycle (years)	40	200
Thinning cycle (years)	20	10
Prescribed fire cycle (years)	20	20
Harvest frequency (years)	80	40
Slope steepness (%)	30	60
Buffer width (m)	30	60
Harvesting disturbance assumptions	85% cover on harvested area in year 1, increasing to 100% in year 5	
Wildfire disturbance	45% cover in the year following the fire, increasing to 100% by year 10	
Road density	4.0 km/km^2	

Montana conditions were a relatively dry forest (average precipitation is 1021 mm), a 40-year fire cycle, an 80-year harvest cycle and 30% slope steepness. The assumptions for the Oregon conditions (average precipitation is 2640 mm) were a 200-year fire cycle, a 40-year harvest cycle and 60% slope steepness common on these less eroded mountains. Both scenarios assumed a 200-m long slope. The two management scenarios were chosen to demonstrate the utility of the prediction tool and the erosion risks associated with fuel management and harvest activities. USDA Forest Service soil quality standards state that timber harvest, thinning and prescribed fire will expose a maximum of 15% mineral soil (Page-Dumroese *et al.*, 2000), so these values were used for their respective runs.

For each of the disturbances, 50 years of typical climate were generated by the FS-WEPP interface, to give 50 possible erosion rates. The interface calculated the average erosion rates and the 5-year return period erosion rates. Runs were carried out for wildfire, prescribed fire, harvesting and thinning. The hillside impact for thinning was assumed the same as harvesting. All scenarios except wildfire assumed an undisturbed buffer. A buffer in this case refers to a strip of vegetation along either side of an ephemeral or perennial stream to reduce delivery of upland eroded sediments. Buffers frequently include much of the forest stream riparian zone. Buffer widths used for this exercise are noted in Table 17.1. In the years following each disturbance, the hillslope was modelled as sequentially recovering, as recommended in the online documentation (Elliot *et al.*, 2002), for 5 years following thinning and prescribed fire, and 10 years following wildfire.

Modelling Results

The results from the two scenarios are presented in Fig. 17.3. Figure 17.3a illustrates the wetter climate in the Oregon Cascade Range and Fig. 17.3b the drier climate in the Bitterroot Mountains in Montana. There are several striking features on these two figures. The erosion following wildfire is more than two orders of magnitude greater than before the fire, and more than a magnitude greater than following a major forest operation with a buffer. Also, the erosion rate in the Cascades is an order of magnitude greater than that in the Bitterroots.

Thinning occurs at a greater frequency in the wetter climate in the Cascades than in the Bitterroot Range (Fig. 17.3a), contributing to the higher overall erosion rate (Table 17.2). Figure 17.3b shows that the erosion rate following a prescribed fire, assuming a 15% mineral soil exposure, is 15 times greater than the erosion rate following a harvest operation.

Table 17.3 presents the runoff predictions expected the year following each respective disturbance from rainfall, and from winter events which include snowmelt and combined rainfall and snowmelt events. Winter processes dominate in both climates, but account for a higher proportion of the total runoff in the Bitterroot climate.

Weather is highly variable year to year. If the year following a fire or other disturbance is drier than normal, sediment yields are low to none. If the year has some major runoff-generating storms, then erosion rates will be high. Figure 17.3 only shows the average predicted sediment yields from 50 different years of weather patterns for each point. Table 17.4 shows the probability that the sediment yield will be non-zero in the year of disturbance, and the sediment yield average from 50 different years. Table 17.4 also shows the sediment yield that may occur if the year following the disturbance is the most erosive year in five.

Table 17.4 appears to have a mismatch of data at first glance because the ‘average’ sediment yields are greater than the greatest yields in 5 years for both harvest scenarios, and for the Bitterroot climate after fire. This is because in these scenarios the only time that sediment was delivered was from a small number of highly erosive years. Hence, the most erosive year in five did not generate any sediment in the Bitterroot climate following harvesting, and was not sufficiently erosive to generate as much sediment as a 50-year average, for the Cascade harvest results, and the Bitterroot wildfire results. Understanding the reliance of extreme events as the causes of hillslope erosion in forests is critical when interpreting erosion research studies, or erosion modelling results.

Compared to the Bitterroot Range, there is a much greater likelihood that there will be

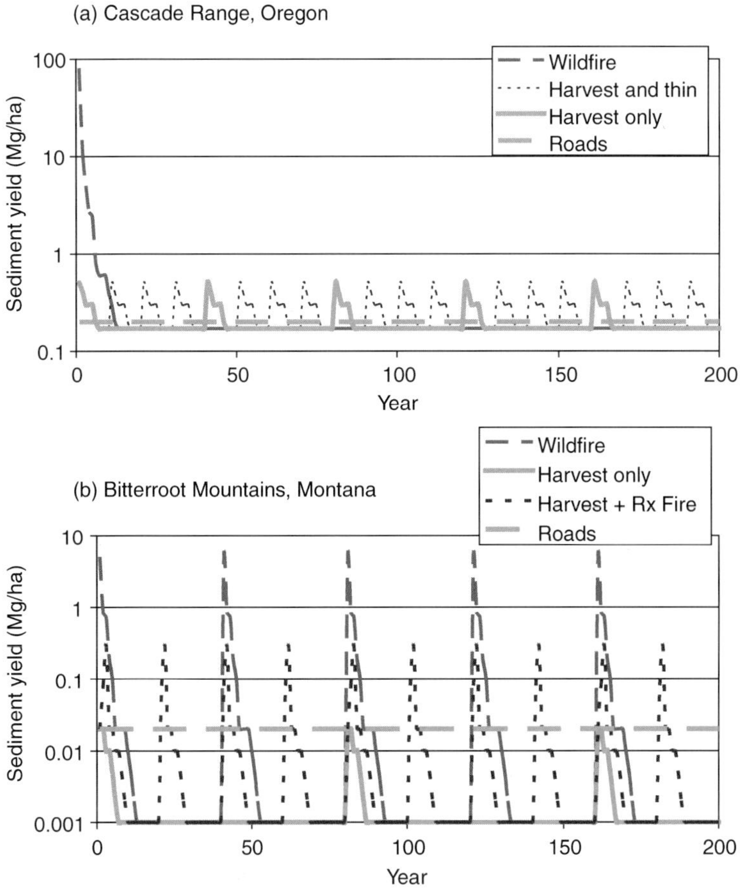

Fig. 17.3. Predicted annual hillslope sediment yield for an 'average' weather pattern vs year for different management conditions for the two scenarios described in Table 17.1 using the WEPP model.

Table 17.2. Predicted average annual sediment delivery rates over 200 years for the two scenarios presented in Table 17.1 and Fig. 17.3.

	Average annual delivery during 200 years (Mg/ha)	
	Bitterroot Range	Cascade Range
Wildfire	0.175	0.69
Harvest only	0.002	0.19
Harvest with thinning	0.004	0.27
Harvest with prescribed fire	0.023	0.28
Roads with density of 4 km/km^2	0.02	0.2

sediment delivered in the Cascade Range following a disturbance, and sediment delivery rates are much higher. In the Bitterroot Range, it was predicted that there is only a 6% chance of sediment delivery in the year following a forest operation (Table 17.4) following current federal guidelines to limit mineral soil exposure to only 15%. Consequently, if a field study is

Table 17.3. Predicted sources of runoff (rain or winter events) for 50 years of run for each of the disturbances for each climate.

	Runoff in year following disturbance (mm)			
	Bitterroot Range		Cascade Range	
Disturbance	Rain	Winter	Rain	Winter
Wildfire	1.8	7.1	23.8	56.5
Harvest or thinning	0.0	0.0	0.1	1.1
Prescribed fire	0.1	0.1	1.0	1.7

Table 17.4. The average predicted sediment yields the year following a disturbance from 50 different annual weather sequences, and the erosion resulting from the most erosive year in five.

	Bitterroot Range	Cascade Range
Precipitation (mm)		
Average	1021	2640
Greatest in 5 years	1138	2868
Sediment yield first year after harvest only		
Probability > 0 (%)	6	34
Average (Mg/ha)	0.02	0.49
Greatest in 5 years (Mg/ha)	0.0	0.14
Sediment yield first year after wildfire		
Probability > 0 (%)	82	100
Average (Mg/ha)	4.98	81
Greatest in 5 years (Mg/ha)	4.42	116

installed to measure sediment yields following a forest operation, there is only a 6% chance that any sediment will be collected, and a 94% chance that there will be no observed sediment yield. In the wetter and steeper Cascade Range scenario, there is a 34% chance of sediment delivery across a buffer in the year following harvesting or thinning.

Sediment yield from road networks depends on a number of watershed attributes, including climate, road design, construction methods and use, and topography. Elliot and Miller (2002) estimated road contributions (per km of road) for a wide range of western USA ecoregions. Table 17.2 provides an average estimate of road network sediment yields for the two sites, assuming a road density of 4 km/km^2. In the absence of traffic associated with fuel management, low-use roads are likely to yield only about 10–20% of the sediment expected from high-use roads. Table 17.2 and Fig. 17.3 both show the higher traffic road sediment rates, as most fuel reduction or harvesting scenarios will result in higher levels of traffic on much of a watershed's road network every year, assuming the watershed area is limited to about 5–10 km^2.

Discussion

On Fig. 17.3 and Tables 17.2–17.4, even though the difference in precipitation is only about a factor of 2.5 and the slope is twice as steep, the differences in runoff and erosion after the disturbance is more than ten times as great. Compared to the Cascade Range, a higher proportion of the runoff in the Bitterroot site is from winter processes (Table 17.3). Snow and snowmelt rates (typically 1 mm/h) are generally

Table 17.5. Observed erosion rates following harvest and wildfire in or near the Bitterroot Range.

	Year	Precipitation (mm)	Sediment yield (Mg/ha)
Sediment yield first year after harvest only[a]			
Dry year	1994	221	0.0
Sediment yield first year after wildfire			
Below avg. precip. year	2001	599[b]	0.5[c]
Above avg. precip. year	2002	1036[b]	10.0[c]

[a]Covert (2003); [b]USDA-NRCS (2005); [c]Elliot and Robichaud (2004).

much lower than rainfall intensities (typically up to 25 mm/h). These differences in amount and form of precipitation tend to bridge some of the thresholds that are common in erosion processes. Snowmelt rates are generally well below forest infiltration rates until soils are saturated. Low runoff rates from snowmelt or low intensity precipitation events frequently do not exceed critical shear values of forest soils, so erosion is limited to raindrop splash erosion. Hence, an increase in precipitation of 2.5 and the steeper slopes cause these types of threshold values to be exceeded at the hillslope scale, which causes the disproportionate increase in predicted erosion rates in Fig. 17.3 and Table 17.4.

The predicted results from the Bitterroot site can be compared to observed erosion rates from studies over the past 10 years. Table 17.5 shows the results from three separate studies, one after thinning and prescribed fire, with a dry year following, one after wildfire with a moderate year following, and one following a wildfire with a wet year following. The results in Table 17.5 support the WEPP model predictions presented in Table 17.4, with observed erosion rates well below average in dry years, and well above average in a wet year containing several high intensity summer storms (Elliot and Robichaud, 2004).

The magnitude of the results from the two climates presented in Fig. 17.3 and Tables 17.2–17.4 cannot be directly compared to each other because of the differences in climate, slope and management. What is apparent, however, is that the same principles of considering erosion following disturbances and frequencies of those disturbances are critical to the watershed planning process.

Results from this study suggest that using an average erosion rate may not accurately reflect the impacts of forest disturbances on watersheds, particularly when attempting to characterize watershed impacts over a short time period. Following a forest disturbance, the greatest amount of sediment is delivered in the first year, and after several years, delivery is below the level of detection. The amount of sediment delivered is highly dependent on the first year's climate. Eroded sediments following wildfire are not likely to be routed through the stream network for a number of years, or even decades. In the interim, sediments are stored in the alluvial deposits of forest streams. Watershed managers need to better understand risks associated with the different levels and temporal nature of sediment yields, and use that knowledge to develop forest management strategies, like timing and frequency of prescribed fire or thinning. Erosion from roads must be considered in any management strategy to estimate the total impact of management activities (Conroy, 2001; Elliot and Miller, 2002). As previously discussed, managers need to exercise caution when dealing with average values, as variability and outliers frequently dominate hydrologic processes.

The results of the modelling analysis presented in Fig. 17.3 and Tables 17.2–17.4 raise a number of important issues for further discussion on the impacts of timber harvest or fuel management on sediment movement in forested watersheds. Erosion following wildfire is much greater than erosion due to forestry operations, despite the higher frequency of such operations. Erosion from wildfire, however, is a natural phenomenon, which has driven the development of forest and associated stream ecosystems. Occasional high upland erosion rates and large sediment yields have played an important role in shaping landscapes and introducing fresh material into our stream systems

(Kirchner *et al.*, 2001). In the last century, scientists have concluded that fire was important for ecosystem health, and that fire exclusion has resulted in a decline of the health of many forests (Schmidt *et al.*, 2002; Conard and Hilburner, 2003). Currently, scientists and forest managers are trying to determine if wildfire severity can be reduced with fuel management practices (Conard and Hilburner, 2003). If wildfire occurrence is reduced, will the large runoff and erosion events that follow lead to a decline in the health of hydrologic and aquatic ecosystems? A related question is: can erosion associated with severe wildfires be reduced with fuel management practices? These questions will require significant interdisciplinary research to increase understanding of the relationships among wildfire, fuel management and watershed health.

Forest roads contribute sediment to stream systems in most years, and in many years generate more sediment than forested hillsides (Fig. 17.3). Modelling results from this study assumed that forest operations may contribute low levels of sediment to stream systems more frequently than the natural wildfire cycle (Fig. 17.3). Sediment delivered from roads and forestry operations are likely to have a finer texture than sediment from wildfire. They are less likely to contribute cobbles to the stream beds that are preferred by many aquatic organisms. Further research is needed to evaluate the importance of large runoff events following fires for delivering coarse sediments to streams, while flushing fine-textured sediments through the stream system.

Summary

The WEPP model was used to compare sediment yields from forested hillslopes following wildfire to those from the same slopes following forestry operations. Sediment yields following forestry operations are much lower than those following wildfire in both the year following the disturbance and when averaged over two centuries. It is not known, however, if reducing large sediment yields that follow wildfire will result in improved watershed health in the long term.

Field work and modelling results lead to the following conclusions:

- Sediment delivery following forest operations and prescribed fire with forested buffers are an order of magnitude or more lower than following wildfire.
- Roads can generate a significant proportion of sediment in a forested watershed.
- Additional research is needed to determine long-term effects of fuel management practices on the occurrence and severity of wildfire, and on watershed health.

References

Conard, S. and Hilbruner, M. (2003) *Influence of Forest Structure on Wildfire Behavior and the Severity of its Effects*. USDA Forest Service, Washington, DC.

Conroy, W.J. (2001) Use of WEPP modeling in watershed analysis and timber harvest planning. Paper no. 01-8005. Presented at the *American Society of Agricultural Engineers Annual International Meeting*, 20 July–1 August, Sacramento, California. American Society of Agricultural Engineers, St Joseph, Michigan.

Covert, S.A. (2003) Accuracy assessment of WEPP-based erosion models on three small, harvested and burned forest watersheds. Unpublished MSc thesis. University of Idaho, Moscow, Idaho.

Elliot, W.J. (2004) WEPP Internet interfaces for forest erosion prediction. *Journal of the American Water Resources Association* 40, 299–309.

Elliot, W.J. and Foltz, M. (2001) Validation of the FSWEPP interfaces for forest roads and disturbances. Paper no. 01-8009. Presented at the *American Society of Agricultural Engineers Annual International Meeting*, 30 July–1 August, Sacramento, California. American Society of Agricultural Engineers, St Joseph, Michigan.

Elliot, W.J. and Hall, D.E. (1997) *Water Erosion Prediction Project (WEPP) Forest Applications*. General Technical Report INT-GTR-365. USDA Forest Service, Rocky Mountain Research Station, Ogden, Utah.

Elliot, W.J. and Miller, I.S. (2002) Estimating erosion impacts from implementing the National Fire Plan. Paper no. 02-5011. Presented at the *American Society of Agricultural Engineers Annual International Meeting*, 28 July–31 July, Chicago, Illinois. American Society of Agricultural Engineers, St Joseph, Michigan.

Elliot, W.J. and Robichaud, P.R. (2004) The effectiveness of postfire mitigation treatments. Presented at the 2004 *Burned Area Emergency Rehabilitation Training Workshop*. April. Denver, Colorado.

Elliot, W.J., Hall, D.E. and Scheele, D.L. (2002) Disturbed WEPP: WEPP interface for disturbed forest and range runoff, erosion and sediment delivery. http://forest.moscowfsl.wsu.edu/fswepp/docs/distweppdoc.html (Accessed February 2004.)

Flanagan, D.C. and Livingston, S.J. (eds) (1995) *WEPP User Summary. NSERL Report No. 11*. USDA Agriculture Research Service, National Soil Erosion Research Laboratory, W. Lafayette, Indiana.

Foltz, R.B. (1998) Traffic and no-traffic on an aggregate surfaced road: sediment production differences. *Proceedings of the Seminar on Environmentally Sound Forest Roads and Wood Transport*, 17–22 June 1996. Food and Agriculture Organization, Rome.

Kirchner, J.W., Finkel, R.C., Riebe, C.S., Granger, D.E., Clayton, J.L., King, J.G. and Megahan, W.F. (2001) Mountain erosion over 10 yr, 10 k.y., and 10 m.y. time scales. *Geology* 29, 591–594.

McClelland, D.E., Foltz, R.B., Wilson, W.D., Heinemann, R., Cundy, T.W., Saurbier, J.A. and Schuster, R.L. (1997) *Assessment of the 1995 and 1996 Floods and Landslides on the Clearwater National Forest.* Region 1. USDA Forest Service, Missoula, Montana.

Page-Dumroese, D., Jurgensen, M., Elliot, W., Rice, T., Nesser, J., Collins, T. and Meurisse, R. (2000) Soil quality standards and guidelines for forest sustainability in northwestern North America. *Forest Ecology and Management* 138, 445–462.

Robichaud, P.R. and Brown, R.E. (1999) What happened after the smoke cleared: Onsite erosion rates after a wildfire in Eastern Oregon. (Revised) In: Olsen, D.S. and Potyondy, J.P. (eds) *Proceedings AWRA Specialty Conference: Wildland Hydrology*, Nov. 2000, Bozeman, Montana. AWRA, Herndon, Virginia, pp. 419–426.

Robichaud, P.R., Luce, C.H. and Brown, R.E. (1993) Variation among different surface conditions in timber harvest sites in the Southern Appalachians. In: Larionov, J.A. and Nearing, M.A. (eds) *Proceedings from the Russia, U.S. and Ukraine International Workshop on Quantitative Assessment of Soil Erosion*. Moscow, Russia, 20–24 September. The Center of Technology Transfer and Pollution Prevention, Purdue University, West Lafayette, Indiana, pp. 231–241.

Schmidt, K.M., Menakis, J.P., Hardy, C.C., Hann, W.J. and Bunnell, D.L. (2002) *Development of Coarse-scale Spatial Data for Wildland Fire and Fuel Management*. General Technical Report RMRS-GTR-87. US Department of Agriculture, Forest Service, Rocky Mountain Research Station, Fort Collins, Colorado.

Trimble, S.W. (1999) Decreased rates of alluvial sediment storage in the Coon Creek Basin, Wisconsin, 1975–1993. *Science* 285, 1244–1246.

USDA-NRCS (2005) Montana SNOTEL Sites. United States Dept. of Agriculture Natural Resource Conservation Service. http://www.wcc.nrcs.usda.gov/snotel/Montana/montana.html (Accessed February 2005.)

18 Runoff and Erosion Modelling by WEPP in an Experimental Mediterranean Watershed

F. Licciardello,[1] E. Amore,[1] M.A. Nearing[2] and S.M. Zimbone[3]
[1]*University of Catania, Italy;* [2]*USDA-ARS Southwest Watershed Research Center, USA;* [3]*Mediterranean University of Reggio Calabria, Italy*

Introduction

In recent decades several simulation models have been developed to estimate and analyse the impact of water erosion at watershed scales (Renard *et al.*, 1982; Singh, 1995; Singh and Frevert, 2002), but more work is needed to test and improve their applicability and efficiency in environmental situations that differ from those where the models were developed (Goodrich and Simanton, 1995; Soto and Díaz-Fierros, 1998; Duiker *et al.*, 2001).

The Water Erosion Prediction Project (WEPP, Nearing *et al.*, 1989) is a physically based, distributed parameter model that has been widely applied around the world (Laflen *et al.*, 1994; Klik *et al.*, 1995; Liu *et al.*, 1997; Flanagan *et al.*, 1998; Hebel and Siegrist, 1998; Kincaid, 2002) to simulate the main physical processes related to infiltration, percolation, runoff and soil erosion phenomena at hillslope and watershed scales. A linkage between WEPP and geographical databases, called GeoWEPP, is under development to automate slope, soil and management parameterization (Renschler and Harbor, 2002; Renschler *et al.*, 2002).

Several tests of the WEPP model successfully conducted in the USA on both field experimental plots (Zhang *et al.*, 1996; Nearing and Nicks, 1997; Flanagan *et al.*, 1998; Tiwari *et al.*, 2000) and small watersheds (0.34–18.20 ha) (Savabi *et al.*, 1996; Liu *et al.*, 1997) have shown results comparable with those produced by other models (Tiwari *et al.*, 2000; Bhuyan *et al.*, 2002). Some efforts at model calibration at the plot scale in European conditions have produced acceptable results (Klik *et al.*, 1995; Hebel and Siegrist, 1998; Vlnasova *et al.*, 1998).

A few applications of WEPP have been carried out in Mediterranean conditions. Simulations of soil water content, runoff and erosion by WEPP for experimental plots in north-west Spain have shown reasonable agreement with observed values (Soto and Díaz-Fierros, 1998). An overestimation of interrill erodibility by the model was found for Mediterranean soils with stable aggregation in southern Spain (Duiker *et al.*, 2001). Results of comparison with measurements of deposited sediment in three Sicilian reservoirs with drainage watersheds of 115–570 km^2 (Santoro *et al.*, 2002) have shown that the greater the amount of eroded sediment, the smaller were the relative errors that resulted using the WEPP model.

In order to assess of the performance of erosion models in Mediterranean conditions, a monitoring programme in a small mountainous watershed was initiated in eastern Sicily (Italy) 7 years ago. In this paper the results of applications of WEPP to the monitored watershed model are analysed in order to draw conclusions on model implementation and performance in the experimental conditions studied.

Materials and Methods

Main characteristics of the experimental watershed

The model was applied to data sets collected in eastern Sicily from a small mountainous watershed, called Cannata, which is a tributary, ephemeral in flow, of the Flascio River. The watershed (Fig. 18.1), covering about 130 ha, is equipped with the meteorological station A, recording rainfall, temperature, wind, solar radiation and pan evaporation, two additional rainfall gauges indicated by B and C, as well as a hydrometrograph connected to a runoff water automatic sampler (for the control of sediment concentration in the flow) (D and E).

Topsoil characteristics were investigated by a field survey at 57 sites within the watershed. Clay-loam (USDA classification) was the dominant texture (63% of spatially distributed samples). Guelph permeameter measurements yielded low to medium values of the saturated hydraulic conductivity (0.2–17.6 mm/h; $N = 57$; CV = 103%). Land use monitoring has highlighted the persisting prevalence of pasture areas (ranging between 87% and 92% of the watershed area) with different vegetation complexes (each grouping up to 15 species) and ground covers. In particular, four soil cover conditions can be distinguished: a high-density herbaceous vegetation (eventually subjected to tillage operations); a medium density herbaceous vegetation; sparse shrubs; and cultivated winter wheat with a wheat–fallow rotation. More detailed information about the watershed characteristics and the monitoring equipment were reported in a previous paper (Licciardello *et al.*, 2001).

Model parameterization

Morphological discretization of the watershed

GeoWEPP was used for the discretization of the watershed into a number of subwatersheds (groups of hillslopes) contributing to channels (Fig. 18.2). A Digital Terrain Model was arranged over a grid of 5 × 5 m cells using ARCVIEW 3.2 by digitizing 2-m elevation contour lines. The Critical Source Area (the threshold area at which a permanent channel begins) and the Minimum Source Channel Length (the minimum length of a channel segment) were set to 1.25 ha and 100 m, respectively, in order to optimize the reproduction of the watershed morphology.

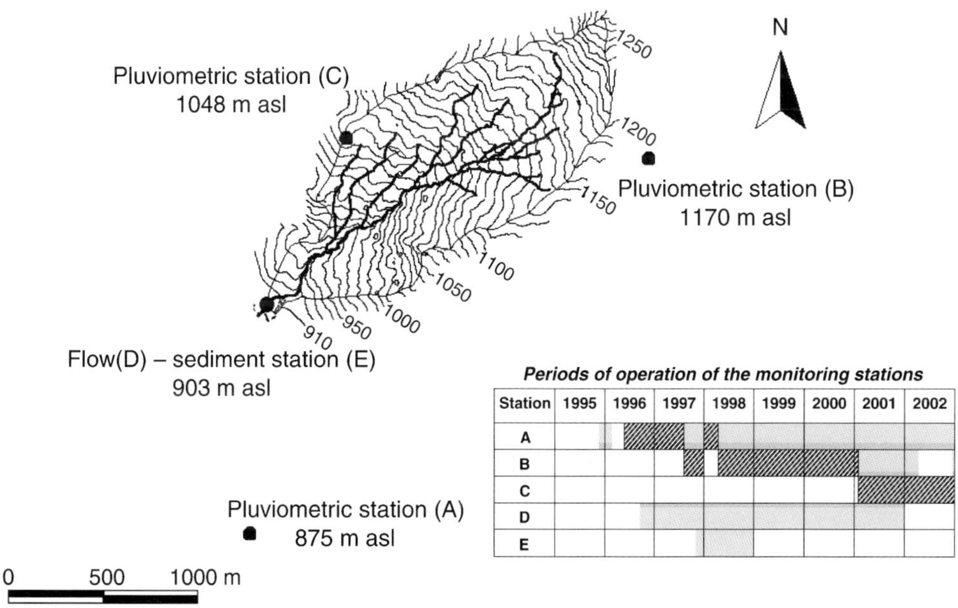

Fig. 18.1. Location and operating periods of the monitoring stations at Cannata watershed, Sicily.

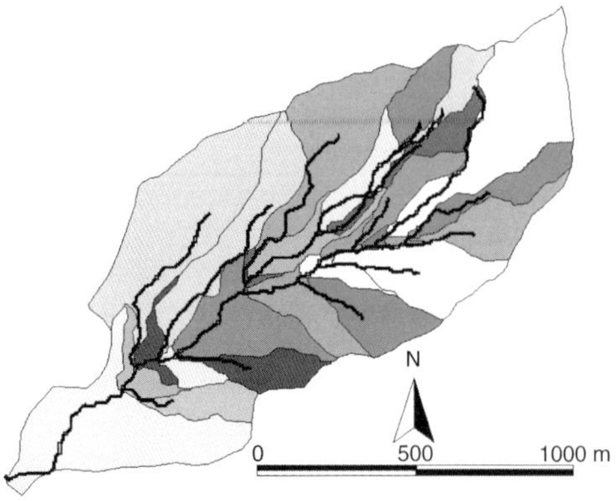

Fig. 18.2. Layout of subwatersheds and drainage network after GeoWEPP application to Cannata watershed.

This resulted in 27 subwatersheds (0.32–16.15 ha with two to three hillslopes), 68 hillslopes (0.01–11.49 ha) and 27 channels. Due to the morphology of the watershed, 30% of the obtained hillslopes were longer than 100 m (common recommended limit; Baffaut *et al.*, 1997).

The Watershed Project (i.e. the morphological schematization of the watershed for input to WEPP) was built through the WEPP Windows Interface using the morphological information on hillslopes and channel network taken from GeoWEPP. ARCVIEW 3.2 was used to overlay soil texture and land use of hillslopes in order to set the size and position of the Overland Flow Elements (OFEs) over each hillslope. Upper channels were treated as ditches, while lower channels were tested as ungraded channels. Twenty different types of OFEs, with a maximum number of ten in a single hillslope, were identified depending on land cover and soil texture combinations.

Construction of input files

The WEPP watershed version was applied on a continuous basis to the observation period from June 1996 to December 2002; the period June–October 1996 was used to train the model and build up initial soil moisture conditions. The Breakpoint Climate Data Generator (BPCDG, Zeleke *et al.*, 1999) was used to build the climate file. Climatic data (daily values of maximum and minimum air temperature, relative humidity, solar radiation and wind velocity and direction at 8:00 a.m. and 6:00 p.m.) were taken from meteorological station A; information on the rainfall pattern (depth and mean intensity in time definite range) were taken from the pluviometric station B in Fig. 18.1, which, based on previous rainfall–runoff data analysis, appeared more representative of true precipitation conditions. Daily values of dew point temperature were calculated on the basis of daily values of air temperature and relative humidity. Uniform soil profiles were assumed. For each of the five soil textures in the watershed, the data were derived by averaging the sand (particle diameter 0.1–2.0 mm), clay, very fine sand (particle diameter 0.05–0.1 mm), organic matter and rock content (particle diameter > 2 mm), cation exchange capacity (CEC), and bulk density from the 57 field samples (up to 36 for each type of soil). Three simulation series were performed using three different sets of the effective hydraulic conductivity inputs, K_e (Table 18.1), to which model outputs have shown a high sensitivity in previous work (Nearing *et al.*, 1990). Numerous attempts have been carried out to improve model simulations by setting the K_e values as a function of physical characteristics (Kidwell *et al.*, 1997; Kincaid, 2002), by a calibration of the runoff data (Hebel and Siegrist, 1998; Savabi, 2001), using measured infiltration data (Savabi, 2001) and by a non-linear regression relationship between K_e and SCS Curve Number (USDA, 1972) for fallow and cropped conditions (Nearing *et al.*, 1996).

In this study, in simulation series I, the K_e input values were internally calculated by WEPP based upon sand and clay content and CEC of the soil. In simulation series II, the K_e values were based on the median field saturated conductivity for each soil type (Bouwer, 1969), resulting in values in the range of 0.4–4.7 mm/h. In simulation series III, K_e values for cropland (1.9–5.8 mm/h) were estimated based on the relationship developed by Nearing *et al.* (1996). WEPP was run using both constant and internally adjusted by the model K_e values for the three simulations series. The interrill erodibility (K_i), the rill erodibility (K_r) and the critical hydraulic shear (τ_c), calculated for the hillslopes as recommended in the WEPP User Summary (Flanagan and Livingstone, 1995), were in the suggested range both for cropland and rangeland areas (Table 18.2).

Table 18.1. Set-up method of effective soil hydraulic conductivity (K_e) in the simulations by WEPP at Cannata watershed.

Simulation series			
I	II	III	
K_e internally calculated by WEPP based upon sand and clay content and CEC of the soil	K_e = 0.5 field saturated conductivity[a] measured by the Guelph permeameter	Rangeland K_e as in the simulation series II	Cropland $K_e = f$ (Curve Number)[b]

[a]As proposed by Bower (1969); [b]as proposed by Nearing *et al.* (1996).

Table 18.2. Interrill erodibility (K_i), rill erodibility (K_r) and critical hydraulic shear (τ_c) input values set according to the equations in the WEPP User Summary (Flanagan and Livingstone, 1995).

	K_i (10^3 kg/s/m^4)		K_r (10^{-3} s/m)		τ_c (N/m^2)	
Soil texture[a]	Rangeland	Cropland	Rangeland	Cropland	Rangeland	Cropland
Clay	796.18	3708.22	0.91	10.0	2.25	3.50
Loam	334.00	4293.62	0.36	4.0	0.49	3.57
Clay loam	603.70	4101.52	0.68	4.0	1.49	4.38
Sandy loam	228.21	4508.77	0.17	5.0	0.19	3.10
Sandy clay loam	276.99	4059.25	0.33	4.0	0.36	3.71

[a]USDA classification.

In the ungraded channels the critical hydraulic shear was set as a function of the stream bed material size, while the erodibility parameters were set to the default value in the WEPP database. For the ditches the default values are used for both parameters. The values for the soil albedo parameter (i.e. the fraction of the solar radiation which is reflected back to the atmosphere) were computed using the Baumer equation (Flanagan and Livingstone, 1995) and ranged between 0.08 and 0.22 for the loam and clay textures, respectively.

For each land use, information about the specific plants and the management practices were designated in the plant/management files. Some studies have reported the difficulty in representing complex plant ecosystems on rangelands (Laflen *et al.*, 1994) and the importance of spatial and temporal variation of vegetation for interrill erosion processes, particularly in semiarid conditions (Blackburn and Pierson, 1994). In this study, the different vegetation complexes have been represented using the plants (up to three) in the WEPP database that better fit the dominant species in the field. Thus the pasture areas of the watershed were modelled using fescue, bluegrass and big sagebrush (the last for the sparse shrubs) from the WEPP database for rangeland and lucerne from the database for cropland. For the crop cultivation, which ranged from 8% to 13% of the watershed area, it was necessary to modify several parameters of the model's default winter wheat database, including planting and harvest dates, type and dates of tillage, and crop rotations.

Furthermore, as the cropland area was characterized by broadcast sown wheat, it was necessary to modify the row width and the distance between plants. For the channels covered with vegetation (ditches), the total Manning roughness (n) coefficient was set as proposed by Knisel (1980) in the Chemicals, Runoff and Erosion from Agricultural Management Systems (CREAMS) manual. For the ungraded channels the default value of Manning's n was used.

Results and Discussion

Statistics of measured and simulated storm runoff depth are reported in Table 18.3 for the 50

Table 18.3. Storm runoff depth statistics for the observation period for the three simulation series performed by WEPP at Cannata watershed.

	Mean (mm)	Median (mm)	Minimum (mm)	Maximum (mm)	SD	r^2	E[a]
Measured	7.8	4.0	1.0	54.0	11.1	–	–
Simulated							
I	1.3	0.0	0.0	3.0	3.7	0.20	–0.34
II	3.3	0.4	0.0	21.2	6.1	0.83	0.54
III	3.0	0.3	0.0	25.5	5.5	0.82	0.47

[a]Nash and Sutcliffe (1970).

Table 18.4. Measured and simulated annual runoff depth during the observation period in the Cannata watershed.

	Annual runoff depth (mm)					
	1997	1998	1999	2000	2001	2002
Measured[a]	62.9	30.7	104.4	65.5	37.5	83.7
Simulated						
I	12.4	3.7	4.7	1.0	8.3	2.8
II	73.8	45.6	44.1	40.6	43.7	53.6
III	68.0	42.2	40.0	34.1	40.3	49.1

[a]Rainfall depth recorded at station A.

Table 18.5. Measured and simulated runoff coefficients during the observation period in the Cannata watershed.

	Annual runoff coefficient (%)					
	1997	1998	1999	2000	2001	2002
Measured[a]	8.8	5.4	17.7	11.3	6.0	10.3
Simulated						
I	1.8	0.7	0.9	0.2	1.6	0.4
II	10.5	8.7	8.8	7.9	8.3	7.3
III	9.7	8.0	7.9	6.6	7.6	6.7

[a]Rainfall depth recorded at station A.

daily values of not less than 1 mm measured during the simulation period. These results were obtained for the computer runs with constant values. The results did not improve using values of K_e that were internally adjusted by the model. WEPP storm runoff depths were better correlated ($r^2 > 0.78$) to the measurements in simulation series II and III (Fig. 18.3), with coefficient of determination and standard error values similar to those found by Savabi *et al.* (1995). These results were also characterized by positive values of model efficiency (Nash and Sutcliffe, 1970). Storm runoff depth was underestimated for nine out of the ten events with runoff over 10 mm. An underestimation also resulted for the smallest events, runoff being zero in 40% of the cases (with a recorded precipitation of 5.2–24.8 mm). A similar behaviour for events with observed runoff depths less than 1 mm was reported by Soto and

Díaz-Fierros (1998). In the simulation series II and III there were more than 200 simulated events during the entire period of simulation. Consequently the annual runoff depth was underestimated between 36% and 62% for 3 of the 6 years (Table 18.4). The annual runoff coefficient values are shown in Table 18.5. These underestimations were similar to those reported by Savabi *et al.* (1996). The regression analysis of simulated vs observed peak runoff gave an r^2 of 0.63 for simulation series II and III (Fig. 18.3) with a model efficiency coefficient less than zero in both cases.

Statistics of storm sediment yields are reported in Table 18.6. High correlations between storm sediment yields were found for simulation series II ($r^2 = 0.92$) and III ($r^2 = 0.77$), but model efficiency coefficients were negative in both cases. Sediment yield in simulation series III was overestimated in seven out of 14 events because of high sediment concentrations in the generated overland flow (Table 18.7). Consequently, the cumulative sediment yield ($N = 14$ events) resulting from simulation series III and II was a factor of 2.3 and 3.2 greater, respectively, than the one estimated through field measurements (Table 18.7). Similar values were found by Liu *et al.* (1997) for the three Holly Springs watersheds (Mississippi, USA). The overestimation of erosion at sites with low erosion rates similar to Cannata, producing between 0.004 and 0.6 t/ha for an event, is supported by numerous examples in the literature (Liu *et al.*, 1997; Nearing, 1998, 2000; Nearing *et al.*, 1999; Tiwari *et al.*, 2000; Santoro *et al.*, 2002).

Ninety-five per cent of the average annual simulated sediment yield was produced from seven hillslopes covering approximately 32 ha, or 25% of the total area. This included the entire cultivated area of approximately 11 ha. The simulated sediment delivery ratio was 0.48 with a coefficient of variation of approximately 11%.

There are two factors that can explain the differences between the measured and predicted values of soil erosion. The first was related to the creation of excessively long slope lengths by GeoWEPP. WEPP overpredicts erosion when slope lengths exceed the recommended value of 100 m (Baffaut *et al.*, 1997). A second reason for the difference between measured and predicted erosion may be due

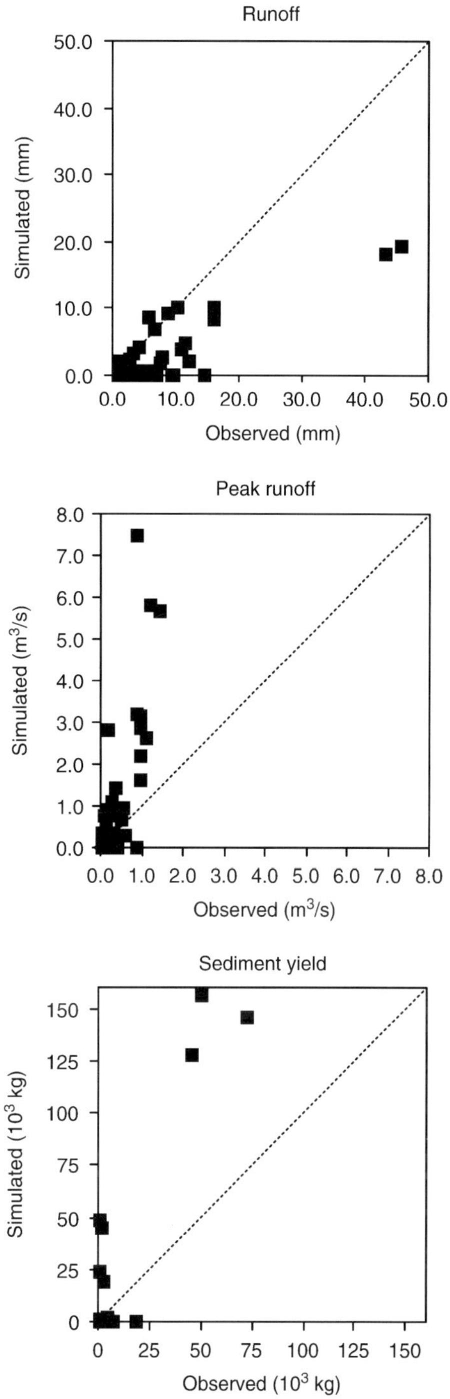

Fig. 18.3. Simulated (by WEPP) vs observed storm runoff ($N = 50$), peak runoff ($N = 45$) and sediment yield ($N = 14$) for simulation series III in the Cannata watershed.

Table 18.6. Storm sediment yield statistics for the observation period for the three simulation series performed by WEPP in the Cannata watershed.

	Mean (10^3 kg)	Median (10^3 kg)	Minimum (10^3 kg)	Maximum (10^3 kg)	SD (10^3 kg)	r^2	E[a]
Measured	17.8	6.3	0.6	72.5	22.6	–	–
Simulated							
I	5.6	0.0	0.0	56.2	15.3	0.54	–0.1
II	57.4	15.4	0.0	287.6	89.7	0.92	–11.6
III	40.6	10.5	0.0	156.3	58.2	0.77	–3.1

[a]Nash and Sutcliffe (1970).

Table 18.7. Runoff volume, sediment yield and average sediment concentration for 14 events in the observation period at the Cannata watershed.

	Runoff volume D (10^3 m^3)	Cumulated sediment yield P (10^3 kg)	Average sediment concentration[a] P/D (g/l)
Measured	227.5	209.8	1.10
Simulated by WEPP			
I	7.9	79.0	9.96
II	75.5	681.0	9.02
III	68.9	457.9	6.64

[a]Computed as the ratio between cumulated sediment yield and cumulated runoff.

simply to natural variation in soil erosion at low rates (Nearing, 1998, 2000; Nearing *et al.*, 1999).

Conclusions

Predicted values of runoff were better correlated to the measurements of runoff in the simulation series with constant K_e values (during the whole period of simulation) set by the user. Storm runoff depth was generally underestimated for both large and small rainfall events. The annual runoff depth was underestimated for 3 of the 6 years. The results suggest possibilities of improvement for the WEPP and GeoWEPP models. The definition of excessively long slope lengths by GeoWEPP needs to be corrected, or alternatively, some modification of the WEPP model is needed to prevent the overprediction of erosion rates from these long slopes. Secondly, in semiarid conditions such as those in this study, spatial variability in rainfall is an important problem that is not currently represented in WEPP. This spatial variability is undoubtedly important in terms of accurate predictions of both runoff and sediment yield. Lastly, it was found time-consuming to generate data input files for the WEPP model because of a lack of model parameters related to the vegetation species typical of Mediterranean areas. None the less, in spite of the difficulties encountered and the limitations of the model, and given the relatively low rates of erosion (with which is associated large natural variation), the results were reasonable with discrepancies within the order of magnitude found in other studies (Liu *et al.*, 1997; Nearing, 1998, 2000; Nearing *et al.*, 1999; Tiwari *et al.*, 2000; Santoro *et al.*, 2002).

References

Baffaut, C., Nearing, M.A., Ascough II, J.C. and Liu, B.Y. (1997) The WEPP watershed model: II. Sensitivity analysis and discretization on small watersheds. *Transactions of the American Society of Agricultural Engineers* 40, 935–943.

Bhuyan, S.J., Kalita, P.K., Janssen, K.A. and Barnes, P.L. (2002) Soil loss predictions with three erosion simulation models. *Environmental Modelling and Software* 17, 137–146.

Blackburn, W.H. and Pierson, F.B. (1994) Sources of variation in interrill erosion on rangelands. In: Blackburn, W.H., Pierson, F.B. Jr, Schuman, G.E. and Zartman, R. (eds) *Variability in Rangeland Water Erosion Processes*. Soil Science Society of America, Madison, Wisconsin, pp. 1–10.

Bouwer, H. (1969) Infiltration of water into non-uniform soil. *Journal of Irrigation and Drainage Division ASCE* 95 (IR4), 451–462.

Duiker, S.W., Flanagan, D.C. and Lal, R. (2001) Erodibility and infiltration characteristics of five major soils of southwest Spain. *Catena* 45, 103–121.

Flanagan, D.C. and Livingstone, S.J. (1995) *USDA Water Erosion Prediction Project: User Summary*. NSERL Report 11. USDA-ARS National Soil Erosion Research Laboratory, West Lafayette, Indiana.

Flanagan, D.C., Laflen, J.M., Nearing, M.A. and Norton, L.D. (1998) The USDA Water Erosion Prediction Project (WEPP) Model. In: *Experience with Soil Erosion Models*. Universitat fur Bodenkultur Wien, Vienna, pp. 53–62.

Goodrich, D.C. and Simanton, J.R. (1995) Water research and management in semiarid environments. *Journal of Soil and Water Conservation* 50, 416–419.

Hebel, B. and Siegrist, S. (1998) Plot size application and validation of WEPP for Swiss conditions. In: *Experience with Soil Erosion Models*. Universität für Bodenkultur Wien, Vienna, pp. 149–160.

Kidwell, M.R., Weltz, M.A. and Guertin, D.P. (1997) Estimation of Green-Ampt effective hydraulic conductivity for rangelands. *Journal of Range Management* 50, 290–299.

Kincaid, D.C. (2002) The WEPP model for runoff and erosion prediction under sprinkler irrigation. *Transactions of the American Society of Agricultural Engineers* 45, 67–72.

Klik, A., Savabi, M.R., Norton, L.D. and Baumer, O. (1995) Application of WEPP hillslope model in Austria. *Proceeding of American Water Resources Association Conference*, 5–9 November 1995, Houston, Texas. American Water Resources Association, Middleburg, Virginia, pp. 313–322.

Knisel, W.G. (1980) *CREAMS: A Field-scale Model for Chemicals, Runoff, and Erosion from Agricultural Management Systems*. USDA Conservation Research Report 26. US Department of Agriculture, Washington, DC.

Laflen, J.M., Flanagan, D.C., Weltz, M.A., Stone, J.J. and Ascoug II, J.C. (1994) The WEPP model and its applicability for predicting erosion on rangelands. In: Blackburn, W.H., Pierson, F.B. Jr, Schuman, G.E. and Zartman, R. (eds) *Variability in Rangeland Water Erosion Processes*. Soil Science Society of America, Madison, Wisconsin, pp. 11–22.

Licciardello, F., Rizzo, N., Somma, F. and Zimbone, S.M. (2001) Runoff and soil erosion in a small Sicilian watershed. *Proceedings of International Symposium: Erosion Research for the 21st Century*, 3–5 January 2001, Honolulu, Hawaii, pp. 363–366.

Liu, B.Y., Nearing, M.A., Baffaut, C. and Ascough II, J.C. (1997) The WEPP Watershed Model. III. Comparison to measured data from small watershed. *Transactions of the American Society of Agricultural Engineers* 40, 945–952.

Nash, J.E. and Sutcliffe, J.V. (1970) River flow forecasting through conceptual models, 1. A discussion of principles. *Journal of Hydrology* 10, 282–290.

Nearing, M.A. (1998) Why soil erosion models over-predict small soil losses and under-predict large soil losses. *Catena* 32, 15–22.

Nearing, M.A. (2000) Evaluating soil erosion models using measured plot data: accounting for variability in the data. *Earth Surface Processes and Landforms* 25, 1035–1043.

Nearing, M.A. and Nicks, A.D. (1997) Evaluation of WEPP: hillslopes and small watersheds. In: Boardman, J. and Favis-Mortlock, D.T. (eds) *Global Change: Modelling Soil Erosion by Water*. NATO-ASI Global Change Series. Springer, Berlin, Germany, pp. 43–53.

Nearing, M.A., Foster, G.F., Lane, L.J. and Finker, S.C. (1989) A process-based soil erosion model for USDA-Water Erosion Prediction Project Technology. *Transactions of the American Society of Agricultural Engineers* 32, 1587–1593.

Nearing, M.A., Deer-Ascough, L. and Laflen, J.M. (1990) Sensitivity analysis of the WEPP hillslope profile erosion model. *Transactions of the American Society of Agricultural Engineers* 33, 839–849.

Nearing, M.A., Liu, B.Y., Risse, L.M. and Zhang, X. (1996) Curve numbers and Green-Ampt effective hydraulic conductivities. *Water Resources Bulletin-American Water Resources Association* 32, 125–136.

Nearing, M.A., Govers, G. and Norton, L.D. (1999) Variability in soil erosion data from replicated plots. *Soil Science Society of America Journal* 63, 1829–1835.

Renard, K.G., Rawls, W.J. and Fogel, M.M. (1982) Currently available models. In: Haan, C.T., Johnson, H.P. and Brakensiek, D.L. (eds) *Hydrologic Modelling of Small Watersheds*. American Society of Agricultural Engineers, St Joseph, Missouri, pp. 507–522.

Renschler, C.S. and Harbor, J. (2002) Soil erosion assessment tools from point to regional scales – the role of geomorphologists in land management research and implementation. *Geomorphology* 47, 189–209.

Renschler, C.S., Flanagan, D.C., Engel, B.A. and Frankenberger, J.R. (2002) GeoWEPP – the Geo-spatial interface for the Water Erosion Prediction Project. *Proceedings of the 95th Annual International Meeting of ASAE – XV World Congress of CIGR*, 28–31 July 2002, Chicago, Illinois. American Society of Agricultural Engineers, St Joseph, Missouri.

Santoro, V., Amore, E., Modica, C. and Nearing, M.A. (2002) Application of two soil erosion models to a large Sicilian basin. In: Rubio, J.L., Morgan, R.C.P., Asins, S. and Andreu, V. (eds) *Proceedings of III International Congress of European Society for Soil Conservation, Man and Soil at the III Millennium*. Geoforma, Valencia, Spain, pp. 1445–1458.

Savabi, M.R. (2001) Determining soil water characteristics for application of WEPP model in south Florida. *Transactions of the American Society of Agricultural Engineers* 44, 59–70.

Savabi, M.R., Flanagan, D.C., Hebel, B. and Engel, B.A. (1995) Application of WEPP and GIS-GRASS to a small watershed in Indiana. *Journal of Soil and Water Conservation* 50, 477–483.

Savabi, M.R., Klik, A., Grulich, K., Mitchell, J.K. and Nearing, M.A. (1996) Application of WEPP and GIS on small watershed in USA and Austria. In: Kovar, K. and Nachtnebel, H.P. (eds) *Proceedings of HydroGIS 96: Application of Geographic Information System in Hydrology and Water Resources Management*. IAHS Publication 235. IAHS Press, Wallingford, UK, pp. 469–476.

Singh, V.P. (1995) *Computer Models of Watershed Hydrology*. Water Resources Publications, Littleton, Colorado.

Singh, V.P. and Frevert, D.K. (2002) *Mathematical Models of Small Watershed Hydrology and Applications*. Water Resources Publications, Littleton, Colorado.

Soto, B. and Díaz-Fierros, F. (1998) Runoff and soil erosion from areas of burnt scrub: comparison of experimental results with those predicted by the WEPP model. *Catena* 31, 257–270.

Tiwari, A.K., Risse, L.M. and Nearing, M.A. (2000) Evaluation of WEPP and its comparison with USLE and RUSLE. *Transactions of the American Society of Agricultural Engineers* 43, 1129–1135.

USDA – Soil Conservation Service (1972) *National Engineering Handbook, Hydrology, Section 4*. USDA Soil Conservation Service, Washington, DC.

Vlnasova, K., Klik, A., Hebel, B. and Zarl, A. (1998) Validation and comparison of EUROSEM, SMODERP and WEPP hillslope models with measured data from erosion plots in Austria and the Czech Republic. In: *Experience with Soil Erosion Models*. Universität für Bodenkultur Wien, Vienna, pp. 287–293.

Zeleke, G., Winter, T. and Flanagan, D. (1999) BPCDG: Breakpoint Climate Data Generator for WEPP using observed climate standard weather data sets. http://topsoil.nserl.purdue.edu/nserlweb/weppmain/BPCDG.html

Zhang, X.C., Nearing, M.A., Risse, L.M. and McGregor, K.C. (1996) Evaluation of runoff and soil loss predictions using natural runoff plot data. *Transactions of the American Society of Agricultural Engineers* 39, 855–863.

19 Spatial Modelling of Ephemeral Gully Incision: a Combined Empirical and Physical Approach

V. Jetten,[1] J. Poesen,[2] J. Nachtergaele[2] and D. van de Vlag[3]
[1]*Utrecht University, The Netherlands;* [2]*Katholieke Universiteit Leuven, Belgium;*
[3]*International Institute for Geoinformation Science and Earth Observation, The Netherlands*

Introduction

Erosion forms caused by flowing water are usually classified as interrill erosion (which can be seen as a combination of splash erosion and sheet flow erosion or wash), rill erosion and gully erosion. The distinction between rills and gullies is a practical one: the first consists of numerous small channels that can be obliterated by normal tillage operations, while the second cannot be obliterated nor crossed with farm equipment (Hutchinson and Pritchard, 1976). Furthermore the formation of rills is associated with micro-relief generated by tillage or land-forming operations (Haan *et al.*, 1994), whereas gullies are permanent erosion features formed along natural concentration flow lines such as depressions and thalwegs. There are problems with these definitions: first, there is a lot of evidence that gullies can form along artificial flow lines such as field boundaries and, second, they are not always permanent features in the landscape but may be removed by the farmer, although maybe not every growing season. Foster (1986) therefore discussed the term 'ephemeral gully' to indicate channels that are formed in agricultural fields in natural concentration lines but that are removed in tillage operations. These are distinguished from rills by the fact that they may reoccur each year in the same location. Furthermore, Poesen (1993) proposes a cross section of 1 ft^2 (or 929 cm^2) to distinguish between rills and ephemeral gullies. Analysing data collected in different parts of the world, Poesen *et al.* (2003) show that soil loss rates by gully erosion represent from 10% up to 94% of total sediment yield caused by water erosion. Poesen *et al.* (1996, 2003) show that ephemeral gullies are not a rare feature caused only by exceptionally severe rainstorms: the relative importance of ephemeral gullies in Belgium as contributors to the total sediment budget is highest for the more frequent rainfall events. In spite of this importance, the explicit modelling of gullies lags far behind compared to the progress made with catchment-based erosion modelling. In a recent review of gully erosion and environmental change, Poesen *et al.* (2003) conclude that there is a great need for monitoring, experimental and modelling studies of gully erosion as a basis for predicting the effects of environmental change (climatic and land use changes) on gully erosion rates. All physically deterministic models simulate splash erosion, interrill erosion and rill erosion in some way, sometimes distinguishing flow erosion in the interrill zone and in the rills (e.g. EUROSEM, Morgan *et al.*, 1998). Gully erosion is explicitly

 Soil Erosion and Sediment Redistribution in River Catchments (eds P.N. Owens and A.J. Collins)

modelled by CREAMS (Knisel, 1980) and WEPP (Flanagan and Nearing, 1995), but these models do not account for a change of the hillslope morphology when gully erosion occurs. Other successfully used gully models are Ephemeral Gully Erosion Model (EGEM) (Woodward, 1999) and Gully Thermoerosion and Erosion Model (GULTEM) (Sidorchuk, 1999). Both are two-dimensional models that simulate incision along a transect, the former using an empirical approach and the latter a purely physical approach. A similar approach is taken by Casalí *et al.* (2003), who adapted a river channel model to simulate the change of width, depth and form of an ephemeral gully. Recently, Souchère *et al.* (2003) simulated ephemeral gullies quite well using the 'STREAM ephemeral gully' model, which is based on the calculation of the sensitivity for erosion of the main branches of a collector network.

In 1998 and 1999 several ephemeral gullies were monitored in the Ganspoel catchment near Leuven (Belgium). In this study, a modelling framework to simulate the incision and formation of ephemeral gullies is proposed and tested with the gullies measured in Belgium. However, although gully dimensions and exact locations were measured, soil properties needed for the erosion model were not mapped. Thus the main objective is to investigate if the model is in principle capable of simulating these gullies, using a calibration data set that is within the values normally found in the comparable loess catchments. The system is evaluated by looking at the feasibility of obtaining the field data needed to make adequate predictions with this model.

Methodology

For the simulation of gully incision and development of its dimensions over the course of a season, there are basically three approaches possible: a complete physical approach, an empirical approach or a combination of both. A purely physical approach would have to compare all flow shear forces applied on the bottom and sidewalls of a gully with the shear strength in three dimensions. Although the theoretical framework has been formulated (Torri and Borselli, 2003), we did not have the data needed to implement and test such a model. On the other hand, an empirical approach would also need a large quantity of gully data to set up statistical relations between, for instance, gully volume and rainfall. The method adopted here combines the event-based spatially distributed erosion model Limburg Soil Erosion Model (LISEM) (Jetten and De Roo, 2001) with several empirical algorithms that determine the possible area and dimensions of gully incisions. LISEM is a spatially distributed model for small catchments, which has splash erosion based on kinetic energy of the rainfall and flow erosion and deposition based on the transport capacity that is calculated with the unit stream power. Erosion is based on the transport capacity surplus mitigated with an efficiency factor based on the inverse of the soil cohesion. A kinematic wave module distributes water and sediment over the landscape using a raster GIS-based flow network. While LISEM produces flow erosion, there is no distinction made in the original version between rills and gullies and the digital elevation model (DEM) is not altered because of erosion and deposition processes. The approach taken here was to remain close to the principles followed in LISEM, so that the same input variables can be used as for a normal simulation, but allowing the calculated erosion to create incisions in the DEM according to the following set of rules:

1. Detachment by concentrated flow was translated to incision only in certain parts of the landscape. These so-called critical areas are determined using a combination of contributing area above a given point of the concentrated flow line in the landscape and the local slope of the soil surface at that point.
2. Gullies are assumed to have a rectangular cross section and erosion is distributed equally over the gully perimeter. In a second stage various other distributions are tested.
3. If a subsurface soil layer with a higher bulk density exists, lateral erosion will take place relative to the partitioning of soil strength along the perimeter of the gully.
4. Once there is an incision, the water in the gridcells that are part of a gully is routed with a separate kinematic wave using the hydraulic radius calculated from the gully flow width and

depth. The fraction f of overland flow conveyed to the gully is based on the overland flow velocity u in the cell: $f = (u\,dt)/(0.5(dx - w))$, where dx is the cell width, w is the gully width and dt is the time step (the gully is assumed to be in the middle of the cell hence the factor 0.5).

5. A simulated incision is considered to be a gully if its cross section is larger than a square foot (929 cm^2; Poesen, 1993; Vandekerckhove *et al.*, 2000).

Sensitive areas

Field observations in Central Belgium (e.g. Vandaele *et al.*, 1996; Desmet *et al.*, 1999; Nachtergaele *et al.*, 2002), Spain (e.g. Vandekerckhove *et al.*, 2000) and Australia (e.g. Moore *et al.*, 1988) show that gully incision is likely to occur when certain combinations of contributing area and local slope exceed a given threshold. In this study five algorithms are tested, of which three are based on field data from the loess area in Central Belgium, one on data from Southern Spain and one from Australia. Five algorithms for critical threshold F_c determination were selected based on slope S (m/m), upstream area A (m^2) and normalized for flow width w (m):

Vandaele *et al.* (1996):

$$F_c = S \cdot \left(A/w\right)^{0.4} > 0.5 \quad (19.1)$$

Moore *et al.* (1988):

$$F_c = S \cdot \left(A/w\right) > 18 \text{ and } \ln\left(\frac{A/w}{S}\right) > 6.8 \quad (19.2)$$

Vandaele *et al.* (1996):

$$F_c = S > 0.025 \cdot \left(A/10000\right)^{-0.4} \quad (19.3)$$

Desmet and Govers (1997):

$$F_c = S \cdot \left(A/w\right)^{0.4} > 0.72 \quad (19.4)$$

Vandaele *et al.* (1997):

$$F_c = S \cdot \left(A/w\right) > 40 \text{ and } \ln\left(\frac{A/w}{S}\right) > 9.8 \quad (19.5)$$

The term $S(A/w)^b$ is explained by O'Loughlin (1986) as being a measure for Hortonian overland flow intensity, as it resembles the LS factor of the USLE type models which are also based on the product of power functions of slope angle and slope length. The term $\ln(A/S)$ resembles the wetness index (cf. Beven and Kirkby, 1979) and is a measure for saturation overland flow and generally produces 'wetter' areas (large values) in the valley floors and dry areas (lower values) towards the water divides. This was shown by Moore *et al.* (1988) who analysed shallow soils which were relatively easily saturated. A second threshold is needed to predict the 'lower end' of a gully, because the critical area algorithms will predict an unlimited continuation of a gully. It appeared from the ephemeral gully data in Belgium that the lower threshold is best represented by a minimum slope angle 3–4% below which all observed ephemeral gullies ended (Nachtergaele *et al.*, 2001). In the GIS PCRaster (Wesseling *et al.*, 1996) a flow network can be generated from the digital elevation model (DEM) using the steepest slope angle, or the GIS is capable of generating a tillage direction based flow network (including dead furrows and headlands) after the method described by Takken *et al.* (2001). Both networks were used in this study for different periods in the year. The five algorithms are easily implemented in PCRaster using the flow networks and the following rules:

- A critical zone is made according to the gully initiation algorithms and the minimum slope angle of 4%.
- The map edge is excluded from the zone because otherwise a flow path will be created there.
- Isolated 'critical' pixels are removed.
- Isolated 'non-critical' pixels are classified as 'critical' if the pixels upstream and downstream are also 'critical' (assuming that the runoff has enough momentum to overcome small non-critical areas).

Gully width

In a homogeneous soil, detachment by concentrated runoff will occur both on the bottom and at the sidewalls of the gully. In a first approach we assume a rectangular cross section and a distribution of the eroded volume over the width and depth according to the ratio of gully width and depth to its perimeter. The increase in width Δw (m) over a depth d (m) on both sides of the gully is then:

$$2d\Delta w = \frac{2d}{P}V \rightarrow \Delta w = \frac{V}{P} \qquad (19.6)$$

and the change of depth Δd (m) over the gully width w (m) is:

$$w\Delta d = \frac{w}{P}V \rightarrow \Delta d = \frac{V}{P} \qquad (19.7)$$

where V is the total eroded soil volume per unit length (m^3/m), i.e. the cross section of the eroded volume. The gully width is initialized on the average measured initial width: 0.2 m.

Second layer

The effect of soil strength is expressed in LISEM as a dimensionless detachment efficiency parameter Y, based on the inverse of cohesion c(kPa): $Y = 1/(0.89 + 0.56*c)$. In case a second soil layer is specified with a different bulk density and cohesion, and the incision reaches the second layer, the detachment is based on an apparent efficiency Y' calculated from the distribution of the efficiencies of the first and second layers Y_1 and Y_2 along the perimeter P (m) of the gully:

$$Y' = Y_1\frac{2d_1}{P} + Y_2\left(1 - \frac{2d_1}{P}\right) \qquad (19.8)$$

in which d_1 is the depth to the first layer (m). Likewise, the volume of net erosion is calculated with an apparent bulk density ρ' (kg/m^3) based on the two bulk densities of the first and second layers (ρ_1 and ρ_2):

$$\rho' = \rho_1\frac{2d_1}{P} + \rho_2\left(1 - \frac{2d_1}{P}\right) \qquad (19.9)$$

When the incision reaches a second soil layer with a higher cohesion, it is assumed that the width and depth are changed in the same way as Eqns 19.6 and 19.7, but weighted with the different efficiencies Y_1 and Y_2. If d_1 is the depth to the first layer and d_2 is the incision depth into the second layer (i.e. gully depth $d = d_1 + d_2$), the change in width and depth become:

$$2d\Delta w = \frac{2d_1Y_1 + 2d_2Y_2}{2d_1Y_1 + 2d_2Y_2 + wY_2} * V$$
$$w\Delta d = \frac{wY_2}{2d_1Y_1 + 2d_2Y_2 + wY_2} * V \qquad (19.10)$$

When the incision has not yet reached the second layer it is calculated as specified in the 'Gully width' section above. It should be noted that due to the limitations of a raster system with one flow line per grid cell, the gully width cannot become larger than the grid cell width.

The data set

The model was tested on a selection of three narrow winter gullies (formed in November 1998) and three wide and shallow summer gullies (formed in spring and summer 1998) which were recorded in the Kinderveld catchment in Central Belgium (Nachtergaele, 2001). Examples of winter and summer gullies are shown in Fig. 19.1. The summer gullies were adjacent in a sub-catchment of 13.7 ha. The winter gullies were at different locations. From the photographs it can be seen that the assumption of a rectangular cross section is reasonable: the summer gullies are very wide and shallow and the winter gullies are rectangular or sometimes even have overhanging walls.

The dimensions of the gullies were recorded using measuring tape and differential GPS. Although breakpoint rainfall was recoded, unfortunately very few soil physical parameters were measured, nor was the net soil loss from the catchment known for the events that led to the gully formation. Therefore the input parameters for LISEM were taken from research in catchments in the vicinity (the Ganspoel catchment, see Takken *et al.*, 1999). Grid cell size was 5 m × 5 m for all maps.

It is important to note that the only variables for which the spatial patterns are known are those derived from the elevation (DEM, slope, flow pattern). All other variables had to be assumed homogeneous because of lack of

Fig. 19.1. Example of a winter gully (left, photo Laboratory of Experimental Geomorphology, Leuven, Belgium), and a summer gully (right, photo Waterboard Roer en Overmaas, The Netherlands).

spatial data. During the simulations, however, differences in k_{sat}-values had to be assumed for the winter gullies, which are located far apart in different fields in the catchment, because a single set of calibration values that fit all three winter gullies could not be found. The summer gullies were located near each other and therefore a single calibration set could be found.

Results

Critical areas and spatial precision

Figure 19.2 shows the DEM and the critical areas derived from Eqns 19.1–19.5 for the summer gullies. It can be seen that the measured gullies (black line) are inside most of the critical areas except from the one defined by Eqn 19.5: the critical area starts considerably lower on the slope than the measured gully. Also the gullies appear to end higher on the slope than predicted by the critical areas (areas steeper than 4%). Figure 19.3 shows the effect of using a critical area: the map on the left shows simulated gullies (incisions larger than 0.0929 m^2) when the whole area is considered critical, on the right are shown the simulated gullies for critical area predicted with Eqn 19.4. It is clear that when LISEM gully predictions are not limited to a critical area, too many incisions are created. Comparing the right hand map in Fig. 19.3 with the measured gullies in Fig. 19.2, it seems that LISEM still produces more gullies than those observed in the field. However, the solid lines in Fig. 19.2 only represent the recorded gullies, while many rills were also observed on the slope, but these were not all separately recorded. In a raster cell rills cannot be represented separately and LISEM will calculate the total eroded soil volume, which it will then assume to be a gully if the total eroded cross section in a cell is larger than 0.0929 m^2. Thus the total simulated gully volume is likely to be

Fig. 19.2. The Belgium test area. Top: DEM and critical threshold area predicted from Eqn 19.1 (white pixels represent zones where gullies might develop, black line represents observed gullies in the field), middle: critical threshold areas predicted using Eqns 19.2 and 19.3, bottom: critical threshold areas predicted using Eqns 19.4 and 19.5. Pixels are 5 m × 5 m.

more than the recorded volume because of the discretization in raster cells.

A second problem arises from the quality of the DEM. The critical areas derived from Eqns 19.1–19.5 depend on the overland flow network that is derived from the DEM, which in turn is digitized from a topographic map. While the gullies are recorded with a differential GPS,

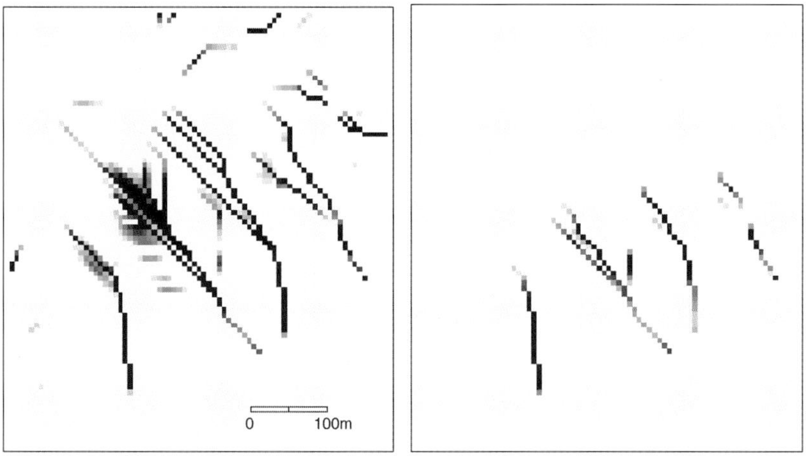

Fig. 19.3. Effect of the use of critical area on the position of simulated summer gullies when the whole slope is considered to be a critical area (left) and gullies formed only within a critical area predicted using Eqn 19.4 (right).

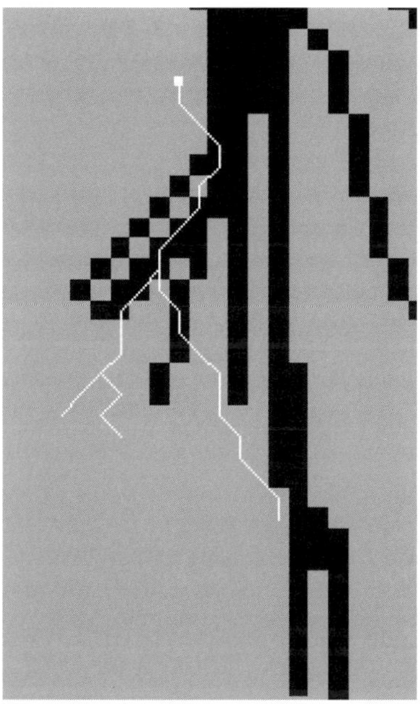

Fig. 19.4. One of the measured winter gullies (indicated by white line, transformed to the raster GIS) compared to the critical area (indicated by black pixels) based on the flow network derived from the DEM. Pixel size is 5 m × 5 m.

the concentrated flow paths in the GIS are derived from the DEM, which itself is generated from a digitized topographic map. The concentrated flow lines derived from the DEM are therefore not exactly in the same location but often several pixels apart, and the shape of the flow path is also slightly different (see Fig. 19.4). This makes a comparison between measured and simulated values very difficult. Unfortunately, the only way to solve this is to force the flow network derived from the DEM through the measured gully positions. This was only done for the winter gullies as the summer gully locations coincided better with the DEM. For the simulations below, Eqn 19.4 is selected to define the critical areas, based on the fact that with Eqns 19.1 and 19.3 the areas seem too large and with Eqn 19.5 the area seems too small. From the remaining equations, only Eqn 19.4 is derived from data in central Belgium.

Winter gullies

Figure 19.5 shows the results for the winter gullies using the first approach, i.e. equal distribution of the eroded soil volume over the perimeter of the gully cross section and an initial width of 0.2 m. The gullies are located far apart and not in the same field as is the case with the summer gullies (discussed below). Therefore, it was not possible to find a single set of calibration parameters that would accurately simulate all gullies: gully 1 is simulated with a saturated hydraulic conductivity (k_{sat}) of 17 mm/h, gully 2

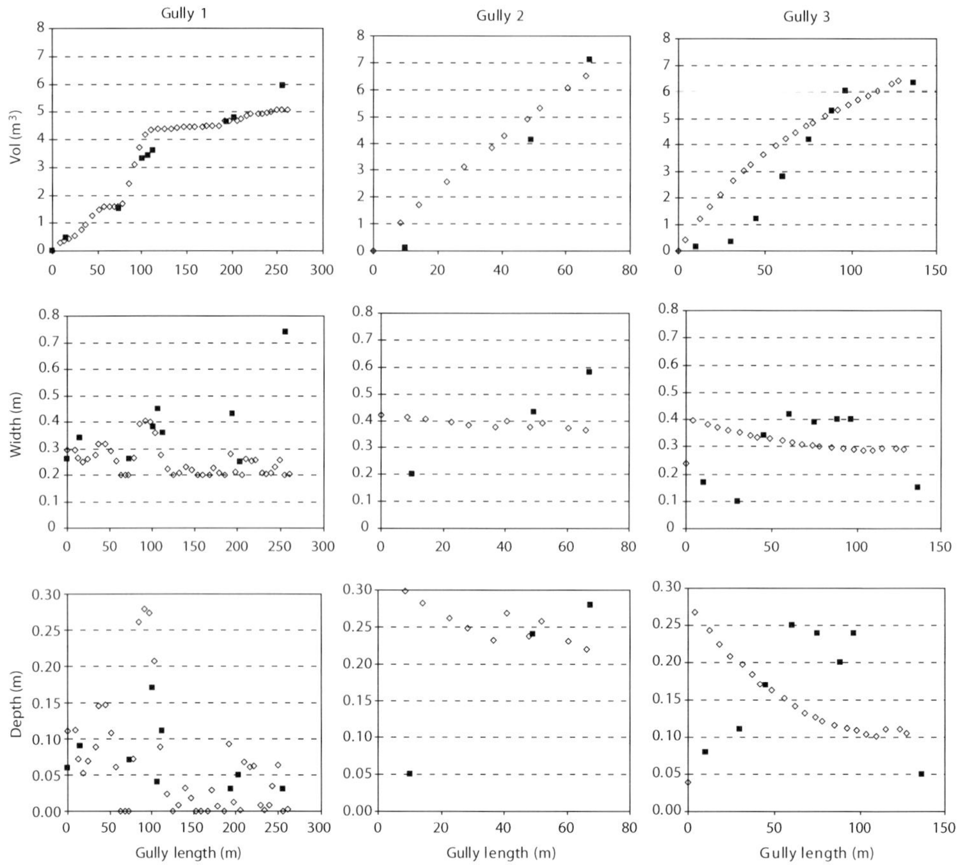

Fig. 19.5. Simulated (open symbols) vs measured (solid symbols) dimensions of winter gullies in the Kinderveld catchment. Three different k_{sat} values are used for gullies 1, 2 and 3: respectively 17 mm/h, 5 mm/h and 20 mm/h. Soil cohesion for all gullies = 4 kPa, bulk density = 1200 kg/m^3, initial gully width = 0.3 m. Critical area is based on Eqn 19.4.

with a k_{sat} of 5 mm/h and gully 3 with a k_{sat} of 20 mm/h, causing different runoff volumes. The results show that the gully volumes can be fairly well predicted. However, the prediction of the variation of width and depth along the gully length is not good. The predicted widths are not as variable as the measured ones, while the simulated depths have a different variability as well, although they have the same range as the measured depths. Because LISEM predicts the eroded cross section as one value, an over or under prediction in increase of gully width is compensated by an under or over prediction of the increase in gully depth. The fairly good prediction of volume along the gully length therefore suggests that the hydraulics part of the model and the prediction of eroded volume are adequate; it is the distribution of the eroded soil volume over width and depth that is not correct. The main reason for this is probably that the spatial variability of the input parameters was not known and therefore was assumed to be homogeneous. Torri and Borselli (2003) assume that one must apply different efficiency parameters for the shear stress applied at the gully walls and the gully bottom, to account for processes such as sidewall collapse. If this is the case, sufficient lateral erosion could probably be simulated, but in the absence of sufficient data this was not done.

Summer gullies

For the simulation of the summer gullies a single calibration set was used, as the gullies are located close together. Two soil layers were defined with different bulk densities and strengths. The differences in parameter values were exaggerated to create as much widening as possible using Eqns 19.9–19.10. The first layer had a depth of 4 cm, a cohesion of 1 kPa and a bulk density of 1100 kg/m^3 (a loose seedbed); the second layer a cohesion of 8 kPa and a bulk density of 1400 kg/m^3. This resulted in simulated widths that remained less than 1 m for all gullies (not shown here), while the measured widths ranged from 2 to more than 8 m (measured values shown in Fig. 19.6). When starting with an initial width of 2 m the gully widths remained below 2.5 m and the variability in width was not well predicted.

A second approach was taken for the summer gullies, using a flow discharge–width relationship to determine the width of the incision. Nachtergaele *et al.* (2002) analysed a series of both winter and summer gullies and found that the best fit is given by:

$$W = 2.51 \cdot Q^{0.412} \tag{19.11}$$

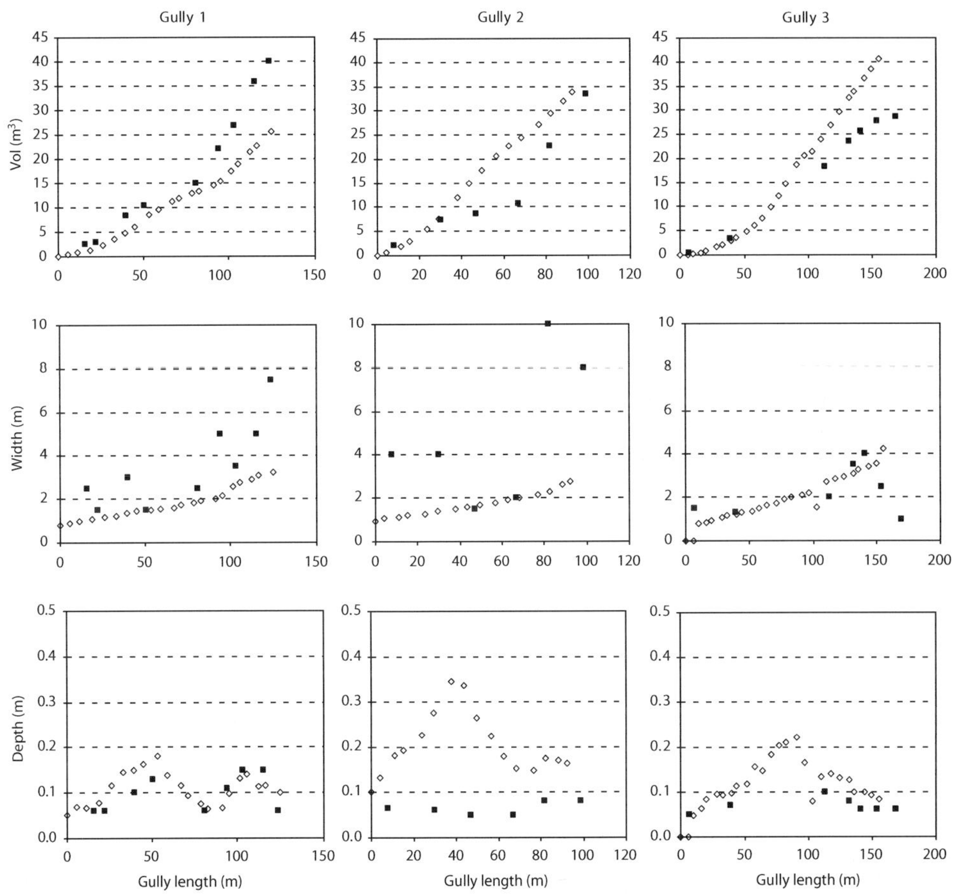

Fig. 19.6. Simulated (open symbols) vs measured (solid symbols) dimensions of three summer gullies in the Kinderveld catchment. Calibration data for this simulation are: k_{sat} = 7 mm/h; 1st layer depth = 4 cm, soil cohesion = 1 kPa, bulk density = 1100 kg/m^3; 2nd layer soil cohesion = 8 kPa, bulk density = 1400 kg/m^3. Width is based on Eqn 19.11 and allowed to grow (see text). Critical area is based on Eqn 19.4.

which is based on data of 67 large rills and gullies from various environments ($r^2 = 0.72$). Equation 19.11 is only used to widen the gully. If the runoff discharge decreases after the rainfall event, the gully obviously should not shrink. Additionally, when the runoff discharge decreases but erosion still takes place, the gully continues to widen using Eqns 19.8–19.10.

This method gives better results, but the gullies still remain too narrow. A reason may be the data set on which the equation is based, because it contains both winter and summer gullies and therefore Eqn 19.11 represents the average *Q–w* relation, which is not suitable for either of the specific situations. Nevertheless, the variation in channel width resembles the measured variation more closely. In contrast to gully 2, the change in depth of the gullies 1 and 3 is quite good compared to the measured values. Gully 2 is much wider in the field than could be simulated, which is then compensated by depth that is too large. A disadvantage of this method is that no assumption is made on the mechanism for the detachment and widening of the summer gullies. Apparently the washing away of the relatively weak and initially air-dry seedbed (due to slaking effects, Nachtergaele and Poesen, 2002) is significantly different from the more 'normal' incision in wet top soils during the winter period.

Discussion and Conclusions

Modelling of ephemeral gully erosion with the LISEM-Gully framework shows mixed results:

- Prediction of the critical areas (i.e. areas prone to ephemeral gully erosion) is necessary to limit the number of incisions and to obtain a better coincidence between predicted gullies and those observed in the field. Still the number of gullies formed is too large. This is partly caused by the discretization of the flow between the raster grid cells in a flow network: the generated network will have parallel flow lines, e.g. 5 m apart, while in reality such an area will have one concentrated flow line. In other words, the raster environment artificially creates more flow lines than are found in reality. The reverse is also true: parallel rills over short distances will be seen as one gully in LISEM. This makes a comparison of simulated and measured incisions difficult.
- A high-quality DEM is needed for this model to work, and the position of the measured gullies has to be mapped very accurately. If this is not available, simulated and measured gullies will not have the same location and they cannot be compared. Comparing winter to summer gullies it seems that gullies in small homogeneous areas can be predicted better with LISEM-gully than gullies in large areas. The quality of the DEM is a weak point of the method: first, it is questionable whether DEMs of such detail are available and, secondly, even on smooth and relatively flat areas the raster GIS will use the steepest slope angle to 'invent' a likely flow network, which then become rather arbitrary because the algorithm reacts to small differences in elevation. The resulting patterns may still not concur with measured incisions even for a high quality DEM.
- The gully volumes are fairly well simulated with the calibration sets used, which are perfectly acceptable for the loess soils in winter and summer circumstances. This means that the volume of soil detachment predicted by LISEM and therefore the gully hydraulics are well simulated. Moreover, not many extra variables are needed above the regular ones used by LISEM.
- The distribution of channel widths and depths, however, does not show the same changes along the gully length as measured in the field, although they do have the same range as the measured values. The framework as described causes insufficient widening of the gullies in both summer and winter situations. Possibly a different 'soil erodibility' must be used for the sidewalls and the bottom of the gully. Clearly more research in this subject is necessary.
- The summer gullies most probably represent areas with a washed away seedbed or a layer with shallow root development and the strength of such a layer is not well represented in this system, and not well modelled by the transport capacity surplus mechanism of LISEM. The distribution of shear

stresses along the perimeter of the summer gullies needs more research in order to predict their development well. For the time being, better results are obtained with a runoff discharge–width relationship, but these are still not satisfactory.

Summarizing, it seems that the LISEM-gully framework has sufficient degrees of freedom to model the ephemeral gully development for small areas, provided an accurate DEM is available. More data are needed on the processes of gully widening versus deepening to improve the simulation.

Acknowledgement

This research was funded by the European Commission, within the framework of the project 'Modelling Within Storm Erosion Dynamics' (MWISED, ENV4 CT97-0887).

References

Beven, K.J. and Kirkby, M.J. (1979) A physically based, variable contributing area model of basin hydrology. *Hydrological Sciences Bulletin* 24, 43–69.

Casalí, J., López, J.J. and Giráldez, J.V. (2003) A process-based model for channel degradation: application to ephemeral gully erosion. *Catena* 50, 435–447.

Desmet, P.J.J. and Govers, G. (1997) Two-dimensional modelling of the within-field variation in rill and gully geometry and location related to topography. *Catena* 29, 283–306.

Desmet, P.J.J., Poesen, J., Govers, G. and Vandaele, K. (1999) Importance of slope gradient and contributing area for optimal prediction of the initiation and trajectory of ephemeral gullies. *Catena* 37, 377–392.

Flanagan, D.C. and Nearing, M.A. (eds) (1995) *USDA-Water Erosion Prediction Project: Hillslope Profile and Watershed Model Documentation*. NSERL report no. 10. NSERL, West Lafayette, Indiana.

Foster, G.R. (1986) Understanding ephemeral gully erosion. *Soil Conservation, vol. 2*. National Academy of Science Press, Washington, DC, pp. 90–125.

Haan, C.T., Barfield, B.J. and Hayes, J.C. (1994) *Design Hydrology and Sedimentology for Small Catchments*. Academic Press, London.

Hutchinson, D.E. and Pritchard, H.W. (1976) Resource conservation glossary. *Journal of Soil and Water Conservation* 31, 1–63.

Jetten, V. and de Roo, A.P.J. (2001) Spatial analysis of erosion conservation measures with LISEM. In: Harmon, R. (ed.) *Landscape Erosion and Evolution Modelling*. Kluwer Academic Press, Dordrecht, The Netherlands, pp. 429–445.

Knisel, W.G. (ed.) (1980) *CREAMS: A Field Scaled Model for Chemicals, Runoff, and Erosion from Agricultural Management Systems*. Conservation Report No. 26, Vol. 1: Model Documentation, Chapter 3. USDA-SEA, Washington, DC.

Moore, I.D., O'Loughlin, E.M. and Burch, G.J. (1988) A contour-based topographic model for hydrological and ecological applications. *Earth Surface Processes and Landforms* 13, 305–320.

Morgan, R.P.C., Quinton, J.N., Smith, R.E., Govers, G., Poesen, J.W.A., Auerswald, K., Chisci, G., Torri, D. and Styczen, M.E. (1998) The European Soil Erosion Model (EUROSEM): a dynamic approach for predicting sediment transport from fields and small catchments. *Earth Surface Processes and Landforms* 23, 527–544.

Nachtergaele, J. (2001) A spatial and temporal analysis of the characteristics, importance and prediction of ephemeral gully erosion. Unpublished PhD thesis. K.U. Leuven, Belgium.

Nachtergaele, J. and Poesen, J. (2002) Spatial and temporal variations in resistance of loess-derived soils to ephemeral gully erosion. *European Journal of Soil Science* 53, 449–464.

Nachtergaele, J., Poesen, J., Steegen, A., Takken, I., Beuselinck, L., Vandekerckhove, L. and Govers, G. (2001) The value of a physically based model versus an empirical approach in the prediction of ephemeral gully erosion for loess-derived soils. *Geomorphology* 40, 237–252.

Nachtergaele, J., Poesen, J., Sidorchuk, A. and Torri, D. (2002) Prediction of concentrated flow width in ephemeral gully channels. *Hydrological Processes* 16, 1935–1953.

O'Loughlin, E.M. (1986) Prediction of surface saturation zones in natural catchments by topographic analysis. *Water Resources Research* 22, 794–804.

Poesen, J. (1993) Gully typology and gully control measures in the European loess belt. In: Wicherek, S. (ed.) *Farm Land Erosion in Temperate Plains Environment and Hills*. Elsevier Science, Amsterdam, pp. 221–239.

Poesen, J., Vandaele, K. and van Wesemael, B. (1996) Contribution of gully erosion to sediment production in cultivated lands and rangelands. In: Walling, D.E. and Webb, B.W. (eds) *Erosion and Sediment Yield: Global and Regional Perspectives*. IAHS Publication 236. IAHS Press, Wallingford, UK, pp. 251–266.

Poesen, J., Nachtergaele, J., Verstraeten, G. and Valentin, C. (2003) Gully erosion and environmental change: importance and research needs. *Catena* 50, 91–133.

Sidorchuk, A. (1999) Dynamic and static models of gully erosion. *Catena* 37, 401–414.

Souchère, V., Cerdan, O., Ludwig, B., Le Bissonnais, Y., Couturier, A. and Papy, F. (2003) Modeling ephemeral gully erosion in small cultivated catchments. *Catena* 50, 489–505.

Takken, I., Beuselinck, L., Nachtergaele, J., Govers, G., Poesen, J. and Degraer, G. (1999) Spatial evaluation of a physically based distributed erosion model (LISEM). *Catena* 37, 431–447.

Takken, I., Jetten, V., Govers, G., Nachtergaele, J. and Steegen, A. (2001) The effect of tillage-induced roughness on runoff and erosion patterns. *Geomorphology* 37, 1–14.

Torri, D. and Borselli, L. (2003) Equation for high-rate gully erosion. *Catena* 50, 449–467.

Vandaele, K., Poesen, J., Govers, G. and van Wesemael, B. (1996) Geomorphic threshold conditions for ephemeral gully incision. *Geomorphology* 16, 161–173.

Vandaele, K., Poesen, J., Marques da Silva, J.R., Govers, G. and Desmet, P. (1997) Assessment of factors controlling ephemeral gully erosion in southern Portugal and central Belgium using aerial photographs. *Zeitschrift für Geomorphologie* 41, 273–287.

Vandekerckhove, L., Poesen, J., Oostwoud Wijdenes, D., Nachtergaele, J., Kosmas, C., Roxo, M.J. and De Figueiredo, T. (2000) Thresholds for gully initiation and sedimentation in Mediterranean Europe. *Earth Surface Processes and Landforms* 25, 1201–1220.

Wesseling, C.G., Karssenberg, D., Van Deursen, W.P.A. and Burrough, P.A. (1996) Integrating dynamic environmental models in GIS: the development of a Dynamic Modelling language. *Transactions in GIS* 1, 40–48.

Woodward, D.E. (1999) Method to predict cropland ephemeral gully erosion. *Catena* 37, 393–399.

20 Simulating Fine Sediment Delivery in Lowland Catchments: Model Development and Application of INCA-Sed

N.P. Jarritt[1] and D.S.L. Lawrence[2]
[1]*University of Reading, UK.* Present address *Entec UK Ltd, London, UK;*
[2]*University of Reading, UK*

Introduction

A variety of fully distributed, physically based models describing sediment erosion and transport have been developed during the past decade, e.g. WEPP (Laflen *et al.*, 1991), LISEM (de Roo *et al.*, 1996), and EUROSEM (Morgan *et al.*, 1998). Much of this work was motivated initially by detailed plot-scale investigations of soil erosion processes, and these physically based model components have been extended to the catchment scale to generate linked hillslope/channel models for sediment flux. This modelling strategy provides a powerful tool for investigating physical processes in conjunction with field experiments or detailed monitoring programmes; however, the large number of input parameters often limits the model applications to small-scale, research catchments (Jetten *et al.*, 1999). Fully distributed models also usually discretize catchment slope processes to a grid or mesh-scale resolution, which can result in long run times and numerical errors and instabilities, undermining the viability of routine application. These limitations highlight the need for 'integrated' catchment sediment delivery models that incorporate more readily available environmental data characterizing hydrometeorological conditions, land use, soil erodibility, and catchment and stream channel morphology at an appropriate level of spatial aggregation. In this chapter we present a new model, INCA-Sed, which simulates suspended sediment concentrations in streams using a semi-distributed representation of catchment variables. The semi-distributed approach has been used for several years in hydrological modelling (Hughes and Sami, 1994) and more recently in the modelling of sediment yield (Liden *et al.*, 2001). The principal objective motivating the development of Integrated Catchment Model of Sediment (INCA-Sed) is to characterize the 'response' of the fine sediment regime to hydrologic forcing, rather than to fully quantify the functioning of the sediment delivery system. The model structure and key equations comprising INCA-Sed are presented in the following sections, and the model is then applied to two lowland catchments in southern England, the Lambourn and the Enborne.

INCA-Sed Model Structure

INCA-Sed uses the same model structure as the INCA-N (Whitehead *et al.*, 1998) and INCA-P (Wade *et al.*, 2001, 2002) models, both of which incorporate a semi-distributed representation of the catchment system. Catchment physical features are recognized at three spatial levels, rather than on a grid cell basis. At the first level, the main river channel is divided into a series of reaches. The land area that drains into each of these reaches is then defined as a 'sub-catchment' using a digital terrain model and Geographical Information System (GIS) algorithms (Morris and Flavin, 1994). At the second level, each sub-catchment is further divided into a maximum of six land use classes. This is achieved by overlaying the sub-catchment boundaries on a land use map and calculating the percentage of each land use type within each sub-catchment. At the third spatial level, a generic cell of unit area is applied to each land use type. A parameter set is derived for the cell by averaging the spatial parameters and this is used in processing the model equations. Sub-catchment totals are calculated by summing the results for each land use type.

The in-stream component of the model treats each reach as a fully-mixed reservoir, with inputs from upstream and the sub-catchment for the reach, and an output to the reach immediately downstream. The INCA models run on a daily timestep, producing output as daily averages for each sub-catchment and stream reach. The model equations are solved simultaneously using a fourth-order Runge-Kutta numerical technique to ensure that no individual process takes precedence over any other (Wade *et al.*, 2002).

Figure 20.1 illustrates the linkages between the processes driving the sub-catchment generation and delivery of sediment in INCA-Sed. The set of processes is considerably simplified relative to soil erosion models such as EUROSEM. For each sub-catchment of the model, material for transport (available sediment) is generated on the catchment slopes. Given sufficient direct runoff, this material is transported from the land to the in-stream phase of the model. Direct runoff can also further erode sediment from the surface once this supply is depleted. The sediment concentration in the direct runoff is, therefore, a combination of the sediment stored on the slope and that generated by flow erosion. The simplifications relative to more detailed soil erosion models include the omissions of: (i) an interception component; (ii) explicit modelling of runoff flow hydraulics; and (iii) a distinction between rill and interrill erosion. Within the channel, the average uniform conditions generated by the hydrological model are used to represent the average bulk movement of sediment within the channel (Fig. 20.2). The bulk entrainment and deposition of sediment is governed by the flow capacity, specified in terms of stream power. Suspended sediment concentration in the flow increases with stream power, given the presence of available material either from the sub-catchments or entrained from the bed. With decreasing stream power, the sediment in suspension will settle and be deposited on the stream bed, but is available for subsequent in-stream resuspension as stream power increases.

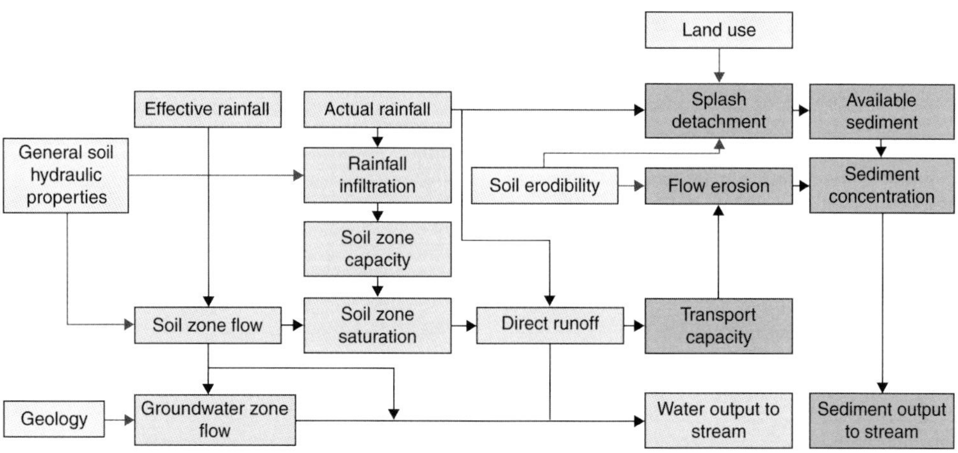

Fig. 20.1. Process components and linkages for sediment generation and transport on sub-catchment hillslopes.

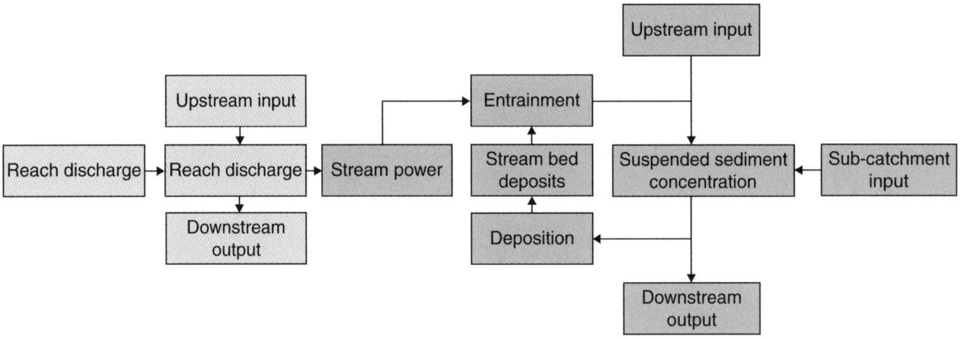

Fig. 20.2. Process components and linkages for sediment transfer in sub-catchment channel reaches.

INCA-Sed Model Equations

The INCA hydrological model is driven by a single input time series, the daily effective rainfall, derived from the Meteorological Office Rainfall and Evaporation Calculation System (MORECS) soil moisture and evaporation accounting model. The MORECS model produces estimates of evapotranspiration, soil moisture deficit and hydrologically effective rainfall on a 40 km × 40 km grid basis (Gardner and Field, 1983). The hydrological model consists of two components, a land phase for the sub-catchment zones and an in-stream phase for the river reaches. The equations for these phases are presented below using the variable and parameter names and units of Wade *et al.* (2002).

The land phase of the hydrological model considers three principal pathways within a catchment: direct runoff, shallow soil zone drainage and flow through the groundwater zone, i.e. baseflow generation. The flux through each of these zones is modelled using mass balance equations. The hydrological system is driven by an effective rainfall flux (p_{eff}, $m^3/s/m^2$) into the soil zone:

$$\frac{dq_{sw}}{dt} = \frac{p_{\text{eff}} - q_{sw}}{T_2} \qquad (20.1)$$

where q_{sw} ($m^3/s/m^2$) is the soil zone flow and T_2 (days) is the soil water zone residence time. A portion of the soil zone flow percolates into the groundwater zone, as controlled by the baseflow index, c_3 (dimensionless):

$$\frac{dq_{gw}}{dt} = \frac{c_3 q_{sw} - q_{gw}}{T_3} \qquad (20.2)$$

where q_{gw} ($m^3/s/m^2$) is the groundwater zone flow and T_3 (days) is the residence time. In the INCA-Sed model, direct runoff (overland flow) can be generated as either saturation-excess overland flow or infiltration-limited overland flow. Direct runoff, q_{dr}, derived from a saturation excess is represented as a proportion, c_1, of the soil zone flow in excess of the saturation threshold, q_{sat}, as given by:

$$\frac{dq_{dr}}{dt} = \frac{c_1(q_{sw} - q_{sat}) - q_{dr}}{T_1} \qquad (20.3)$$

where T_1 is the surface runoff residence time. The soil zone saturation flow, q_{sat}, is related to the soil type, a semi-distributed parameter within INCA-Sed. The proportion of the excess soil zone flow that does not contribute to direct runoff input is assumed to be lost to surface depressions and subsequent evaporation. This water loss is therefore given by:

$$(1 - c_1)(q_{sw} - q_{sat}) \qquad (20.4)$$

Direct runoff is also generated when the rainfall rate exceeds the infiltration rate (i.e. infiltration-limited overland flow) such that:

$$\frac{dq_{dr}}{dt} = \frac{c_2(p - i) - q_{dr}}{T_1} \qquad (20.5)$$

where p ($m^3/s/m^2$) is the rainfall rate, i ($m^3/s/m^2$) is the variable infiltration rate and c_2 (dimensionless) is the proportion of the rainfall excess that becomes direct runoff. The infiltration rate is directly proportional to the hydraulic conductivity of the soil and inversely proportional to the water content of the soil. INCA-Sed does not model the volumes of water within each of the sub-catchment flow zones, but rather calculates the flux rates within each zone. To accommodate this, the effective degree of saturation is calculated from the relative magnitudes of the soil zone flow and the saturation threshold flow. The full equation governing the direct runoff within INCA-Sed is obtained by combining Eqns 20.3 and 20.5 to yield:

$$\frac{dq_{dr}}{dt} = \frac{c_1\langle q_{sw} - q_{sat}\rangle + c_2(p - i) - q_{dr}}{T_1} \tag{20.6}$$

where $\langle\,\rangle$ is used to indicate that this quantity only takes on a value when $q_{sw} > q_{sat}$ and is zero in all other cases. The total discharge from the sub-catchment into the reach (Q_{SC}, m^3/s) is then given by:

$$Q_{SC} = A(q_{dr} + (1 - c_3)\, q_{sw} + q_{gw} \tag{20.7}$$

where A is the sub-catchment area. The discharge within the channel is governed by the equations presented by Whitehead *et al.* (1998) and the reader is referred to that work for details.

As illustrated in Fig. 20.1, sediment is generated by splash detachment and erosion by direct runoff, and equations must be developed to represent these processes. The transport capacity of the direct runoff also needs to be specified, as does a mass balance for the sediment in each sub-catchment. The detachment of soil particles by raindrop impact is a function of the energy imparted to the soil surface by the individual drops (Sharma *et al.*, 1993), and in terms of available data, the best proxy for this is the daily rainfall total. Splash detachment (S_{SP}, kg/m^2/s) is therefore modelled as a function of the rainfall (p, m/s), a soil erodibility parameter spatially linked to soil type, E (kg/m^2/s) and a vegetation cover index linked to land use, V (dimensionless):

$$S_{SP} = pE^{\frac{p}{V}} \tag{20.8}$$

The transport capacity of surface runoff is critical in simulating sediment delivery in models that are driven by direct runoff (Ferro, 1998), as it acts as an upper limit to the potential contribution of each sub-catchment to sediment concentrations in the channel. In INCA-Sed, the sediment transport capacity, S_{TC}, is modelled as a simple power law relationship given by:

$$S_{TC} = a_4\left(\frac{Aq_{dr}}{L} - a_5\right)^{a_6} \tag{20.9}$$

where a_4 (kg/m^2), a_5 (m^2/s) and a_6 (dimensionless) are calibration parameters, and L is the length of the channel reach for the sub-catchment such that A/L provides an estimate of the average slope length. The erosion of sediment by direct runoff is represented in a similar manner, although the erosive potential of the flow is given in terms of the total transport capacity, less the current sediment load. Direct runoff flow erosion S_{FL} (kg/s) is therefore given by:

$$S_{FL} = a_1 E \frac{(S_{TC} - S_C)}{S_{TC}}\left(\frac{Aq_{dr}}{L} - a_2\right)^{a_3} \tag{20.10}$$

where a_1 (dimensionless), a_2 (m/s) and a_3 (dimensionless) are calibration parameters and S_C (kg/s) is the sediment transport rate. Finally, a mass balance accounting is used for each sub-catchment to determine the mass of sediment remaining on the slope and removed to the channel during each time step.

The suspended sediment flux within the stream channel can be conceptualized as having two components: (i) lateral downstream transfer through reaches; and (ii) vertical exchange with the bed material. Downstream movement can be readily linked to the hydrologic flux; however, the vertical exchange is much more complex, as sediment is potentially entrained, deposited and reentrained during its migration through channel reaches. In order to accommodate this, the relative rates of entrainment from and deposition to the bed must be linked to the hydrologic equations. A key variable controlling this is the grain size distribution of the sediment, and INCA-Sed incorporates the effects of sediment grain size by calculating a mass balance for each of five grain size classes (with boundaries at 2, 60, 200 and 600 μm). The grain size of the material delivered to the stream from the slopes is assumed to be related to the soil texture of the sub-catchment contributing to a reach and this texture, together with the distribution of soil types within each sub-catchment is used to estimate the proportion of each of the five grain size classes delivered to the stream. Soil texture is used in lieu of the effective grain size delivered to the stream, as data quantifying effective grain size are not routinely available for large-scale hydrologic modelling. However, where such observations are available, they could be incorporated into INCA-Sed and would provide better control on the partitioning of grain size distributions along flow pathways.

Entrainment is driven by the boundary shear velocity, υ_*, which for equilibrium flow in a wide rectangular channel is given by:

$$\upsilon_* \approx \sqrt{gd \sin\theta} \qquad (20.11)$$

where d is the flow depth, θ is the slope of the water surface and g is the gravitational acceleration. Within INCA-Sed, this is applied in the form:

$$\upsilon_* \approx \sqrt{gda_7 \sin\theta} \qquad (20.12)$$

where a_7 (dimensionless) is a calibration parameter introduced to accommodate the departure from ideal conditions in natural channels. Inman (1949) related the sediment grain diameter to the threshold shear velocity required to entrain a particle from the bed, and the following approximation to the Inman curve is applied in INCA-Sed to determine when particular grain size classes are vulnerable to entrainment:

$$D_{max} = x_1 \upsilon_*^{x_2} \qquad (20.13)$$

where D_{max} is the maximum grain diameter that can be entrained at a given shear velocity, and x_1 and x_2 are constants of regression with values of 9.99 and 2.52, respectively. Once suspended, sediment particles will tend to fall towards the bed under gravity, although the turbulent structure of the flow can locally enhance the suspension. As a first approximation, though, INCA-Sed calculates a potential deposition rate for each grain size class based on the terminal settling velocity for the median grain size of each class.

The volume of sediment stored on the reach bed is increased by deposition from the flow and is decreased by the material entrainment. Therefore, for a timestep of length T seconds, the mass balance equation is:

$$\frac{dm_{bed}}{dt} = T(m_{dep} - m_{ent}) \qquad (20.14)$$

where m_{bed} (kg) is the mass of sediment on the bed, m_{dep} (kg) and m_{ent} (kg), are the masses deposited and entrained to and from the bed, respectively, during time t (s). The mass balance for sediment in suspension within a reach is more complex. In addition to local exchange with the bed, the suspended 'store' in each reach receives sediment from upstream and from the contributing sub-catchment and also releases suspended sediment downstream. The mass balance equation, for a timestep of length T, is thus:

$$\frac{dM_{sus}}{dt} = T\big(M_{out} + LW(m_{ent} - m_{dep}) + M_{up} - Qm_{sus}\big) \qquad (20.15)$$

where M_{sus} (kg) is the total mass of sediment in suspension in the reach, M_{up} (kg/s) is the rate of input from upstream, M_{out} (kg/s) is the rate of input from the sub-catchment to the reach downstream, W (m) is the channel width, Q (m^3/s) is the flow discharge and m_{sus} (kg/m^3) is the suspended sediment concentration (SSC) given as:

$$m_{sus} = \frac{M_{sus}}{Vol} \qquad (20.16)$$

where Vol is the volume of water (m^3) in the reach.

Model Demonstration

To demonstrate model performance, INCA-Sed was applied to two neighbouring catchments, the Lambourn and the Enborne, both tributary to the River Kennet in Berkshire, England. The Lambourn and Enborne catchments have areas of 234 and 153 km^2, respectively. The Lambourn is a highly permeable, baseflow-dominated system, being underlain almost entirely by chalk, while the Enborne is overlain by low-permeability Tertiary clays and gravels throughout most of the catchment. Land use in both catchments largely comprises arable farming and grazing. To apply INCA-Sed, the catchments were partitioned into contributing sub-catchments using a digital terrain map for the area. The percentage distribution of soil type, land use and geology were derived for each sub-catchment by using GIS overlays for each of these physical characteristics. The reach lengths of the main river channel were also derived from the GIS database. The discharge time series were obtained from 15-min records from the Environment Agency gauging stations at Boxford and Brimpton, for the Lambourn and Enborne, respectively. The daily suspended sediment time series used for the model demonstration extend from 21 March 1999 to 20 March 2000 and were sampled approximately 2 km north of Boxford for the Lambourn and a few hundred metres upstream of the gauging station at Brimpton for the Enborne. The samples were taken using an EPIC automatic sampler and the water samples were processed using vacuum filtration with 0.45 μm Whatman filter papers (Evans *et al.*, 2003). Two daily samples were available for each site and the average of those two values was used.

Comparisons between the observed suspended sediment concentrations and the model results are illustrated in Figs 20.3 and 20.4 for the two catchments. The observed values vary significantly between the two sites, with annual average SSCs of 8.5 and 34.1 mg/l for the Lambourn and Enborne, respectively. The Enborne also exhibits much greater variability in concentrations, with a standard deviation of 64.3 mg/l, relative to the 3.1 mg/l observed for the Lambourn. The model very effectively reproduces the flashy sediment response trends seen at the Enborne (Fig. 20.4), although the actual peak value is difficult to model precisely and would require a recalibration of the model with this objective as the target. The Nash-Sutcliffe values (Nash and Sutcliffe, 1970) for the Enborne model are 0.723 and 0.32 for the catchment hydrology and sediment concentration (Fig. 20.4), respectively. The model appears to perform more poorly for the Lambourn (Fig. 20.3), although this is largely due to the scale at which the data are displayed, and the Nash-Sutcliffe values for the model fit are 0.83 and 0.36 for the hydrology and sediment concentrations (Fig. 20.3). However, the Lambourn data series exhibits a lower degree of autocorrelation and only during the period between November 1999 and January 2000 is a consistent peak response observed. This is not surprising, given the baseflow-dominated hydrology. Much of the sediment suspended in the water column is likely the product of resuspended in-channel sources, including organic material derived from stream vegetation, rather than from direct runoff. During the period in which there is a distinct sediment response, the model in fact reproduces the observed trend reasonably well. Given the low values of suspended sediment in this stream, the average values and range are fairly well characterized by

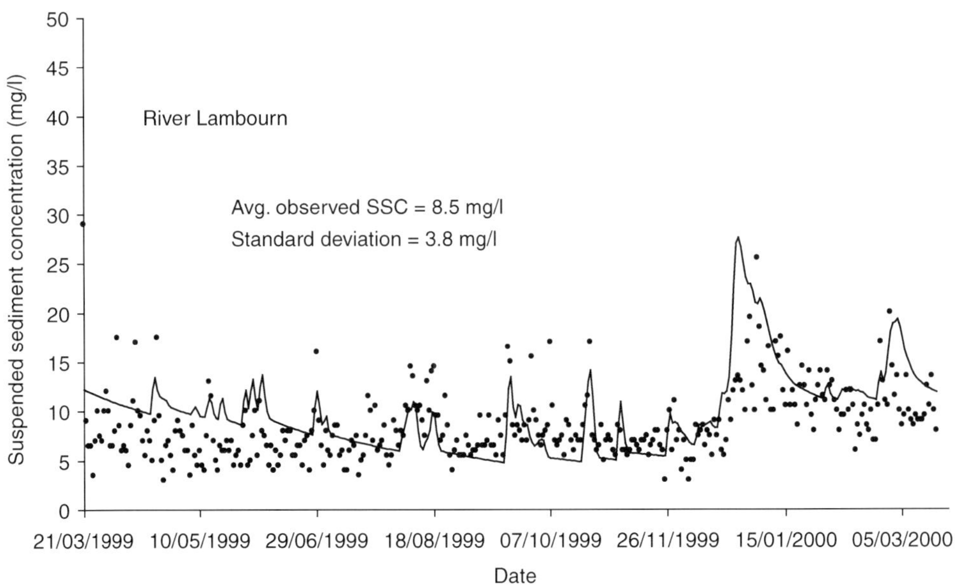

Fig. 20.3. Observed versus simulated suspended sediment concentrations for the River Lambourn.

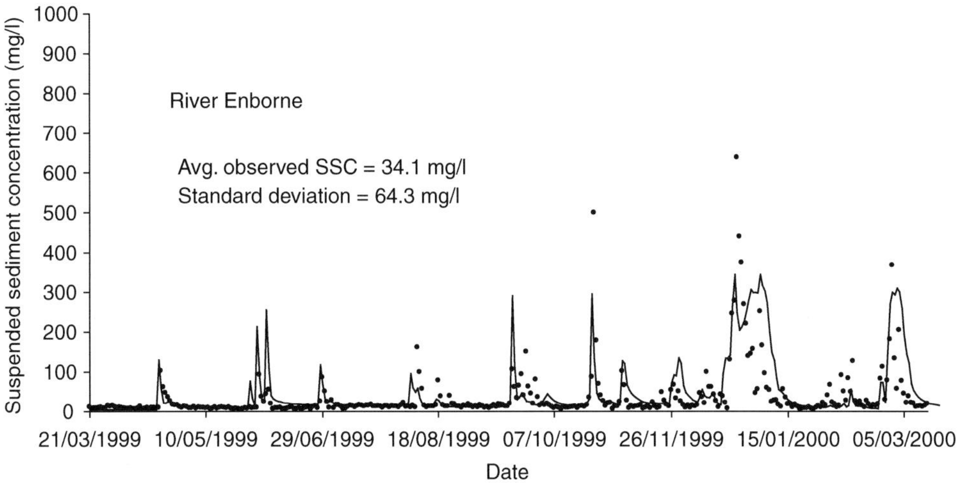

Fig. 20.4. Observed versus simulated suspended sediment concentrations for the River Enborne.

the model results, although the temporal structure of the response and the role of background 'noise' at these low concentrations require further investigation.

The model demonstration illustrates the application of a semi-distributed linked hillslope-channel model for characterizing the fine sediment response to hydrologic forcing in small lowland catchments. The principal advantages of the modelling approach are the reliance on readily available catchment data and the relatively simple numerical structure for representing catchment spatial properties. The results suggest that INCA-Sed can be used to simulate time series of suspended sediment concentration in two catchments with very different hydrologic and sediment responses to a fair level of correspondence with observed concentrations. Significant discrepancies between observed and simulated concentrations, though, are present and may be a consequence of several factors including: (i) the lack of explicit modelling of in-stream sediment sources derived from biological activity and bank erosion; and (ii) the limited spatial structure of the hillslope component of the model, which relies on percentages of land use and soil type for each sub-catchment rather than mapping out the connectivity of these sediment generators and stores in the landscape. Further development of INCA-Sed will investigate the potential role of these factors and strategies for incorporating them into the simulation of sediment delivery.

Acknowledgements

Dr Dan Evans is gratefully acknowledged for the use of his SSC data for the Lambourn and Enborne catchments. This work was supported by a NERC postgraduate research studentship to Nick Jarritt.

References

De Roo, A.P.J., Wesseling, C.G. and Ritsema, C.J. (1996) LISEM: a single-event based physically-based hydrological and soil erosion model for drainage basins. I. Theory, input and output. *Hydrological Processes* 10, 1119–1126.

Evans, D.J., Johnes, P.J. and Lawrence, D.S.L. (2003) Suspended and bed load sediment transport dynamics in two lowland UK streams – storm integrated monitoring. In: Bogen, J., Fergus, T. and Walling, D.E. (eds) *Erosion and Sediment Transport Measurement in Rivers: Technological and Methodological Advances.*

IAHS Publication 283. IAHS Press, Wallingford, UK, pp. 103–110.
Ferro, V. (1998) Evaluating overland flow sediment transport capacity. *Hydrological Processes* 12, 929–946.
Gardner, C.M.K. and Field, M. (1983) An evaluation of the success of MORECS, a meteorological model, in estimating soil-moisture deficits. *Agricultural Meteorology* 29, 269–284.
Hughes, D.A. and Sami, K. (1994) A semi-distributed, variable time-interval model of catchment hydrology – structure and parameter-estimation procedures. *Journal of Hydrology* 155, 265–291.
Inman, D.L. (1949) Sorting of sediments in the light of fluid mechanics. *Journal of Sedimentary Petrology* 19, 51–70.
Jetten, V., de Roo, A., Favis-Mortlock, D. (1999) Evaluation of field-scale and catchment-scale soil erosion models. *Catena* 37, 521–541.
Laflen, J.M., Lane, L.J. and Foster, G.R. (1991) WEPP – a new generation of erosion prediction technology. *Journal of Soil and Water Conservation* 46, 34–38.
Liden, R., Harlin, J., Karlsson, M. and Rahmberg, M. (2001) Hydrological modelling of fine sediments in Odzi River, Zimbabwe. *Water SA* 27, 303–314.
Morgan, R.P.C., Quinton, J.N., Smith, R.E., Govers, G., Poesen, J.W.A., Auerswald, K., Chisci, G., Torri, D. and Styczen, M.E. (1998) The European Soil Erosion Model (EUROSEM): a dynamic approach for predicting sediment transport from fields and small catchments. *Earth Surface Processes and Landforms* 23, 527–554.
Morris, D.G. and Flavin, R.W. (1994) *Sub-set of the UK 50 m by 50 m Hydrological Digital Terrain Model Grids.* NERC, Institute of Hydrology, Wallingford, UK.
Nash, J.E. and Sutcliffe, J.V. (1970) River flow forecasting through conceptual models. *Journal of Hydrology* 10, 282–290.
Sharma, P.P., Gupta, S.C. and Foster, G.R. (1993) Predicting soil detachment by raindrops. *Soil Science Society of America Journal* 57, 674–680.
Wade, A.J., Hornberger, G.M., Whitehead, P.G. and Jarvie, H.P. (2001) On modeling the mechanisms that control in-stream phosphorus, macrophyte and epiphyte dynamics: an assessment of a new model using general sensitivity analysis. *Water Resources Research* 37, 2777–2792.
Wade, A.J., Whitehead, P.G. and Butterfield, D. (2002) The Integrated Catchments model of Phosphorus dynamics (INCA-P), a new approach for multiple source assessment in heterogeneous rivers: model structure and equations. *Hydrology and Earth System Science* 6, 583–606.
Whitehead, P.G., Wilson, E. and Butterfield, D. (1998) A semi-distributed integrated nitrogen model for multiple source assessment in catchments (INCA): Part I – model structure and process equations and Part II – application to large river basins in south Wales. *The Science of the Total Environment* 210/211, 547–583.

IV

Management

Sorting and treatment of sediment dredged from the River Elbe, Hamburg, Germany (photo: P.N. Owens).

21 Estimating Sediment Generation from Hillslopes in England and Wales: Development of a Management Planning Tool

G.A. Wood,[1] M. McHugh,[1] R.P.C. Morgan[1] and A. Williamson[2]
[1]National Soil Resources Institute, UK; [2]Environment Agency, UK

Introduction

The accelerated erosion of land in England and Wales has been an issue of concern for many years (Evans, 1971; Morgan, 1980, 1985; Evans and Skinner, 1987; Robinson and Blackman, 1990; Evans and Boardman, 1994) because of its agricultural (Evans, 1996), economic (Butcher *et al.*, 1992) and ecological (Theurer *et al.*, 1998) impacts. Overall costs of erosion to the UK economy may be as high as £90 million per annum (Environment Agency, 2002). There are, therefore, strong arguments to support the need for erosion control. Before resources can be effectively focused, however, it is necessary to identify target areas for resource investment, especially as erosion mitigation has previously been on an *ad hoc* basis, with problems being dealt with locally. This research addressed the need for a national assessment of soil erosion risk for England and Wales by producing a national map of soil erosion taking account of soils, slopes and present-day land use. It formed part of a larger study that incorporated consideration of slope–channel connectivity (Walling and Zhang, 2004) to account for sediment delivery to watercourses (McHugh *et al.*, 2002).

Methodology

Estimating erosion probability

Between 1996 and 2002, various surveys of soil erosion carried out in England and Wales, at subset field sites of the National Soil Inventory (NSI), formed the basis of three extensive erosion monitoring projects for lowland arable soils (Harrod, 1998), upland soils (Harrod *et al.*, 2000; McHugh, 2002) and lowland grassland soils (Harrod *et al.*, 2000). The NSI database contains comprehensive data on soils for over 6000 field sites located at 5 km intervals across England and Wales. Through repeated summertime visits to arable sites (in 1996, 1997, 1998), upland sites (in 1997, 1999, 2002) and grassland sites (in 1998), these studies recorded area and volumetric soil loss from channelled erosion (rills and gullies) and macro-scale areas of bare soil on all sites. They did not account for soil loss through tillage, wind or sheet erosion processes. Whilst these data may not provide a suitable base for estimating long-term erosion rates, they represent the most comprehensive erosion database available for England and Wales.

The erosion data were analysed to determine the probability of erosion of a certain magnitude occurring for given soil and slope combinations. The numbers of observations when grouping the soils data into soil series were too small for robust statistical analysis; consequently, data were analysed at the soil subgroup level (Avery, 1980), although in some cases it was necessary to combine subgroups and aggregate further to the soil group level.

Using slope gradient measurements made on-site with a clinometer, each field site was assigned to a slope class. For arable and grassland land cover these were: 0–2.99°, 3–6.99° and > 7°. The boundary at 3° was chosen because it represents the critical angle at which, for many soils in northern Europe, rill erosion begins (De Ploey, 1984). The boundary at 7° represents the upper limit of land considered suitable for arable farming in many land capability classifications (Hudson, 1981) and is the upper limit of land in the highest capability class in the UK (Bibby and Mackney, 1969). In the upland areas, field sites were classed as either < 12° or > 12°. The 12° boundary was selected as the slope gradient in the uplands at which erosion was found to decrease significantly compared with that on shallower slopes (McHugh, 2000). It is also the upper limit at which ploughing can be carried out safely using standard equipment. For some soil subgroups it was necessary to group slope classes together either because the number of observations in a given slope class was very small or because there was no difference in the frequency of erosion between slope classes.

For each combination of soil subgroup and slope class, frequency histograms were produced showing the number of instances of erosion of different magnitudes: Fig. 21.1 shows two examples from the lowland arable data set. As expected (Boardman, 1990; Evans, 1998), these show that the volumes of soil eroded are strongly negatively skewed. Such a Poisson distribution provides a better basis than the normal distribution for estimating the probability of rare events with a low probability of occurrence. Table 21.1 gives the cumulative probability values of erosion for each magnitude class for the two data sets illustrated in Fig. 21.1. From these, it is possible to estimate the magnitude of erosion that will occur for probability levels $P = 0.99$, $P = 0.995$ and $P = 0.999$. By adopting the ergodic principle of statistical mechanics, spatial probabilities can be used to approximate temporal probabilities (Scheidegger and Langbein, 1966), based on the assumption that 'in a steady state, the ensemble of all possible configurations of the system at a given time is identical to the ensemble obtained by watching the system evolve through all times t_1, t_2 . . . t_i' (Scheidegger, 1970) and that, in geographical and geomorphological situations, 'ensembles' can be 'replaced by sets of spatial points or areas' (Paine, 1985). Hence these probabilities obtained from spatial data sets can be used as surrogates for the probability of events occurring over time. Events with annual probabilities of 0.99, 0.998 and 0.999 correspond to events

Table 21.1. Cumulative probability distributions based on Poisson distribution for selected soil subgroups under lowland arable.

	Erosion class	Cumulative probability		Erosion class	Cumulative probability
A	0	0.348351	B	0	0.206192
	0.01–0.04	0.715702		0.01–0.04	0.531758
Subgroup 5.71	0.05–0.09	0.909397	Subgroup 5.72	0.05–0.09	0.788784
Slopes ≥ 3°	0.10–0.49	0.977483	Slopes ≥ 3°	0.10–0.49	0.924061
($\lambda = 1.0546$)	0.50–0.99	0.995434	($\lambda = 1.57895$)	0.50–0.99	0.977460
	1.00–4.99	0.999219		1.00–4.99	0.994323
	5.00–9.99	0.999885		5.00–9.99	0.998760
	10.00–24.99	0.999985		10.00–24.99	0.999761
	25.00–49.99	0.999998		25.00–49.99	0.999959
				50.00–99.99	0.999994
				100.00–199.99	0.999999

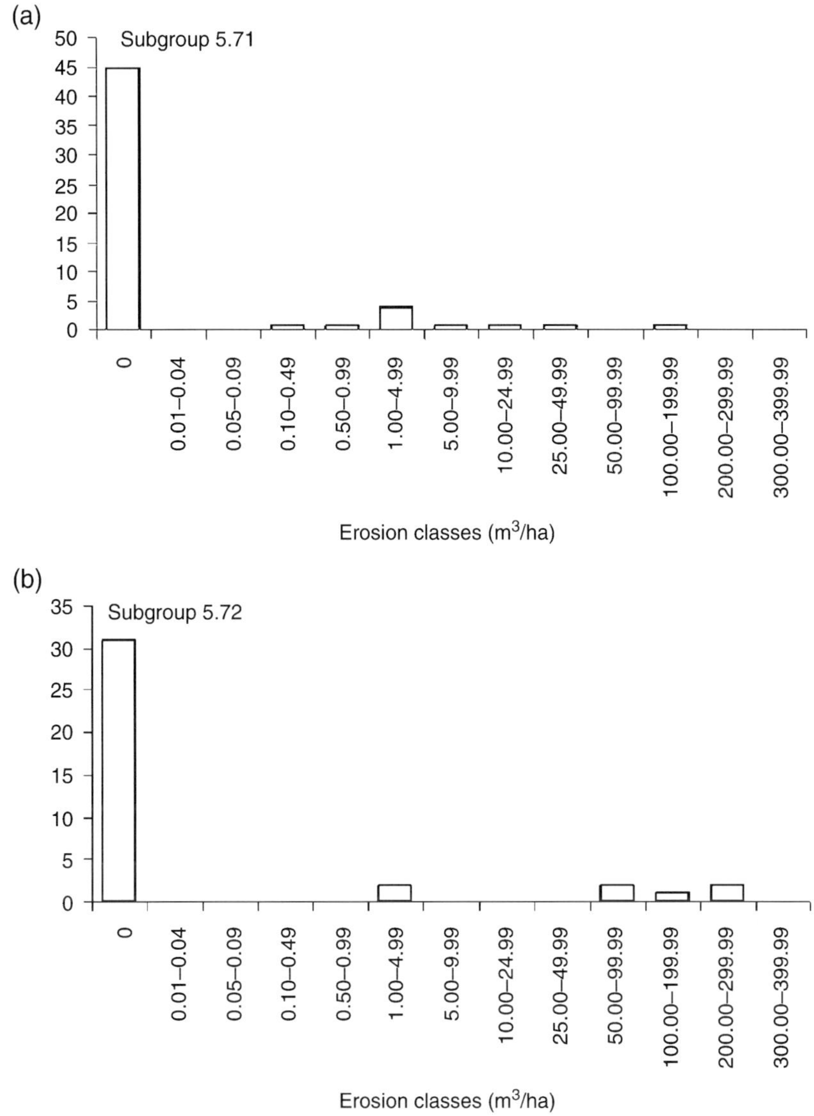

Fig. 21.1. Frequency of erosion classes for soil subgroups 5.71 and 5.72 under lowland arable on slopes ≥ 3°.

with respective recurrence intervals of once in 1, 5 and 10 years.

Geographical Extrapolation over England and Wales

A GIS-based approach was used to:

1. Determine a spatially averaged slope class for each 1 × 1 km pixel using data from the Ordnance Survey 50 m PANORAMA digital elevation model.

2. Describe the soil composition of each 1 × 1 km pixel according to the areal proportion of the dominant, second and third most common soil subgroups, based on the National Soil Map of England and Wales.

3. Determine the predominant land use of each 1 × 1 km pixel according to the broad classification of lowland arable, lowland grassland

and upland, based on the 1990 Land Cover Map of Great Britain produced by the Centre for Ecology and Hydrology (CEH).

4. Calculate the annual erosion rate expected with a 10-year return period for each 1 × 1 km pixel, according to land use, and spatially averaged over the combination of soil and slope classes.

Spatial characterization of slope gradient

A typical landscape, bounded by a 1 km square, will contain a range of slopes that cannot be sensibly characterized using an average value of slope gradient, and certainly not without loss of important detail. Consequently, slope gradients calculated using a 50 m Digital Elevation Model were derived, and the frequency distribution of slope gradient classes were stored as a histogram for each 1 km pixel.

Soil distribution

The NSRI National Soil Map (NatMap) provided a summary of the soil types within a 1 km grid for England and Wales with proportions of each soil, by soil association, available in descending order of dominance within each 1 km square. Within this work, the three most predominant soil associations, which, on average, represented 95% of the soil area in each 1 km pixel, were used to characterize each 1 km pixel.

Land cover mapping

The Landcover Map of Great Britain (LCMGB) (Fuller *et al.*, 1994) uses a classification of Landsat Thematic Mapper to map 25 cover types, including water, beaches and bare ground, developed and arable land, and 18 types of semi-natural vegetation. The map has an overall mapping accuracy of 67%, consistent with known mapping accuracies of satellite-image-derived thematic maps (Richards and Jia, 1999). LCMGB data were reclassified into the arable, grassland and upland landcover categories used in the erosion surveys. Comparison of erosion sites with coincident LCMGB locations indicated that the LCMGB arable map class was in agreement with 89% of erosion arable sites (of the remaining 11%, 10% were grassland). Only 44% of lowland grassland sites agreed with LCMBG locations, while upland areas had an accuracy of 81%. The classification of the remaining areas was equally divided between grassland and arable classes. These levels of uncertainty have a direct impact on the final estimates of sediment generation potential.

Erosion modelling

Total erosion amount within each 1 km pixel was estimated by calculating a spatially weighted sum of all soil–slope class combinations present in any one pixel, according to the predominant land use. Combinations of different soil–slope classes will normally occur in varying proportions. When summarized as 1 km pixels, only the relative proportions of each slope and soil class are known. Without being able to co-locate soils with their slope class, it had to be assumed that each soil subgroup contained slope classes in proportion to their weighted averages within the 1 km pixel. This is not what one would expect in reality because soils occupy landscape positions according to their pedogenic origins and may be expected, therefore, to have a specific spatial relationship to different slope classes. With these assumptions, the total erosion for each 1 km pixel was estimated using Eqn 21.1.

$$R = \sum_{t=1}^{m} \omega_t . \sum_{s=1}^{n} \omega_s . E_{ts} \qquad (21.1)$$

where R is the estimate of total channelled erosion (m^3/ha); ω_s is the slope-area weighting factor (%); ω_t is the soil group area weighting factor (%) and E_{ts} is the estimated erosion rate (m^3/ha) in association with slope class s and soil group t as presented in Tables 21.2–21.4.

Results

Erosion risk on lowland arable soils

The annual erosion expected with given probabilities indicated (Table 21.2) that the brown earths were the major soil group with the highest

Table 21.2. Estimated annual soil erosion (m^3/ha) for lowland arable soils with given return periods.

	1-in-1 year			1-in-5 years			1-in-10 years		
	Slope			Slope			Slope		
Soil subgroup	0–2°	3–6°	≥ 7°	0–2°	3–6°	≥ 7°	0–2°	3–6°	≥ 7°
3.43	0.00	0.00	0.00	0.00	0.00	0.03	0.00	0.00	0.05
4.11	0.00	0.07	0.07	0.00	0.10	0.10	0.00	0.50	0.50
5.11	0.00	0.02	0.02	0.00	0.05	0.05	0.00	0.05	0.05
5.13	0.06	0.06	0.06	0.10	0.10	0.10	0.50	0.50	0.50
5.41 + 5.47	0.04	0.09	0.07	0.30	0.42	0.42	0.42	0.49	0.50
5.51, 5.52, 5.54	0.00	0.19	0.19	0.00	0.53	0.53	0.00	0.81	0.81
5.71	0.04	0.38	0.38	0.06	0.84	0.84	0.08	0.97	0.97
5.72	0.05	0.87	0.87	0.09	4.29	4.29	0.10	6.00	6.00
5.81	0.18	0.18	0.18	0.50	0.50	0.50	0.79	0.79	0.79
7.11, 7.12, 7.13	0.02	0.07	0.07	0.05	0.10	0.10	0.05	0.32	0.32
8.31	0.05	0.05	0.05	0.10	0.10	0.10	0.14	0.14	0.14

Table 21.3. Estimated annual soil erosion (m^3/ha) for upland soils with given return periods.

	1-in-1 year		1-in-5 years		1-in-10 years	
	Slope		Slope		Slope	
Soil subgroup	≤ 12°	> 12°	≤ 12°	> 12°	≤ 12°	> 12°
3.11	0.02	0.02	0.04	0.04	0.09	0.09
5.41	0.03	0.03	0.04	0.04	0.12	0.12
6.11, 6.12, 6.21, 6.22, 6.31, 6.32, 6.42, 6.43, 6.51, 6.52, 6.54	0.04	0.06	0.16	0.20	0.19	0.31
7.21	0.01	0.01	0.02	0.02	0.03	0.03
10.11, 10.13	0.07	0.09	0.36	0.39	0.40	1.20

Table 21.4. Estimated annual soil erosion (m^3/ha) for lowland grassland soils with given return periods.

	1-in-1 year			1-in-5 years			1-in-10 years		
	Slope			Slope			Slope		
Soil group	0–2°	3–6°	≥ 7°	0–2°	3–6°	≥ 7°	0–2°	3–6°	≥ 7°
5.41	0.00	0.03	0.03	0.00	0.17	0.17	0.00	0.19	0.19
5.72	0.01	0.01	0.01	0.03	0.03	0.03	0.03	0.03	0.03
7.13	0.08	0.08	0.08	0.20	0.20	0.20	0.28	0.28	0.28

vulnerability overall. The Soil Survey of England and Wales (1983) recognized soils in subgroups 5.41 (typical brown earths), 5.51 (typical brown sands), 5.54 (argillic brown sands) and 5.71 (typical argillic brown earths) as having a risk of water erosion. It was not surprising that these subgroups figured strongly in the data set with 140 observations of 5.41,

46 observations of 5.51, 5.52 and 5.54 combined, and 87 observations of 5.71. In the 1997–1999 survey, there were 68 observations of subgroup 5.72 (stagnogleyic argillic brown earths) with two occurrences > 100 m^3/ha. The probabilities of erosion may be overestimated for typical paleo-argillic brown earths (5.81), colluvial brown earths (5.47) and stagnogleyic brown calcareous earths (5.13) where, in each case, two instances were recorded in data sets of ≤ 15 observations.

Erosion risk on upland soils

The annual erosion expected with given probabilities for upland soils (Table 21.3) revealed that, as expected (Bower, 1962; Phillips *et al.*, 1981; Tallis, 1987; Evans, 1990), peat soils (subgroups 10.11 – raw oligo-fibrous peat soils; 10.13 – raw oligo-amorphous peat soils) were the most at risk. The most vulnerable mineral soils were the podzols, particularly the wetter stagnopodzols (Group 6.5).

Erosion risk on lowland grassland soils

Erosion on lowland grassland soils is most commonly associated with either heavily grazed permanent pasture or short-term rotational grass or leys (Heathwaite *et al.*, 1990a,b). Ley grass is commonly resown every 7–10 years and any erosion usually occurs in the time between when the land is prepared for resowing and when grass cover has become established. For both heavily grazed land and rotational grass, it is reasonable to assume that any erosion observed in the field has occurred some time within the previous 1–4 years. In order to provide a temporal resolution to the data, the measured values of soil loss were divided by three (Table 21.4).

Discussion

The results for the 1-in-1 year probability showed erosion rates that were extremely low. In contrast, the rates shown for the 1-in-10 year probability (Fig. 21.2) indicated parts of England and Wales where erosion exceeded 1.0 m^3/ha/ year, equivalent to 1.4 t/ha/year on the assumption that the bulk density of the soil is 1.4 Mg/m^3. It is recognized that annual erosion rates in excess of 1 t/ha can give rise to unacceptable problems of diffuse pollution and sedimentation (Moldenhauer and Onstad, 1975; Morgan, 1995).

The marked differences in rates between 1-in-1 year and 1-in-10 year probabilities are expected. Data on rates of erosion for individual events or individual years are strongly negatively skewed, indicating that erosion rates in very frequent events, such as 1-in-1 year, are low. Serious erosion events, defined as > 10 m^3/ha, have return periods of 1-in-25 years or more. When such events are sampled within short-term monitoring periods, calculations of mean annual erosion are biased and unrealistically high (see Boardman, 1998; Evans, 1998). The use of probability analysis helps to reduce this bias, although some data sets were small and not ideally suited to fitting a probability distribution because of gaps (zeros or lack of occurrences) in moderate to high erosion events.

The 1-in-10 year probability is proposed as a reasonable indicator of events that are likely to cause environmental problems. Although the rates shown on the map for erodible areas concur with those measured by Evans (1998) and Boardman (1998), they do not contain occurrences > 10 m^3/ha/year. It should be noted that the data set included five instances of measured field erosion > 100 m^3/ha, all on arable land, two under winter cereals, two under potatoes and one where the intended crop was not recorded. As indicated earlier, such high values have return periods of greater than 1-in-10 years. They also usually relate to field sizes of between 1 and 3 ha, whereas the data presented here represent averages calculated for areas of 1 km^2 (Fig. 21.2) scaled up to catchments up to 1273 km^2 in size (Fig. 21.3). The impact of a single large event in one field will be considerably reduced by this procedure. When selecting rates for tolerable erosion, allowance must be made for the size of the area being considered and the fact that sediment yields per unit area generally decrease with the 0.2 power of the basin area (Roehl, 1962). Against this background, the rate of 0.7 m^3/ha seems reasonable as an indicator of catchments with a potential erosion problem.

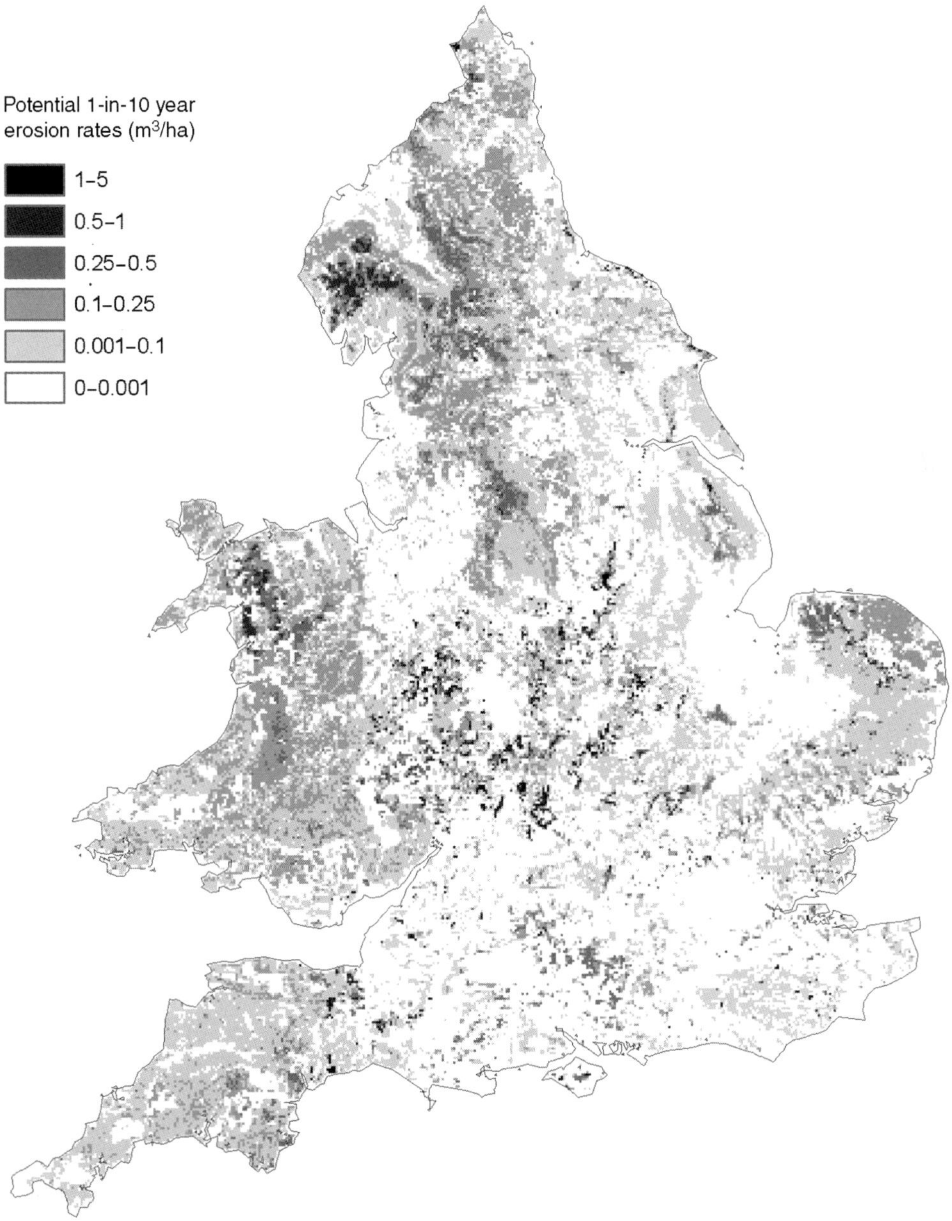

Fig. 21.2. Distribution of estimated annual erosion in England and Wales. Erosion rates represent estimates of the 1-in-10 year erosion events.

The appearance and interpretability of the sediment generation map in Fig. 21.2 is fractured due to the high degree of variability between adjacent pixels. These data have been re-presented in order to identify catchments that are most at risk from sediment generation. To do this it was necessary to compare the values of soil erosion with threshold values. There are no universally accepted definitions of tolerable or allowable erosion rates. However, scientists have proposed that annual soil erosion rates of greater than 1 t/ha at a field scale

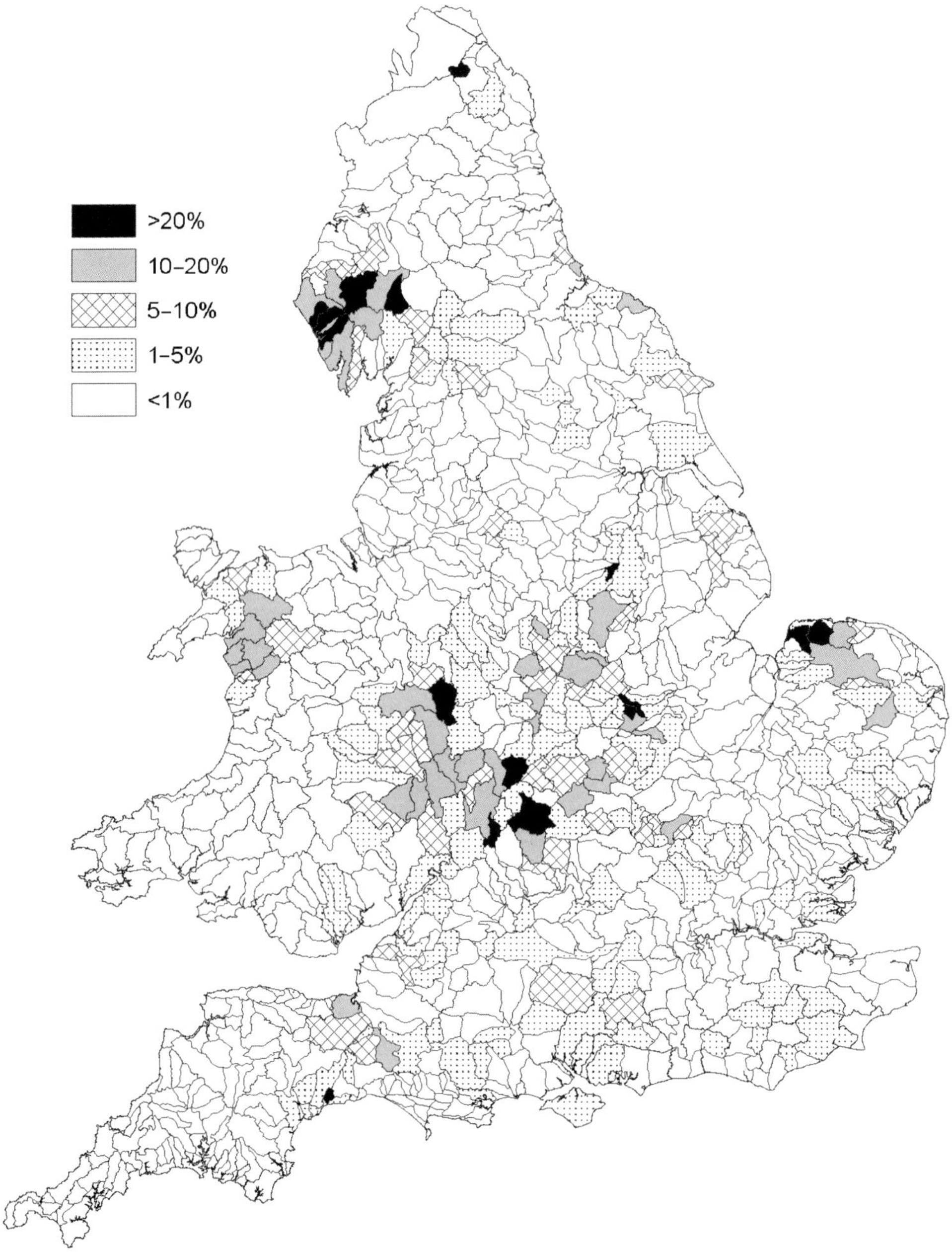

Fig. 21.3. Percentage area of catchment with a probability of a 1-in-10 year erosion event ≥ 0.7 m^3/ha.

are sufficient to cause erosion damage and pollution off-site (Moldenhauer and Onstad, 1975). Assuming a bulk density of 1.4 Mg/m^3, this would equate to 0.7 m^3/ha. Soil erosion rates ≥ 0.7 m^3/ha have therefore been used here as indicators of vulnerable conditions. The mapped output in Fig. 21.3 shows the percentage area of each catchment where these values are equalled or exceeded. It provides a more readily interpretable product for policy use or when targeting catchments for more detailed survey.

Although the maps have not been validated, overall they confirm the observations of Evans (1998) that erosion in lowland arable areas is associated with soil associations 3.43, 5.41, 5.51 and 5.71 and that there is an erosion problem in Somerset, the Isle of Wight, Norfolk, the West Midlands and Nottinghamshire. Surprisingly, the area of the South Downs, shown by Boardman (1990) to be subject to serious erosion, is not clearly represented on the map, although some individual grid squares designated in the third highest category of erosion probably contain many of his eroded sites.

The probability analysis enabled erosion rates to be obtained for arable, upland and lowland grassland soils with the potential to integrate the three land cover types into a single map. However, this presumes that land cover can be objectively mapped with an acceptable level of accuracy. In this work, the only national coverage of land cover available had a mapping accuracy of *c.* 65% and the accuracy of distinction between lowland grassland and arable was a particular problem. As in some 1-km squares, the same soils could be managed in different ways, leading to contrasting erosion risks, this land cover misclassification has implications for the accuracy of the final output. Also, as land use is temporally dynamic, these maps will not represent erosion at later dates.

Conclusions and Recommendations

Although the erosion model is based on the most comprehensive erosion data available for England and Wales, it is not sufficient for quantifying the risk of erosion from different soils at the soil series level and it is questionable how representative erosion data from the sampling period (1996–2002) are of long-term soil erosion rates. Validation of the mapped output is also required. Extrapolation of the erosion data across land use types nationally, using GIS, requires more accurate maps of vegetation cover and specific land uses than those currently available. Although the probability of erosion may be overestimated for soil subgroups 5.13, 5.47 and 5.81, generally erosion rates are likely to be underestimates because the methodology used to measure erosion did not account for rain-splash or unchannelled flow. The erosion analysis is also based on the assumption that spatial probabilities can be used to approximate temporal probabilities.

The final maps of vulnerability to erosion represent the first consolidated assessment at a national level, allowing comparisons of the situation across the range of slope, soil and land use combinations. Despite their limitations and the need for further work, the maps can be used to identify broad regions where more detailed local investigations can be targeted, and hence represent an important management tool.

The Environment Agency of England and Wales (EA) has used the maps developed in this paper to identify combinations of factors contributing to an increased likelihood of soil erosion – and subsequently to sediment delivery to watercourses. The broad geographical areas where it is estimated that there is the greatest erosion risk have contributed to prioritization exercises relating to the Water Framework Directive (WFD) (European Parliament, 2000). The EA have identified areas that may be at risk of failing to achieve good ecological status, in accordance with the WFD, because of a number of pressures, including sediment from land sources. This work is providing decision support for regulatory and advisory visits to farms.

References

Avery, B.W. (1980) *Soil Classification for England and Wales (Higher Categories)*. Soil Survey Technical Monograph No. 14. Soil Survey of England and Wales, Harpenden, UK.

Bibby, J.S. and Mackney, D. (1969) *Land Use Capability Classification*. Soil Survey Technical Monograph No. 1. Soil Survey of England and Wales, Harpenden, UK.

Boardman, J. (1990) Soil erosion on the South Downs: a review. In: Boardman, J., Foster, I.D.L. and Dearing, J.A. (eds) *Soil Erosion on Agricultural Land*. Wiley, Chichester, UK, pp. 87–105.

Boardman, J. (1998) Modelling soil erosion in real landscapes: a Western European perspective. In: Boardman, J. and Favis-Mortlock, D. (eds) *Modelling Soil Erosion by Water*. Springer, Berlin, Germany, pp. 17–29.

Bower, M.M. (1962) The cause of erosion in blanket peat bogs. A review of evidence in the light of recent work in the Pennines. *Scottish Geographical Magazine* 78, 33–43.

Butcher, D.P., Claydon, J., Labdaz, J.C., Pattinson, V.A., Potter, A.W.R. and White, P. (1992) Reservoir sedimentation and colour problems in Southern Pennine reservoirs. *Journal of the Institute of Water and Environmental Management* 6, 418–431.

De Ploey, J. (1984) Hydraulics of runoff and loess loam deposition. *Earth Surface Processes and Landforms* 9, 533–539.

Environment Agency (2002) *Agriculture and Natural Resources: Benefits, Costs and Potential Solutions*. Report of the Environment Agency, May 2002. Environment Agency, Bristol, UK.

European Parliament (2000) Establishing a framework for Community action in the field of water policy. Directive 2000/60/EC of the European Parliament and of the Council. *Official Journal of the European Communities*, 23 October 2000.

Evans, R. (1971) The need for soil conservation. *Area* 3, 20–23.

Evans, R. (1990) Erosion studies in the Dark Peak. *Journal of Northern Soils Discussion Group* 24, 39–61.

Evans, R. (1996) *Soil Erosion and its Impacts in England and Wales*. Friends of the Earth Trust, London.

Evans, R. (1998) Field data and erosion models. In: Boardman, J. and Favis-Mortlock, D. (eds) *Modelling Soil Erosion by Water*. Springer, Berlin, Germany, pp. 313–327.

Evans, R. and Boardman, J. (1994) Assessment of water erosion in farmers' fields in the UK. In: Rickson, R.J. (ed.) *Conserving Soil Resources: European Perspectives*. CAB International, Wallingford, UK, pp. 13–24.

Evans, R. and Skinner, D. (1987) A survey of soil erosion. *Soil and Water* 15, 28–31.

Fuller, R.M., Groom, G.B. and Jones, A.R. (1994) The Land Cover Map of Great Britain: an automated classification of Landsat Thematic Mapper data. *Photogrammetric Engineering and Remote Sensing* 60, 553–562.

Harrod, T.R. (1998) *A Systematic Approach to National Budgets of Phosphorus Loss Through Soil Erosion and Surface Runoff at National Soil Inventory (NSI) Nodes*. Final Report to MAFF. MAFF project NT1014. MAFF (DEFRA), London.

Harrod, T.R., McHugh, M., Appleby, P.G., Evans, R., George, D.G., Haworth, E.Y., Hewitt, D., Hornung, M., Housen, G., Leeks, G., Morgan, R.P.C. and Tipping, E. (2000) *Research on the Quantification and Causes of Upland Erosion*. Final report to MAFF. MAFF Project SP0402. MAFF (DEFRA), London.

Heathwaite, A.L., Burt, T.P. and Trudgill, S.T. (1990a) The effect of land use on nitrogen, phosphorus and suspended sediment delivery to streams in a small catchment in Southwest England. In: Thornes, J.B. (ed.) *Vegetation and Erosion*. Wiley, Chichester, UK, pp. 161–177.

Heathwaite, A.L., Burt, T.P. and Trudgill, S.T. (1990b) Land-use controls on sediment production in a lowland catchment, south-west England. In: Boardman, J., Foster, I.D.L. and Dearing, J.A. (eds) *Soil Erosion on Agricultural Land*. Wiley, Chichester, UK, pp. 69–86.

Hudson, N.W. (1981) *Soil Conservation*. Batsford, London.

McHugh, M. (2000) Extent, causes and rates of upland soil erosion in England and Wales. PhD thesis. Cranfield University, UK.

McHugh, M. (2002) *NSI Erosion Survey*. Final Report to DEFRA, Project SP0407. DEFRA, London.

McHugh, M., Morgan, R.P.C., Walling, D.E., Wood, G.A., Zhang, Y., Anthony, S. and Hutchins, M. (2002) *Prediction of Sediment Delivery to Watercourses from Land. Phase II*. Final Report to Environment Agency. R&D Project P2-209. Environment Agency, Bristol, UK.

Moldenhauer, W.C. and Onstad, C.A. (1975) Achieving specified soil loss levels. *Journal of Soil and Water Conservation* 30, 166–168.

Morgan, R.P.C. (1980) Soil erosion and conservation in Britain. *Progress in Physical Geography* 4, 24–47.

Morgan, R.P.C. (1985) Assessment of soil erosion risk in England and Wales. *Soil Use and Management* 1, 127–131.

Morgan, R.P.C. (1995) *Soil Erosion and Conservation*, 2nd edn. Longman, London.

Paine, A.D.M. (1985) 'Ergodic' reasoning in geomorphology: time for a review of the term. *Progress in Physical Geography* 9, 1–15.

Phillips, J., Yalden, D. and Tallis, J. (1981) *Moorland Erosion Project, Phase 1 Report*. Peak Park Planning Board, Bakewell, UK.

Richards, J.A. and Jia, X. (1999) *Remote Sensing Digital Image Analysis – An Introduction*, 3rd edn. Springer, Berlin, Germany, pp. 259–290.

Robinson, D.A. and Blackman, J.D. (1990) Some costs and consequences of soil erosion and flooding around Brighton and Hove, Autumn 1987. In: Boardman, J., Foster, I.D.L. and Dearing, J.A. (eds) *Soil Erosion on Agricultural Land*. Wiley, Chichester, UK, pp. 369–382.

Roehl, J.W. (1962) Sediment source areas, delivery ratios and influencing morphological factors. In: *Symposium of Bari. 1–8 October 1962. Commission of Land Erosion*. IAHS Publication 59. IAHS Press, Wallingford, UK, pp. 202–213.

Scheidegger, A.E. (1970) *Theoretical Geomorphology*. Springer, New York.

Scheidegger, A.E. and Langbein, W.B. (1966) Probability concepts in geomorphology. United States Geological Survey Professional Paper No. 500C.

Soil Survey of England and Wales (1983) *Legend for the 1:250,000 Soil Map of England and Wales*. Soil Survey of England and Wales, Harpenden, UK.

Tallis, J.H. (1987) Fire and flood at Holme Moss: erosion processes in an upland blanket mire. *Journal of Ecology* 75, 1099–1129.

Theurer, F.D., Harrod, T.R. and Theurer, M. (1998) *Sedimentation and Salmonids in England and Wales*. Environment Agency Technical Report P194. Environment Agency, Bristol, UK.

Walling, D.E. and Zhang, Y. (2004) Predicting slope–channel connectivity: a national scale approach. In: Golosov, V., Belyaev, V. and Walling, D.E. (eds) *Sediment Transfer through the Fluvial System*. IAHS Publication 288. IAHS Press, Wallingford, UK, pp. 107–114.

22 Management of Sediment Production and Prevention in River Catchments: a Matter of Scale?

R.J. Rickson
National Soil Resources Institute, UK

Introduction

This chapter addresses some of the issues in soil erosion and sedimentation research that are related to scale and scale effects, with particular reference to spatial scales. First, a brief review of soil erosion research is given, illustrating the diverse spatial scales at which investigations have been carried out by erosion scientists and researchers. The wealth of experience and knowledge gained from these studies serves a very wide audience, ranging from developers of sub-process level, physically based erosion models through to regional planners and policy makers.

However, in many cases, the results of soil loss and sediment delivery obtained at one spatial scale are not consistent with those found at another. There has been little work aimed at understanding and then explaining these apparent anomalies in erosion data generated at different spatial scales. This is illustrated by some experimental data on the performance of erosion control blankets and mats. The implication of these anomalies is that the results at one scale cannot/should not be extrapolated to another, although in reality such extrapolation does take place.

To understand the effect of spatial scale on erosion processes, there is a need to investigate the links between different scales. This can be achieved by designing and applying experimental methods which encompass a range of spatial scales. This is ambitious at a time when funding for erosion research is severely limited. However, an EU LIFE Environment/Syngenta funded project (SOWAP – Soil and water protection measures in North and Central Europe) does consider soil erosion and sediment production at a variety of scales, from the process of individual soil aggregate breakdown at the smallest spatial scale, through rainfall simulation plot trials, to field scale erosion and runoff plots, representing the sub-catchment scale. Attention is paid to the changes in erosion and sediment production rates at the varying scales under investigation. It is hoped that the observations of soil loss at these different scales will provide insights to help explain the variability of erosion due to spatial scale effects. Also, this research aims to provide data which can provide direct links between the diverse range of interested parties – from scientists, interested in sub-process level modelling, to agencies involved in implementing EU land management policies.

The Matter of Scale in Soil Erosion Research

Looking back at the history of soil erosion research reveals that studies have been carried out at varying spatial and temporal scales. Appreciation of the mechanics of the erosion process is only possible at a small spatial scale (mm^2 or cm^2), because the primary agent of erosion by water is the individual raindrop, which initiates the first phase of erosion – detachment of soil particles or aggregates from the soil mass. Studies of individual soil aggregate breakdown may use aggregates of only a few millimetres in diameter (Rahim, 1990; Simmons, 1998). Morgan *et al.* (1988) used aluminium canisters 15 mm in diameter to measure rain splash detachment of standard laboratory sand underneath a plant canopy. Ellison splash cups (Ellison, 1944), 7 cm in diameter, are used specifically to isolate soil erosion caused by rain splash detachment and transport from that caused by overland flow. This level of detail is essential if the mechanics of erosion are to be investigated and understood.

Table 22.1 shows some selected studies of soil erosion at various spatial scales (as classified by Wickenkamp *et al.*, 2000), illustrating the range of scales at which erosion studies have been carried out. There are examples of erosion studies carried out at scales even greater than those shown in Table 22.1. Mushala *et al.* (1994) report on the extent of soil erosion at the regional scale (e.g. Middleveld of Swaziland) over time. Investigations of the extent and degree of soil erosion at the national scale can be found in Boardman and Evans (1994) and van der Knijff *et al.* (2000), for example. De Ploey (1989) assessed the extent of soil erosion and land degradation for the continent of Europe. Global scale mapping of soil erosion can be found in research programmes such as GLASOD (Oldeman *et al.*, 1991), which considers the extent and degree of land degradation, including soil erosion at the 1 : 1,000,000 scale. Beyond our own world, erosion studies have been carried out on other planets, such as an investigation of wind erosion on Mars (Hynek and Phillips, 2001).

This wealth of data serves many audiences well – including developers of physically based erosion models (such as EUROSEM (Morgan *et al.*, 1998) and WEPP (Nearing *et al.*, 1989) etc.), who strive to simulate realistic processes mathematically as white-box (as opposed to grey- or black-box) erosion models.

At a larger spatial scale, regional planners (for example, in national government departments and ministries) may wish to use erosion studies at the regional scale to identify vulnerable areas with high soil erosion risk. Once identified, these areas can be targeted for design and implementation of soil conservation programmes. This approach is described in Mushala *et al.* (1994), where land systems analysis identifies land facets in the Middleveld of Swaziland, which are characterized as having high risk of erosion. The reasons for the high erosion risk are investigated further, in order to avoid land management practices that may increase erosion risk in the future.

At the largest spatial scale, erosion studies will be relevant to national or international policy makers who wish to identify regions where soil erosion is leading to irreversible land degradation. This is illustrated by the efforts made by the United Nations Environment Programme (UNEP) in attempting to arrest the expansion of desertification in North Africa and the Mediterranean region (http://www.unccd.int/convention/menu.php). At this scale, the socio-economic impacts associated with soil degradation can have significant effects on land resources and human livelihoods.

Ciesiolka and Rose (1998) observe that smaller scale studies tend to focus on 'on-site' impacts of soil erosion, whilst larger spatial-scale studies concentrate on the 'off-site' impacts. This may also determine the type of audience interested in any given study.

Scale Effects on Erosion Processes and Erosion Control Practices: the Challenges

Despite the enormous range of scales at which erosion has been studied and the resultant massive knowledge- and data-bases, there is concern that very little work has been done either on linking these different scales or ensuring that different results, which may be a function of scale, can be explained. There has been some work on converting field-scale to catchment-scale erosion data, based on the concept of sediment delivery ratios (Osterkamp and Toy, 1997;

Table 22.1. Range of spatial scales of soil erosion research.

Erosion research technique	Area	Scale	Dimension descriptors (Wickenkamp *et al.*, 2000)		Dominant processes operating	Reference (examples)
Splash cup	mm^2	< 1 : 5k	Nanoscale	Subtope	Rain splash dominant; overland flow/deposition limited. No gullies, stream bank erosion or mass movements.	Ellison (1944); Morgan *et al.* (1988)
Laboratory tray	cm^2	< 1 : 5k	Nanoscale	Subtope	Rain splash dominant? Overland flow/deposition limited. No gullies, stream bank erosion or mass movements.	Idowu (1996)
Runoff rig	m^2	1 : 5k–10k	Microscale	Tope	Rain splash and overland flow; some deposition possible. No gullies, stream bank erosion or mass movements.	Kamalu (1993)
Field plot	m^2	1 : 5k–10k	Microscale	Tope	Rain splash and overland flow; some deposition. Some gullying and mass movements possible; no stream bank erosion.	Wischmeier and Smith (1978); Ciesiolka and Rose (1998); Pierson *et al.* (1994)
Field	ha	1 : 10k–25k	Mesoscale	Chore	Rain splash, overland flow and deposition. Gullying and mass movements possible. No stream bank erosion.	Evans and Boardman (1994)
Sub-catchment	ha – km^2	1 : 10k–25k	Mesoscale	Chore	Rain splash, overland flow and deposition. Gullying possible. Some stream bank erosion.	Rapp *et al.* (1972); Hudson (1981)
Catchment landscape	km^2	1 : 25k–50k	Macroscale	Region	Rain splash, overland flow and deposition. Some gullying and mass movement possible. Stream bank erosion.	Dickinson and Collins (1998)

Imeson and Cammeraat, 1998; Walling, Chapter 2, this volume; Wood *et al.*, Chapter 21, this volume). However, connections between different scales are still poorly understood, although the issue is addressed by Kirkby (2001).

These limited studies tend to concentrate on soil erosion and sediment production, with less emphasis on the problems of extrapolating the performance of soil erosion control or soil conservation methods to different spatial scales. For example, whilst there has been considerable research into the effects of the morphology and architecture of individual plants on erosion processes (e.g. Morgan, 1985; Styczen and Høgh-Schmidt, 1988), little attempt has been made to scale up these results to field stands of the same crop.

The C (cropping) factor in the Universal Soil Loss Equation (Wischmeier and Smith, 1978) does express the effect of field-scale crop stand on soil loss, but here there has been very limited attempts to explain how the physiology of the individual crop affects the processes of erosion at this broader scale. At best, the C factor effect on erosion has been sub-divided into three components: Ci (above ground canopy effects), Cii (at/on ground canopy effects) and Ciii (residual effects of land use, such as the influence of soil structure, organic matter content, soil bulk density, tillage, roots, subsurface stems and biological activity in soil) (Wischmeier, 1975). This partitioning of the C factor has been considered in soil erosion models (e.g. Beasley *et al.*, 1980; Park *et al.*, 1982), but still there is little consideration of the interaction effects between Ci, Cii and Ciii. Despite this partitioning of the C factor in this way, modelling the effect of vegetation on soil erosion processes using the C factor still represents a 'black box' approach.

More comprehensive data on erosion control effectiveness at different spatial scales can be found, notably regarding the use of geotextiles for soil erosion control. Geotextiles are defined as 'permeable textiles or fabrics used in conjunction with soil, foundation, rock, earth or any geotechnical engineering related material, as an integral part of a man-made project' (John, 1987, p. 1). Whilst these materials are used for numerous applications (including drainage, filtration, separation of two distinct ground layers, slope stabilization and vegetation management), the example given in this chapter only considers their use for soil erosion control.

A review of the literature on this subject reveals that there have been many studies at different scales, often using the same geotextile product, under similar or at least comparable environmental conditions (soil type, rainfall intensity, slope gradient and length, etc.; Rickson, 2000). Table 22.2 describes some of the different erosion control geotextiles that have been investigated at varying spatial scales.

Table 22.3 shows the erosion control performance of these products at different spatial scales. These data were selected because the experimental conditions used (soil type, rainfall intensity and slope) are similar. Any variations due to experimental conditions will be accounted for, as the results have been standardized in the following manner. The bare soil control (with no geotextile treatment) is always assigned an erosion control effectiveness factor of 1, and the results of the geotextile treatments

Table 22.2. Examples of geotextiles used in soil erosion control.

Geotextile name	Weight (g/m^2)	Description	% Area of geotextile
Geojute	500	Woven jute bionet	54
Fine Geojute	292	Woven jute bionet	80
Enviromat	450	Biomat of wood shavings held in a polypropylene mesh	94
Enkamat – surface laid	265	Nylon, 3 dimensional geomat	40
Enkamat – buried	265	Nylon, 3 dimensional geomat	15
Tensarmat	450	Polypropylene 3 dimensional geomat	62*/10**
Bachbettgewebe	700	Woven coir bionet	58

*% Area of geotextile when surface applied. **% Area of geotextile when buried.

Table 22.3. Relative erosion control performance by selected geotextiles, as a function of scale.

Source	Field or Lab. based	Plot size (m^2)	Treatments							
			Bare soil	Geojute	Fine Geojute	Enviromat	Enkamat (S)	Bachbett-gewebe	Enkamat (B)	Tensarmat
Rickson (2000) – Splash cups	L	0.005	1.00	0.140	0.184	0.250	0.350	0.296	1.30	1.24
Sutherland and Ziegler (1996) – soil tray	L	0.18	1.00	0.017	*	< 0.001	*	*	*	*
Thomson and Ingold (1986) – trial plots	F	0.9	1.00	0.08	0.700	0.08	*	0.100	0.600	*
Cazzuffi *et al.* (1991) – artificial slope	L	1.5	1.00	1.059	*	*	*	*	0.824	0.882
Rickson (2000) – artificial slope	L	2	1.00	0.228	0.147	0.360	0.512	0.228	0.763	0.995

*Data not collected.

are expressed relative to the bare soil control, all other factors being equal.

The selected data show that erosion control effectiveness is affected by the scale at which the experimental trial takes place. For example, the relative soil loss results for the Geojute product vary considerably. Geojute is a woven mat of 500 g/m^2 weight, made from yarns of spun jute fibres. This product's performance ranges from producing only 1.7% of the soil loss measured from the bare soil control (at a scale of 0.18 m^2), up to 105.9% (i.e. more than that observed from the bare soil plot) at a plot size of 1.5 m^2. Conversely, the buried synthetic product, Enkamat, which resembles a mesh of randomly spaced, thin nylon filaments reduces soil loss to 76% of that from the bare soil control at 2 m^2, but at the smaller scale, soil loss is greater under this treatment (at 130%) than that yielded by the bare soil plot. Some products appear to perform best at the small spatial scales (e.g. surface-laid Enkamat), some at the intermediate scale (e.g. Geojute, Enviromat, Bachbettgewebe and buried Enkamat), and some at the larger spatial scale (fine Geojute and Tensarmat).

These are confusing messages for geotextile designers, manufacturers, specifiers and end-users, as it is unclear as to what is the most effective product overall. The results in Table 22.3 imply that the effectiveness of each product varies with site-specific conditions, which requires further investigation. A number of manufacturers quote similar data in promotional material to illustrate that their own product performs most effectively, but this conclusion may be based on very carefully selected (and thus biased) data.

Implications of Scale Effects in Erosion and Erosion Control Studies

The data shown in Table 22.3 imply that erosion control effectiveness demonstrated at one scale cannot/should not be extrapolated to another. Simple 'scaling up' is not possible – it appears that the mean value of erosion per unit area will change when the sample size is increased, all other factors being equal. Unfortunately, this scaling up (or down) cannot be predicted because, according to van Noordwijk *et al.* (1998), there are no 'scaling rules'. In other words, linear additivity is not valid for erosion studies (Pierson *et al.*, 1994), as shown by Smith and Quinton (2000), when comparing erosion rates for different slope lengths (Table 22.4). Despite these apparent anomalies and errors, Smith and Quinton (2000) report that such scale extrapolations are commonly applied in erosion and sedimentation modelling.

There has been some attempt to explain why differences in erosion rates take place at different spatial scales. For example, it is not possible to simulate certain erosion processes at small spatial scales, such as under laboratory conditions (e.g. gullying, mass movements, stream bank erosion). Idowu (1996) compared soil erosion generated from laboratory-based soil trays (0.375 m^2) with that measured for the same soil in the field (plot size = 4.5 m^2), and found that differences in slope length (0.75 m versus 3 m for laboratory and field plots, respectively), soil preparation and soil depth (0.05 m versus 0.20 m, respectively) significantly affected the rates of erosion measured. Bryan (1981) also reports on the variability in erosion rates under simulated rainfall seen when comparing laboratory and field plots.

Fragmenting or deconstructing processes at small spatial scales may omit synergy with other processes, but may also gain something from the greater resolution of simulation. Van Noordwijk *et al.* (1998) express this concept in terms of the 'whole being greater than sum of the parts'.

This concept helps to explain some of the anomalies shown in Table 22.3. At the small-scale experiments, the process of rainsplash detachment (especially) and transport will dominate the erosion process. As the spatial scale increases, overland flow becomes the dominant agent of erosion. Some geotextile

Table 22.4. Non-linearity in soil erosion rate as a function of slope length (Smith and Quinton, 2000).

Slope length (m)	Erosion rate (kg/ha)
5	446
20	632
200	1095

products are more effective at controlling erosion caused by rainsplash than they are at controlling erosion caused by overland flow. For example, the buried Enkamat product offers no surface protection from raindrop impact (and consequent detachment and transport). Hence the erosion control effectiveness of this product at the small scale (where rainsplash processes dominate, not least because there is insufficient slope length to generate surface flow) is very poor. Indeed, this product yields even more soil loss than the bare soil control, because backfilling the soil into the structure of the Enkamat destroys any soil structure, making it highly susceptible to erosion. However, at the larger spatial scale, surface flow can be generated, so that erosion is now dominated by detachment and transport by runoff. As the buried product encourages infiltration, through its simulation of a dense network of vegetative roots, there is less generation of runoff and thus less detachment and transport by overland flow from this product, and soil loss is only 76% of that observed for the bare soil control.

Clearly, we need to understand the connectivity between different scales in soil erosion research, and yet the number of studies looking at different scales is very limited. Small-scale, often laboratory-based studies tend to dominate in erosion research, not least because compared with larger scale studies (e.g. field plots) they are cheaper, carrying out multiple simulations and replications is quicker and easier, there is better control of variables and the impact of failure is less significant. However, in terms of evaluating soil erosion control practices, small-scale research has a major flaw in that it takes little notice of socio-economic factors, which often dominate uptake of sediment control practices. According to Turkelboom and Trebuil, 'The traditional biophysical approach relying on researcher-controlled runoff plots is no longer adapted to the actual agronomic, economic, social and policy circumstances, faced by farmers' (1998, p. 51).

The Way Forward. . .

The preceding discussion establishes that there is a need for specific erosion studies to consider a broader range of scales than is currently the case. The aim would be to address both the need for understanding the processes operating *and* assessing the practical viability of the erosion control practice under evaluation. This is an ambitious remit, but there have been some studies whose experimental design has allowed these objectives to be met.

Hudson (1993) reports on the 'nested catchments' approach in soil erosion research. He reports on a project based in Sri Lanka where measurements of erosion and runoff are taken at the plot (*c.* 200 m^2), field (1 ha), sub-catchment (7–284 ha) and catchment (5000 ha) scale (Hudson, 1981). This approach was developed from biological research, investigating biodiversity and species richness at different scales. Turkelboom and Trebuil (1998) developed a methodology for erosion process analysis at the field, farm and catchment levels, and ways of linking these different scales. Their approach is summarized in a diagram of the multiscale approach, involving the physical, economic and social aspects affecting erosion.

Kirkby (2001) describes the hierarchical MEDRUSH model, which simulates erosion and runoff processes operating at a scale of 1 m^2 in the first instance. Interactions at this scale between atmosphere, vegetation, surface characteristics and soil determine runoff and sediment generation. These results are then 'nested' or 'embedded' within representative 'flow strips' of up to 100 m wide, oriented up/down the slope. The water and sediment generated at this scale are then 'routed' via computed linear transfer functions into the next scale up – the sub-catchment (1–10 km^2). Output from this scale then feeds the main catchment-scale channel network, which may be up to 2500 km^2 in area. Kirkby argues that the MEDRUSH model demonstrates that there can be 'a proper physical basis for constructing coarse scale models, and that coarse and fine scaled models can be linked together consistently with a sound physical basis' (2001, p. 11).

One example of current work which considers the issues of connectivity between scales in erosion research is the EU LIFE Environment/Syngenta funded 3-year project, SOWAP (Soil and water protection in North and Central Europe) (http://www.sowap.org/index.htm). SOWAP's intention is to be a 'truly integrated programme, meeting the demands of scientific experiments and modelling, farmers, and policy makers at national and international level'. SOWAP's objectives are to demonstrate:

- the viability and effectiveness of 'conservation oriented' arable land management systems in protecting soil resources, improving catchment water quality and promoting biodiversity;
- the environmental, ecological, economic and social benefits of 'conservation oriented' land use practices;
- the environmental impacts associated with 'conventional' arable land use practices, where intensive soil management can lead to degradation of soil resources, water pollution, reduced biodiversity and less carbon sequestration;
- how an environmentally sound land use policy can be implemented, as recommended by the EU 6th Environment Action Programme and the EC Communication on Soil Protection;
- how a unique database can be disseminated successfully at the local, regional, national and EU level via workshops, multi-media, field visits, publications and the Internet.

SOWAP is based at four field sites – in Belgium, Hungary and two in the UK (in Leicestershire and Somerset), where the effects of different arable cultivation and soil management practices on soil erosion and runoff are investigated at a range of spatial scales. The intended audiences range from soil science researchers who are interested in the effects of soil management practice on erosion processes through to EU-level policy makers, who wish to know how policies advocated in the 6th Environmental Action Programme can be implemented successfully at the farm scale.

At the micro scale, the stability of individual soil aggregates when subjected to rainfall impact is being quantified as a function of soil management. The results will be explained in terms of physical, chemical and/or biological factors affecting soil stability and resistance to disruption. At the next scale, rainfall simulation is being carried out at the plot scale (1 m^2) to generate soil loss and runoff. This is to ensure the simulation of erosion caused by rain splash and overland flow. Finally, at the field scale, erosion plots have been installed to measure runoff and sediment losses under the different soil management treatments. The results from these field-scale plots will be used as direct indicators of sediment delivery to the catchment and any off-site consequences of this, notably the effect of soil loss on water quality and thus biological impacts. As an EU-supported project, SOWAP has an international component, whereby the erosion rates generated in the three partner countries (using the same experimental methodologies and data collection protocols) are compared, taking into account the unique environmental conditions of each site.

The results of soil erosion rates at the different scales will be compared and explained in terms of factors affecting the processes operating and the delivery mechanisms at the three scales. It is hoped that this project allows the connections between these three scales to be better understood.

In this holistic approach, the results of the erosion component are then applied at different scales too – ranging from the impact of erosion on microbial community structure and function, through to informing policy makers at the EU level about Best Management Practices, and how Directives such as the Water Framework Directive and the forthcoming Soil Protection Directive can be implemented successfully.

Conclusions: Which is the Right Scale?

Studying soil erosion processes at the process and even sub-process level is essential if we are to understand fully and simulate realistically the mechanics of soil erosion and sediment production. Greater experimental control at smaller spatial scales makes it easier to reduce plot variability (e.g. soil, rainfall and slope characteristics), but then these plots are less representative of the more heterogeneous landscapes. De Coursey and Meyer note that 'by minimising the variability, we greatly reduce our ability to extrapolate to other areas' (1977, p. 194). Kirkby concludes that fine-scale models 'can never be suitable on their own for grappling with resource issues' (2001, p. 11).

Acknowledging these limitations, the Australian Centre for International Agricultural Research (ACIAR) for example, has moved away from research at the plot and farm level to the catchment/regional level (Lal *et al.*, 1998). This is defined as the 'appropriate scale of research activity', mainly because small scale, process-based investigations cannot adequately evaluate the technical, economic nor social performance of proposed land management policies, which may include sediment control practices such as agroforestry.

By means of a compromise, Cielsiolka and Rose argue that it is important to consider the wide range of scales at which erosion research can take place: 'In order to develop sound and effective integrated catchment management strategies, it is desirable to understand hydrological, sediment and chemical transport processes at a number of landscape scales' (1998, p. 300). Until we know and understand the connections between the different spatial scales, there is a need to encourage soil erosion research to encompass as wide a range of scales as possible. This has the multiple benefits of:

- linking soil erosion rates generated at varying spatial scales;
- supplying knowledge which will be of interest to many parties, from physically based erosion modellers through to policy makers;
- identifying if there are any rules to be applied when upscaling or downscaling the results of soil erosion research.

References

Beasley, D.B., Huggins, L.F. and Monke, E.J. (1980) ANSWERS: A model for watershed planning. *Transactions of the American Society of Agricultural Engineers* 23, 938–944.

Boardman, J. and Evans, R. (1994) Soil erosion in Britain: a review. In: Rickson, R.J. (ed.) *Conserving Soil Resources: European Perspectives*. CAB International, Wallingford, UK, pp. 3–12.

Bryan, R.B. (1981) Soil erosion under simulated rainfall in the field and laboratory. In: *Erosion and Sediment Transport Measurement.* IAHS Publication 133. IAHS Press, Wallingford, UK.

Cazzuffi, D., Monti, R. and Rimoldi, P. (1991) Geosynthetics subjected to different conditions of rain and runoff in erosion control application: a laboratory investigation. In: *Erosion Control: A Global Perspective*, Proceedings of the XXII IECA Annual Conference, 1991, Orlando, Florida. International Erosion Control Association, Steamboat Springs, Colorado, pp. 193–208.

Ciesiolka, C.A.A. and Rose, C.W. (1998) The measurement of soil erosion. In: Penning de Vries, F.W.T., Agus, F. and Kerr, J. (eds) *Soil Erosion at Multiple Scales: Principles and Methods for Assessing Causes and Impact.* CAB International, Wallingford, UK, pp. 287–301.

De Coursey, D.G. and Meyer, L.D. (1977) Philosophy of erosion simulation for land use management. In: Foster, G.R. (ed.) *Soil Erosion: Prediction and Control.* Proceedings of a national conference on soil erosion, Purdue University, West Lafayette, Indiana. Soil Conservation Society of America, Ankeny, Iowa, pp. 193–200.

De Ploey, J. (1989) Erosional systems and perspectives for erosion control in European loess areas. *Soil Technology Series* 1, 93–102.

Dickinson, A. and Collins, R. (1998) Predicting erosion and sediment yield at the catchment scale. In: Penning de Vries, F.W.T., Agus, F. and Kerr, J. (eds) *Soil Erosion at Multiple Scales: Principles and Methods for Assessing Causes and Impact.* CAB International, Wallingford, UK, pp. 317–342.

Ellison, W.D. (1944) Studies of raindrop erosion. *Agricultural Engineering* 25, 131–136, 181–182.

Evans, R. and Boardman, J. (1994) Assessment of water erosion in farmers' fields in the UK. In: Rickson, R.J. (ed.) *Conserving Soil Resources: European Perspectives*. CAB International, Wallingford, UK, pp. 13–24.

Hudson, N.W. (1981) A research project on hydrology and soil erosion in mountain watersheds in Sri Lanka. In: Lal, R. and Russell, E.W. (eds) *Tropical Agricultural Hydrology*. Wiley, New York, pp. 311–324.

Hudson, N.W. (1993) *Field Measurement of Soil Erosion and Runoff.* FAO Soils Bulletin No. 68. FAO, Rome.

Hynek, B.M. and Phillips, R.J. (2001) Evidence for extensive denudation of the Martian highlands. *Geology* 29, 407–410.

Idowu, J. (1996) Tillage management for the control of soil and water losses from a soil with an impermeable subsurface pan. Unpublished PhD thesis. Cranfield University, UK.

Imeson, A.C. and Cammeraat, L.H. (1998) Upscaling erosion processes from the plot to the catchment scale. *Proceedings of the GCTE Soil Erosion Network Conference*, Utrecht, The Netherlands, 14–18 April 1997.

John, N.W.M. (1987) *Geotextiles*. Blackie and Son, Glasgow, UK.

Kamalu, C. (1993) A laboratory study of soil erosion on a model road shoulder due to simulated rainfall and runoff. PhD thesis. Cranfield Institute of Technology, UK.

Kirkby, M. (2001) From plot to continent: reconciling fine and coarse scale erosion models. In: Stott, D.E., Mohtar, R.H. and Steinhardt, G.C. (eds) *Sustaining the Global Farm*. Selected papers from the 10th International Soil Conservation Organisation meeting, held 24–29 May 1999 at Purdue University and USDA-ARS National Soil Erosion Research Laboratory. http://topsoil.nserl.purdue.edu/nserlweb/isco99/pdf/ISCOdisc/SustainingTheGlobalFarm/P271-Kirkby.pdf

Lal, P., Couglan, K. and Menz, K. (1998) Designing and implementing outcome oriented soil conservation research. In: Penning de Vries, F.W.T., Agus, F. and Kerr, J. (eds) *Soil Erosion at Multiple Scales: Principles and Methods for Assessing Causes and Impact.* CAB International, Wallingford, UK, pp. 357–363.

Morgan, R.P.C. (1985) Effect of corn and soybean canopy on soil detachment by rainfall. *Transactions of the American Society of Agricultural Engineers* 28, 1135–1140.

Morgan, R.P.C., Brandt, C.J. and Studdard, M.J. (1988) The role of plant covers in controlling soil detachment by rainfall. Silsoe College pamphlet.

Morgan, R.P.C, Quinton, J.N., Smith, R.E., Govers, G., Poesen, J.W.A., Auerswald, K., Chisci, G., Torri, D. and Styczen, M.E. (1998) The European Soil Erosion Model (EUROSEM): a dynamic approach for predicting sediment transport from fields and small catchments. *Earth Surface Processes and Landforms* 23, 527–544.

Mushala, H.M., Scholten, T., Felix Henningsen, P., Morgan, R.P.C. and Rickson, R.J. (1994) *Soil Erosion and River Sedimentation in Swaziland.* Final report to the EC. Contract No. TS2-CT90-0324. University of Swaziland, Kwaluseni.

Nearing, M.A., Foster, G.R., Lane, L.J. and Finkner, S.C. (1989) A process based soil erosion model for USDA water erosion prediction project technology. *Transactions of the American Society of Agricultural Engineers* 32, 1587–1593.

Oldeman, L.R., Hakkeling, R.T.A. and Sombroek, W.G. (1991) *World Map of the Status of Human-induced Soil Degradation: An Explanatory Note*, 2nd edn. ISRIC, Wageningen and UNEP, Nairobi.

Osterkamp, W.R. and Toy, T.J. (1997) Geomorphic considerations for erosion prediction. *Environmental Geology* 29, 152–157.

Park, S.W., Mitchell, J.K. and Scarborough, J.N. (1982) Soil erosion simulation on small watersheds: a modified ANSWERS model. *Transactions of the American Society of Agricultural Engineers* 25.

Pierson Jr, F.B., van Vactor, S.S., Blackburn, W.H. and Wood, J.C. (1994). Incorporating small scale variability into predictions of hydrologic response on sagebrush rangelands. In: Blackburn, W.H., Pierson Jr, F.B., Schuman, G.E. and Zartman, R. (eds) *Variability in Rangeland Water Erosion Processes*. Proceedings of a symposium of the Soil Science Society of America, Minneapolis, Minnesota, November, 1992. SSA Special Publication No. 38. SSSA, Madison, Wisconsin.

Rahim, S.E. (1990) Spatial and seasonal variations of soil erodibility. Unpublished PhD thesis. Cranfield University, UK.

Rapp, A., Murray-Rust, D.H., Christiansson, C. and Berry, L. (1972) Soil erosion and sedimentation in four catchments near Dodoma, Tanzania. *Geografiska Annaler* 54A, 255–318.

Rickson, R.J. (2000) The use of geotextiles for soil erosion control. Unpublished PhD thesis. Cranfield University, UK.

Simmons, R.W. (1998) The effect of polyacrilamide based soil conditioners on structural sealing at a sub-process level. Unpublished PhD thesis. University of Kent at Canterbury, UK.

Smith, R and Quinton, J.N. (2000) Dynamics and scale in simulating erosion by water. In: Schmidt, J. (ed.) *Soil Erosion: Application of Physically Based Models*. Springer, Berlin, Germany, pp. 282–294.

Styczen, M. and Høgh-Schmidt, K. (1988) A new description of splash erosion in relation to raindrop sizes and vegetation. In: Morgan, R.P.C. and Rickson, R.J. (eds) *Erosion Assessment and Modelling*. Commission of the European Communities, Directorate General for Agriculture DG VI, Luxembourg.

Sutherland, R.A. and Ziegler, A.D. (1996) Geotextile effectiveness in reducing interrill runoff and sediment flux. In: *Erosion Control Technology. . .Bringing it Home*. Proceedings of the XXVII IECA Conference, 27 February–1 March, 1996, Seattle, Washington. International Erosion Control Association, Steamboat Springs, Colorado, pp. 393–406.

Thomson, J. and Ingold, T. (1986) Use of jute fabrics in erosion control. Report prepared for the ITC, UNCTADD/GATT, Geneva, January.

Turkelboom, F. and Trebuil, G. (1998) A multi-scale approach for on-farm erosion research: application to Northern Thailand Highlands. In: Penning de Vries, F.W.T., Agus, F. and Kerr, J. (eds) *Soil Erosion at Multiple Scales: Principles and Methods for Assessing Causes and Impact*. CAB International, Wallingford, UK, pp. 51–72.

Van der Knijff, J., Jones, R.J.A. and Montanarella, L. (2000) *Soil Erosion Risk Assessment in Europe*. EUR 19044 EN. Office for Official Publications of the European Communities, Luxembourg.

Van Noordwijk, M., van Roode, M., McCaillie, E.L. and Lusiana, B. (1998) Erosion and sedimentation as multi-scale, fractal processes: implications for models, experiment and the real world. In: Penning de Vries, F.W.T., Agus, F. and Kerr, J. (eds) *Soil Erosion at Multiple Scales: Principles and Methods for Assessing Causes and Impact.* CAB International, Wallingford, UK, pp. 223–253.

Wickenkamp, V., Duttmann, R. and Mosimann, T. (2000) A multi-scale approach to predicting soil erosion on cropland using empirical and physically based soil erosion models in a Geographic Information System. In: Schmidt, J. (ed.) *Soil Erosion: Application of Physically Based Models*. Springer, Berlin, Germany, pp. 109–133.

Wischmeier, W.H. (1975) Estimating the soil loss equation's cover and management factor for undisturbed areas. In: USDA (ed.) *Present and Prospective Technology for Predicting Sediment Yields and Sources*. US Agricultural Research Service, ARS-S-40, Washington, DC, pp. 118–124.

Wischmeier, W.H. and Smith, D.D. (1978) *Predicting Rainfall Erosion Losses*. USDA Agricultural Handbook 537. USDA, Washington, DC.

23 Changes in the Spatial Distribution of Erosion within a Selectively Logged Rainforest Catchment in Borneo 1988–2003

R.P.D. Walsh,[1] M.A. Clarke,[2] K. Bidin,[3] W.H. Blake,[4] N.A. Chappell,[5] I. Douglas,[6] N. Ramli,[3] A.M. Sayer,[1] W. Sinun,[7] J. Larenus[8] and J. Hanapi[8]

[1]*University of Wales Swansea, UK;* [2]*National Soil Resources Institute, UK;* [3]*Universiti Malaysia Sabah, Malaysia;* [4]*University of Plymouth, UK;* [5]*Lancaster University, UK;* [6]*Manchester University, UK;* [7]*Innoprise Corporation Sdn. Bhd., Malaysia;* [8]*Danum Valley Field Centre, Malaysia*

Introduction

Despite many assessments of the immediate erosional and hydrological impacts of rainforest logging and replacement land uses (for reviews see Anderson and Spencer, 1991; Douglas, 1999), few have addressed longer term recovery or sediment source and delivery issues linking slope and catchment scales. Furthermore, long-term studies in the humid tropics are sparse (those of Larsen *et al.* (1999) in Puerto Rico and Malmer (2004) in southeastern Sabah being notable exceptions). This chapter synthesizes results of a 15-year study assessing changes in slope and catchment erosion and sediment sources within a small catchment in Sabah that was selectively logged using a combination of tractor and high-lead logging techniques between December 1988 and June 1989 and then left to regenerate naturally. Comparisons are drawn with data on slope erosion rates and sediment sources in adjacent primary forest catchments.

Study Area

The study area lies in the Segama catchment in eastern Sabah, Borneo (Fig. 23.1A). Study catchments and erosion monitoring sites are located close to the Danum Valley Field Centre (4° 58′N, 117° 48′E) in regenerating forest selectively logged in 1988–1989 and in primary rainforest of the Danum Valley Conservation Area (DVCA) (Fig. 23.1B). The DVCA is a 438 km^2 area of undisturbed lowland dipterocarp rainforest that, together with the Maliau Basin Conservation Area farther west, is nested within a large forest area under rotational selective logging management by the Sabah Foundation (Fig. 23.1A).

The climate at Danum is equatorial (Walsh and Newbery, 1999). Mean annual temperature (1985–2003) is 26.8°C, with annual and diurnal ranges of 1.8°C and 8.6°C respectively. Mean annual rainfall (July 1985–June 2004) is 2829 mm, with monthly means ranging from 158 mm (April) to 313 mm (January). Highest rainfall tends to occur following the equinoxes (May–June and October–November)

 Soil Erosion and Sediment Redistribution in River Catchments (eds P.N. Owens and A.J. Collins)

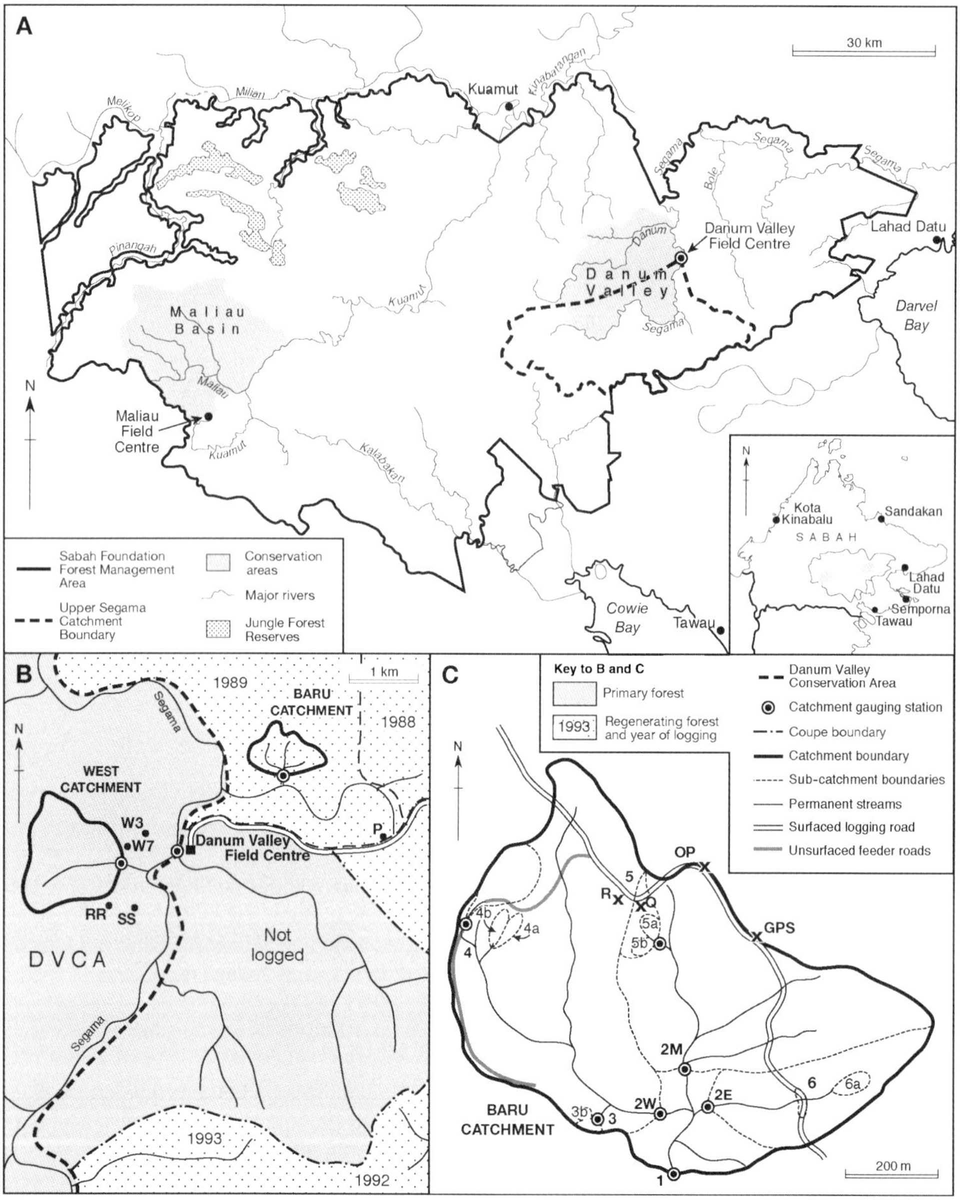

Fig. 23.1. Location maps of the study area: (A) The position of the Upper Segama catchment and Danum Valley Conservation Area within the Sabah Foundation Natural Forest Management Area. (B) The study catchments and other erosion monitoring sites at Danum. (C) The selectively logged Baru catchment.

and during the northerly monsoon months of December–January. There are on average each year 218 days with rain and 9.3 and 0.9 days with at least 50 mm and 100 mm respectively. The highest recorded fall to date is 177.4 mm on 29 March 1988.

The geology of the study area mostly comprises rocks of the Kuamut Formation, which consists of a melange of slumped sedimentary and volcanic rocks with interbedded sandstone, mudstone and tuffs, known collectively as slump breccia (Leong, 1974; Marsh and

Greer, 1992). The topography of the study area is hilly (130–260 m in altitude) and finely dissected with a drainage density (derived by field mapping) of 20–22 km/km^2. Maximum slope angles range from 10° to 30°. The soils developed on the Kuamut Formation belong to the Bang Association (Wright, 1975) and classify as Ultisols in the USDA system and Acrisol-Alisol groups in the FAO classification (FAO/UNESCO, 1990). In the study area, loams and sandy loams predominate (Chappell *et al.*, 1999a), but sandy loams occur on some steeper slopes (Clarke, 2002). Depths are generally at least 1.5 m, but often much deeper.

Research Design and Methods

Catchment erosion

Changes in catchment sediment yield during and following selective logging were assessed by monitoring the flow and sediment transport of two small instrumented catchments (Fig. 23.1B) from mid-1988 to 2003. The Baru catchment (area 0.44 km^2) was monitored while still primary forest, during the various phases of selective logging operations from December 1988 to June 1989 and then through post-logging time; the West (or W8S5) catchment (area 1.7 km^2) remained under primary forest throughout. River stage was recorded continuously using water level recorders (later linked to data loggers); rating curves were used to convert stage data to flows (Douglas *et al.*, 1992). Suspended sediment was sampled using flood-activated automatic water samplers. The Baru monitoring system was upgraded and extended in 1995 (Chappell *et al.*, 1999b) by installing V-notch weirs, Partech IR15C turbidity sensors and data loggers at the main catchment site, on the principal tributaries (West, Middle and East), on a rutted gully at Site 4 and on headwater ephemeral channels at Sites 3, 5 and 6 (Fig. 23.1C). This within-catchment network operated from July 1995 to June 1996; records since were maintained only at the main site. In 2002–2003, the Baru and West gauging stations were again upgraded by installing PDCR1830 Campbell Scientific pressure transducers and 195-Analite turbidity sensors coupled to CR10X Campbell Scientific data loggers. The data loggers recorded mean stage and turbidity every 15 min from readings taken at 1-min intervals. Turbidity (T) was converted to suspended sediment concentration (SSC) using a calibration equation (SSC (mg/l) = 0.674 T (in NTUs)), derived by filtering and weighing suspended sediment samples covering a wide range of recorded turbidities. The West Stream site was upgraded further by construction of a 120° V-notch weir. The new systems were also installed in November 2002 on two piped mini-catchments (W3 and W7 in Fig. 23.1B) to assess soil pipe hydrology and sediment transport (Sayer *et al.*, 2004, in press) and on the Segama River at Danum (catchment area 721 km^2), monitored since 1985 (Douglas *et al.*, 1999).

Slope and channel erosion

Slope and channel erosion has been measured since 1990 at networks of sites within the 1988–1989 selectively logged catchment and adjacent area (Fig. 23.1C) and primary forest close to the West catchment (Fig. 23.1B). Erosion was assessed primarily by periodic (6-monthly or yearly) remeasurement of ground level at a series of points across slope (or gully) transects *between* two permanent stakes using either an erosion bridge (a type of microprofiler) (Shakesby *et al.*, 1991; Shakesby, 1993), a simple wooden bar or a stretched tape as the datum line. These methods avoid problems of interference with the natural movement of soil or sediment associated with erosion pins or stakes. Erosion bridges with spans of 1.1 m (37 points, 25 mm apart), 2.7 m (up to 27 points, 100 mm apart) and 3.0 m width (up to 28 points, 100 mm apart) were used. The wooden bar and stretched tape methods were used for very wide and deeply gullied cross-sections.

In the primary forest, 29 erosion bridge and erosion bar transects were established in June 1990 on slopes and ephemeral channel heads in the W3 area of the DVCA (Fig. 23.1B). In 1997–1998 an additional 50 erosion bridge transects, arranged down seven slope profiles, were established in areas of steeper relief (15–25° and 25–32° maximum slope angle) at sites SS and RR (Fig. 23.1B).

In the logged forest, the aim was to assess erosion of different elements of the post-logging regenerating forest mosaic. This comprises (in addition to intact and lightly disturbed forest fragments): (i) heavily disturbed terrain (including heavily compacted log-landing areas); (ii) skid trails (tracks made by tractors, called 'skidders', when dragging individual logs to the logging roads); (iii) unsurfaced feeder (or spur) and surfaced main logging roads; and (iv) landslides and their scars. Over 20 transects were installed in 1990, focusing on a log-landing area with incipient gullies at Site 5 in the Baru catchment; and a skid trail and a deeply rutted, unsurfaced feeder road at Site P. The principal aim was to assess whether incipient gullies developed in different situations during logging enlarged or infilled as forest regeneration proceeded. An additional 32 transects were installed in the Baru catchment in 1994–1995. These included five along a gullied unsurfaced feeder road and two on skid trails at Site 4; eight on heavily disturbed slopes at Sites 5 and 6; three on the scar of a road-related landslip of December 1994 at Site 6; and 14 channel cross-sections. In 1997–1998, six transects were established on steep, moderately logged slopes at Site 5 and 10 at two heavily compacted and still (8 years post-logging) unvegetated locations at the GPS and OP sites (Fig. 23.1C).

Results and Analysis

Rainfall 1988–2003

Rainfall during the study (Table 23.1) was particularly high in 1989, 1994–1996 and 1999–2003 and low during the ENSO drought events of 1991–1993 and 1997–1998. Large falls of at least 50 mm were particular frequent (11–17 per year) in 1999–2003 and very few (3–6 per year) in 1996–1998. Highest daily falls were 162.5 mm on 19 January 1996; 140.6 mm on 8 January 2001 and 139.8 mm on 31 January 2000.

Erosion at catchment and sub-catchment scales

In the selectively logged Baru catchment, sediment yield (Table 23.2) rose sharply to an

Table 23.1. Rainfall at Danum Valley Field Centre during the study period 1988–2003.

		Number of days with:		Maximum daily fall and date	
Year	Annual rainfall (mm)	> 50 mm	> 100 mm	mm	Date
1988	2938	7	2	177.2**	29 Mar
1989	3205	10	0	80.1	22 Feb
1990	2729	10	1	135.0	4 Jan
1991	2609	10	1	114.0	3 Oct
1992	2366	8	0	77.3	19 Jun
1993	2501	8	0	92.2	9 Jan
1994	2978	7	3	122.5	2 Apr
1995	3294	11	0	98.6	20 Oct
1996	2989	6	1	162.5	19 Jan
1997	1918	3	0	93.0	21 Feb
1998	2139	6	2	124.3	27 Sep
1999	3382	15	0	89.6	17 Dec
2000	3501	13	3	139.8*	31 Jan
2001	3075	11	1	140.6	8 Jan
2002	2728	12	0	95.2	10 May
2003	3539	17	2	119.5	14 Dec

*183 mm fell in a 20-h period straddling 30/31 January 2000.
** Prior to the catchment monitoring period.

Table 23.2. Changes in the sediment yield of the Baru catchment 1988–2003. All yields are converted to annual rates.

Period	Years after logging ceased	Sediment yield ($t/km^2/year$)	Remarks
Jun–Sep 1988	Pre-logging period	342	West stream yield = 114 $t/km^2/year$, but > Baru yield in Jun and Aug
Oct–Dec 1988	Affected by logging road construction	1372	West stream yield = 68 $t/km^2/year$
Jan–Dec 1989	During and up to 0.5 years post-logging	1633	May–June = main logging phase; West stream yield = 118 $t/km^2/year$
Jan–Dec 1990	0.5–1.5 years post-logging	1017	West stream yield = 117 $t/km^2/year$
Jan–Jul 1991	1.5–2.1 years post-logging	193	West stream yield = 79 $t/km^2/year$; Baru yields decline further 1992–1994
Jul 1995–Jun 1996	6.0–7.0 years post-logging	592	Mostly in two storms of 99 mm (22 Oct 1995) and 163 mm (19 Jan 1996)
Feb–Sep 2003	14.5–15.5 years post-logging	277	Baru yield less than that of West stream in all months

annual rate of 1367 $t/km^2/year$ in the final 3 months of 1988 following the construction of the main logging road in the east of the catchment (Fig. 23.1C), reached a peak of 1633 $t/km^2/year$ in 1989 during and following logging in January–June 1989 and remained very high throughout 1990. Peak monthly yields were 766 t/km^2 in October 1989 and 696 t/km^2 in May 1990. Annual yields were 13.9 and 8.7 times those of the primary forest West catchment in 1989 and 1990, respectively. Baru sediment yield fell sharply in 1991 to an annual rate (193 $t/km^2/year$) only twice as high as for West Stream (79 $t/km^2/year$) and yields declined further in 1992–1994.

A second peak in sediment yield of 592 $t/km^2/year$ occurred in 1995–1996, 40% of which was accounted for by the single extreme event of 19 January 1996, following a rainstorm of 162.5 mm. Within the catchment (Table 23.3), yields were twice as high (1467 $t/km^2/year$) in East Tributary (affected by landsliding in the event) than in Middle Tributary and four times as high as in West Tributary (unaffected by landslides and little affected by the main logging road) (Fig. 23.1C). The proportion of annual load transported in the January storm was also much lower for the West Tributary (15.6%) than for the other two sub-catchments (41.0 and 49.2%).

Sediment yields declined for a second time since 1995–1996 and in 2003 monthly yields (Table 23.4) and peak suspended sediment concentrations (SSCs) in storm events (Fig. 23.2) were mostly lower in the Baru than in the primary forest West Stream. Thus peak SSCs of 4136 mg/l on 2 March and 588 mg/l on 30 June 2003 were exceeded by the 4492 mg/l and 2147 mg/l values respectively recorded for West Stream. Nevertheless SSCs remained high (> 100 mg/l) in events for longer in the Baru than in West Stream, despite the latter's larger catchment area.

Changes in erosion rates and sediment sources within the Baru catchment

Table 23.5 summarizes changes in ground lowering and erosion rates for different elements of the post-logging mosaic. Erosion rates on skid trails and low-angle (< 25°) heavily disturbed slopes (including log-landings) fell from high levels in 1990–1994 to values often lower than in primary forest in most years thereafter, with rises in ground level being recorded since 1999. At skid trail transect P-2 (Fig. 23.3) the early active phase up to 1996 involved 15–20 cm infill of the incipient gully that had formed and 10–15 cm erosion of the upper terrain, thus recreating a planar slope. A similar tendency towards infill of incipient rills and gullies was recorded at most cross-sections on low-angle log-landing terrain at Site 5. The sharp declines

Table 23.3. Spatial variations in erosion rate and the proportion of erosion accounted for by the 19 January 1996 extreme event within the Baru catchment in the year July 1995–June 1996 (modified after Chappell *et al.*, 1999, 2004). For locations see Fig. 23.1C.

Catchment unit	Unit description	Area (ha)	Annual load (t)	Sediment yield ($t/km^2/year$)	19 Jan 1996 storm load as % of annual load
Baru catchment	Entire catchment	44.1	261	592	40.3
Sub-catchments:					
East tributary	Affected by road-linked 1994 and 1996 landslides	4.6	67	1467	49.2
Middle tributary	Affected by main logging road but no major landslides in 1994–1996	14.3	98	685	41.0
West tributary	Contains eroding unsurfaced logging road at Site 4, but less affected by the main logging road and unaffected by landslides	19.0	69	361	15.6
Ephemeral headwater units:					
Site 4	Unsurfaced logging road gully (but above waterfall section)	1.3	8.4	643	12.8
Site 5	Stream below heavily logged and log-landing terrain	0.75	0.107	14	11.5
Zero-order ephemeral streams or slope units:					
Site 3	Primary forest slope	0.06	0.015*	24*	no data
Site 4a	Skid trail	0.16	0.126*	81*	no data
Site 5a	Channel head below lightly affected slope	0.15	0.022*	15*	no data
Site 5b	Gully below log-landing area	0.14	0.058	41	34.1
Site 6b	Channel head on steep lightly logged slope	0.03	0.030	99	31.1

*Values are underestimates as they do not include records for the 19 January storm.

Table 23.4. Monthly sediment yields (t/km²/month) for the Baru (15 years post-logging) and the undisturbed West (W8S5) catchment in 2003.

Month	Baru	West (W8S5)	Difference
Feb	20.0	32.0	12.0
Mar	60.2	77.4	17.2
Apr	23.2	27.1	3.9
May	2.3	28.8	26.5
Jun	3.4	16.3	12.9
Jul	31.0	832.1*	801.1*
Aug	0.1	1.8	1.7
Sep	44.2	84.5	40.3
Mean	23.1	137.5	114.4

*Provisional only. Includes exceptional events that overtopped the weir; estimates for these events may change when a more accurate stage-discharge relationship for over-top weir flows has been developed.

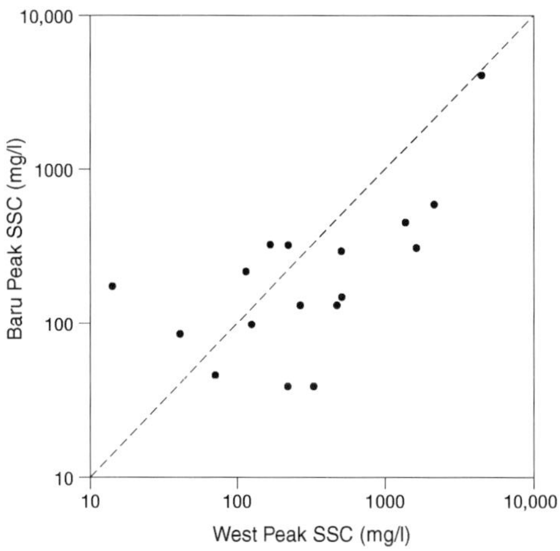

Fig. 23.2. Comparison of storm peak suspended sediment concentrations for the selectively logged Baru catchment and the primary forest West catchment in January–June 2003. A 45° line of equivalence is shown to aid the comparison. Number of storm events is 17.

in erosion rate were linked to revegetation at some sites, but to formation of protective stone lags at other sites where revegetation was slow. Rises in ground level since 1999 may be due to two factors: re-expansion of compacted soil aided by revegetation and soil faunal activity; and soil expansion in the very wet weather of 1999–2002. The shorter record for high-angle (> 25°) heavily disturbed terrain indicates very high erosion in the 19 January 1996 extreme storm, but lower rates since.

In contrast, unsurfaced logging roads have remained important sediment sources. The two monitored rutted gully systems at Site 4 and Site P show similar patterns. Whereas gullied ruts in upper sections of systems tended to stabilize or infill with progressive revegetation of the roads and adjacent contributing slopes and reductions in magnitudes and erosional effectiveness of overland flow, erosion became concentrated in lower 'waterfall' sections where road runoff descended steep road embankments to the pre-logging stream network. Thus at Site 4 in the Baru catchment, such a waterfall retreated via slot erosion over 6 m between 1995 and 2000. Vertical incision at Transect 4-6 (Fig. 23.3) totalled 20 cm between February 1995 and April 1998, but was followed by around 40 cm in 1998–1999 and over 1.6 m in 1999–2000 as headward erosion of the waterfall reached the cross-section. In total, the site has contributed around 30 m³ (over 30 t) of sediment to the Baru system since 1995, an amount that is much higher than the annual load of 0.126 t of 1995–1996 from the same road gully upstream (Table 23.4). Sediment from the waterfall site is currently (2000–2004) being stored behind a debris dam a short distance downstream. At Site P, headward erosion of 5 m and vertical incision of over 2 m was recorded between 1990 and 1996, but rates have since slowed considerably.

Table 23.5. Changes between 1990 and 2002 in ground lowering and erosion rates for different elements of the post-logging terrain mosaic of the Baru catchment and for primary forest slopes. Dry bulk densities (g/cm^3) used to convert ground lowering to erosion rates were: primary forest = 0.88; skid trails = 0.94; landslide scars = 1.34; others = 1.33. Measurement times are in summer except in 2002 (December for primary forest; February for other categories).

(a) Ground lowering rates (GL) mm/year (negative = lowering; positive = ground level rise; *n* = number of transects)

	Primary forest		Skid trails		Unsurfaced logging roads		Landslide scars		Heavily disturbed slopes < 25°		Heavily disturbed slopes > 25°	
Period	GL	*n*	GL	*n*	GL	*n*	GL	*n*	GL	*n*	GL	*n*
1990–1992	–2.03	13	–14.4	3	–75.1	5	n/a		–5.0	11	no data	
1992–1994	+1.91	13	–17.1	3	–9.6	4	n/a		–9.3	1	no data	
1994–1995	–1.78	13	–1.7	3	–53.5	10	–136.5	1	–1.6	5	no data	
1995–1996	–3.84	13	+6.3	5	–1.9	10	–101.4	3	–6.8	5	–106.1	3
1996–1997	–0.02	13	–6.9	5	–8.7	10	–19.9	3	–0.5	5	+20.7	3
1997–1999	+2.11	13	–0.5	5	+7.2	10	–3.0	3	–0.3	15	+1.3	8
1999–2000	–2.60	13	+6.8	4	–56.7	7	–109.4	1	+14.7	15	+29.4	7
2000–2002	–0.07	13	no data	no data	no data	+8.5	9	–1.7	5			
1990–2002	–0.36											

(b) Erosion rates t/km^2/year (negative = erosion; positive = ground level rise)

Period	Primary forest	Skid trails	Unsurfaced logging roads	Landslide scars	Heavily disturbed slopes < 25°	Heavily disturbed slopes > 25°
1990–1992	–1,786	–13,536	–99,883	n/a	–7,315	no data
1992–1994	+1,681	–16,074	–12,768	n/a	–12,369	no data
1994–1995	–1,566	–1,598	–71,155	–183,044	–2,128	no data
1995–1996	–3,379	+5,922	–2,527	–135,876	–9,044	–141,113
1996–1997	–18	–6,486	–11,571	–26,666	–665	+27,531
1997–1999	+1,857	–470	+9,576	–4,020	–399	+1,729
1999–2000	–2,288	+6,392	–75,411	–146,596	+19,551	+39,102
2000–2002	–62	no data	no data	no data	+11,305	–2,261
1990–2002	–314					

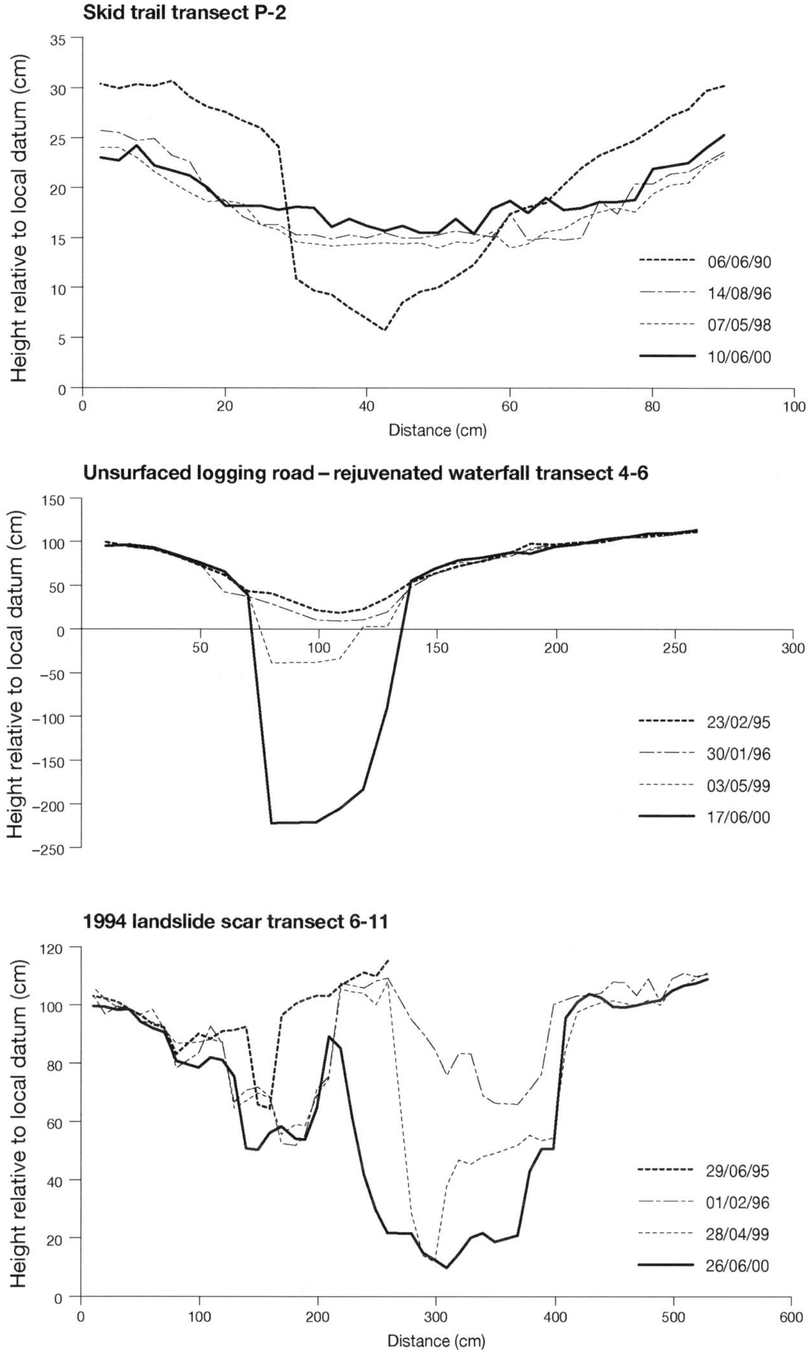

Fig. 23.3. Long-term changes in ground level at erosion bridge transects of different post-logging mosaic landscape elements. Top: skid trail Transect P-2 over the period 1990–2000; middle: gullied unsurfaced logging feeder road Transect 4-6 over the period 1995–2000; bottom: Transect 6-11 on the scar of the December 1994 landslide at Site 6 (see Fig. 23.4) over the period 1995–2000.

Table 23.6. The sequence of events in the Site 6 area of the East sub-catchment of the Baru from 1994 to 2003 (for locations see Fig. 23.4).

Date/period	Events
1994	Instrumentation of headwater ephemeral stream downslope of surfaced logging road
1994 (14 Dec)	Landslide including road and road embankment material; toe deposits and formation of debris dam in ephemeral stream channel at the base of the landslide
	Cutting of initial gully in scar deposits by time of installation of erosion bridge site 6-10
1996 (19 Jan)	Major rainstorm (163 mm in a day) results in (in chronological order)
	1. New landslide blocks valley 200 m downstream of first landslide
	2. Lake forms behind the valley dam
	3. Upstream debris dam of first landslide bursts
	4. Enlargement of gullies in scar and toe deposits of first landslide (see Fig. 23.3)
	5. Coarse dam-burst material is deposited in the channel and floodplain drowned by the new lake
1996–1999	Deposition of progressively finer material in the lake behind lower dam
	Progressive draining of lake (complete by 1998) as outlet height is lowered by fluvial incision
	Re-establishment of stream channel course through lake deposits
2000 (31 Jan)	Marked re-excavation of original ephemeral channel through debris of the first landslide (see Fig. 23.3)
2000–2003	Continued enlargement (re-excavation) of original ephemeral channel at 6-11 and downstream including exposure of weir and other instrumentation buried in 1996

Although there was some landsliding in 1989–1990 during and immediately after logging, landslides, and subsequently their scars and toe deposits, became a major sediment source in the Baru catchment since 1994 and were largely responsible for the second sediment yield peak recorded in 1995–1996 (Table 23.2). The first major landslide occurred on 14 December 1994 at Site 6 (where there had been some previous activity in 1989–1990). Another 10 landslides were recorded along the same surfaced logging road (Fig. 23.1C) during the 162.5 mm rainstorm of 19 January 1996. Mechanisms involved include: (i) bridge collapse with the decay of the logs from which they are constructed; (ii) the failure of artificially steep embankment slopes downslope of roads; (iii) interference to drainage from upslope (especially when culverts become blocked); and (iv) undermining of the strength of road material by pipe development. Most of these mechanisms require time after cessation of logging to develop and this may explain why they did not occur until over 5 years after logging.

The sequence of events and consequences for downstream sediment movement of two of these landslides in the headwaters of the East Tributary are summarized in Table 23.6 and Fig. 23.4. Although both landslides led to major immediate sediment inputs to and downstream pulses of sediment along East Tributary and the Baru, much material from the first landslide was stored as a debris dam and channel and valley fill. In the January 1996 event, the 1994 debris dam burst and there was significant gullying of the 1994 landslide scar, as shown by Transect 6-11 (Fig. 23.3). Much of this material (and that of subsequent events) was deposited as a 10–20 cm basal deposit in a temporary 80 m^2 lake that formed behind a dam created by the second landslide earlier in the rainstorm event. Although landslide scar erosion tends to decline with revegetation after the first 2 years, gullying and re-excavation of channels can continue episodically for much longer. Thus re-excavation of the pre-landslide ephemeral channel at Transect 6-11 (Fig. 23.4) was still episodically occurring 6–10 years after the 1994 landslide, with massive enlargement in the January 2000

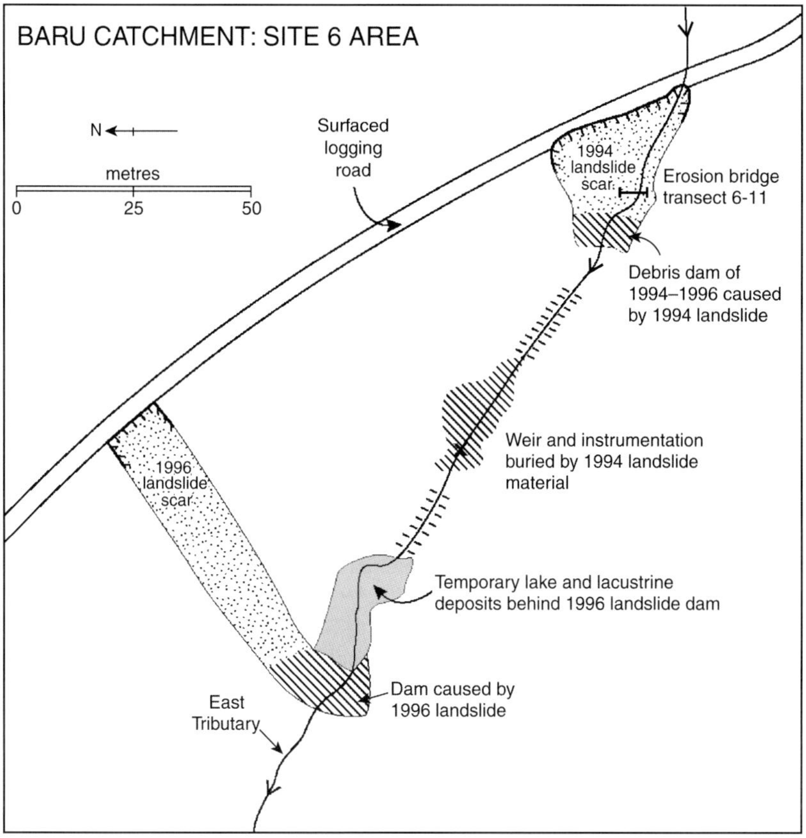

Fig. 23.4. Sketch map of landslides of 14 December 1994 and 19 January 1996 at Site 6 and their impacts on the East Tributary of the Baru catchment. See Table 23.6 for the sequence of events.

event (Fig. 23.3) and also subsequently, with erosion of one of the fixed stakes in 2003. The importance of landslides is demonstrated by the fact that the sub-catchment (East Tributary) containing them contributed 25.8% of the 1995–1996 annual suspended sediment yield of the Baru, despite occupying only 10.4% of the surface area (Table 23.3). Thus the two landslips not only provided sediment direct to stream channels during the event itself, but also acted as important sediment regulators and sources during succeeding years.

Erosion rates and sediment sources in the primary forest

Natural rates of catchment erosion in the Danum area are high for a rainforest environment, a fact linked by Douglas *et al.* (1992) to the erodible geology and silt-loam soils. Slopewash rates at the long-term erosion bridge sites (Table 23.5) averaged 0.36 mm/year (314 t/km^2/year) over the 12.5-year period June 1990–December 2002. This rate is much higher than the 24 t/km^2/year value recorded using an unbounded erosion plot on an undisturbed forest slope at Site 3 in the Baru catchment in 1995–1996 (Table 23.3; Chappell *et al.*, 1999b). This is in part because the unbounded plot record missed much of the January 1996 storm event because of burial of the turbidity sensor by sediment and leaves, whereas the erosion bridge record indicated the event to have been the most erosive in the entire 12.5-year record (Table 23.5). It may also reflect, however, a bias towards eroding sites in the bridge network through a lack of slope-base sites.

The rather high slopewash rates may be linked to overland flow occurrence in the Danum area. Evidence from networks of simple overland flow recorders (Clarke, 2002; Sayer *et al.*, 2004) and rainfall simulations (Clarke *et al.*, 2002), when added to previous runoff plot results (Sinun *et al.*, 1992; Chappell *et al.*, 1999b), indicate that overland flow (albeit only 2–5% of rainfall) is more widespread and frequent at Danum than formerly thought. Measured saturated hydraulic conductivities for Danum surface soil horizons are considerably higher than storm rainfall intensities (Chappell *et al.*, 1998) and the depth and storage capacity of upper permeable soil suggest that widespread saturation overland flow is unlikely. The reason for limited but widespread overland flow may be a degree of surface impermeability produced by the lower, decaying component of leaf litter and/or the surface fine root mat (which permeameter experiments tend to miss). This would explain why Hortonian overland flow was produced by rainfall simulation experiments (Clarke *et al.*, 2002), as they involved minimal disturbance to the surface.

Current research at Danum indicates the importance of pipeflow as a storm runoff process and pipe erosion as a sediment source. A single pipe, for example, was found to contribute around 47% of stream discharge and 22% (43 t/km^2/year) of the sediment yield (200 t/km^2/year) of the largely ephemeral W3 stream over 16 months of monitoring between December 2002 and July 2004 (see also Sayer *et al.*, 2004, in press). As the W3 catchment also received considerable sediment from unmonitored pipes, the contribution from pipe erosion is likely to be much higher. The efficacy of pipe erosion was also demonstrated by around 2 m retreat of the channel head and roof collapse of a downstream piped section of a mainly pipe-fed ephemeral channel in undisturbed forest at Site 3 in the Baru catchment during 2003–2004.

Evidence presented elsewhere (Spencer *et al.*, 1990; Douglas *et al.*, 1999) demonstrated the significance of debris dams, channel banks and landslides as important, but temporally highly variant sources and regulators of sediment transport on the West Stream. The result is that suspended sediment responses to similar-sized rainstorms vary considerably depending on antecedent rainfall and discharge patterns, whether a landslide occurs or has recently occurred, and whether debris dams are in a decaying or constructional phase upstream. Whereas the most extreme events, including the 19 January 1996 storm, have been recorded as clearing the channel upstream of existing debris dams (Douglas *et al.*, 1999), lesser events are more varied in their suspended sediment response. Thus in storm responses of West Stream in January–June 2003, the highest SSC of 4492 mg/l was recorded in a relatively modest event (peak stage 89 cm), whereas peak concentrations in the eight larger hydrographs with peak stages of 92–135 cm varied from 293 to 2147 mg/l.

Synthesis

Several points emerge from the results. First, it is clear that the impacts of selective logging are more complex and longer-term than previously thought. The temporal patterns of erosion recorded by the erosion bridge network for the various elements of the post-logging mosaic, when combined with the long-term Baru and 1995–1996 multi-scale catchment record, suggest a multi-phase model of erosional impact and recovery involving changes in both the spatial pattern and relative importance of sediment sources (Table 23.7). Some sources (skid trails, heavily disturbed terrain including log-landings, and gullies along unsurfaced logging roads) associated with the main erosional peak immediately after logging decline in importance with revegetation of terrain and a degree of sediment exhaustion and surface armouring where vegetation is slow to recover. Incipient gullies tend to stabilize or aggrade, except where gullies connect via steep sections (e.g. road embankments or lower convexities of slopes) to the pre-existing stream network, leading to the creation of retreating knickpoints or waterfall sections (as at Site 4 and Site P).

The second peak in erosion and sediment transport in the Baru catchment in 1995–1996 over 6 years after logging ceased saw a switch in relative importance of sources to road-linked landsliding, landslide scars, rejuvenated sections of road gullies as well as debris dam bursts

Table 23.7. Changes in principal sediment sources in the Baru catchment 1988–2003.

Phase	Principal sediment sources	Storage/lag mechanisms
Pre-logging/primary forest	Slopewash Pipe erosion Landslides (where slope angles are sufficiently steep) Channel bed and banks (including debris dams)	Natural debris dams
During logging and up to 2 years post-logging	Building of surfaced logging roads Creation and gullying/rutting of unsurfaced logging roads Log-landing areas Skid trails Heavily logged slopes Enhanced debris dam bursts Only limited mass movement activity	Enhanced debris dam creation Choking of headwater channels Aggradation of channels
2–5 years post-logging	Headward retreat of rejuvenated sections of unsurfaced roads Log-landing areas – reducing Skid trails – rapidly reducing Heavily logged slopes – rapidly reducing Enhanced debris dam bursts No known landslide activity	Debris dam activity Aggradation and re-excavation of channels
5–12 years post-logging	Road-related landslides (5–7 years post-logging) Gullying of landslide scars and toe deposits Headward retreat of rejuvenated sections of unsurfaced roads Erosion of old, and new landslide-related, debris dams in extreme events Re-excavation of aggraded channel reaches	Continued debris dam activity enhanced by creation of landslide-related dams Re-excavation of aggraded channels
13–15 years post-logging	As 5–12 years post-logging, but progressive reduction in activity (?) and no further landslides	Continued debris dam activity Re-excavation of aggraded channels

and re-excavation of valley fill. Because landslides occurred mainly in the southeast of the catchment, where the main logging road cuts across the middle of a slope, this phase also saw a spatial switch of sources to the East sub-catchment. Although linked to extreme rainstorms, the underlying reason why the enhanced landslide activity – and to some extent debris dam bursts – did not occur sooner after logging is biogenic; it takes time for logs in bridges, culverts and in logging-enhanced debris dams to decay sufficiently for landslides or debris dam collapse to ensue. Complex sequences of sediment supply, dam creation and decay, in-channel storage and release continue to be as characteristic of this second erosional phase as in the immediate post-logging phase, as the sequence of events in East Tributary demonstrates. Landslides are particularly important in providing both immediate and subsequent large inputs of sediment to the stream and in creating dams either at their base or in the form of debris dams downstream.

Whether the current situation represents a return to pre-logging conditions is unclear. Although peak SSCs in the Baru catchment in February–September 2003 were close to or lower than those recorded in the primary forest West Stream, one should await results from more extreme storms (exceeding 100 mm/day) than those so far included. Sediment loads in the January 1996 event were disproportionately high in the East and Middle Tributaries (more affected by landsliding) than in the West Tributary (unaffected by post-logging landslides) or a primary catchment. The recovery of the Baru has been described by Douglas *et al.* (1999) as a punctuated equilibrium, in which occasional large events interrupt quiescent periods (such as much of 1991–1994) by setting off new pulses of sediment movement; February–September 2003 may be part of one such quiescent period. The continued marked erosion at the 1994 landslide scar and the current storage of sediment from the actively eroding Site 4 knickpoint behind a new debris dam on West Tributary suggest that a new sediment pulse may be initiated by future extreme events.

Conclusions

1. The impact that selective logging has on slope erosion and sediment transport is longer term and more complex than formerly thought. A second phase of enhanced erosion 5–12 years after logging is identified, associated with biological decay of logs: (i) in bridges and culverts along a logging road section aligned along the contour in a mid-slope position; and (ii) in debris dams within the stream system; this leads to road-linked landslides and debris dam bursts in extreme events.

2. The spatial distribution and sources of sediment change with time since logging, with the early sources (skid trails, rutted unsurfaced roads and heavily disturbed terrain) being replaced in the later phase by landslides, landslide scars, knickpoints along road drainage gullies and re-excavation of valley and channel fill.

3. Extreme events play enhanced roles in the sediment budgets of logged catchments compared with those in primary forest.

4. Soil pipes are major sediment sources for primary forest streams in the study area.

5. Erosional impacts of selective logging can be reduced significantly by: (i) avoiding aligning logging roads across the middle of slopes; (ii) selective destruction of bridges and culverts (sites of potential landslides) at the end of logging operations; and (iii) avoiding creation of steep road drain sections down road embankments.

Acknowledgements

Immense thanks are due to Mr Jonny Larenus and Mr Jamil Hanapi and many others for their help in the fieldwork. We also thank Nicola Jones and Anna Ratcliffe for drawing the diagrams and maps. The authors would like to thank the Danum Valley Management Committee, the Economic Planning Unit of the Prime Minister's Department of Malaysia and the Sabah Chief Minister's Department for permission to carry out research at Danum Valley. This is paper A/440 of the Royal Society SE Asia Rain Forest Research Programme.

References

Anderson, J.M. and Spencer, T. (1991) *Carbon, Nutrient and Water Balances of Tropical Rainforest Ecosystems Subject to Disturbance: Management Implications and Research Proposals*. MAB Digest 7. UNESCO, Paris.

Chappell, N.A., Franks, S.W. and Larenus, J. (1998) Multi-scale permeability estimation for a tropical catchment. *Hydrological Processes* 12, 1507–1523.

Chappell, N.A., Ternan, J.L. and Bidin, K. (1999a) Correlation of physicochemical properties and sub-erosional landforms with aggregate stability variations in a tropical Ultisol disturbed by forestry operations. *Soil and Tillage Research* 50, 55–71.

Chappell, N.A., McKenna, P., Bidin, K., Douglas, I. and Walsh, R.P.D. (1999b) Parsimonious modelling of water and suspended sediment flux from nested catchments affected by selective forestry. *Philosophical Transactions of the Royal Society of London B* 354, 1831–1846.

Chappell, N.A., Douglas, I., Bidin, K. and Tych, W. (2004) Sources of sediment within a tropical catchment recovering from selective logging. *Hydrological Processes* 18, 685–701.

Clarke, M.A., Walsh, R.P.D., Sinun, W., Larenus, J. and Hanapi, J. (2002) Soil erosion and surface roughness dynamics in relation to slope angle in primary and logged tropical rainforest, eastern Sabah. In: Rubio, J.L., Morgan, R.P.C., Asins, S. and Andreu, V. (eds) *Man and Soil at the Third Millennium*, Vol. II. Geoforma Ediciones, Logroño, Spain, pp. 1471–1484.

Clarke, M.A. (2002) Slope angle and soil erosion in primary and disturbed tropical rainforest in South-East Asia. Unpublished PhD thesis. University of Wales Swansea, UK.

Douglas, I. (1999) Hydrological investigations of forest disturbance and land cover impacts in South-East Asia: a review. *Philosophical Transactions of the Royal Society of London B* 354, 1725–1738.

Douglas, I., Spencer, T., Greer, T., Bidin, K., Sinun, W. and Wong, W.M. (1992) The impact of selective commercial logging on stream hydrology, chemistry and sediment loads in the Ula Segama rain forest, Sabah, Malaysia. *Philosophical Transactions of the Royal Society of London B* 335, 397–406.

Douglas, I., Bidin, K., Balamurugan, G., Chappell, N.A., Walsh, R.P.D., Greer, T. and Sinun, W. (1999) The role of extreme events in the impacts of selective tropical forestry on erosion during harvesting and recovery phases at Danum Valley, Sabah. *Philosophical Transactions of the Royal Society of London B* 354, 1749–1761.

FAO/UNESCO (1990) *Soil Map of the World. Revised Legend*. World Soil Resources Report 60. FAO, Rome.

Larsen, M.C., Torres-Sánchez, A.J. and Concepción, I.M. (1999) Slopewash, surface runoff and fine-litter transport in forest and landslide scars in humid-tropical steeplands, Luquillo Experimental Forest, Puerto Rico. *Earth Surface Processes and Landforms* 24, 481–502.

Leong, K.M. (1974) *The Geology and Mineral Resources of the Upper Segama Valley and Darvey Bay Area, Sabah, Malaysia*. Memoir 4 (revised). Government Printer, Kuching, Malaysia.

Malmer, A. (2004) Streamwater quality as affected by wild fires in natural and manmade vegetation in Malaysian Borneo. *Hydrological Processes* 18, 853–864.

Marsh, C.W. and Greer, A.G. (1992) Forest land-use in Sabah, Malaysia: an introduction to Danum Valley. *Philosophical Transactions of the Royal Society of London B* 335, 331–339.

Sayer, A.M., Walsh, R.P.D., Clarke, M.A. and Bidin, K. (2004) The role of pipe erosion and slopewash in sediment redistribution in small rainforest catchments, Sabah, Malaysia. In: Gosolov, V., Belyaev, V. and Walling, D.E. (eds) *Sediment Transfer through the Fluvial System*. IAHS Publication 288. IAHS Press, Wallingford, UK, pp. 29–36.

Sayer, A.M., Walsh, R.P.D. and Bidin, K. (in press) Pipeflow suspended sediment dynamics and their contribution to stream sediment budgets in small rainforest catchments in Sabah, Malaysia. *Forest Ecology and Management*.

Shakesby, R.A. (1993) The soil erosion bridge: a device for micro-profiling soil surfaces. *Earth Surface Processes and Landforms* 18, 823–827.

Shakesby, R.A., Walsh, R.P.D. and Coelho, C.O.A. (1991) New developments in techniques for measuring soil erosion in burned and unburned forested catchments, Portugal. *Zeitschrift für Geomorphologie, N.F. Supplement-Band* 83, 161–174.

Sinun, W., Meng W.W., Douglas, I. and Spencer, T. (1992) Throughfall, stemflow, overland flow and throughflow in the Ulu Segama rain forest, Sabah, Malaysia. *Philosophical Transactions of the Royal Society of London B* 335, 389–395.

Spencer, T., Douglas, I., Greer, T. and Sinun, W. (1990) Vegetation and fluvial geomorphic processes in south-east Asian tropical rainforests. In: Thornes, J.B. (ed.) *Vegetation and Erosion*. Wiley, Chichester, UK, pp. 451–469.

Walsh, R.P.D. and Newbery, D.M. (1999) The ecoclimatology of Danum, Sabah, in the context of the world's rainforest regions, with particular reference to dry periods and their impact. *Philosophical Transactions of the Royal Society of London B* 354, 1869–1883.

Wright, P.S. (1975) *The Soils of Sabah. Volume 3. Western Parts of Tawau and Lahad Datu Districts*. Ministry for Overseas Development, UK.

24 Erosion and Deposition Rates on 'Headlands' in Low-gradient Sugarcane Land in Australia

F. Visser
Katholieke Universiteit Leuven, Belgium

Introduction

Sugarcane cultivation is the most important type of land use on the floodplains of rivers along the tropical North Queensland coast in Australia. Recent concern about the potentially negative impact of agricultural activities on the Great Barrier Reef Marine Park has focused attention on sediment and nutrient runoff from the cane-growing region (Bramley and Roth, 2002; Neil *et al.*, 2002). To allow sugarcane production under the wet tropical rainfall conditions (> 2000 mm/year), the low gradient floodplain landscape has been strongly modified, for example through construction of dense drainage networks. Although floodplains are generally considered areas of sediment storage, such highly modified landscapes could still comprise significant sources of sediment and nutrients.

A recent study by Visser *et al.* (2002) showed that under certain conditions the sugarcane land is indeed a net source of sediment. In this previous study a sediment budget was used to identify which landscape elements in cane land are the most important sediment sources. Table 24.1 summarizes the results of the budget, which is based on direct measurements of erosion and deposition during a wet summer season from December 1999 to June 2000. A net amount of 1509 t of sediment was lost from the study area, which comprises 3.9 t/ha/year from the cultivated surface area.

Only one landscape element functioned as a net sediment sink during the study period. This was the element 'headlands', which is usually present in the low-gradient sugarcane landscape. This chapter presents the detailed results of the estimates of erosion and deposition rates on the landscape element 'headlands', from which the input for the 1999/2000 sediment budget was calculated. It also presents the estimates for the 2000/01 season. Different ways to obtain the erosion and deposition rates are compared. The results are discussed along with a selection of qualitative observations. Based on the results and observations, recommendations are made for headland management practices in low-gradient sugarcane land, which can help reduce sediment and nutrient export to sensitive downstream environments.

Study Area

The study area is a 5.4 km^2 section of the Ripple Creek catchment. Ripple Creek is a tributary of the Herbert River, one of the major rivers along the North Queensland coast (Fig. 24.1). The low-gradient area in the Ripple Creek catchment

Table 24.1. Overview of the sediment budget calculation for sugarcane land over the period 1 December 1999 to 31 May 2000 (from Visser *et al.*, 2002).

Landscape element	Input *I* (t)	Storage ΔS (t)	Net input ($I - \Delta S$) (t)
2nd–4th year crop with trash blanket (ratoon)	157*	–	157
1st year crop without trash blanket (plant cane)	666*	–	666
Water furrows	738	369	369
Drains	533	412	121
Headlands	299	597	–298
Uncultivated upland	269*	–	269
Total	*2662*	*1378*	*1284*
Net input from all landscape elements ($I - \Delta S$)			1284
Output from study area (*O*)			1509
Budget difference (($I - \Delta S$) – *O*)			225

*Measurement techniques used in these landscape elements only provide net soil loss values.

forms part of the Herbert River floodplain. It becomes completely inundated during floods with a ± 25-year return period, when the river overtops its bank. Several times each year large parts of the floodplain also become inundated by local heavy rainfall and runoff from the surrounding uplands. Most of the rainfall results from tropical cyclonic disturbances during the wet summer season from November to May. The sugarcane crop tolerates the very wet conditions and can even withstand temporary inundation.

The subject of this study, the headlands, are strips of grassland along the sugarcane fields. They are located throughout the study area in between the boundaries of sugarcane fields and the edges of the numerous drains, as schematically drawn in Fig. 24.2. The 3–5 m wide strips provide space to turn the large cane harvesters and are used to access fields in areas without official roads or tracks. In total they take up 6.9% of the cultivated area. All headlands have a grass cover, though cover density significantly varies due to differences in maintenance practices.

Methods

Because it was not known in advance whether the headlands were a sediment source or sink, both erosion and deposition rates needed to be quantified as input for the sediment budget. The erosion pin method seemed most useful for this purpose, despite its sensitivity to measurement errors (Loughran, 1990). Several other issues complicated estimation of total erosion and deposition rates on headlands:

- The elongated shape of the surface area hinders the application of common sampling strategies and interpolation methods.
- There is a possibility of runoff and sediment supply on to the headlands from the fields as well as from overbank flow by the drains.
- Cane growers need vehicle access to most headlands.
- Installing, measuring and removing plots are time consuming.

In addition to this there is a likely variation in erosion and deposition rates on headlands throughout the study area due to a number of spatially variable conditions:

- Soil texture: The texture of the soil in the catchment varies from clay to sand. Various studies identified texture as an important erosion-controlling factor (Morgan, 1995).
- Type of drain along which the headland stretch is located: Overbank flow from Ripple Drain and major tributaries has high velocities and will be more erosive or transport more sediment than flow from minor tributaries.
- Crop cover on the fields bordering the headlands: More sediment supply is expected from plant cane fields than from ratoon fields (Prove *et al.*, 1995).

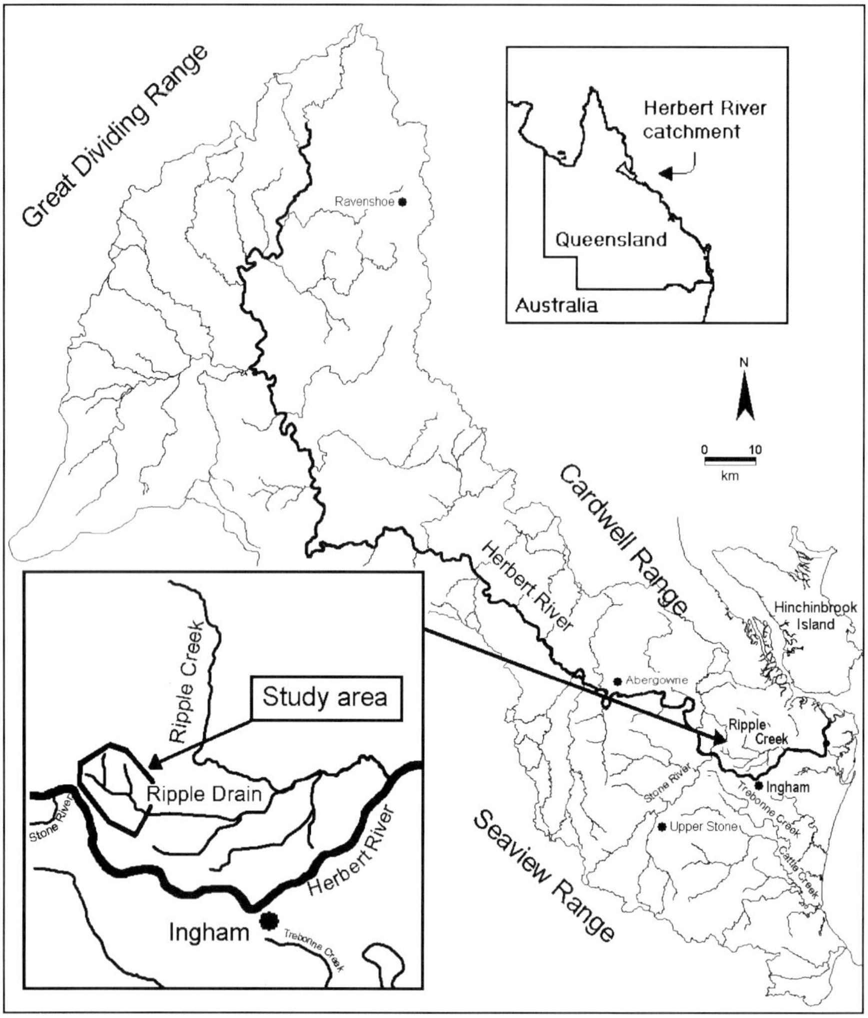

Fig. 24.1. Location of the study area in the Herbert River catchment, North Queensland, Australia.

- Vegetation cover: Preliminary observations indicated that sites with clear signs of surface erosion often have low vegetation cover.

The set-up of the pin plots was designed based on the above considerations. In order to best quantify the erosion and deposition rates for the total headland surface area, different ways of data analysis were applied and compared.

General set-up

Erosion pin plots were positioned as transects between the edge of the fields and the drains (Fig. 24.2). Each plot was five pins wide and seven to nine pins long, depending on the width of the headland. The distance between pins in a plot was 50 cm. The pins consisted of 350 mm long steel rods of 5 mm diameter, with 12 mm washers. The distance from the top of the erosion pin to the surface of the washer was measured with a digital calliper that has 0.1 mm precision. For each plot, erosion, deposition and net surface level change rates were calculated. Net surface level change (mm/year) was obtained by averaging height values of all pins. Erosion and deposition rates were obtained by separately summing positive and

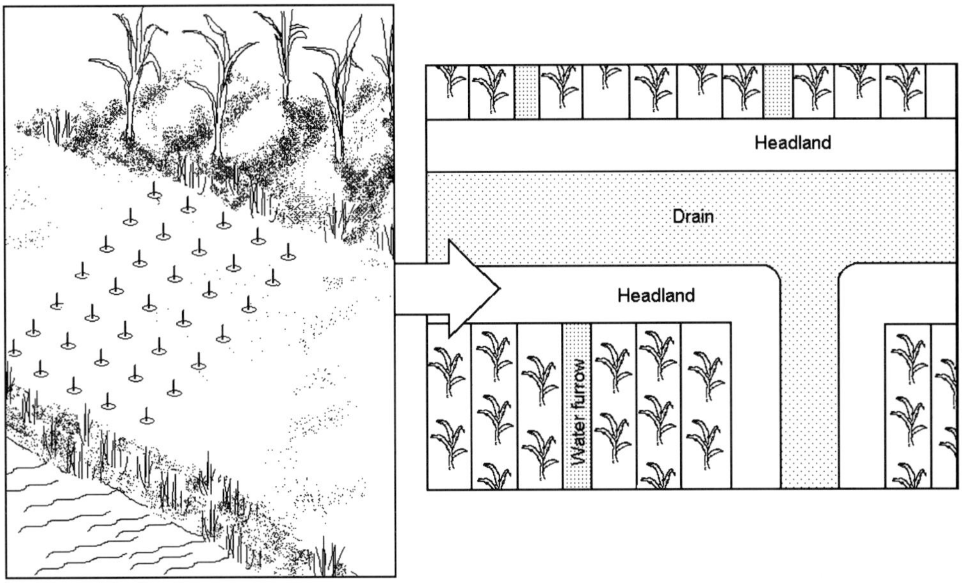

Fig. 24.2. Schematic drawing of sugarcane landscape components with example of erosion pin plot set-up.

negative values and dividing them by the total number of pins in a plot.

Effect of headland conditions

The pin plots covered the full width of a head land and therefore sufficiently represented spatial variation in this direction. Capturing variation along the length of the headlands throughout the catchment was more difficult. Application of spatial interpolation techniques based on random sampling was impractical and simple averaging of a number of plots would provide little additional information on processes in the cane land. Therefore plots were distributed through the study area so that they covered variation in surface conditions such as soil texture, drain type and crop cover as pointed out above. A non-parametric test was used to indicate whether any of the surface level change rates varied significantly between plots with different surface conditions. When calculating total amounts of erosion and storage as input for the sediment budget of the whole area, significant median rates could be applied to sub-areas with similar surface conditions and yield more reliable results than simple total averages.

Effect of vegetation cover

Vegetation cover of the headland surface was not taken into account in the distribution of the pin plots. Cover percentage is a continuous variable with a high spatial variability. It can be studied in a different manner and was therefore used to obtain an alternative estimate of surface level change rates.

At the end of each budget period the vegetation cover percentage was estimated for each pin plot. An estimate consisted of the average result of a visual assessment by two people. In February of the 1999/2000 field season a vegetation cover survey was done in the same way for 75% of the headland surface area. With the regression equation for the relationship between headland surface cover and surface level change and the information from the vegetation cover survey, alternative estimates of total headland erosion, deposition and net surface level change were calculated.

Results and Discussion

The original average erosion, deposition and net surface level change for each pin plot and

Table 24.2. Data from erosion pin plots on headlands (see Table 24.3 for coding of the surface conditions).

Plot	Erosion (mm/year)	Deposition (mm/year)	Net (mm/year)	Surface conditions			
				Cover (%)	Drain type	Soil type	Crop cover
Pin plots 1999/2000 season							
A	3.3	1.4	−1.9	75	maj	g	R
B	0.4	2.0	1.6	98	maj	g	P
C	0.7	5.4	5.7	70	min	c	P
D	0.6	5.4	7.3	70	min	c	P
E	0.0	2.4	2.4	90	maj	g	R
F	0.0	3.2	3.2	100	maj	g	P
G	13.3	9.7	−3.6	20	rip	s	R
H	3.0	3.0	0.0	65	rip	s	R
I	0.3	1.3	1.0	90	rip	s	R
J	0.2	2.2	2.0	95	rip	s	R
K	0.1	2.2	2.1	90	rip	s	P
L	0.3	2.4	2.1	80	rip	s	P
M	0.4	2.7	2.2	70	min	s	P
Pin plots 2000/01 season							
A	2.0	0.3	−1.7	45	min	c	P
D	1.1	0.8	−0.3	65	min	c	P
E	0.6	1.3	0.7	85	maj	g	R
F	0.3	1.3	1.1	75	maj	g	P
G	3.5	1.2	−2.4	30	rip	s	R
J	4.0	0.2	−3.8	99	rip	s	R
M	0.7	0.6	−0.1	65	min	s	P

each season are listed in Table 24.2. Further analysis is based on these values.

Surface level change rates based on headland surface conditions

Thirteen erosion pin plots were installed in the 1999/2000 season. The number of plots representing each of the headland surface conditions is listed in Table 24.3. In the 2000/01 season only seven plots were installed, to save operating time. Where possible the same sites were used to allow comparison between the seasons. They did, however, not cover all conditions. In the second season none of the sites contained a plant cane crop due to a reduction in total plant cane area: 20% compared to 34% in the 1999/2000 season.

All sample distributions were heavy-tailed or skewed, therefore the non-parametric Kruskal-Wallis test was used to study the effects of the surface conditions. The results only indicate a significant variation in net surface level change between plots along different crop types ($\alpha = 0.05$) (Table 24.4). There is a significantly higher positive surface level change on plots adjacent to plant cane fields compared to ratoon fields. No conclusions can be drawn from the results of the other variables. The rate of net surface level change calculated from separate median values for headlands along fields with ratoon and plant cane crop is 2.8 mm/year.

For the 2000/01 field season the test does not indicate significant differences in erosion and deposition rates ($\alpha = 0.05$) for any of the headland surface conditions (Table 24.4). No conclusions can be drawn from these results. The sample sizes are small and probably do not sufficiently represent the headland conditions.

Table 24.3. Distribution of pin plots across headland sites with different surface conditions.

Surface condition		1999/2000 season	2000/01 season
Soil type	Silty clay (s)	7	3
	Grey sand (g)	4	2
	Clay (c)	2	2
Drain type	Ripple Drain (rip)	6	2
	Major tributary (maj)	4	2
	Minor tributary (min)	3	3
Crop type	Plant cane (p)	7	–
	Ratoon (r)	6	7

Table 24.4. Results of Kruskal-Wallis tests for differences in surface level change due to differences in headland surface conditions.

Headland surface conditions	Type of surface level change	*P* for 1999/2000 data	*P* for 2000/01 data
Soil type	Net	0.08	0.12
	Erosion	0.50	0.14
	Deposition	0.17	0.15
Drain type	Net	0.06	0.07
	Erosion	0.55	0.07
	Deposition	0.21	0.15
Crop type	Net	0.04*	–
	Erosion	0.51	–
	Deposition	0.72	–

*Significant ($\alpha = 0.05$).

Surface level change rates based on vegetation cover

For each season the pin plot values for net surface level change (NSLC) are plotted against the vegetation cover estimates (VC) (Fig. 24.3). Both seasons show an increase in surface level (deposition) with increasing vegetation cover. The trend in the 1999/2000 season data is, however, not significant ($\alpha = 0.05$), because two outliers (plots C and D) affect the regression. If the data from these plots are excluded, the regressions for both seasons become very similar. This is not necessarily expected. Surface level changes will be different between years because of different weather conditions. The following equations were derived from the regressions:

Eqn 24.1: regression for 1999/2000 data:

$$\text{NSLC} = 0.06 * \text{VC} - 2.8 \quad (r^2 = 0.2, \text{ not significant}) \qquad (24.1)$$

Eqn 24.2: regression for 2000/01 data:

$$\text{NSLC} = 0.06 * \text{VC} - 4.3 \quad (r^2 = 0.9) \qquad (24.2)$$

In the 1999/2000 season the vegetation cover was surveyed for 75% of the headlands. For each surveyed headland stretch an estimate of total surface level change rates can be calculated from the surface area and the net surface level change calculated from the cover percentage and Eqn 24.1. For the total surveyed headland area this results in an average net surface level change of 1.7 mm (deposition). This value is thought to be representative of the whole study area.

The surface level change value obtained this way is not reliable because it is based on the 1999/2000 regression, which is not significant. The relationship for the 2000/01 season is significant, but the headland cover survey was not repeated. For this season an estimate is therefore made based on the 1999/2000

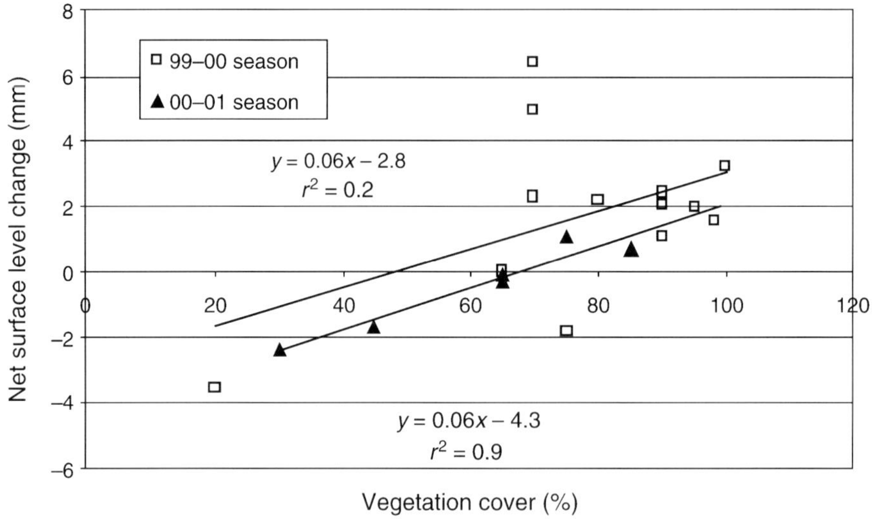

Fig. 24.3. Scatter diagram of vegetation cover percentage and net surface level change (mm), with separate regressions for the 1999/2000 and 2000/01 data ($n = 13$, $P = 0.09$, and $n = 6$, $P < 0.01$).

Table 24.5. Different estimates of erosion rates, deposition rates and net surface level change on headlands.

Rates in mm/year	Average	Log transformed	Median	Separate rates for crop type	Vegetation cover based
1999/2000 season					
Erosion	1.7	0.4	0.4	–	–
Deposition	3.3	2.8	2.4	–	–
Net	1.7	–	2.1	2.8	1.7*
2000/01 season					
Erosion	1.4	–	1.1	–	–
Deposition	0.9	–	0.8	–	–
Net	–0.5	–	–0.2	–	0.7**

*Calculated from insignificant regression equation.
**Estimated using 1999/2000 cover survey data.

cover data, assuming that the overall cover was similar for both seasons. There is no reason to expect a significant difference between the seasons, although local variation between the seasons was observed. The estimate results in a net surface level change of 0.7 mm (deposition) for the 2000/01 season.

Estimates of total surface level change rates

To compose the sediment budget, erosion and deposition rates for all of the headland surface area are needed. The discussion above shows some difficulties for obtaining these rates. Table 24.5 summarizes the results of the different methods, including the average and median values of the samples.

It is not recommended to use the average to calculate the surface level change rates, because the samples were not randomly chosen from the area. In the case of the erosion and deposition rates an additional problem occurs. The way the sample values were obtained from net surface level change results in a lognormal distribution. Thus, because of the low sample numbers, the average values can be strongly affected by outliers. Log-transformed data are also presented in Table 24.5 for the 1999/2000 season. Due to extremely low sample numbers transformation did not seem appropriate for the 2000/01 data.

The different analyses of the 1999/2000 data show consistent positive net surface level changes. When the erosion and deposition rates are used as input for the sediment budget, the log-transformed or median values will give the most reliable results because these rates are least affected by outliers. Although analysis of the measured data for the 2000/01 season suggests net erosion on the headlands, the estimate based on vegetation cover indicates that there should still be overall deposition. Because of the low and biased sample numbers this could be possible. The way that input for the sediment budget is calculated from the erosion and deposition rates is explained in Visser (2003). The results in Table 24.5 may not correspond with the results of the sediment budget presented in Table 24.1, because calculations have been revised since Visser *et al.* (2002).

Qualitative observations: spatial variation

Each sample value in Table 24.2 includes the average information of 25–35 individual erosion pins. By presenting only these average values, much interesting spatial information is lost. Although no further quantitative analysis was done, the more detailed observations clearly suggested that a highly variable pattern of erosion and deposition across the headlands is caused by the combined effect of two types of water flow (Fig. 24.4). After overbank flow events relatively coarse (sandy) sediment was deposited in the vegetation along the edge of the drains and some pins showed signs of scouring in the direction of the flow. Locally, rills were created where concentrated runoff from the inter-row area flowed on to the headland surface. Where the runoff was rich in sediment, flow deposition occurred rather than erosion and fan-shaped deposits were observed at the margins of fields. In exceptional cases pins showed erosion and subsequent deposition within one season.

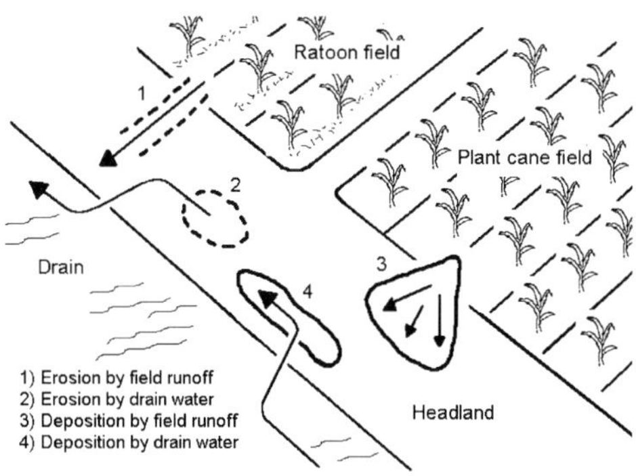

Fig. 24.4. Schematic drawing of erosion and deposition processes observed on headlands.

Representativeness: temporal variation

Because the data from this study cover only 2 years, the observed total erosion and deposition rates may not be representative for longer term averages. Both measurement seasons experienced above average rainfall (Table 24.6). The higher rainfall in the first season clearly caused more frequent overbank flow on to the headlands. This may have resulted in an above average deposition rate. On longer timescales the effect of large-scale flooding of the Herbert River may cause deposition or erosion effects of a different order of magnitude and reduce the values from this study to background noise. Such major events were not observed, however.

Conclusions and Recommendations

Analysis of the erosion pin plot data from the 1999/2000 season indicated that headlands collect sediment and act as sinks in the sugarcane landscape. The data from the second season (2000/01) does not confirm this and suggest that headlands are a net sediment source. However, the reliability of the data for this season is questionable due to the limited sample numbers and biased pin plot distribution. An alternative estimate based on the headland vegetation cover also contradicts the result and, because the erosion rate of the second season is smaller than deposition during the first season, headlands appear to be only a secondary sediment source.

Table 24.6. Yearly and 'total wet season' rainfall (Oct–May) data from Ingham weather station (Australian Bureau of Meteorology).

Year	1999		2000		2001	1968–2004*
Annual rainfall (mm)	2845		3375		1600	2047
Wet season rainfall Oct–May (mm)		3335		2288		1896

*Long-term average.

When the erosion and deposition processes are studied in more detail, observations of the first season show significantly higher net deposition rates on headlands along fields with a plant cane crop compared to fields with a ratoon crop. This result supports the values of these landscape components in the sediment budget, which show higher sediment export from unprotected plant cane fields.

Further quantitative interpretation of the results should be done with caution, for example because of high uncertainties around the actual measurements (which have not been discussed in this chapter; see Visser, 2003). The qualitative observations, however, confirmed the quantitative data. They also stress the complex connectivity between sediment sources in the sugarcane landscape, which is strongly influenced by flood hydrology. Full understanding of the erosion and deposition processes on the headlands requires further study, which should involve hydrological modelling. This can also help with estimating the effect of rare major flood events compared to the yearly local flooding, in particular regarding downstream sediment export.

Recommendations for soil management

Despite the limitations of the applied method and the questionable reliability of the data for the longer term, the results of this study provide important information for soil management practices in sugarcane land. Since headlands are directly connected with the fields, they can, under suitable conditions, act as buffer strips and trap sediment in the runoff from the field before it reaches the drains. In addition to this, the grass filters sediment from overbank flow. On the contrary, because of their strong connectivity with the drainage system, which provides a direct route for sediment export, headlands can become an effective sediment source when their surface is degraded.

Suitable headland conditions are obtained through design and maintenance. Headlands should, for example, be wide enough to effectively trap the sediment and sufficient vegetation cover should be maintained (see Fig. 24.3). Farmers can achieve the latter by reducing the frequency of slashing and repairing damaged headlands with fast-growing vegetation. Through implementation of appropriate headland management measures, farmers will contribute to a decrease of sediment and nutrient losses from sugarcane land and thus help to reduce their potential impact on downstream river ecosystems and near-shore reefs.

Acknowledgements

The author of this paper is grateful to Sugar Cane Research and Development Corporation, CSIRO, Bureau of Sugar Experiment Stations and the Australian National University for providing funding and assistance.

References

Bramley, R.G.V. and Roth, C.H. (2002) Land use impact on water quality in an intensively managed catchment in the Australian humid tropics. *Marine and Freshwater Research* 53, 931–940.

Loughran, R.J. (1990) The measurement of soil erosion. *Progress in Physical Geography* 13, 216–233.
Morgan, R.P.C. (1995) *Soil Erosion and Conservation*, 2nd edn. Longman, Harlow, UK.
Neil, D.T., Orpin, A.R., Ridd, P.V. and Yu, B. (2002) Sediment yield and impacts from river catchments to the Great Barrier Reef lagoon. *Marine and Freshwater Research* 53, 733–752.
Prove, B.G., Doogan, V.J. and Truong, P.N.V. (1995) Nature and magnitude of soil erosion in sugarcane land on the wet tropical coast of north-eastern Queensland. *Australian Journal of Experimental Agriculture* 35, 641–649.
Visser, F. (2003) Sediment budget for cane land on the Lower Herbert River floodplain, North Queensland, Australia. Unpublished PhD thesis. Australian National University, Canberra, Australia.
Visser, F., Roth, C.H., Wasson, R.J. and Prosser, I.P. (2002) Quantifying sediment sources in low-lying sugarcane land: a sediment budget approach. In: Dyer, F.J., Thoms, M.C. and Olley, J.M. (eds) *The Structure, Functioning and Management Implications of Fluvial Sedimentary Systems.* IAHS Publication 276. IAHS Press, Wallingford, UK, pp. 169–175.

25 Land Use Change, Sediment Fluxes and Reef Conservation in Belize, Central America

R.S. Nunny,[1] P. Stone[2] and D.E. Walling[3]
[1]Ambios Ltd, UK; [2]PM Atkins Water, UK; [3]University of Exeter, UK

Background

The Watershed Reef Interconnectivity Scientific Study (WRIScS) was a 4-year (1997–2002) research project with the objective of promoting a balance between sustainable land use development and reef conservation in Belize, Central America. The project was based on field data collection and concerned with the transport of fine sediment in river systems and the coastal zone in the Stann Creek District of southern Belize (Fig. 25.1).

Belize has the second largest barrier reef in the world and the largest in the Western Hemisphere. Changing conditions within river basins represent one of four critical causes of degradation of the coastal zone of this region, compounding the effects of climate change, tourist pressures and over-fishing. Altered land use practices, in particular the clearing of land for agriculture, has led to increased concern about the potential impact of soil erosion on contaminant transfer and sediment loads in rivers and consequent impacts on the coastal zone and the barrier reef. The Stann Creek area could be considered a high-risk region in this respect, due to its high rainfall (2–3 m/year), hilly terrain (up to 1120 m above sea level) and erodible soils, coupled with extensive agricultural development and the continued expansion of such development into marginal lands.

The Stann Creek area supports most of the citrus and banana production of the country. Clearing of land for citrus and banana farming reduces the protection against soil erosion offered by the canopy cover of natural forest. The erosion of hillslopes in farmed areas increases the amounts of fine sediment washed into the rivers. By this means increased sediment and associated contaminants (e.g. agricultural pesticides) may be delivered to the sea and carried to the barrier reef.

Clearance of forest potentially poses a threat to the globally important ecosystem of the Belize barrier reef. The fishery and tourism industries of Belize depend directly on this resource. Belize is in the process of developing strategies to manage land use development in such a way that it does not threaten the coastal resources, the associated ecosystems and the industries that depend on them. However, a prerequisite for environmental management and sustainable development is a good understanding of the natural and human systems involved. Such information is scarce in Belize. The WRIScS project therefore attempted to provide sound, scientific information to identify and quantify the processes at work in order to define the land–sea links and effects, thereby providing guidance for sustainable and integrated land use development and coastal zone management.

The purpose of the WRIScS project was to promote sustainable land use development consistent with reef conservation. In pursuing this purpose the project aimed to determine whether farming activity in targeted river valleys has led

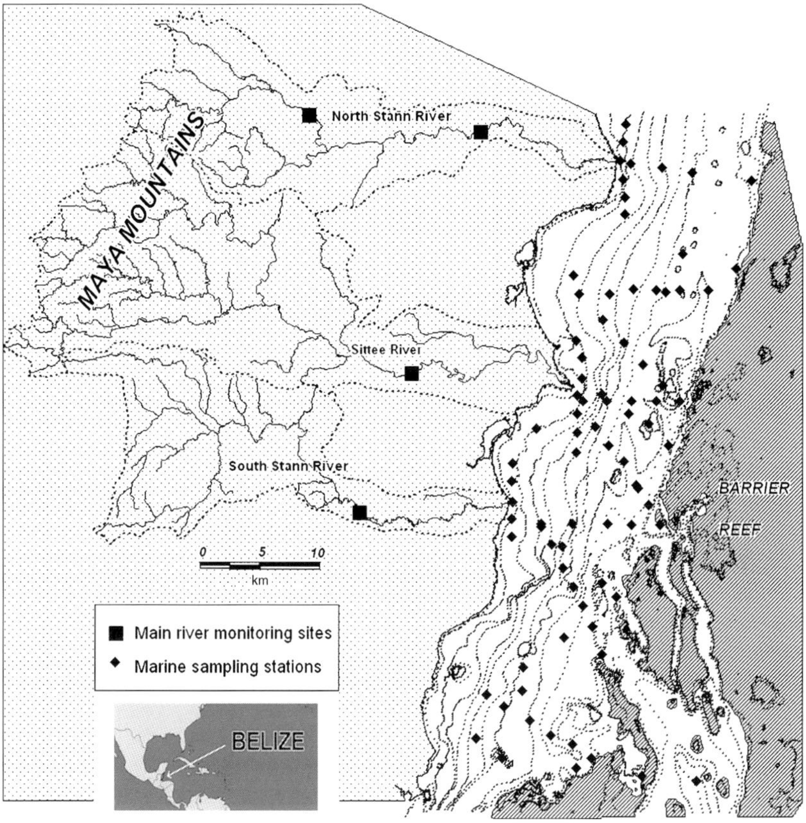

Fig. 25.1. Study locations.

to an increase in the amount of sediment and associated contaminants transported to the reef. In accomplishing its purpose and aim, the specific objectives of the project were as follows:

1. To quantify the effects of agricultural land use activity in the Stann Creek area on the transfer of sediment to rivers (North Stann Creek, Sittee River and South Stann Creek) and subsequently to the sea.
2. To examine the dispersion of sediment and any associated contaminants entering the sea, investigating in particular the ability of such discharges to impact upon the inner margins of the barrier reef in the sea area between Dangriga and Placentia.
3. To identify, through a programme of seabed sediment studies, changes in the quantities and quality of river sediment affecting the barrier reef, and to lay the foundation for future monitoring of the same.
4. To integrate the knowledge gained during the study into its socio-economic context, and to provide advice relating to the prevention of any environmental degradation identified.
5. To promote an awareness of the project and to share knowledge of scientific methodologies used within the Belize environmental community at large.

This chapter presents a brief overview of these activities, with emphasis on objectives 1–3, and the conclusions and recommendations that emerged in relation to objectives 4 and 5.

Methods

The three major river basins of the study area (Fig. 25.1) were instrumented in their mid to lower reaches to generate continuous data (2 min averages every 15 min) for water level, turbidity, water temperature and specific conductance over a 2-year period. Current meter measurements were made to develop stage-discharge rating curves at each site. Suspended sediment samples were collected both on regular servicing trips and through storm events to calibrate the turbidity readings in mg/l. Three sub-catchments within two of the basins were also instrumented over shorter periods (duration of the rainy season), to identify the effects of specific land usage.

Bulk samples of suspended sediment were collected during river flood conditions. Samples were taken of over-bank deposits on the river floodplains, within both forest and farm environments. These samples were subject to fingerprint analyses (Walling *et al.*, 1999) to determine the sources of sediments being transported by the rivers.

Observations in the coastal zone were made from temporary bases set up by Raleigh International expeditions at remote cays on the barrier reef (Fig. 25.1) during March–September in two consecutive years. Water quality observations (salinity, temperature and turbidity) were made through the area at 1–2-week intervals. Suspended sediment traps were established at a range of sites and emptied and cleaned at 3-week intervals. Recording current meters were deployed at three sites (mid-channel, off the river mouths) over a 12-month period. Sea-bed surveys were conducted using SCUBA, acoustic and grabbing/coring methods (five detailed surveys of local areas and a broad-scale survey).

Determination of marine and river suspended sediment concentrations (gravimetric), basic marine sediment analyses and pre-processing of all sediment samples were undertaken at a laboratory set up in the Coastal Zone Management Institute in Belize. Pre-processed samples were returned to the UK for determination of dating (based on radionuclide profiles, ^{137}Cs, ^{210}Pb, undertaken for a limited number of floodplain cores) and 'fingerprinting' parameters (C, N, P) and for particle size, trace metal, hydrocarbon and pesticide analyses.

Results

River systems

The measurement of the water and sediment discharged from the three main catchments of the WRIScS study area (Fig. 25.2) have shown

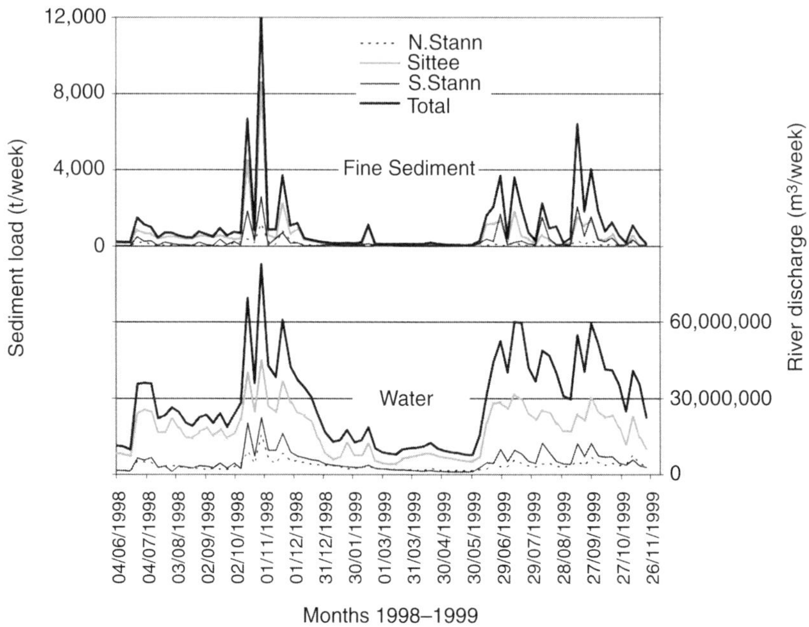

Fig. 25.2. Water and sediment discharges from the three major catchments during the wet seasons of 1998 and 1999.

Table 25.1. Suspended sediment loads measured in the main rivers of the study area (Fig. 25.1), during 1998 and 1999. Note 1998 contained the unusual conditions of Hurricane Mitch.

	Suspended sediment load (t/rainy season)		Specific suspended sediment load (t/km^2/rainy season)	
Site	1998	1999	1998	1999
North Stann upper	4,612	2,158	49.7	23.3
North Stann lower	–	11,207	–	45.3
Sittee	28,458	14,270	71.2	35.7
South Stann	8,948	11,012	40.6	49.9

that, on regional and global scales (Walling, 1996), delivery of water is high (attaining 1800 mm runoff annually) and suspended sediment yields are low (~ 50 t/km^2/year specific suspended sediment yield). This is a natural phenomenon, reflecting the rainfall and catchment characteristics (notably the topography, geology and vegetation cover) of the watersheds. Table 25.1 shows both total load and specific load for the main catchments per rainy season (the rainy season produces 75–95% of the total annual load in these rivers).

The three river basins studied drain the Maya Mountains, with the rivers crossing a flat coastal plain in their lower courses. Approximately 20–30% of the area of the catchments is flat land. A range of forest types occupy the uncultivated lands. Humans are modifying this situation with large-scale clearances of land for plantation-type cultivation of citrus fruits and bananas. Only a small proportion of the watersheds is cultivated, with citrus plantations accounting for 10% of the total catchment area, and bananas about 1%. The clearance of land for citrus development extends up the length of the valleys occupying both flat and foot-slope areas, whilst that for banana is confined to the flat coastal plain area. This development began nearly a century ago, but most clearance and planting has occurred over the past 25 years.

By comparing specific suspended sediment yields from forested and cultivated areas (sub-catchments), results from the study can be interpreted to show that change in land use can produce an order of magnitude or greater increase in fine sediment input to the rivers. On average, the delivery of suspended sediment to the sea from the study catchments is estimated to have increased about four- or fivefold as a result of land use change during the 20th century (from comparison of actual loads with loads that would be expected if all the land was naturally forested). This low net increase is primarily due to the small percentage of farmed land (~10%) within the total watershed areas. As a consequence, runoff from the large areas of natural forest that still prevail is having a significant diluting effect on human impact. Weekly mean suspended sediment concentrations measured on the main rivers during the project never exceeded 200 mg/l. Sediment is delivered to the sea mostly during the rainy season (June–November, Fig. 25.2), in a series of river flood events. The total fine sediment load carried to the sea each year is about 50,000 t, of which 90% is carried during flood events. Human impact has increased the amount of sediment carried in suspension during the peaks of these floods, as shown by source-tracing techniques which identify increased contributions from farmed land as a flood event develops. Table 25.2 shows the relative contribution of each of the different land uses to the catchment suspended sediment yields. The study found no evidence that human activity is changing the particle size of the suspended sediment load.

Marine systems

The channel (shelf lagoon of Perkins, 1983) between the mainland and the reef in the study area is dominated by marine conditions throughout most of the year. Observed salinity rarely fell below 25. Some 30% of the freshwater found in the channel is estimated to be derived from the three major local rivers, which deliver some 1500×10^6 m^3 of freshwater per year. The other

Table 25.2. The relative contribution (%) of each of the potential sources of suspended sediment to catchment suspended sediment yields. 1998 and 1999 data.

	Source type			
Site	Riverbank (%)	Citrus (%)	Forest (%)	Banana (%)
North Stann upper	0.7	66.7	32.6	–
North Stann lower	2.2	78.3	19.5	–
Sittee	5.5	40.0	54.5	–
South Stann	0.7	26.7	55.3	17.3

70% must come from river sources both to the north and south of the study area. This observation underlines the importance of water and fine sediment sources from the wider region to processes active in the study area.

The receiving waters of the coastal zone are characterized by low energy levels. The reef itself provides effective shelter from ocean waves. Currents produced by the combined effects of regional circulation systems, wind and tide effects are very slow, with peak flows rarely exceeding 0.25 m/s. Currents inside the reef are mostly southward flowing, but were observed on occasion to reverse to produce northerly flow, persistently during the months of September and October. Regional (western Caribbean) circulation is thought to be the primary control of flow inside the reef, with wind, tide and river discharge playing secondary roles.

The study has provided a large body of data on the turbidity of the coastal zone in the Stann Creek District. Waters are very clear in most places for most of the time (suspended sediment concentration ~ 1 mg/l). Temporal and spatial variation in turbidity does occur, with muddier waters (5–10 mg/l) being generated in response to both river inflow and the seabed-stirring effects of wave action. River inputs and wave resuspension are equally effective as sources of turbidity, shown by the persistence of turbidity during periods of low river sediment discharge (Fig. 25.3) and particularly during the windiest months of the year (notably April). Application of simple box models to the data generated indicates that the annual river-borne suspended sediment input (~ 50,000 t) mixes into a transient suspended sediment load (of about a further 100,000 t) carried intermittently within the coastal water body. Importantly, the mode of delivery of the bulk of the river sediment (at high concentrations during short-lived flood events) encourages sediment particle flocculation and settling on reaching the sea, and mathematical modelling of dilution and particle settling shows that most sediment accumulates rapidly on the seabed within a few kilometres of the river mouths. These model predictions are consistent with extensive field measurement (turbidity and sediment trap data) made in the coastal zone during the study.

All sediment accumulating in shallow waters close to the mainland shore is susceptible to further erosion and onward transport as a result of wave action. However, the wave-energy and current regimes, which determine the capacity of this second phase of sediment transport, operate independently of river processes and are unaffected by local human activity. Thus any transport of river sediment to the reef involving periods of accumulation on the seabed effectively de-couples river-mouth and reef-margin turbidity regimes. Turbidity levels at the reef resulting from this transport process therefore remain essentially unaffected by increases in river sediment inputs due to land use change.

A small proportion of the sediment introduced by the rivers may not undergo temporary accumulation on the seabed, and can be carried directly to the reef. Current-meter data indicate that conditions conducive to this transport prevail on several (~ 5) occasions each year, with minimum travel times to the reef of about 3 days. Residual turbidity in these plumes is, however, very low and plume duration is short. Model outputs indicated that during these short-lived events, suspended sediment concentrations may have increased two- or threefold as a result of land use changes over the past century. These predictions are again consistent

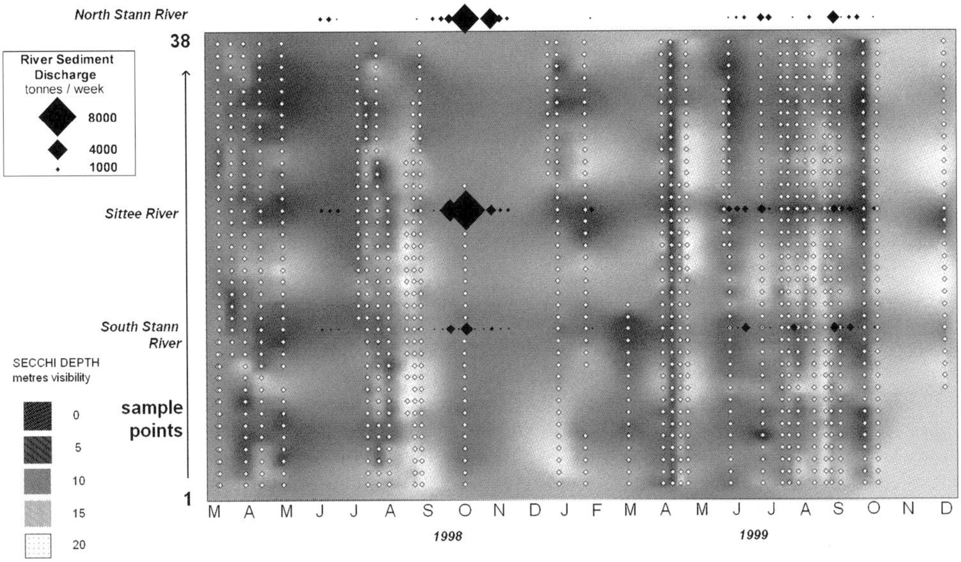

Fig. 25.3. Spatial and temporal variability in the turbidity of the coastal waters within the study areas, as measured using secchi discs (note turbidity is inversely proportional to secchi depth). Black diamonds note periods of high river suspended sediment load. Sample point 1 is in the south, point 38 is in the north.

with field observations. Absolute concentrations during these events remain low, and potential effects have to be evaluated in the context of the natural turbidity regime of the reef margin areas, where persistent periods of wave action commonly generate suspended sediment concentrations of similar and higher values (Fig. 25.3). In consequence, it is reasonable to conclude that stresses on coral populations are unlikely to have been significantly increased as a result of augmented sediment delivery related to land use change.

The bulk of additional sediment entering the sea as a result of changed land use is accumulating on the bed of the channel separating the mainland from the reef. Accumulation rates are very low (around 10 mm per century) and thus deposition is not causing problems or significantly altering water depths (hence not changing the potential for wave-induced sediment resuspension). The muddy sediment on the floor of the channel between the mainland and the reef is of both reef and terrigenous origin, and active sediment transport processes relate to both sources. At the barrier reef, river-borne accumulation comprises less than 25% of the seabed sediment (based on carbonate content analyses of bed sediments undertaken during the project).

Analyses of soil and sediment samples undertaken during the study indicate that low levels of contamination (pesticides, metals) can be identified on cultivated land surfaces, but these are not traceable into the marine environment due to high levels of dilution with uncontaminated sediments. By global standards, the coastal zone sedimentary environment is pristine. Only a few point locations of modest metal concentrations, and low levels of copper (Fig. 25.4) across wider zones, can arguably be attributed to anthropogenic sources. Copper is extensively used as an additive micro-nutrient in crop sprays in the region.

The coral biotopes found within the channel zone and along the inner reef margins show a gradient in characteristics, which apparently relates to the effects of intermittent fluvial influences. Such variation in habitat and assemblages would be expected as a natural feature of this transitional environment. The coral surveys undertaken would have been insensitive to subtle anthropogenic effects over short timescales, but provide a good benchmark against which future changes can be assessed.

From the findings of the WRIScS project a clear-cut series of conclusions were formulated relating to the natural resource management

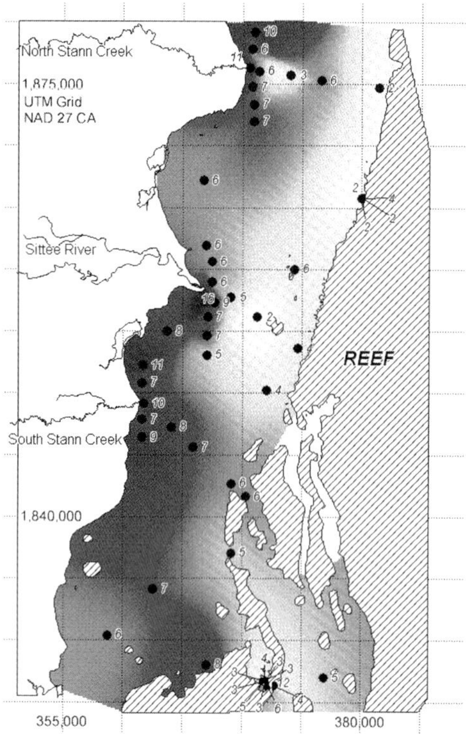

Fig. 25.4. Concentration of copper (μg/g) in seabed surface sediment of the coastal zone in the study area.

issues. These are contained in a series of reports presented to Belizean institutions, and can be accessed at http://www.ambios.net/wriscs/archive.htm

Conclusions and Management Implications

The scientific findings of the WRIScS project are comprehensive and to some extent unexpected. The dataset on which the findings are based represents the first complex series of inter-related measurements of hydrological and marine sedimentary processes for Belize, if not the tropical world.

- Incidents of localized, high intensity rainfall in the Maya Mountains during the rainy season produce large volumes of runoff, but there is little soil erosion in these areas due to the high density of the natural forest cover.
- Land use changes have increased the amount of fine sediment transported by rivers in the study area. Land farmed for citrus could generate up to 15 times more fine sediment than land under natural conditions.
- Although only less than one-fifth of the study area is farmed for citrus and banana, this land generates over half the sediment carried in suspension by the rivers.
- Land use change from forest to citrus and banana through the 20th century is estimated to have caused a net four- to five-fold increase in the fine sediment load at the river mouths within the study area.
- The total volume of river water flowing to the sea from local catchments is typically about 1500×10^6 m^3 per year. This flow carries about 50,000 t of sediment to the sea annually. In global terms the runoff from the study area rivers is relatively high but the sediment yield is low. This low sediment yield may be an important factor contributing to the existence of the Belize barrier reef.
- When river sediment reaches the sea it is effectively diluted by mixing processes, and most is initially deposited on the seabed within a few kilometres of the river mouths.
- On only about five occasions each year do currents prevail that can carry river sediment plumes directly to the reef. Transport to the reef takes 2–15 days, with the plumes persisting over the reef for only a few days.
- As the settling characteristics of the sediment load carried by the rivers seem unaffected by the land use changes, and as accumulation rates in the channel between the land and the reef remain very low (~ 0.1 mm/year), the nearshore zone is providing an effective sink for the increased amounts of sediment released as a result of land use change.
- Wind–wave effects are as important as river discharge in influencing coastal water turbidity. Wave-induced resuspension of bed sediment allows longer term transport of river sediment away from river-mouth areas, potentially toward the reef. The capacity for this wave-induced onward transport and the resulting turbidity are, however,

independent of the processes delivering terrestrial sediment to the coastal zone.

- A range of contaminants (trace metals, pesticides, polyaromatic hydrocarbons) were detectable in soil samples and river sediments, but concentrations were well below globally acknowledged safe levels. No significantly elevated concentrations of any contaminants were found in marine sediments.

In conclusion, there is no evidence to suggest that changed sedimentary processes resulting from farming activity in the study area are having a negative impact on the barrier reef. Although sediment and contaminant delivery to the coast has increased as a result of land use change, the channel separating the mainland river mouths from the Belize barrier reef provides an effective buffer, absorbing the effects of change to date and preventing impact on the reef. The natural coastal system is effective in dealing with the impact of increased sediment yield and sediment contaminant loading produced by current land use.

Based on these conclusions and also relating to actual and perceived shortcomings of the WRIScS project, a series of recommendations were made in relation to future natural resource management activity in Belize. These related to:

- the value of future farm scale investigations of soil erosion;
- the need to provide public maps of areas undergoing and also sensitive to river bank and soil erosion;
- the requirement to study freshwater ecology in the rivers of the study area;
- the need for study of effects on the coastal zone of nutrients contained in river discharges;
- the need for studies similar to that undertaken by WRIScS to be completed elsewhere in Belize, particularly in relation to the sugar cane growing areas in the north;
- the value of future oceanographic studies to understand water circulation in the coastal zone; and
- the setting up of a National Marine Environmental Quality Monitoring Programme, which should be a simple, effective, inexpensive and sustainable monitoring system, established immediately and based on existing resources supported by national funding.

As well as addressing the research tasks, the project was seen from the outset as a demonstration vehicle, and effort was put in to stimulating environmental monitoring in Belize, not just by delivering equipment and data to institutions, but through a sharing of European and Caribbean research experiences. This sharing took place on both formal and informal levels, in workshops, in seminars, in hands-on teaching sessions and in the field. One of the final reports of the study was dedicated solely to methodology and includes manuals and training materials produced specifically for use in Belize. This aspect of the project proved very fruitful and significant amounts of 'confidence and awareness' were gained at both institutional and individual levels.

References

Perkins, J.S. (1983) *The Belize Barrier Reef Ecosystem: An Assessment of its Resources, Conservation Status and Management*. New York Zoological Society, and Yale School of Forestry and Environmental Studies, New York.

Walling, D.E. (1996) Suspended sediment transport by rivers: a geomorphological and hydrological perspective. *Archiv für Hydrobiologie Special Issue, Advances in Limnology* 47, 1–27.

Walling, D.E, Owens, P.N. and Leeks, G.J.L. (1999) Fingerprinting suspended sediment sources in the catchment of the River Ouse, Yorkshire, UK. *Hydrological Processes* 13, 955–975.

26 Understanding the Distribution, Structure and Behaviour of Urban Sediments and Associated Metals Towards Improving Water Management Strategies

I.G. Droppo,[1] K.N. Irvine,[2] K.J. Curran,[3] E. Carrigan,[4] S. Mayo,[4] C. Jaskot[1] and B. Trapp[1]

[1]National Water Research Institute, Canada; [2]State University of New York, Buffalo, USA; [3]Dalhousie University, Canada; [4]McMaster University, Canada

Introduction

Within urban watersheds, sediments and their associated contaminants follow multiple pathways that are unique to the urban system, but analogous to the natural fluvial system. Droppo *et al.* (2002) described the transport of sediments and associated contaminants within an urban watershed as an urban continuum with linkages of pathways or compartments for the conveyance of sediments and contaminants from a depositional or eroded surface to a treatment system or receiving water body. While there are similarities between urban watersheds and those of a more naturalized system, the complexity of anthropogenic factors within an urban system may result in substantive changes in the physical, chemical and biological characteristics of sediment during relatively short transport distances and times. Within natural river basins, spatial and temporal changes in sediment structure and behaviour (physical, chemical and biological) are generally more gradual due to moderate changes in land use or water quality (with the exception where a significant point source discharge may impact water chemistry).

Runoff and sediment transport within urbanized areas is complex spatially and temporally due to variations in surface types and land use. Pervious and impervious surfaces are mixed within the urban environment, creating a spatially diverse rate of infiltration and runoff. Land use (industrial, residential, commercial) also influences runoff dynamics and the nature of the particles removed. The array of land uses and land types combined with long and short range atmospheric deposition result in a wide variety of sediment sources, particle types, sizes and structures available for transport through the urban continuum (Vermette *et al.*, 1987; Roberts *et al.*, 1988; Droppo *et al.*, 2002). Frequently the finer fraction (silts and clays) will be present in the form of conglomerated particles called flocs or aggregates (Droppo *et al.*, 2002). Often roadways are the repositories of sediment and associated contaminants derived from adjacent land uses as the surface runoff flows to the sewer systems (Droppo *et al.*, 2002). Not all sediment and associated contaminants deposited within the street gutters are delivered to receiving water bodies or sewage treatment plants (STPs) due to urban management practices (e.g. street sweeping) and

 Soil Erosion and Sediment Redistribution in River Catchments (eds P.N. Owens and A.J. Collins)

hydraulic sorting of particles during storm events (Droppo *et al.*, 2002).

Surface washoff channelled along street gutters is generally collected within stormwater sewers or within combined sewer systems. Unlike fluvial systems, combined sewer system flow is dependent on human (sanitary flow), industrial effluent discharge and surface runoff during storm events. Storm sewer flow is only dependent on stormwater runoff. Given the variability in water usage (combined sewer) and flows (storm and combined sewer), sewers are subject to sudden variations in water and sediment supply (Mark *et al.*, 1996). The effective management of urban water systems, therefore, requires increased knowledge of the temporal variations in flow for sewer systems and the origin (spatial distribution), the structure and chemical composition of the sediments entering and being transported within a sewer system to an STP or receiving water body. The aim of this chapter is to examine the spatial distribution of sediments and associated contaminants within the Kenilworth sewershed of Hamilton, Ontario, Canada and to discuss their impact on the management of urban water systems. The specific objectives are to: (i) evaluate the physical structure and grain size distribution of street sediment samples and how it changes during washoff events to better understand sediment transport behaviour; (ii) evaluate the selective partitioning of metals within the matrix of the sediment particles to assess the metal bioavailability in receiving waters; and (iii) map using GIS the spatial distribution of these sediments and contaminants to identify trends and areas of concern within the sewershed.

Materials and Methods

Study site

The Kenilworth sewershed (Fig. 26.1) has a contributing area of 256.5 ha and is serviced by a combined sewer system which discharges to the Woodward STP or to Hamilton Harbour, Lake Ontario via a single combined sewer overflow (CSO) during rainfall events of greater than 5 mm (Paul Theil Associates and Beak Consultants, 1991). The sewershed is relatively flat (1% slope on average), with the exception of

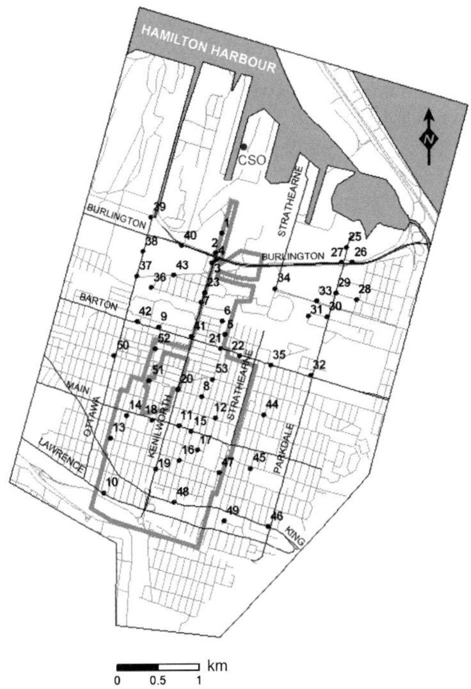

Fig. 26.1. Sampling sites in and around the Kenilworth sewershed of Hamilton, Ontario, Canada. Grey line delineates the Kenilworth sewershed contribution area.

the Niagara Escarpment, and can be divided into two sections. The upper portion of the sewershed contains approximately 52% impervious land and is dominated by older residential, single family dwellings, with commercial ribbons along major streets. The lower portion of the sewershed (approximately 9% of the contributing area) has mixed industrial practices with steel manufacturing being dominant. Sixty-six per cent of the industrial area can be considered impervious with extensive unpaved industrial storage lots. It is estimated that the total overflow volume from the sewershed for a typical rainfall year is 311,000 m^3, with a total sediment load of greater than 6×10^4 kg (Irvine *et al.*, 1998).

Dry surface street sediment samples were collected from the gutters at 52 sites located throughout the sewershed and external to its boundaries (Fig. 26.1) on 14 May 2001. Sixteen represented industrial land use sites which had at least one side of the street adjacent to an industry or an industrial storage lot regardless of traffic volume. Eighteen sites were classified as high traffic volume commercial/residential sites (13,560 and 70,137 vehicles per 24-h period). Eighteen sites were defined as low traffic volume (< 13,560 vehicles per 24-h period) commercial/residential sites.

Sample collection and particle sizing

All bulk dry street sediment samples were gathered with a polyethylene scoop which was washed with acetone and distilled water prior to each sample collection and placed in a 500 ml polyethylene container. Samples were air dried and split into two samples using a rotating V-splitter run for 2 min. One sample was analysed for bulk fractionated metal concentrations while the other was further fractionated into five size classes (> 2000, 500–2000, 250–500, 63–250, < 63 μm) using standard sieving methods prior to metal fractionation analysis.

To assess the impact of hydraulic sorting and enrichment of contaminants entering the sewer system, surface washoff was collected during storm events by holding a sample bottle below street level within the sewer catch basin (gully pots). Suspended sediment (floc) subsamples were then immediately taken with a wide mouth pipette (3.74 mm) and transferred into a 25 ml plankton chamber partially filled with tap water (surrogate for particle-free sewer water). This procedure is not considered destructive to natural sediment flocs (Gibbs and Konwar, 1982). Given that the flocs settle within the plankton chamber where they are sized following the image analysis method of Droppo *et al.* (1997), it is assumed that representative floc sizes are obtained. The Droppo *et al.* (1997) method allows for structural observations and determination of floc grain size distributions contained in the plankton chambers by employing an inverted microscope interfaced with a CCD video camera and computer image analysis system.

Metals analysis

All 52 grab samples of street sediment were analysed using the sequential extraction procedure of Tessier *et al.* (1979) with a Hitachi 180-80 Polarized Zeeman Atomic Absorption Spectrophotometer. This method separates the metals into the following operationally defined fractions: fraction 1: exchangeable; fraction 2: bound to carbonates; fraction 3: bound to Fe/Mn oxides; fraction 4: bound to organic matter: fraction 5: residual. While metal fractionation data alone can not be used to directly address bioavailability, the sequence of extraction (fraction 1 to fraction 4) can be viewed as an inverse scale of the relative availability of metals (fraction 5 is considered not available) (Stone and Droppo, 1996). The sediment metals standard WQB-1 (NWRI, 1990) was analysed with each run of samples providing a measure of QA/QC.

Results and Discussion

Structural and transport characteristics of street sediment

Grain size and spatial distribution of street sediment

Kenilworth sewershed street sediments have an average median grain size of 327 μm with a range of 124–665 μm ($n = 52$). Similar size spectrums were measured by Sartor and Boyd (1972), Klemetson (1985) and Vermette *et al.* (1987). Urban sediment distributions are generally poorly sorted and positively skewed, demonstrating the dominance of larger single-grained particles within the distribution. The street sediment in the industrial sector is typically finer than the commercial/residential sector (Table 26.1). The fine sediment within the industrial sector is consistent with increased local atmospheric particulate levels from steel mill emissions. Larger particles within the commercial/residential sector likely are related to the degradation of building materials such as concrete surfaces (Vermette *et al.*, 1987; Roberts *et al.*, 1988).

Structural changes and hydraulic sorting during rain events

An assessment of street sediment particle structure using optical microscopy and computer image analysis revealed that the majority of sediment (by mass), regardless of site, contains large individual solid mineral particles with minimal aggregated particles present (Fig. 26.2a). Using a modified Imhoff Cone, Droppo *et al.* (2002)

Table 26.1. Mean contribution of grain size fractions (%) for bulk street sediment samples of the industrial (Industrial), high traffic volume commercial/residential (comm/res – H) and low traffic volume commercial/residential (comm/res – L) sectors.

	Size range (μm)				
Sector	> 2000	2000–500	500–250	250–63	< 63
Industrial	8	19	20	41	11
comm/res – H	10	26	23	32	8
comm/res – L	11	28	19	30	12

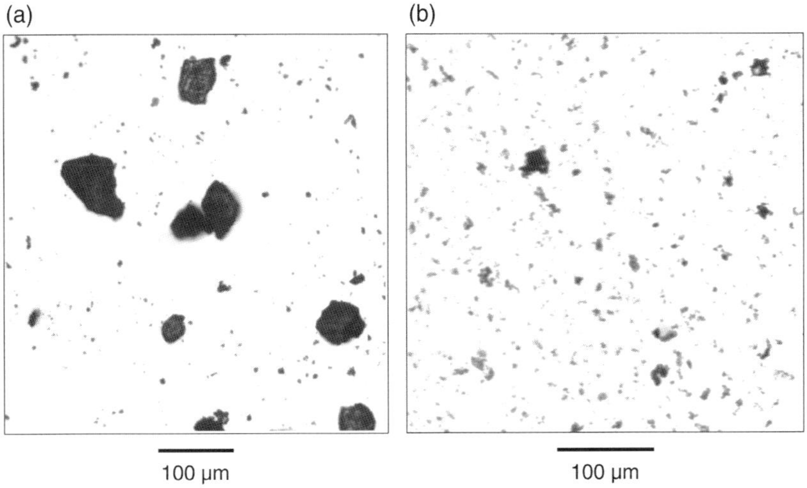

Fig. 26.2. Micrographs showing representative urban sediments: (a) dry street dust and (b) surface washoff.

found particles from Site 1 to settle within the Stokes' region and to have a density of between 2.0 and 2.65 g/cm^3. Similar densities were found by Butler *et al.* (1992). Settling experiments within the Imhoff Cone (Droppo *et al.*, 2002), however, were performed on the bulk samples and do not take into account the influence of hydraulic sorting on the settling and transport results. During a rain event, the surface runoff will only entrain a given size of particles (competence of flow). The size entrained is dependent on particle shape, particle density, particle composition, surface roughness, flow characteristics (depth and velocity), rainfall intensity, duration between rain events, source area contributions and flocculation and/or deflocculation processes. Finer sediments generally are selectively transported to the sewer systems (Droppo *et al.*, 2002; Sutherland, 2003).

Figure 26.2b shows that finer particles in runoff entering the sewer system exhibit the classical floc structure of multiple particles aggregated together. From multiple samples, the median size for stormwater runoff ranged from 25 to 50 μm, which is an order of magnitude lower than dry street sediment samples. This range is comparable to other urban areas (Chebbo *et al.*, 1990; Verbanck *et al.*, 1990; Chebbo and Bachoc, 1992). Verbanck *et al.* (1990) reported that 75% of the stormwater sediment mass was finer than 100 μm with a median diameter between 25 and 44 μm.

Bulk sediment chemistry

Spatial distribution of metals

Spatial analysis using a GIS and bulk street sediment (Kriging analysis in ArcInfo 8.3 used for spatial metal mapping) was used to assess variations in contaminant concentrations over the sample area. Mean total metal concentration and metal speciation of street sediments are compared to the effects level of the Ontario Ministry for the Environment and Energy (MOEE) *Guidelines for the Protection and Management of Aquatic Sediment Quality* (1993) (discussed below) of each sector. With the exception of Zn, the highest bulk chemistry values were present in the industrial sites, followed by the commercial/residential high traffic volume and the commercial/residential low traffic volume sites. This is not surprising given the nature of the industry and heavy industrial vehicles in the area. The industrial sector has the highest standard deviations, which are likely related to land use as well as vehicle volume and type. For example, the high standard deviation value for Pb in the industrial sector is largely related to site 3, which had a concentration of 1344 μg/g due to its location next to a Pb recycler at the time of sampling. The spatial distribution and 95% confidence interval of metal concentrations are shown in Fig. 26.3. The figures show there is less confidence in the data on the outskirts of the sampling region which is related to fewer samples contributing to the assessment. On average there is a threefold increase in Cd, Cu, Fe, Mn and Pb from the commercial/residential (low traffic volume) to the industrial sector (Table 26.2). Zn showed no evidence of a trend. This is consistent with Ellis *et al.* (1987), who observed for a variety of European sites that Zn is more evenly distributed in catchments due to the influence of long-range atmospheric transport and deposition.

Impact assessment of bulk street sediments

To understand the relative severity of the metal contamination in the street sediment and the degree to which management strategies for their remediation, abatement and/or containment are required, the MOEE guidelines were applied. While these guidelines were developed for aquatic sediments, they allow for the assessment of the potential impact that these metals would have if they reached Hamilton Harbour via a combined sewer overflow. Table 26.3 describes these guidelines, which represent a gradient of ecotoxic effects and are based on the chronic, long-term effects of contaminants on benthic organisms. The 52 samples from this study showed that metal concentrations in street sediment are between the SEL and LEL (Table 26.2) and are similar to the smaller data set of Droppo *et al.* (1998). Consequently, if this street sediment were transported through the urban continuum to Hamilton Harbour it will probably have a chronic effect on benthic organisms. For some sites, metal levels were above the SEL of the MOEE guidelines (particularly for Cu, Fe and Mn) (Table 26.3). In these cases, the street sediment could have chronic or acutely toxic effects on benthic organisms. The values within this study are generally within the ranges observed from other street sediment studies summarized by Stone and Marsalek (1996).

Contaminant loading related to hydraulic sorting of street sediment

Bulk analysis of street sediments typically is used for the assessment of contaminant source area identification. Bulk samples are less useful for the assessment of contaminant loads to the sewer system of urban environments because of hydraulic sorting (Sutherland, 2003). Contaminants preferentially bind to smaller particles (Horowitz, 1991), which has significant implications for the loading of contaminants to the sewers and receiving water bodies and for the management of urban water systems.

Table 26.4 provides the per cent contribution of metals by size fraction based on reasonable estimates from field observations and the literature of flow (3 l/min), sediment concentration (30 mg/l), grain size proportions (0% (> 2000 μm), 5% (500–2000 μm), 10% (250–500 μm), 35% (63–250 μm) and 50% (< 63 μm)) and measured metal concentrations provided in Table 26.5. The calculations used in this approach make the assumption that there is no sorption or desorption of metals and that pH is constant. From Table 26.4, it is evident that the finer fraction contributes the greatest load of metals to the sewer system (> 45%). As such, any management strategies must target this fine street sediment, which is a major vector for contaminant transport to urban water systems.

Metal speciation and bioavailability in street sediment

Cadmium was by far the most bioavailable metal. The exchangeable Cd fraction was on

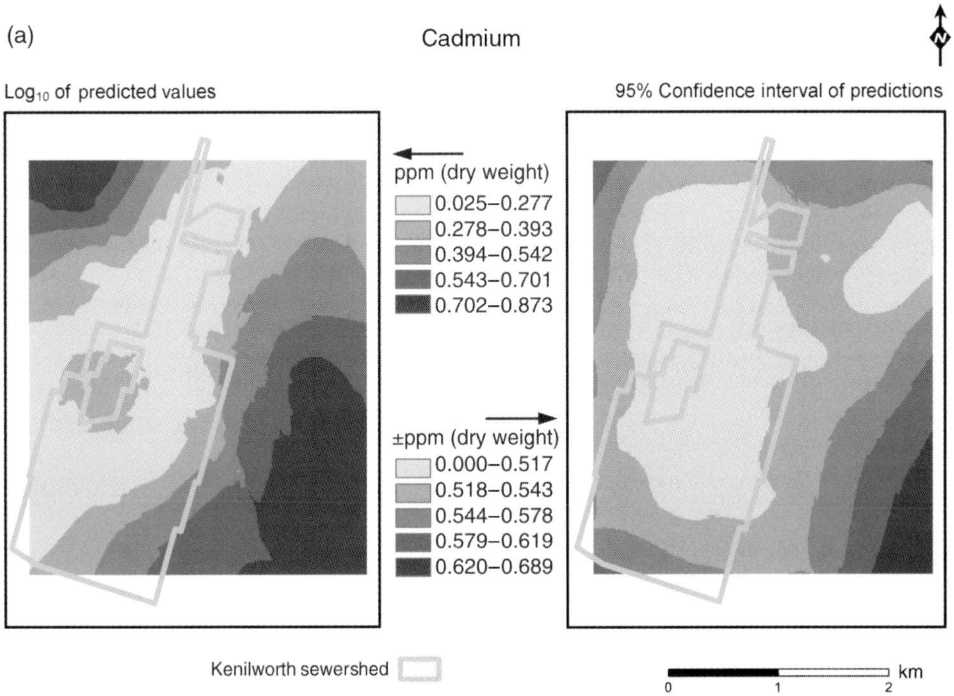

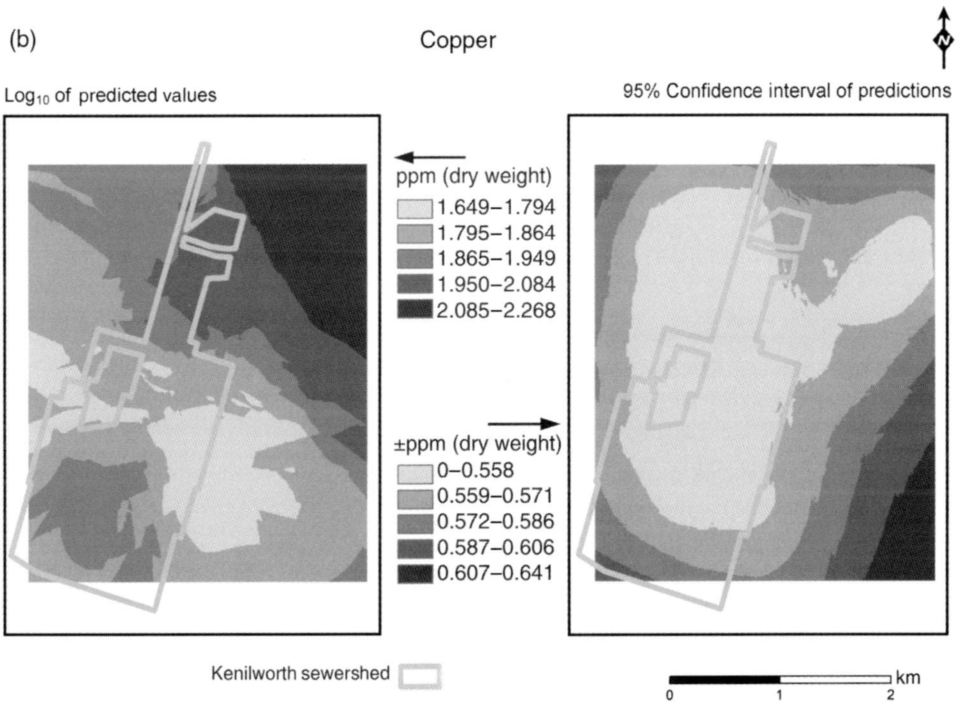

Fig. 26.3. Spatial distributions of metals and confidence intervals for the predicted values. (a) Cd, (b) Cu, (c) Fe, (d) Pb, (e) Mn and (f) Zn.

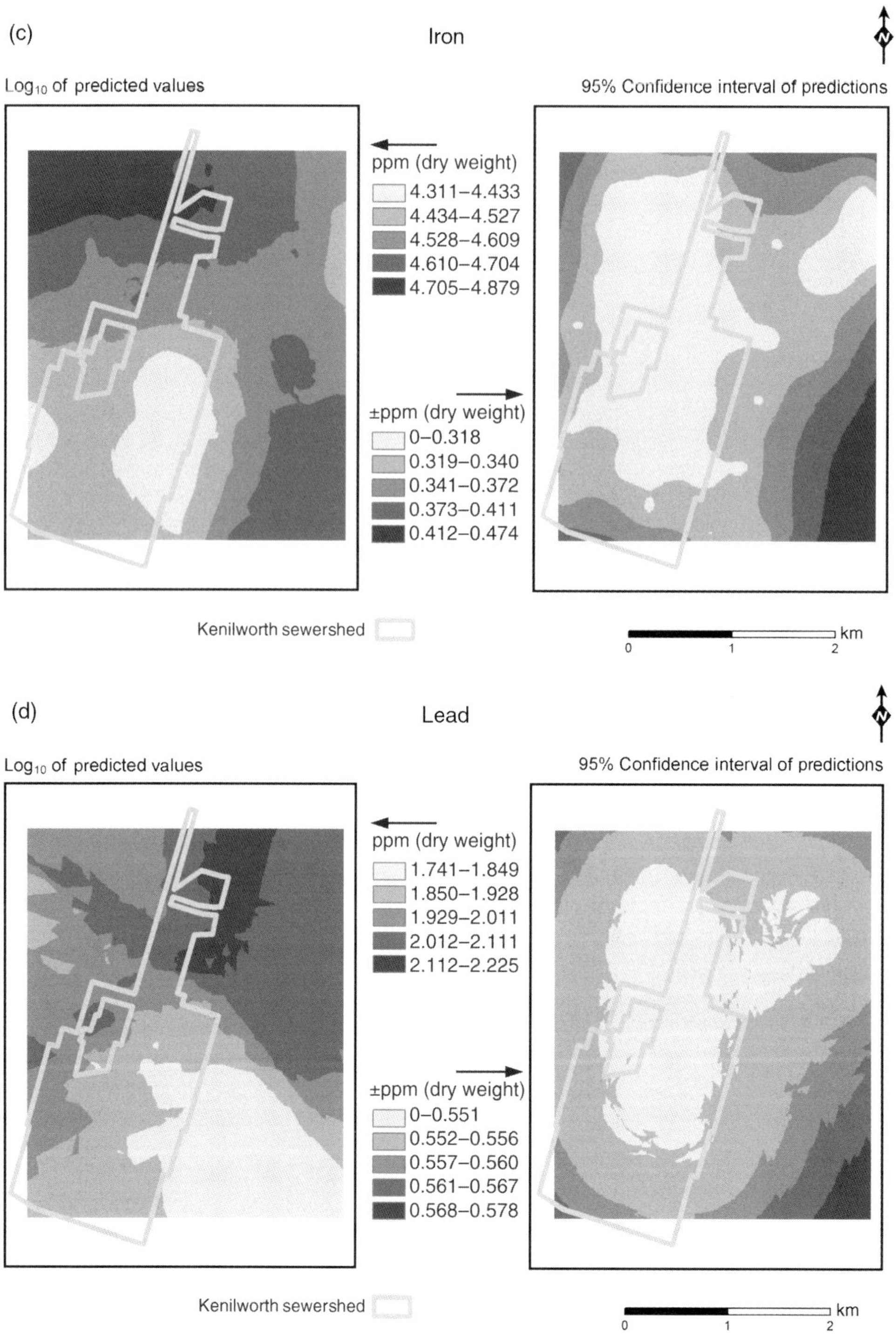

Fig. 26.3. *Continued.*

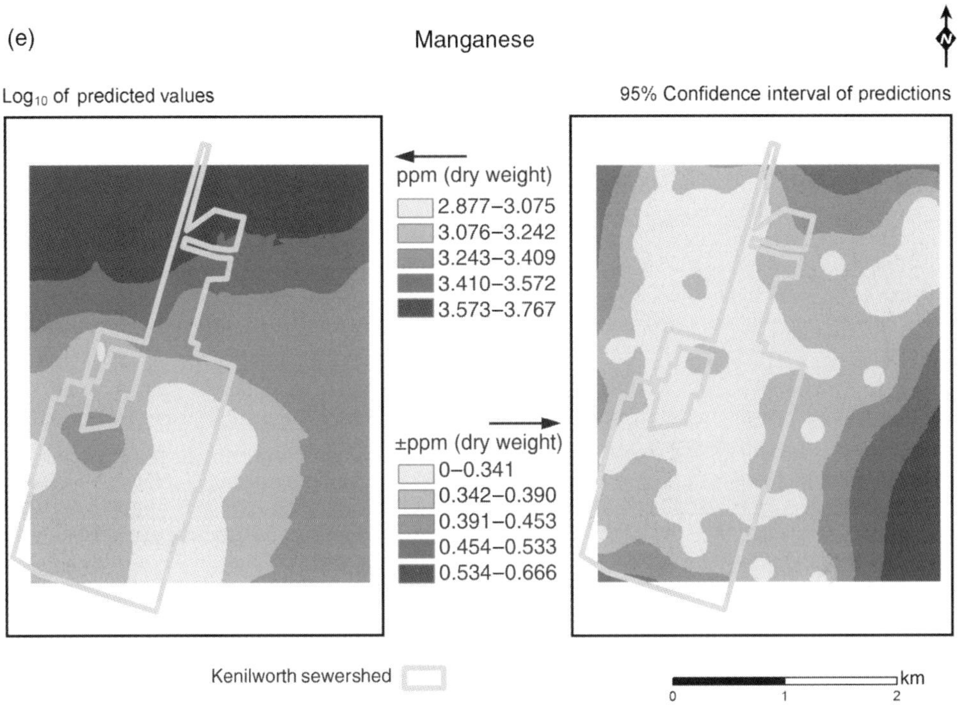

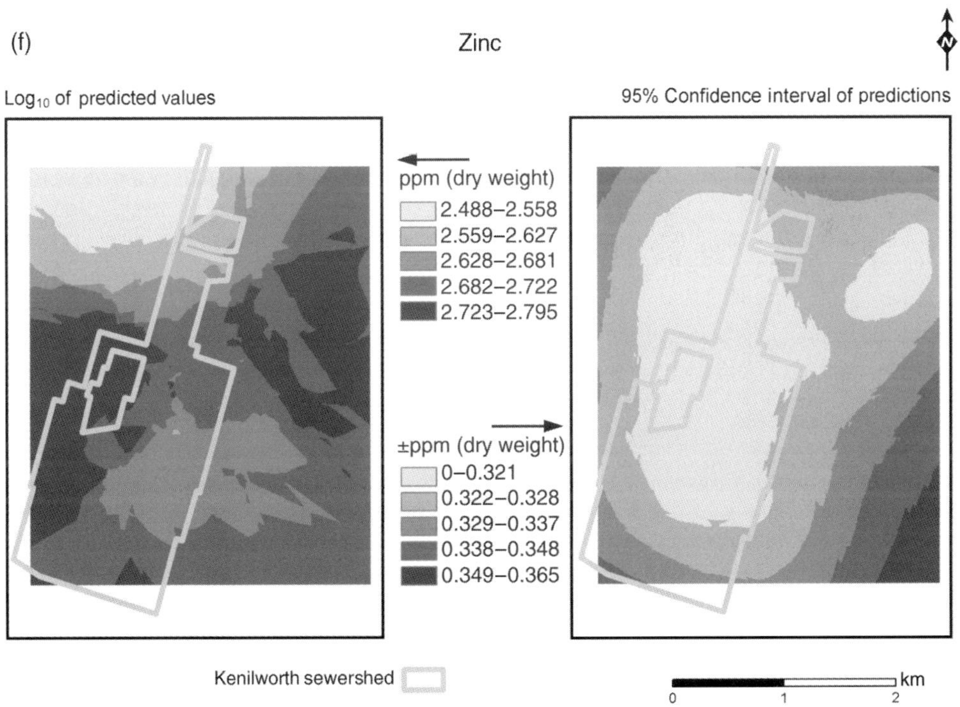

Fig. 26.3. *Continued.*

Table 26.2. Mean total trace metal concentrations, number of sampling sites within MOEE effects levels and per cent metals bound within fractions 1 to 5 by sector.

Trace metal and sector	Total concentration (μg/g)	No. sites > SEL	No. sites < SEL but > LEL	Fraction 1 (%)	Fraction 2 (%)	Fraction 3 (%)	Fraction 4 (%)	Fraction 5 (%)
Cd – Ind	6.06	2	14	7.09	10.91	12.59	22.53	46.89
	(10.69)			(7.64)	(11.12)	(9.73)	(12.72)	(23.85)
Cd – C/R	4.69	1	17	4.48	12.73	13.96	13.53	55.30
high traf.	(7.22)			(3.65)	(7.90)	(8.45)	(10.18)	(22.48)
Cd – C/R	1.93	0	18	6.87	17.99	16.37	11.27	47.49
low traf.	(1.53)			(4.95)	(10.78)	(8.94)	(7.05)	(22.23)
Cu – Ind	162	7	9	1.22	2.53	1.88	38.29	56.07
	(235)			(0.61)	(4.13)	(1.30)	(12.10)	(13.47)
Cu – C/R	119	7	11	1.65	6.88	2.32	48.39	40.75
high traf.	(44)			(0.66)	(4.66)	(0.94)	(15.08)	(16.80)
Cu – C/R	61	1	17	3.03	10.09	6.21	35.69	44.99
low traf.	(65)			(1.46)	(8.25)	(6.86)	(16.50)	(16.41)
Fe – Ind	54,444	12	4	0.02	0.75	8.39	3.36	87.48
	(20,306)			(0.01)	(0.44)	(2.81)	(1.67)	(3.29)
Fe – C/R	43,150	9	9	0.01	0.93	5.63	3.72	89.71
high traf.	(9,650)			(0.01)	(0.73)	(2.66)	(5.15)	(6.27)
Fe – C/R	23,483	2	16	0.02	0.52	6.59	3.34	89.52
low traf.	(8,687)			(0.01)	(0.50)	(3.89)	(1.47)	(4.18)
Mn – Ind	4,642	16	0	0.07	17.86	30.61	12.12	39.36
	(1,314)			(0.04)	(6.08)	(8.69)	(2.89)	(10.69)
Mn – C/R	2,375	18	0	0.13	30.54	24.58	8.56	36.20
high traf.	(703)			(0.05)	(5.00)	(5.13)	(2.90)	(5.55)
Mn – C/R	1,344	7	11	0.35	30.23	31.01	6.87	31.51
low traf.	(855)			(0.22)	(4.36)	(10.50)	(2.99)	(9.18)
Pb – Ind	219	2	14	1.69	7.05	22.01	18.51	50.75
	(344)			(1.12)	(6.06)	(9.99)	(9.18)	(11.44)
Pb – C/R	132	1	17	1.30	10.02	27.10	21.61	39.99
high traf.	(116)			(0.67)	(4.80)	(6.72)	(7.45)	(9.81)
Pb – C/R	73	0	18	2.55	13.67	43.45	18.14	22.19
low traf.	(40)			(1.32)	(5.30)	(16.23)	(6.90)	(14.53)
Zn – Ind	428	1	15	0.34	24.20	25.70	25.23	24.54
	(185)			(0.17)	(11.10)	(12.12)	(8.94)	(17.68)
Zn – C/R	583	2	16	0.50	38.98	13.44	17.01	30.06
high traf.	(177)			(0.27)	(11.99)	(3.89)	(7.28)	(16.06)
Zn – C/R	544	1	17	0.64	23.80	16.08	18.46	41.02
low traf.	(405)			(0.46)	(11.34)	(8.95)	(9.79)	(19.48)

Ind = industrial, C/R = commercial/residential with low and high traffic volumes. SEL = severe effect level, LEL = lowest effect level. Values in parentheses = standard deviation.
Guideline levels (in μg/g): Cd – LEL = 0.6, SEL = 10; Cu – LEL = 16, SEL = 110; Fe – LEL = 2%, SEL = 4%; Mn – LEL = 460, SEL = 1100; Pb – LEL = 31, SEL = 250; Zn – LEL = 120, SEL = 820.

Table 26.3. MOEE *Guidelines for the Protection and Management of Aquatic Sediment Quality* (Ontario Ministry of Environment and Energy, 1993).

Pollutant categories	Sediment quality	Potential impact
> SEL	Grossly polluted	Will significantly affect use of sediment by benthic organisms
> LEL	Marginally–significantly polluted	Will affect sediment use by some benthic organisms
> NEL	Clean–marginally polluted	Potential to affect some sensitive use
< NEL	Clean	No impact on water quality or benthic organisms anticipated

SEL = severe effect level; LEL = lowest effect level; and NEL = no effect level.

Table 26.4. Per cent contribution for metals as related to size fractions.

Size fraction (μm)	Cd load (%)	Cu load (%)	Fe load (%)	Mn load (%)	Pb load (%)	Zn load (%)
> 2000	0	0	0	0	0	0
500–2000	2	3	7	9	3	1
250–500	6	7	11	11	8	8
63–250	31	34	35	35	38	40
< 63	61	57	46	45	51	51

Table 26.5. Trace metal concentrations by size class for Site 1.

Size fraction (μm)	% Size contribution	Cd (μg/g)	Cu (μg/g)	Fe (μg/g)	Mn (μg/g)	Pb (μg/g)	Zn (μg/g)
> 2000	4	N/A	N/A	N/A	N/A	N/A	N/A
500–2000	21	0.33	40.2	15.0	16,331	30.0	143
250–500	20	0.56	43.9	11.8	10,277	45.1	378
63–250	42	0.68	55.1	9.1	8,000	51.9	480
< 63	13	1.07	74.4	9.5	8,422	54.7	488

average 6.1% of the total Cd, with a maximum of 25% measured in the industrial sector. Cadmium is considered more mobile than other metals, with binding generally through cation exchange and easily reducible phases (Salomons and Forstner, 1984). Researchers have found similar high bioavailability of Cd in stormwater runoff (Hamilton *et al.*, 1984; Ellis *et al.*, 1987; Morrison *et al.*, 1987; Stone and Marsalek, 1996). The binding of Cd varied between sectors within the sampling area. Within the commercial/residential low traffic volume areas (generally those close to the escarpment), Cd was bound mostly to carbonates (fraction 2), while in the industrial sector Cd is primarily bound to organics (fraction 4). The intermediate zone of commercial/residential high traffic volume area had similar percentages of Cd between fractions 2 and 4. The importance of organics to the binding of Cd in the industrial sector is likely related to the increased organic carbon content from the coal piles as compared to the increased carbonate content derived from the limestone- and dolomite-based Niagara Escarpment and from building materials containing high concentrations of

calcareous material in the more residential areas (Vermette *et al.*, 1987). With the high traffic commercial/residential areas generally between these two sectors, it is not surprising to find equal proportions among binding sites.

Copper is generally considered a signature element for industrial and vehicular impact on surface sediment quality (Irvine *et al.*, 1989; Andrews and Sutherland, 2004) and binds preferentially to organic coatings on sediment (Stone and Marsalek, 1996). The binding of Cu to the organic fraction is likely related to the presence of organic carbon from coal dust in the industrial sector and fuel combustion in the high traffic areas. Other sources of Cu may be from tyre wear, brake dust and general vehicle wear (Atkins and Hawley, 1978). Away from the industrial and high traffic volume sectors, there is no dominance of organic binding, with the Cu fractions 2–4 relatively proportional.

Iron was almost entirely bound within the non-bioavailable residual fraction which is likely reflective of the Fe being bound within the iron-ore crystalline structure. Manganese was preferentially bound to Fe/Mn oxides and in the commercial/residential sectors also to carbonates. Lead was preferentially bound to Fe/Mn oxides but, surprisingly, the greatest binding in this fraction occurred in the low traffic commercial/residential sector far away from the steel industry. Similar to Cd, Pb showed greater binding with carbonates closer to the escarpment.

In contrast to Cd, the remaining metals had less than 2% of their total concentration in the highly bioavailability exchangeable fraction. If, as in Droppo *et al.* (1998), it is assumed that the sum of fractions 1–4 represent the potential bioavailability of metals, then each metal has a strong potential to impact organisms through bioaccumulation (generally, with the exception of Fe, greater than 50% of metals are bound in fractions 1–4). Such an estimate is, however, highly subjective and non-transferable between environments. This is because the partitioning of metals between fractions, and therefore potential bioavailability, will vary during transport and retention on the surface, in the sewer systems and in the receiving water body due to changes in pH, redox potential, ionic strength, suspended solid concentration, microbial activity and DOC concentrations (Ellis, 1986; Ellis *et al.*, 1987; Morrison *et al.*, 1987; Krantzberg, 1994; Stone and Marsalek, 1996; Brassard *et al.*, 1997; Warren and Haack, 2001). It is likely that the anoxic conditions of much of the bed sediments in Hamilton Harbour will increase the bioavailability of the street sediment should it be deposited within the Harbour via a CSO. When taking into account only the potentially bioavailable fractions (fractions 1–4), the ecological impact of the sediments as related to the MOEE (Ontario Ministry of Environment and Energy, 1993) guidelines for sediments is reduced. For example, Table 26.6 demonstrates the reduction of the mean Cu concentration below the SEL when presented as a bioavailable fraction. A similar result also is observed for Fe, which exhibits substantial reductions in levels when presented as bioavailable concentrations. These examples are further significant if one realizes that the sequentially extracted fractions represent

Table 26.6. Mean total metals concentrations and maximum bioavailable concentrations compared to MOEE guidelines (mean potential bioavailablity = sum of fractions 1–4).

Metal	LEL (μg/g)	SEL (μg/g)	Mean total concentration (μg/g)	Mean potential bioavailability (μg/g)
Cd	0.6	10	4.2	2
Cu	16	110	112	60
Fe	2,100	4,380	39,818	4,378
Mn	460	1,100	2,716	1,752
Pb	31	250	139	87
Zn	120	820	522.5	406

SEL = severe effect level; LEL = lowest effect level.

a gradient of bioavailability and, as such, the true bioavailable trace metals concentrations are less than those provided in Table 26.6 (Tessier *et al.*, 1979; Krantzberg, 1994).

Mitigation and remediation strategies

Best management practices are required for removal of street sediments prior to their delivery to sewer systems and receiving water bodies. Traditional street-sweeping measures are generally ineffective, with only 15% of the fine particles (less than 40 μm) removed (Pitt and Clark, 2003). Studies have shown that such poor efficiency in the removal of the chemically active <63 μm fraction can be equated to only a 5–10% reduction in storm runoff pollutant loadings (for sweeping frequencies of 2 days per week). Increasing sweeping frequencies to more than twice a week does not appreciably reduce solids loading any further (Field and Sullivan, 2003a). Studies have also shown that even new street sweeping/vacuuming technologies perform poorly in the removal of street sediments as they are dependent on dirt loading rates, street texture, litter, moisture, parked car conditions and locations, and equipment operating conditions (Pitt and Clark, 2003). Studies by Sutherland and Jelen (1997) and Sutherland *et al.* (1998) with new generation high-efficiency sweepers, however, suggest that bimonthly to weekly sweepings could reduce total suspended solids loads by 40–80%. This reduction in total suspended solids concentration concurrently was associated with a reduction of total metal (Pb, Zn, Cu) loads by 20–60%. While the effectiveness of street sweeping in the reduction of pollutant loads is in question, it should be used as part of a suite of best management practices (BMPs) (Field and Sullivan, 2003a). Other BMPs which may be used in conjunction with street sweeping for source area control of pollutants are described in Field and Sullivan (2003b) (e.g. grassed swale drains, filter strips, stormwater wetlands, dry and wet detention ponds, infiltration trenches, infiltration basins and porous pavements). Urbonas (2000) evaluated 16 structural BMPs for 13 different performance characteristics that included water quality improvement, as well as hydrologic and hydraulic characteristics. The results showed that minimizing the impervious area directly connected to the sewer, detention basins, retention ponds, wetland basins and porous pavements are amongst the most effective structural BMPs. The minimization of impervious area directly connected to the sewer may be as simple as initiation of downspout disconnection programmes to reduce overland runoff and peak flows in the sewer system (e.g. Totten Sims Hubicki Associates *et al.*, 2001).

Of the above BMPs, the grassed swale often is used and has been demonstrated by Wigington *et al.* (1986) to be an effective BMP for the removal of contaminated street sediment. While there may be concern of long-term accumulation of trace metals within swales, Wigington *et al.* (1986) observed that although trace metals do accumulate within grassed swales over time, they typically exhibit a low degree of leachability, with negligible accumulation of trace metals below 5 cm depth in the swale soil. Proper design of roadside swales is important to ensure that sediments are not removed from roadside surfaces or contribute to sediment loads within washoff during rainfall events. For example, depressed, grass-lined roadside swales assist in runoff accumulation and trace metal storage (Wigington *et al.* 1986), whereas gravel roadside shoulders and elevated grassed areas may act as a source of trace metals within road surface runoff (Sutherland and Tolosa, 2000).

The highest total metals loads in CSO sampling done for this study and in Buffalo, New York generally have been associated with the 'first flush' (Irvine *et al.*, 1998, 2005). However, not all studies concur with the universality of the first flush phenomenon (Saget *et al.*, 1996; Hager, 2001). Hall and Anderson (1988) observed a first flush phenomenon for insoluble metals in stormwater runoff, but a secondary peak of soluble metals that had greater toxicity (as measured by a *Daphnia* 96-h test) also was observed. Hager (2001) noted CSO control in the USA often is designed around the capture of first flush, although there also appears to be some discussion that this is not sufficient. Stormwater detention ponds can be an effective solution for the removal of particulate-bound contaminants through deposition of suspended solids, prior to runoff discharge to aquatic environments. Often, however, due to space restrictions in urban environments, surface detention ponds are not possible and so underground, in-line detention tanks are used to collect the sewer flow and remove solids. However, Van Loon *et al.* (2000) indicate that the design of surface detention facilities needs further consideration due to their potential negative impacts on species that use these detention ponds as habitat. Designs that maximize contaminant removal yet minimize contaminant exposure to organisms are required (Van Loon *et al.*, 2000). This may include the use of sand or biofilm filters, although problems of hydraulic conductivity loss due to sediment accumulation within filter interstices may occur (Lau *et al.*, 2000).

Conclusion

To effectively manage water quality issues within urban environments and urban-impacted receiving water bodies, there must be a comprehensive understanding of the characteristics and processes which control the contaminant make up of such an environment. Within urban environments, roadways represent areas of sediment accumulation (deposition) derived from overland flow from adjacent land uses where sediment is eroded from source area material or from short or long range atmospheric deposition. As such, with storm events, roadways represent a significant source of sediment and contaminants to sewer systems with subsequent potential delivery to receiving water bodies via storm or combined sewer overflows. This chapter has demonstrated the following significant conclusions relative to improving urban water management strategies for the Kenilworth sewershed.

1. Street sediment is generally characterized as coarse, but still possessed a proportion of finer sediments. The industrial sector of the Kenilworth sewershed had a higher proportion of fines than the residential/commercial sector.
2. Hydraulic sorting influences metal loading and mobility by preferentially transporting the chemically active fraction ($< 63\ \mu m$) to the sewer systems.
3. All metal concentrations, if delivered to the bed sediments of Hamilton Harbour via a CSO, were found to potentially have a chronic effect on benthic organisms based on MOEE guidelines. Some sites possessed concentrations high enough in Cu, Fe and Mn to have an acute effect.
4. When comparing metal concentrations to standards, it is important to compare the bioavailable fraction, as this will be more representative of the true impact or risk for the environment.
5. GIS analysis illustrated the significant spatial variation in metal concentrations over the sewershed suggesting there is merit for 'sector-based' management strategies.
6. Management options for the control of urban street sediment migration to receiving water bodies are varied and dependent on site-specific and operational considerations. Regardless, street sediment control needs to be a focus of any urban water management strategy, as it can represent a major source of contamination.

References

Andrews, S. and Sutherland, R.A. (2004) Cu, Pb and Zn contamination in Nuuanu watershed, Oahu, Hawaii. *The Science of the Total Environment* 324, 173–182.

Atkins, E.D. and Hawley, J.R. (1978) *Sources of Metals and Metal Levels in Municipal Wastewaters*. Canada-Ontario Agreement on Great Lakes Water Quality, Research Report No. 80. Environment Canada and Ontario Ministry of the Environment, Minister of Supply and Services, Ottawa, Canada.

Brassard, P., Kramer, J.R. and Collins, P.V. (1997) Dissolved metal concentrations and suspended sediment in Hamilton Harbour. *Journal of Great Lakes Research* 23, 86–96.

Butler, D., Thedchanamoorthy, S. and Payne, J.A. (1992) Aspects of surface sediment characteristics on an urban catchment in London. *Water Science and Technology* 25, 13–19.

Chebbo, G. and Bachoc, A. (1992) Characterization of suspended solids in urban wet weather discharges. *Water Science and Technology* 25, 171–179.

Chebbo, G., Musquerre, P. and Bachoc, A. (1990) Solids transferred into sewers. In: Iwasa, Y. and Sueishi, T. (eds) *Proceedings of the 5th International Conference on Urban Storm Drainage.* Asahi Printing Co., Chuo-ku, Osaka, Japan, pp. 885–890.

Droppo, I.G., Leppard, G.G., Flannigan, D.T. and Liss, S.N. (1997) The freshwater floc: a functional relationship of water and organic and inorganic floc constituents affecting suspended sediment properties. *Water, Air and Soil Pollution* 99, 43–53.

Droppo, I.G., Irvine, K.N., Murphy, T.P. and Jaskot, C. (1998) Fractionated metals in street dust of a mixed land use sewershed, Hamilton, Ontario. In: Wheater, H. and Kirby, C. (eds) *Hydrology in a Changing Environment,* Volume 3. Wiley, Chichester, UK, pp. 383–394.

Droppo, I.G., Irvine, K.N. and Jaskot, C. (2002) Flocculation of cohesive sediments in the urban continuum: implications for stormwater management. *Environmental Technology* 23, 27–41.

Ellis, J.B. (1986) Pollutional aspects of urban runoff. In: *Urban Runoff Pollution.* NATO ASI series, vol. G10. Springer, Berlin, Germany, pp. 1–38.

Ellis, J.B., Revitt, D.M., Harrop, D.O. and Beckwith, P.R. (1987) The contribution of highway surfaces to urban stormwater sediments and metal loadings. *The Science of the Total Environment* 59, 339–349.

Field, R. and Sullivan, D. (2003a) Management of wet weather flow in the urban watershed. In: Field, R. and Sullivan, D. (eds) *Wet-weather Flow in the Urban Watershed: Technology and Management.* Lewis Publishers, Boca Raton, Florida, pp. 1–41.

Field, R. and Sullivan, D. (eds) (2003b) *Wet-weather Flow in the Urban Watershed: Technology and Management.* Lewis Publishers, Boca Raton, Florida.

Gibbs, R.J. and Konwar, L.N. (1982) Effect of pipetting on mineral flocs. *Environmental Science and Technology* 16, 119–121.

Hager, M.C. (2001) Evaluating first-flush. *Storm Water* 2, 1–12.

Hall, K.J. and Anderson, B.C. (1988) The toxicity and chemical composition of urban stormwater runoff. *Canadian Journal of Civil Engineering* 15, 98–106.

Hamilton, R.S., Revitt, D.M. and Warren, R.S. (1984) Levels and physico-chemical associations of Cd, Cu, Pb and Zn in road sediments. *The Science of the Total Environment* 33, 59–74.

Horowitz, A.J. (1991) *A Primer in Sediment-Trace Element Chemistry*, 2nd edn. Lewis, Chelsea, Michigan.

Irvine, K.N., Murray, S.D. and Drake, J.J. (1989) Spatial and temporal variability of dry dustfall and associated trace elements: Hamilton, Canada. *Environmental Technology Letters* 10, 527–540.

Irvine, K.N., Droppo, I.G., Murphy, T.P. and Stirrup, D.M. (1998) Annual loading estimates of selected metals and PAHs in CSOs, Hamilton, Ontario, using a continuous PCSWMM approach. In: James, W. (ed.) *Advances in Modeling the Management of Stormwater Impacts – Volume 6.* Ann Arbor Press, Ann Arbor, Michigan, pp. 383–398.

Irvine, K.N., Caruso, J. and McCorkhill, G. (2005) Consideration of metal levels in identifying CSO abatement options. *Urban Water Journal* 2, 193–200.

Klemetson, S.L. (1985) Factors affecting stream transport of combined sewer overflow sediments. *Water Pollution Control Federation Journal* 57, 390–397.

Krantzberg, G. (1994) Spatial and temporal variability in metal bioavailability and toxicity of sediment from Hamilton Harbour, Lake Ontario. *Environmental Toxicology and Chemistry* 13, 1685–1698.

Lau, L.Y., Marsalek, J. and Rochfort, Q. (2000) Use of a biofilter for treatment of heavy metals in highway runoff. *Water Quality Research Journal of Canada* 13, 1685–1698.

Mark, O., Cerar, U. and Perrusquia, G. (1996) Prediction of locations with sediment deposits in sewers. *Water Science and Technology* 33, 147–154.

Morrison, G.M., Revitt, D.M. and Ellis, J.B. (1987) Heavy metal exceedance of water quality standards during storm events. In: Gujer, W. and Krejci, V. (eds) *Topics in Urban Storm Water Quality, Planning and Management.* International Association for Hydraulic Research, Lausanne, Switzerland, pp. 91–96.

NWRI (1990) *National Water Research Institute Certificate of Analysis: Certified Reference Material WQB-1; Trace Metals in Sediment.* Aquatic Quality Assurance Program, Research and Applications Branch, National Water Research Institute, Burlington, Ontario.

Ontario Ministry of Environment and Energy (MOEE) (1993) *Guidelines for the Protection and Management of Aquatic Sediment Quality in Ontario.* Ontario Ministry of Environment and Energy, Ottawa.

Paul Theil Associates and Beak Consultants (1991) *Regional Municipality of Hamilton-Wentworth Pollution Control Plan.* Paul Theil Associates and Beak Consultants, Hamilton, Ontario.

Pitt, R. and Clark, S. (2003) Emerging stormwater controls for critical source areas. In: Sullivan, D. and Field, R. (eds) *Management of Wet-weather Flow in the Watershed: Technology and Management.* CRC Press, Boca Raton, Florida, pp. 103–139.
Roberts, A.H., Ellis, J.B. and Whalley, W.B. (1988) The size and surface texture of sediment in an urban catchment. *The Science of the Total Environment* 72, 11–27.
Saget, A., Chebbo, G. and Betrand-Krajewski, J.-L. (1996) The first flush in sewer systems. *Water Science and Technology* 33, 101–108.
Salomons, W. and Forstner, U. (1984) *Metals in the Hydrocycle.* Springer, New York.
Sartor, J.D. and Boyd, G.B. (1972) *Water Pollution Aspects of Street Surface Contaminants.* Report EPA-R2- 72- 08. USEPA, Washington, DC.
Stone, M. and Droppo, I.G. (1996) Distribution of lead, copper and zinc in size-fractionated river bed sediment in two agricultural catchments of southern Ontario, Canada. *Environmental Pollution* 3, 353–362.
Stone, M. and Marsalek, J. (1996) Trace metal composition and speciation in street sediments: Sault Ste Marie, Canada. *Water, Air and Soil Pollution* 87, 149–169.
Sutherland, R.A. (2003) Lead in grain size fractions of road-deposited sediment. *Environmental Pollution* 121, 229–237.
Sutherland, R.A. and Tolosa, C.A. (2000) Variation in total and extractable elements with distance from roads in an urban watershed, Honolulu, Hawaii. *Water, Air and Soil Pollution* 127, 315–338.
Sutherland, R.C. and Jelen, S.L. (1997) Contrary to conventional wisdom, street sweeping can be an effective BMP. In: James W. (ed.) *Advances in Modeling the Management of Stormwater Impacts – Volume 5.* Computational Hydraulics International, Guelph, Ontario.
Sutherland, R.C., Jelen, S.L. and Minton, G. (1998) High efficiency sweeping as an alternative to the use of wet vaults for stormwater treatment. In: James W. (ed.) *Advances in Modeling the Management of Stormwater Impact* – Volume 6. Computational Hydraulics International, Guelph, Ontario.
Tessier, A., Campbell, P.G.C. and Bisson, M. (1979) Sequential extraction procedure for the speciation of particulate trace metals. *Analytical Chemistry* 51, 844–850.
Totten Sims Hubicki Associates, Donald G. Weatherbe Associates, Elizabeth Leedham (2001) *Storm Water Pollution Prevention Handbook.* Report for the Ontario Ministry of Energy and Environment, Ottawa, Canada.
Urbonas, B. (2000) Assessment of stormwater best management practice effectiveness. In: Field, R., Heaney, J.P. and Pitt, R. (eds) *Innovative Urban Wet-weather Flow Management Systems.* Technomic Publishing Co., Lancaster, Pennsylvania.
Van Loon, G., Anderson, B.C., Watt, W. and Marsalek, J. (2000) Characterizing stormwater sediments for ecotoxic risk. *Water Quality Research Journal of Canada* 35, 341–364.
Verbanck, M.A., Vanderborght, J.P. and Wollast, P. (1990) Size distributions of suspended particles in combined sewers during dry- and wet-weather. In: Iwasa, Y. and Sueishi, T. (eds) *Proceedings of the 5th International Conference on Urban Storm Drainage.* Asahi Printing Co., Chuo-ku, Osaka, Japan, pp. 891–896.
Vermette, S.J., Irvine, K.N. and Drake, J.J. (1987) Elemental and size distribution characteristics of urban sediments: Hamilton, Canada. *Environmental Technology Letters* 8, 619–634.
Warren, L.A. and Haack, E.A. (2001) Biogeochemical controls on metal behaviour in freshwater environments. *Earth-Science Review* 54, 261–320.
Wigington, P.J. Jr, Randall, C.W. and Grizard, T.J. (1986) Accumulation of selected trace metals in soil of urban runoff swale drains. *Water Resources Bulletin* 22, 73–79.

27 Managing Sediment in the Landscape: Current Practices and Future Vision

R.P.C. Morgan
National Soil Resources Institute, UK

Introduction

The techniques for managing land to control erosion are well known and fall broadly into three groups: structural or engineering methods, of which terracing is the most visible example in the landscape; vegetative or agronomic methods, which involve manipulation of the land cover to provide soil protection; and soil management, concerned particularly with the tillage practices used to cultivate the soil and control weeds. Although these methods date back to the times of early civilization, they have been improved through research and their uptake has been promoted through various conservation or advisory services. The last century saw the establishment of soil conservation services in many countries, their work supported by research and development activities in agricultural research services and universities. Soil conservation was promoted at the level of the individual farmer, with the emphasis on retaining soil *in situ* in the field.

Since the 1950s, the importance of the transfer of sediment from land to watercourses has been increasingly recognized as a result of associated adverse environmental impacts, particularly water pollution and eutrophication. Between 45 and 90% of the phosphorus delivered to watercourses annually is associated with sediment (Sibbesen *et al*., 1994; Sibbesen and Sharpley, 1997). Sediment also contributes to the muddy floods that result in damage to property, sedimentation on roads and substantial clean-up costs to individual householders, local authorities and insurance companies (Boardman *et al*., 1994). Unwanted sedimentation downstream causes the silting of reservoirs, canals and harbours. With greater concern on the off-site effects of erosion, the emphasis has switched from single farmers to the management of the watershed with the need to address not only the source areas of sediment but also the pathways along which the sediment is moved over the landscape to water bodies.

With the focus on off-site damage and pollution, erosion control also became important in non-agricultural areas such as construction sites, highway slopes and pipeline corridors. Erosion is now appreciated as a problem in many recreational areas that come under pressure from an ever-increasing urban population. Urban parks, country parks and national parks within easy reach of urban areas suffer particularly from footpath erosion and damage of habitat. In response to this, an erosion-control industry has grown up, manufacturing and selling products like soil stabilizers, geotextiles, gabions and various types of interlocking concrete cellular blocks to stabilize slopes and control sediment movement. Under increasing environmental pressure for sustainable solutions and avoidance of 'hard' engineering, a range of approaches to erosion control has emerged. These may be briefly summarized as natural adjustment, allowing erosion to stabilize

 Soil Erosion and Sediment Redistribution in River Catchments (eds P.N. Owens and A.J. Collins)

through natural processes over time; bioengineering, making use of the engineering properties of vegetation; biotechnical engineering, using bioengineering to enhance the engineering function of engineering structures; and structural engineering. Much of this work is based on techniques developed in Austria, Switzerland and southern Germany for the control of erosion on road banks (Schiechtl and Stern, 1996).

Against this background, this paper focuses on present-day sediment issues, discusses the problems of current management practices and proposes a way forward for improving the implementation of control measures.

Sediment Issues

The main sources of sediment in agricultural areas are where erosive rains fall on erodible soils with less than 30% ground cover, conditions which can arise from row-crop cultivation, bare-soil fallows, unprotected land beneath tree crops, and overgrazing. In non-agricultural areas, the main sediment sources are simply unprotected land, particularly construction sites, whether for housing, highway development or pipeline installation. The environmental damage that can occur on these sites inhibits the restoration of land once construction work has been completed. In addition, it is now recognized that the breakdown of soil by erosion releases carbon to the atmosphere; it is estimated that, globally, soil erosion increases the atmospheric carbon content by 1.1 Pg/year (Lal, 2002). Although there is considerable uncertainty surrounding this estimate because of the methods used to estimate both erosion rates and the global area of land affected by accelerated erosion, the figure is sufficiently high to indicate that controlling erosion could make an important contribution to carbon sequestration and therefore help to mitigate some of the effects of any global warming.

The drivers for erosion control and sediment management are varied. They range from ensuring global and national food security to support for the welfare of individual farmers and their families. They also include flood control, maintenance of acceptable water quality, protection of the land from environmental damage and restoration of disturbed land. Most recently, there is interest in the potential for carbon sequestration. The biggest driver, however, is the need to reduce the costs that arise from erosion. Failure to control erosion on cut slopes and embankments along highways can lead to sediment washing out on to roads, causing dangerous driving conditions or leading to closure of the road until clean-up is effected. Many local authorities spend large sums of money on sediment removal to keep roads open. Erosion is estimated to cost the UK economy £90 million per annum (Environment Agency, 2002) of which 95% is associated with agriculture. This figure, however, is likely to be an underestimate, since it does not take account of flooding or the additional costs of water purification arising from sedimentation and pollution.

Responsibilities

National governments clearly have a responsibility to ensure sufficiency of food for a country's population and to protect the environment. Through their involvement in international organizations, they also have a commitment to global food security. Traditionally, governments have fulfilled this role by setting up conservation or advisory services to promote erosion control on farms. Service staff surveyed the farm and its erosion risk and then proposed control measures. For many years, terrace and waterway systems were considered an important component of any control because they were fundamental to the approaches promoted in the United States and many conservation services world-wide were modelled on the American system. By the 1970s it was clear that in many areas this 'top-down' approach was not working (Hudson, 1992). From the 1980s, erosion control was no longer seen as an objective in its own right. It was viewed, instead, as a component of overall land husbandry (Shaxson *et al.*, 1988). Emphasis was given to involving farmers in the design and implementation of conservation measures and ensuring that these were compatible with the farmer's objectives. Control measures became strongly agronomically and soil-management based with greater reliance on using traditional techniques. Unfortunately these changes did not necessarily improve the sustained up-take of erosion control by farmers. Adoption took place only where there was a perceived 'economic' benefit to the farmer or incentives were provided by government. Studies in the Kericho area of Kenya showed that where farmers recognized sufficient benefit, they would adopt soil conservation measures themselves without artificial incentives (Tiffen *et al*., 1994).

At the same time, emphasis switched from consideration of single farmers to catchment management. Sediment control thus required collaboration between land users and the establishment of bodies with responsibility for catchments, such as water companies, national park authorities and local government. As a result, the responsibility for erosion control became diffused among many stakeholders including individual farmers and land owners; soil conservation services and other agricultural advisory services; local, regional and national governments; transnational organizations such as the European Union and the Food and Agricultural Organisation of the United Nations; farmer associations; and NGOs. Many soil conservation projects can involve many NGOs and several government departments, particularly those dealing with agriculture, forestry, environment and human welfare. Obtaining collaboration between these groups is often a major challenge, particularly where conflicts of interest arise.

An unresolved issue is who should pay for erosion damage. Should it be the 'polluter', even though the polluter may get no economic benefit from implementing erosion control? Or should it be the people downstream who would gain the benefit, or the community as a whole, using taxes and redistributing the money as a financial incentive to the farmer? The 'polluter pays' principle is clearly unworkable where the 'polluter' is so poor that there is no way in which any payment can be made. An alternative is for national governments to provide incentives to farmers to adopt environmentally-friendly practices. Policies in Iceland to support farmers to engage in environmental projects provided they reduce stocking rates have been a major factor in reducing overall sheep numbers by 50%, with resultant benefit to the quality of the Icelandic rangeland (Arnalds, 1999). Compliance-based financial incentives have been a feature of erosion control programmes in the USA since the 1985 Farm Bill, but research clearly demonstrates that they do not provide a long-term sustainable solution. Once the incentives are withdrawn, farmers return to highly erosive but often, for them, economically productive systems (Napier, 1999). Sustainable soil conservation must rely on methods which are inherently economic for the farmer or land user, which means that the market prices paid by consumers for food must reflect the true costs of production and environmental protection.

In non-agricultural areas, land owners and land users, like highways agencies and construction companies, are more willing to pay the costs of erosion control both to avoid liability for environmental damage and to promote their 'green' credentials. Erosion rates from construction sites can frequently exceed 100 t/ha from individual storms, and there is a need to either control the sediment at source or prevent its passage downslope to areas where it is not wanted, such as settlements, water bodies and ecologically sensitive sites. In many situations, the need for erosion control is temporary. It is not an overall objective but forms a part of the overall programme of land restoration. Erosion can impair restoration work by washing-out seeds and plants and, in some cases, removing the top soil completely. Land restoration and erosion control are seen as complementary.

Learning from Agriculture

At present there is very little contact between organizations with responsibility for erosion control on farmland and those concerned with non-agricultural areas. The erosion-control industry has not found a market for its products with farmers, partly because many of its products provide an all-year control over erosion which would prohibit the use of the land for agriculture. Farmers require erosion control for short periods of time, usually between ploughing and crop establishment and then after harvest. Products such as soil conditioners or soil stabilizers which might be useful in both situations are generally too expensive for farmers to use. Even where similar measures are employed, great differences in practice exist between agricultural and other engineers. An example is found in the pipeline industry where the diverter berms used as a standard method of erosion control along rights-of-way are analogous to the diversion terraces used on arable land. Their purpose is to intercept surface runoff and convey it at a safe velocity to a suitable place of disposal. Pipeline engineers and agricultural engineers use different formulae for calculating spacing of the diverters. Many of those developed by pipeline engineers take account only of slope and soil, ignoring differences in climate. As a result, in areas of intense erosive rainfall, they can lead to spacings too large to control erosion adequately. Further, pipeline engineers tend to construct the channels behind the diverters at gradients of 10%, whereas agricultural engineers use channel grades of 1 : 250 to 1 : 500. It is not surprising that diverter berms often fail to prevent erosion when their channel slopes are sufficient to induce runoff velocities capable of cutting gullies (Morgan and Hann, 2003).

Despite these differences, the principles of erosion control are similar for both agricultural and non-agricultural areas. Construction and civil engineers can learn a lot from agriculture by applying some basic principles (Morgan and Rickson, 1996; Northcutt, 1998), namely:

- minimize soil disturbance – stabilized and protected soil will not erode unless it is disturbed; the soil also contains a seedbank which can provide a foundation for revegetation;
- keep the soil covered – establishment of ground cover is the most important tool in controlling erosion;
- improve or maintain soil quality;
- erosion models can be used to estimate sediment loss and deposition and as design tools to evaluate the effectiveness of different erosion-control scenarios;
- buffer strips can protect a slope from sediment movement and prevent the discharge of sediment into water bodies and can be more cost-effective than concrete dams in river channels.

Research into erosion control in agriculture and forestry has also indicated that a sound understanding of the role of vegetation is needed to use it effectively in erosion control. A uniform ground cover of > 70% should be the aim. Vegetation where the canopy is 2 m or more above the ground or where plants grow in clumps instead of giving a uniform cover can result in higher erosion rates than those from bare ground. The vegetative architecture of buffer strips needs to be carefully considered, since clumpy grasses and 'gappy' bushes will not be effective. If the standards of sediment management are to improve, erosion scientists and practitioners in agriculture and engineering will need to come together and share their methods and experiences.

Community-based Erosion Control

A major operational weakness in present sedimentation management is the lack of a framework within which the large number of stakeholders can work together, share their interests and resolve conflicts. Generally erosion affects the person or persons on whose land it occurs and the various stakeholders downslope or downstream who suffer from environmental damage. Although there are long-term national and international issues in relation to food production and delivery of sediment to the oceans, the majority of the stakeholders are associated with the local community where the erosion takes place. These are the people who will benefit directly from erosion control. They are therefore in the best position to decide how much they are prepared to pay either for rehabilitation of existing damage or prevention of future damage.

The recent approaches to erosion control in agricultural areas based on participatory techniques involving farmers have been more successful than the previously tried 'top-down' approaches. In Australia, these approaches have developed into the Land Care movement, which brings together farmers, other members of the local community, extension services, and the State and Federal Government to address local and regional problems of land management. There are now some 4000 Land Care or similar groups throughout Australia working as essentially 'self-starting, autonomous grassroots bodies' in which the members share aspirations, skills and experiences, seek advice from experts and identify sources of funding from government and business (Roberts, 1992). Land Care groups have the following characteristics (Campbell, 1994; Marston, 1996):

- they tackle a broad range of issues, providing an integrated approach to resource management;
- they are based on neighbourhoods, usually covering catchments with contiguous boundaries, rather than just groups of farmers with a common interest;
- their impetus comes from the community, providing 'ownership' of any programmes for erosion or sediment control and improving coordination, collaboration and communication between various stakeholders;
- they can produce and implement proposals which are economically realistic within the funding support mechanisms available.

In the USA, the Soil Conservation Districts have gradually evolved to be inclusive of local business, industry, recreation and community interest, instead of being strongly farmer-oriented, and are therefore able to perform a similar role to the Land Care groups. They now have a broad-based agenda in natural resources rather than a focus on soil conservation.

The Future

Land Care groups are ideally placed to overcome many of the deficiencies in previous and much current practice. They can bring together not only the stakeholders with respect to erosion damage and benefits of erosion control, but can also provide the forum for liaison between the community, experts in erosion assessment and control from a range of disciplines, and the erosion-control industry. By working at a catchment level, they can contribute to watershed management whilst recognizing that this is best achieved by working with the individual farmers and land users within the catchment.

Specific tasks which Land Care groups can perform with respect to erosion and sediment control include:

- provision of an institutional framework;
- undertaking an erosion audit to establish the baseline condition;
- deciding appropriate performance criteria for acceptable erosion in different parts of the region, reflecting the priorities of different stakeholders ranging from long-term protection of the soil resource to the prevention of erosion damage, flooding and loss of water quality downstream;
- proposing an integrated approach to land management to achieve the performance targets, selecting sites where current management needs to be improved;
- identifying best management practices and determining how they should be applied, installed and maintained;
- monitoring success of the programme by follow-up audits.

In relation to the erosion audit, it is necessary to decide what the audit should cover. For example, it could be restricted to an assessment of the current state of erosion, sediment pathways and sediment delivery to watercourses, covering agricultural, residential, commercial, industrial and institutional sites. Audits could also cover the work of the design engineers, planners and contractors who might be employed by the group in order to assess how well they are performing.

In carrying out their work the Land Care groups will need to operate within current government environmental policies and any legal requirements, such as specified maximum erosion rates, specified threshold sediment concentrations in rivers, compliance with recommended best management practices and penalties for non-compliance, and types and sources of finance available to support different activities. The objective would be for each Land Care group to define an Erosion and Sediment Control Programme with clear objectives, a business plan, a structured programme for implementation and a programme for monitoring.

For Land Care groups to be successful, it is clear that the following issues need to be resolved:

- Who should pay for the benefits of erosion control? How much of the costs should be borne by the taxpayer and how much by the land owner, land user or land developer (potential polluter)? To what extent can potential pollution from one activity on the land be traded against more land-protective activities on another piece of land?
- To what extent is an educational programme needed to make communities

aware of watershed processes of erosion and sediment movement and their consequences on stream hydrology, aquatic ecology, water quality, flooding including functioning of floodplains, and stability of stream banks?

- What should be the methodology for undertaking erosion audits, designing erosion-control measures, evaluating different policy and erosion-control scenarios and providing a base for decision-making? There is a need for integrated decision-support models which can be used to inform stakeholders through visualization of different management scenarios so that the most cost-effective best management practices to meet specific erosion and conservation objectives can be determined.
- What should be the role in Land Care groups for organizations such as DEFRA and the Environment Agency, which have responsibilities for implementing regulations? How might that role be best discharged?
- To whom should Land Care groups be responsible? Should they be part of or independent from existing local authority structures?

Conclusions

Responsibility for erosion and sediment control is presently diverse and involves many different organizations, depending upon the context. There is little liaison between those responsible for erosion control in the agricultural and non-agricultural sectors, which often results in differences in practice. Participatory approaches to watershed management and erosion control are proving more successful than previous 'top-down' approaches. The Land Care movement in Australia has been particularly successful because of its community base and its ability to bring together all the stakeholders involved in erosion and its control. The chapter envisages the establishment of similar groups in most countries of the world, but their success will be dependent on government support politically, philosophically and financially. Do we and our governments have the vision to trust the local community and therefore provide the necessary legislative back-up?

References

Arnalds, A. (1999) Incentives for soil conservation in Iceland. In: Sanders, D.W., Huszar, P.C., Sombatpanit, S. and Enters, T. (eds) *Incentives in Soil Conservation: From Theory into Practice.* Oxford & IBH Publishing, New Delhi, pp. 135–150.

Boardman, J., Ligneau, A., de Roo, A. and Vandaele, K. (1994) Flooding of property by runoff from agricultural land in northwestern Europe. *Geomorphology* 10, 183–196.

Campbell, A. (1994) *Landcare: Community Shaping the Land and the Future*. Allen and Unwin, Sydney.

Environment Agency (2002) *Agriculture and Natural Resources: Benefits, Costs and Potential Solutions*. Environment Agency, Bristol, UK.

Hudson, N. (1992) *Land Husbandry*. Batsford, London.

Lal, R. (2002) Soil conservation and restoration to sequester carbon and mitigate the greenhouse effect. In: Rubio, J.L., Morgan, R.P.C., Asins, S. and Andreu, V. (eds) *Man and Soil at the Third Millennium.* Geoforma Ediciones, Logroño, Spain, pp. 37–51.

Marston, D. (1996) Extending soil conservation from concept to action. In: Sombatpanit, S., Zöbish, M.A., Sanders, D.W. and Cook, M.G. (eds) *Soil Conservation Extension: From Concepts to Adoption.* Soil and Water Conservation Society of Thailand, Bangkok, pp. 27–34.

Morgan, R.P.C. and Hann, M.J. (2003) Design of diverter berms for erosion control and biorestoration along pipeline rights-of-way. Poster paper presented at *Land Reclamation 2003*, 13–16 May, Runcorn, UK.

Morgan, R.P.C. and Rickson, R.J. (1996) Water erosion control. In: Morgan, R.P.C. and Rickson, R.J. (eds) *Slope Stabilization and Erosion Control: a Bioengineering Approach.* E & F N Spon, London, pp. 133–190.

Napier, T.L. (1999) Inadequacies of voluntary soil and water conservation incentives threaten adoption of regulatory approaches. In: Sanders, D.W., Huszar, P.C., Sombatpanit, S. and Enters, T. (eds) *Incentives in Soil Conservation: From Theory into Practice.* Oxford and IBH Publishing, New Delhi, pp. 151–164.

Northcutt, G. (1998) Lessons from the farm. *Erosion Control* 5, 26–39.
Roberts, B. (1992) *Landcare Manual.* New South Wales University Press, Kensington, Australia.
Schiechtl, H.M. and Stern, R. (1996) *Ground Bioengineering Techniques for Slope Protection and Erosion* Control. Blackwell Science, Oxford, UK.
Shaxson, T.F., Hudson, N.W., Sanders, D.W., Roose, E. and Moldenhauer, W.C. (1988) *Land Husbandry, a Framework for Soil and Water Conservation.* Soil and Water Conservation Society, Ankeny, Iowa.
Sibbesen, E. and Sharpley, A. (1997) Setting and justifying upper critical limits for phosphorus in soils. In: Tunney, H., Carlton, O.T., Brookes, P.C. and Johnston, A.E. (eds) *Phosphorus Loss from Soil to Water.* CAB International, Wallingford, UK, pp. 151–176.
Sibbesen, E., Schjønning, P., Hansen, A.C., Nielsen, J.D. and Heidmann, T. (1994) Surface runoff, erosion and loss of phosphorus relative to soil physical factors as influenced by tillage and cropping systems. In: Jensen, H.E., Schjønning, P., Mikkelsen, S.A. and Madsen, H.B. (eds) *Soil Tillage From Crop Production and Protection of the Environment.* ISTRO, Aalborg, Sweden, pp. 245–250.
Tiffen, M., Mortimore, M. and Gichuki, F. (1994) *More People, Less Erosion. Environmental Recovery in Kenya.* Wiley, Chichester, UK.

V

Summary and Outlook

One person's soil erosion is another person's sediment problem. Soil erosion and sediment deposition in adjacent fields separated by a hedge boundary, Devon, England (photo: P.N. Owens).

28 Soil Erosion and Sediment Redistribution in River Catchments: Summary, Outlook and Future Requirements

P.N. Owens[1] and A.J. Collins[2]
[1]*National Soil Resources Institute, UK;* [2]*National Soil Resources Institute, UK;* Present address *Landcare Research, New Zealand*

Introduction

The previous chapters have presented results and ideas relating to research on soil erosion and sediment redistribution in river catchments. This chapter places these in a broader framework of soil–sediment–water systems and attempts to identify future research needs to assist the science in addressing the requirements of catchment managers and policy-makers.

Measurement and Monitoring of Soil Erosion and Sediment Redistribution

Measurement and monitoring techniques

Measurement (determination of the extent, amount or flux of soil erosion or sediment transport and deposition at a point in time and/or its associated impacts) and monitoring (determination of temporal trends in extent or fluxes or impacts through repeat estimation procedures such as measurement) are intimately linked. There are a variety of different ways to measure and thus monitor soil erosion and sediment fluxes. These include direct approaches, such as visual observations and mapping (Morgan, 2005), erosion pins (Peart *et al.*, Chapter 3; Visser, Chapter 24), erosion bridges (Shakesby *et al.*, Chapter 5; Walsh *et al.*, Chapter 23), erosion plots (Fig. 28.1a), volumetric measurements (Belyaev *et al.*, Chapter 4), turbidity sensors (Hejduk *et al.*, Chapter 9; Walsh *et al.*, Chapter 23), and the trapping and collection of fluvial sediment (Farguell and Sala, Chapter 8; Hejduk *et al.*, Chapter 9; Fig. 28.1b). There are also indirect approaches, such as the use of tracers (Foster, 2000), including environmental radionuclides (Walling, Chapter 2; Peart *et al.*, Chapter 3; Belyaev *et al.*, Chapter 4), environmental magnetism (Blake *et al.*, Chapter 6) and bedload tracing techniques (Evans and Gibson, Chapter 10).

Useful and comprehensive reviews of the direct and indirect approaches used to measure and monitor soil erosion, and to identify sediment sources (including soil erosion and bank erosion) are provided by Brazier (2004) and Collins and Walling (2004), respectively. Some additional techniques to measure and monitor soil erosion and sediment (suspended and bedload) fluxes in rivers are described in a series of publications by two international organizations: (i) the International Association of Hydrological Sciences (IAHS), including Bogen *et al.* (1992, 2003), Horowitz and Walling (2005) and Walling and Horowitz (2005);

Fig. 28.1. (a) Erosion plot and Coshocton-wheel, Stuart-Takla Research Forest, British Columbia, Canada (photo: E.L. Petticrew); and (b) Helley-Smith bedload sampler, Ebro River, Spain (photo: P.N. Owens).

and (ii) the International Association for Sediment Water Science (IASWS), including Evans *et al.* (1997), Evans and Evans (2001) and Kronvang (2003). The IAHS and IASWS series of publications provide a thorough record of methodological and technological developments in soil erosion and sediment transport measuring and monitoring techniques and the reader is directed to these for further information.

The approaches described in the chapters in this book and in the other publications mentioned above are important in that they provide valuable information that has improved our understanding of soil erosion and sediment transport processes and rates. While it is important that there are continued efforts to improve these techniques, such methodological advances should be driven by both scientific questions and management needs.

Measurement and monitoring programmes

Soil erosion

While it has been advocated that there may be a shift from monitoring towards tracing sediment movement from source to sink (Walling, Chapter 2), it is clear that long-term measurement and monitoring programmes are still necessary to identify spatial patterns and temporal trends in erosion and sediment fluxes. Measuring and monitoring of soil erosion at the national level is generally limited. In most cases soil erosion is not seen as a priority issue. Time-specific measurements and longer term monitoring of soil erosion tend to be for specific plots or study areas related to a particular research institute/university or research question (e.g. Peart *et al.*, Chapter 3; Belyaev *et al.*, Chapter 4; Shakesby *et al.*, Chapter 5). Other, larger scale studies have tended to focus on certain land uses or biogeoclimatic areas, such as lowlands or uplands, at particular points in time and/or have been reconnaissance-type surveys that have not considered all erosion processes. For England and Wales, for example, such studies include Evans (1993), Chambers and Garwood (2000) and McHugh *et al.* (2002a,b) (also see Brazier, 2004; Evans, Chapter 7; Wood *et al.*, Chapter 21). These studies, however, often provide the best assessment of the extent of soil erosion and its severity, and are thus particularly useful for management and policy-making (e.g. McHugh *et al.*, 2002a,b). For those countries where national estimates of soil erosion and sediment delivery have been determined (e.g. Jones *et al.*, 2004; Wood *et al.*, Chapter 21), in most cases models such as USLE and PESERA have been used and the resultant information (usually expressed in the form of maps) are only snap-shots of the spatial distribution of estimated erosion at a point in time. Such an approach tends to provide an assessment of soil erosion potential or risk.

Measurement and monitoring at the regional and national scale over a period of time appropriate to the main soil erosion processes is an important requirement if we are to assess the nature and extent of soil erosion, and perhaps more importantly, the trends. Within the European Union (EU), recommendations have been made for such a soil erosion measurement and monitoring programme that uses a risk-based assessment approach validated through measurement (Vandekerckhove *et al.*, 2004).

An important issue associated with measurement and monitoring of soil erosion relates to the level of detail required. Lumped estimates of erosion that incorporate all erosion processes provide information on extent and severity, and ideally trends, in erosion but fail to identify the relative importance of the different erosion processes. Lumped or aggregated information on erosion is important as an indicator of the situation or state of the system; however, specific information on each erosion process is often more useful for managers and also for scientists interested in landscape evolution. As such, it is important that all dominant erosion processes are identified and measured. While water, and to some extent wind and tillage, erosion processes are reasonably well-studied, other significant processes, such as soil loss due to crop harvesting (Ruysschaert *et al.*, 2004), have received much less attention. Poesen *et al.* (2001), for example, estimated that mean soil loss due to crop harvesting for sugarbeet in Belgium was 8.7 t/ha per harvest. For national or regional measurement and monitoring programmes to be useful, all soil erosion processes should be assessed, at least initially. A risk-ranking assessment approach can then be used to identify those processes and/or sites that require more comprehensive and long-term monitoring (see Table 28.1).

Sediment fluxes in rivers

Generally, regional and national sediment monitoring programmes are inadequate compared to those for water quantity and quality, due to the poor temporal and spatial resolution of monitoring and sample analysis. An example is the revised National Stream Water Quality Accounting Network (NASQUAN) monitoring programme in the USA, where suspended sediment samples are collected from rivers usually only 12–15 times per year (Horowitz, 2004). In the UK, suspended sediment concentration is routinely measured as part of the Harmonised Monitoring Scheme (HMS) and combined with flow data from the National River Flow Archive (NRFA) to calculate fluxes of suspended sediment (and other determinands). This now

Table 28.1. Types, occurrence and severity of soil erosion at the national level in Europe. Numbers relate to the degree of erosion (1 = minor, 2 = important, 3 = predominant, N = not found and ? = not known) (modified from Jones *et al.*, 2004).

Country	Rill and interrill	Gully	Snowmelt	Bank	Tillage	Animals	Wind	Landslides
Albania	2	2	1	2	1	2	?	2
Austria	2	1	2	2	1	N	?	2
Belgium	2	1	N	1	1	N	1	1
Bosnia-Herzegovina	2	2	1	2	1	1	1	1
Bulgaria	2	2	2	1	1	1	1	1
Croatia	2	2	1	2	2	1	1	2
Cyprus	2	2	1	1	2	?	?	1
Czech R.	3	1	1	1	1	?	?	1
Denmark	3	1	N	1	1	N	2	?
Estonia	2	N	N	?	?	1	1	N
Finland	1	N	2	1	?	1	N	N
France	3	2	2	3		1	1	2
Germany	2	1	1	2	1	?	2	2
Greece	1	3	1	3	1	2	1	1
Hungary	2	2	1	2	2	1	1	1
Iceland	1	2	3	3	N	N	1	2
Ireland	1	N	N	3	1	2	N	N
Italy	3	2	1	1	2	?	1	2
Latvia	2	N	N	?	?	1	?	N
Lithuania	2	N	N	?	?	1	?	N
Luxembourg	1	N	N	1	N	N	N	?
Macedonia	2	2	1	2	1	1	?	1
Malta	1	2	N	N	N	1	N	1
Montenegro	2	2	1	2	1	2	?	1
Norway	1	1	3	1	N	1	1	2
Poland	2	1	1	1	?	?	2	2
Portugal	2	3	N	1	1	?	?	1
Romania	2	2	1	2	1	1	?	1
Serbia	2	2	1	2	1	1	1	1
Slovakia	2	1	?	1	?	?	?	1
Slovenia	2	2	1	1	2	?	?	2
Spain	2	3	1	1	1	1	1	2
Sweden	1	2	1	2	N	1	1	3
Switzerland	1	1	3	2	?	1	?	2
The Netherlands	1	N	N	?	N	?	1	N
UK	2	1	1	2	1	2	1	1

represents part of the UK contribution to the OSPAR (Oslo-Paris) Commission, as agreed under the terms of the 1998 Convention for the Protection of the Marine Environment of the North Atlantic (itself a combination of the Oslo Convention of 1972 and the Paris Convention of 1974). Table 28.2 presents the number of samples collected for each HMS site and annual sediment fluxes for key regions in Great Britain (England, Scotland and Wales) for the period 1974–1994 (Littlewood and Marsh, 2005). Despite the problems associated with the

Table 28.2. Specific annual suspended sediment loads (t/km^2) (1975–1994) for regions of Great Britain (modified from Littlewood and Marsh, 2005).

	Atlantic	Irish Sea	Celtic Sea	English Channel	North Sea
Area (%)[a]	4	14	17	8	57
N[b]	245	408	386	589	344
1975	9.5	22.5	22.4	11.9	9.5
1976	13.5	14.3	18.6	17.5	5.5
1977	16.8	14.1	37.2	18.2	17.4
1978	17.3	15.6	26.0	13.2	9.2
1979	14.0	20.7	41.1	14.7	13.9
1980	34.8	22.9	47.1	12.6	11.7
1981	22.1	28.8	42.9	17.3	16.9
1982	22.2	22.9	44.4	18.1	14.9
1983	15.6	21.3	36.5	9.7	9.0
1984	8.6	21.0	22.9	16.1	9.3
1985	9.5	14.8	25.5	10.8	10.3
1986	20.1	21.5	45.2	20.9	22.8
1987	8.8	16.5	24.8	9.0	9.7
1988	16.2	21.9	25.3	20.9	9.6
1989	8.6	16.6	26.5	11.5	6.0
1990	48.2	17.2	21.5	16.6	7.8
1991	19.8	18.9	19.0	7.8	6.0
1992	20.0	20.4	33.9	10.4	8.7
1993	27.6	22.5	31.2	14.8	9.0
1994	16.5	21.8	45.1	18.5	9.0
Mean	18.5	19.8	31.9	14.5	10.8

[a]Approximate percentage cover of the Harmonised Monitoring Scheme (HMS) (Great Britain) area.
[b]Average number of values per HMS site for period.

sampling and load estimation procedures (see Littlewood and Marsh, 2005), the data illustrate both the regional and temporal variation in sediment fluxes, and provide a very valuable source of information for catchment managers and those concerned with sediment inputs to the marine environment.

With the introduction of more comprehensive policy and legislation in many countries (such as the Water Framework Directive (WFD) in the EU), there is a growing demand for improved river measurement and monitoring programmes, especially for water chemistry. In recognition of the intimate link between sediment quality and water quality, there is an awareness of the need to improve sediment monitoring programmes and to develop these alongside those for water quantity and quality. Historically, sediment monitoring programmes were driven by the need to assemble information on sediment transport and loads for water resource engineering requirements (e.g. for hydroelectric power generation), such as the hydrometric programmes in Canada (Day, 1992) and Iceland (Hardardóttir and Snorrason, 2002). In many cases, it is the quality of the sediment, as opposed to the quantity, and how sediment influences environmental quality that are now the main drivers for monitoring and measurement. The realization of the need for improved sediment monitoring and measurement in relation to water quality and aquatic ecology can be seen, for example, with sediment now being addressed by the WFD Expert Group on Analysis and Monitoring of Priority Substances (AMPS) (Brils, 2004a). Barceló *et al.* (2004) have suggested key recommendations regarding the development of guidelines for monitoring contaminants in sediments, with emphasis on priority substances, for implementation of the WFD.

Perhaps the largest monitoring programme for sediment quality is the National Sediment Quality Survey (NSQS) and associated National Sediment Inventory, run by the US Environmental Protection Agency. This programme involves over 50,000 monitoring stations in the USA. The latest NSQS report (USEPA, 2004) describes the accumulation of chemical contaminants in rivers, lakes, oceans and estuaries and includes a screening-level assessment of the potential for associated adverse effects on human and/or environmental health.

In terms of sediment fluxes in rivers, it is important and timely that we reconsider our current measurement and monitoring programmes. For example, much has been written on the sampling frequencies needed for estimating sediment loads in rivers and on the use of procedures to estimate loads using infrequent sampling programmes (Phillips *et al.*, 1999; Horowitz, 2004; Littlewood and Marsh, 2005). Similar work is needed for sediment-associated contaminants and nutrients (Webb *et al.*, 2000; Horowitz, 2004; Littlewood and Marsh, 2005). Furthermore, recent developments in our understanding of the role and importance of:

- particle flocculation (Petticrew and Droppo, 2000; Droppo, 2001; Phillips and Walling, 2005);
- colloids (Kretzschmar *et al.*, 1999; Heathwaite *et al.*, 2005);
- the organic component (Olley, 2002; Petticrew and Arocena, 2003; Petticrew, 2005; Phillips and Walling, 2005); and
- particle-chemical speciation and transformations (Carter *et al.*, in press; Droppo *et al.*, Chapter 26)

in sediment and chemical transport and storage have questioned what it is we are actually measuring and monitoring, and what is important for understanding system behaviour and dynamics. We may need, therefore, to evaluate both what we are doing and the implications of the information that we have so far obtained. In this respect, measurement and monitoring programmes should be viewed as dynamic in that there is a need for continuous evaluation of the protocols and methods of monitoring and of what is being measured. What is clear, though, is that a certain level of standardization of measurement and monitoring and assessment tools is required. While this may be difficult to achieve globally, due to different objectives, requirements and economic considerations, it should be done at least at the national level or at the river basin level if several countries are involved (see also Collins and Owens, Chapter 1).

Tracing and fingerprinting approaches

The need for improved monitoring of soil erosion on land and sediment fluxes and storage in rivers will no doubt increase in the future. However, it is also becoming increasingly apparent that techniques that are able to trace sediment through catchments and are able to identify the sources of sediment have an important role to play in the management of sediment in river catchments, and perhaps should be incorporated within more comprehensive erosion–sediment measurement and monitoring programmes in river catchments. Good examples of the potential of tracing techniques are given by Wallbrink *et al.* (1998), Foster (2000), Whiting *et al.* (2005) and Walling (Chapter 2). In addition, the chapters by Peart *et al.* (Chapter 3), Belyaev *et al.* (Chapter 4) and Blake *et al.* (Chapter 6) further demonstrate the potential that tracing techniques offer, particularly in that they are able to provide information on erosion and sediment dynamics at the catchment scale (Golosov *et al.*, 1992; Owens *et al.*, 1997; Whiting *et al.*, 2005). The work of Belyaev *et al.* (Chapter 4) is also useful in that it demonstrates the potential of using tracing techniques and more conventional measurement (and modelling) techniques in combination, as part of a broader and more comprehensive assessment approach.

Similarly, sediment source fingerprinting techniques (e.g. Collins and Walling, 2004) are likely to be a growing area of research in the near future as source control options (i.e. control of excessive soil erosion or bank erosion) are likely to assume increased importance as an effective and sustainable management strategy. Robust and reliable techniques are required that give unequivocal results relating to soil erosion and sediment dynamics and provenance. If results are ambiguous and unreliable then confidence will soon be lost by potential users, and indeed there may be issues of liability if results are misleading or incorrect. Efforts to determine the uncertainty associated with fingerprinting results (e.g. Rowan *et al.*, 2000; Small *et al.*, 2002) and to standardize tracing approaches (e.g. for environmental radionuclides see Zapata, 2002) are thus to be welcomed and will no doubt strengthen the acceptance of such techniques within the scientific and management communities.

Improvements in soil–sediment tracing and fingerprinting techniques represent an area where scientists can provide useful tools for catchment managers. Recent developments and applications in the use of tracers such as ^{7}Be (e.g. Blake *et al.*, 2002; Whiting *et al.*, 2005), ^{134}Cs (Syversen *et al.*, 2001), rare earth elements (Polyakov *et al.*, 2004) and fluorescent tracers (Baker, 2002), and in fingerprinting properties such as Sr-Nd-Pb isotopic geochemistry (e.g. Douglas *et al.*, 2003) and C and N isotopes and profiles (e.g. Olley, 2002; Hagedorn *et al.*, 2003; Petticrew, Chapter 11), offer exciting opportunities to advance our understanding of the soil–sediment–water system. Many of these developments are being made in other disciplines, such as soil science, water chemistry and microbiology, and have yet to be fully utilized in soil erosion and sediment transport research.

Palaeolimnological approaches

While measurement (including tracing) and monitoring of contemporary or recent rates of soil erosion and sediment fluxes in rivers are essential in order to provide the necessary information on the current situation, it is often important to place this information within the context of a longer time series so as to establish the representativeness of the contemporary values and to identify longer term trends. The information contained within sedimentary environments such as floodplains, lakes, estuaries and the continental shelf offers much potential. Such records can be used to establish trends in catchment soil erosion and sediment yields (Foster *et al.*, 1985; Zolitschka, 1998), trends in sediment sources (Collins *et al.*, 1997; Owens and Walling, 2002) and trends in the fluxes of sediment-associated contaminants, phosphorus and organic carbon (e.g. Owens and Walling, 2003; Gomez *et al.*, 2004; Leithold *et al.*, 2005). Foster (Chapter 12) provides a good overview of the benefits of the palaeolimnological approach, and states that 'Failure to recognize the contribution that this approach might make to our understanding of sediment [and contaminant] dynamics in river catchments over short to medium timescales would deny us a critical future opportunity for understanding the consequences of environmental change on sediment delivery and the sediment delivery system' (p. 140).

Despite a rapid increase in the number of studies using lake, reservoir, floodplain and marine environments to reconstruct changes in sediment sources and yields, there are still areas where further research is needed to improve the approach. There is clearly a need for improved dating methods, and also improved understanding and accuracy of existing techniques such as ^{14}C, unsupported ^{210}Pb and ^{137}Cs (e.g. He *et al.*, 1996; Appleby *et al.*, 2003). There are also other, relatively new, dating methods that offer potential, such as ^{32}Si (Nijampurkar *et al.*, 1998; Morgenstern *et al.*, 2001), which require further testing and development.

Modelling of Soil Erosion and Sediment Redistribution

The role of models

One of the greatest tasks for the scientific community is to develop models in order to:

- improve our understanding of landscape evolution;
- predict erosion and sediment generation and transport;
- evaluate the likely impacts on receptors (such as water bodies); and
- evaluate the effectiveness of management options.

The modelling chapters in the book have been concerned mainly with the first two of these tasks, in terms of predicting soil erosion on land (e.g. Nearing, Chapter 13; Kuhn, Chapter 14; Sidorchuk *et al.*, Chapter 15; Kinnell, Chapter 16; Elliot, Chapter 17; Licciardello *et al.*, Chapter 18; Jetten *et al.*, Chapter 19), sediment delivery from land to rivers (Jarritt and Lawrence, Chapter 20; Wood *et al.*, Chapter 21) and sediment transport in rivers (Jarritt and Lawrence,

Chapter 20). Outside this volume, studies have used models to evaluate the impacts of soil erosion and sediment transport on receptors, while others have been used to assess the potential effectiveness of management options, particularly through scenario analysis (e.g. Van Rompaey *et al.*, 2002; Asselman *et al.*, 2003; Elliot, Chapter 17; see Table 28.3). Some of the models described in these chapters and papers, and additional soil erosion–sediment transport models (such as AGNPS, ANSWERS, EPIC, EUROSEM, GLEAMS, GSTARS, FLUVIAL, HEC, LISEM, MIRSED, PESERA, RUSLE, SHETRAN, SWAT, USLE) are described further in Harmon and Doe (2001) and Summer and Walling (2002).

Most of the aforementioned chapters have highlighted that there are some important requirements (especially in terms of data) and developments needed if models are to be able to undertake the four tasks listed above. Some of these requirements and developments are described below.

Model requirements and developments

Data sets and process information

It is clear that appropriate data sets are needed in order for models to be able to predict soil erosion and sediment transport processes. Such data are required for model development in terms of improving our understanding of the system (especially for physically based models such as WEPP and EUROSEM), for parameterization, and also for subsequent calibration and validation. It is argued by many, including Brazier (2004), that at present we are lacking the relevant data sets for these needs. As such, the requirement for better and more appropriate data sets identifies a direct link between the needs of modelling and the need for comprehensive measurement and monitoring programmes, as identified in the sections above. Brazier states that 'a return to collection and use of datasets that closely match the output of soil erosion models is crucial for the development of the discipline as a whole' (2004, p. 357). Furthermore, 'the use of high quality observed data that exist should be made to improve reliability of results [from models] and constrain uncertainty prior to reliance upon the accuracy of model predictions within a decision-making framework' (Brazier, 2004, p. 358).

Useful examples of the type of data and process information needed for soil erosion modelling are provided by Kuhn (Chapter 14), Sidorchuk *et al.* (Chapter 15) and Kinnell (Chapter 16) in terms of the concept of soil erodibility and of small-scale physico-chemical characteristics of the soil surface. Kuhn describes problems associated with short-term variations in erodibility as a result of soil properties and the wetting and drying characteristics of soils, amongst other factors. He suggests that either erodibility models have to integrate all of the processes and properties that determine their short-term changes, or that new erosion models with more sophisticated representation

Table 28.3. Simulated changes in soil erosion and sediment yield based on land use changes in the Dijle catchment (820 km^2), Belgium (modified from Van Rompaey *et al.*, 2002). Changes were determined using a coupled soil erosion–sediment transport model (SEDEM) in combination with the analysis of historical maps. Note the non-linear response to land use change of both soil erosion and sediment yield.

Scenario	Forest area (%)	Arable area (%)	Soil erosion (t/ha/year)	Sediment yield (t/year × 10^3)
25% deforestation	9.7	61.3	4.85	78.7
15% deforestation	11.1	59.9	4.67	73.9
5% deforestation	12.3	58.7	4.51	69.9
Present situation	13.0	58.0	4.41	67.5
5% afforestation	15.9	55.1	4.03	58.2
15% afforestation	21.7	49.3	3.50	45.7
25% afforestation	27.5	43.5	2.93	36.3

of the soil properties and processes that control soil resistance to erosion have to be developed. The feasibility of these suggestions for practical purposes is, however, questionable, although Sidorchuk *et al.* (Chapter 15) suggest that stochastic approaches to soil erosion modelling offer a way forward. Again, however, detailed data on soil physical conditions (including the size of the soil particles and aggregates) and on hydrodynamic characteristics are needed. Small-scale variations in erodibility, and modelling of the size of particles detached by raindrop impact and transported by water flow, are further considered by Kinnell (Chapter 16).

It is clear from Chapters 14–16 that there is a need for more realistic representation of temporal and spatial variations in soil erodibility, which in turn requires an improved understanding of processes operating at this scale and the provision of appropriate data. Such information could be provided, for example, through detailed experiments using rainfall simulators and artificial soil slopes (e.g. Kuhn, Chapter 14; Fig. 28.2).

Identification of appropriate spatial and temporal scales

The type of data required for soil erosion and sediment redistribution models largely depends on the spatial (i.e. plot, hillslope, catchment, global) and temporal (i.e. storm-event, annual, decadal, Holocene) scales of the models, and as such are likely to be specific to each model or group of models. As one moves to larger spatial scales, such as from plot to catchment scale, the range of erosion and sediment transport processes increases. Thus there is a need to model all relevant processes and ideally have appropriate data sets for these. Kirkby (1998) has suggested that nested data sets at different scales are needed for modelling systems, in order to assess whether models can address issues of changing dominant processes at different scales.

The issue of the most appropriate spatial scales for models is one that has received much attention recently. For example, Martin and Church (2004) argue that natural breaks in landscape process and morphology define appropriate spatial domains for the modelling of landscape evolution. This may entail a

Fig. 28.2. The 6 m × 3 m tri-axial, hydraulically operated soil slope at the Soil Erosion Laboratory, National Soil Resources Institute, England (photo: NSRI). Located above the slope is a pressure-fed rainfall simulator.

mechanistic approach at small spatial scales and a more generalized approach to process definition in large-scale models. Similarly, for modelling of soil erosion and sediment transport at the catchment scale, Wasson (2002) suggests that we may need to re-evaluate our approach. He advocates a focus on emergent properties and the adoption of a more top-down approach, where emergent properties of catchments are identified as the starting point for modelling at the catchment scale, rather than the traditional approach of continuum mechanics in which processes are modelled and combined, usually in a spatial setting, to produce the emergent property, such as runoff and sediment yield.

It is clear, however, that the most appropriate scale (both in time and space) for modelling should be determined by the objectives of the exercise, and also by the level of data available. Models developed to predict soil erosion on hillslopes during individual rainfall events will operate at different temporal and spatial scales and have different data requirements than those models concerned with soil erosion and sediment delivery in small catchments over a period of years (see chapters in this volume for models that operate at these scales), and compared to models that reconstruct sediment redistribution in catchments over the Holocene epoch (e.g. Coulthard *et al.*, 2002). Whether we can ever couple such modelling approaches remains uncertain.

Variability in input data and modelled results

Critical evaluation of model output is an important, but sometimes neglected, part of modelling. It is important to ascertain and assess what it is that the model is telling us. Nearing (Chapter 13) describes the very considerable variability in results associated with soil erosion models, partly a function of the high variability of input data. In addition, the variability (and hence confidence associated with prediction) is a function of erosion magnitude, being greatest for low erosion rates and less for higher erosion rates. An implication of data variability is that there is a limit in terms of accuracy for models. Thus Nearing (Chapter 13) advocates the use of continuous simulation models in order to account for the complex overlap between temporally and spatially heterogeneous distributions of both the driving force of erosion (e.g. rainfall) and the system resistance (e.g. erodibility).

In recognition of the errors and uncertainty associated with input data, and consequently in the modelled results, much work has recently focused on uncertainty analysis in modelling (Bevan, 1996, 2002). It is likely that such research will develop further and it is suggested here that all models should incorporate an element of uncertainty analysis. A major problem with models (or perhaps more with the scientific community) is that the results that they produce are often taken as fact by those not familiar with the model. Too frequently, the results are presented as values without the errors and uncertainty associated with them, and there is very little information associated with the sensitivity of the model(s) to the model parameters. Many models are either being used by managers and policy-makers or are being specifically designed for them (see sections below). It is important that we make it clear where sources of error and uncertainty lie, and that a comprehensive list of warnings accompany any documentation. Improvements in scientific understanding (i.e. process understanding and improved model structure), data availability (i.e. high resolution soils data and DEMs) and computing capability should see a marked improvement in the values predicted by models.

Models and decision-support systems for management

Examples of the type of modelling tools being increasingly demanded by end-users such as regulators and catchment managers are provided by Van Rompaey *et al.* (2002; Table 28.3), Asselman *et al.* (2003), Elliot (Chapter 17), Jarritt and Lawrence (Chapter 20) and Wood *et al.* (Chapter 21). In the case of the latter, which itself was part of a larger study mapping sediment delivery to rivers (McHugh *et al.*, 2002b), maps of vulnerability to soil erosion for England and Wales were produced. As the authors state, despite the inherent problems and limitations of the approach, the maps do provide an indication of hot-spots for subsequent targeted management options and are ‘providing decision support for regulatory and advisory visits to farms’ (Wood *et al.*, Chapter 21, p. 225).

It is likely that scientists will be increasingly asked to develop such models and tools, creating toolboxes to help managers prioritize and target management options in the implementation of legislation and policy. Toolboxes are themselves usually part of Decision Support Systems (DSS), which are often computer-based information systems developed to assist decision-makers to address semi-structured tasks in a decision domain. Typically there are three main components within a DSS:

- a user interface enabling easy interaction between the user and the system;
- a database containing the raw and processed data of the domain and the study area: for DSS relevant to erosion and sediment dynamics the data are usually derived from measurement and monitoring of the main soil–sediment variables, in addition to input data on climate, soils and land use, etc.; and
- a tool base (or toolbox) with the methods, techniques and software instruments required to work in an effective manner with the domain models and data.

The models within a DSS may be required to characterize the state of the catchment in terms of erosion and sediment transport, and to evaluate how soil erosion and sediment transport may respond to changes in policy, land use and climate, or through management interventions such as the placement of buffer features or dredging activities.

As knowledge progresses, the task of modelling undergoes a process of continual refinement and evolution. It is undeniable that the models described in these chapters are improving in terms of their ability to predict erosion and sediment transport. There clearly needs to be a balance, however, between the level of complexity required to model processes and the ease of use for end-users.

Management of Soil Erosion and Sediment Redistribution

The responsibility of the scientific community

With a move towards more accountable public funding of research, the scientific community is increasingly undertaking research to assist with the sustainable management of the environment. In the case of river catchments, research into soil erosion and sediment redistribution should be targeted in order to maintain soil functions, aquatic ecosystems and human health, as part of sustainable soil–sediment–water resources development at the global scale. In this context, there are also important socio-economic considerations or implications associated with soil erosion and sediment redistribution in rivers that include, amongst other things, agricultural productivity and sediment dredging. Thus, Crosson (1997) estimated for the USA that the annual on-farm costs of soil erosion in terms of losses of net farm income are about US$100 million (US$0.6/ha), while Clark *et al.* (1985) have estimated the annual off-site costs in the USA of about US$6 billion. In Europe, yearly economic losses in agricultural areas affected by soil erosion are estimated at €53/ha, while the off-site costs are about €32/ha (EEA, 2003). At a local level, the city of Hamburg, Germany, has to dredge between 2 and 5×10^6 m^3 of sediment each year from the River Elbe to maintain the port. It is estimated that the costs of this are of the order of €30 million annually, not including personnel and capital costs (Netzband *et al.*, 2002).

As scientists, we should be helping to address these and similar problems by undertaking appropriate research and by developing tools for managers. A series of chapters in this book (Chapters 23–26) has demonstrated how scientific research on the processes of soil erosion and sediment redistribution in contrasting catchments has provided information that could inform and develop appropriate management strategies. Thus, Walsh *et al.* (Chapter 23) illustrate the importance of detailed process measurement and monitoring in identifying the long-term response of soil erosion and sediment transport to selective logging in headwater catchments in Sabah, Borneo. In addition, the main sediment sources are changing (e.g. to road-linked landslides) and the authors suggest some management options to reduce the erosional impact of logging activities. Similarly, Visser (Chapter 24) and Nunny *et al.* (Chapter 25) describe management-driven research to investigate the role of land use changes on soil erosion, storage effects and downstream sedimentation in barrier reefs in Australia and Belize, respectively. In both cases, it was demonstrated that the study catchments have significant capacity to buffer changes in sediment delivery due to land use changes, and this knowledge has enabled the authors to make useful recommendations for soil and sediment management.

In urban environments, targeted, process-based research is also important for management. Thus, Droppo *et al.* (Chapter 26) used a combination of sample collection and measurement to examine the sources and pathways of sediment-associated metals in an urban environment in Canada. Again, recommendations for management were made, in this case also based on detailed information on sediment–metal fractionation, bioavailability and hydraulic sorting. In addition, they used spatial analysis techniques within a GIS to identify potential hotspots of metals and suggested that there was 'merit for sector-based management strategies' (p. 284). Similarly, many of the chapters in this book have described tools and models that have either been designed specifically for management (e.g. Jarritt and Lawrence, Chapter 20; Wood *et al.*, Chapter 21) or have potential applications for management (e.g. Walling, Chapter 2).

As discussed in Collins and Owens (Chapter 1), in order to carry out meaningful research and produce useful products (be it process understanding, data provision or the development of tools) to meet the needs of managers and policy-makers, and society more generally, there are several important considerations and activities which scientists should consider. Two of these considerations are the scale of research that we undertake and the level of complexity of the information that we provide. Rickson (Chapter 22; see also Morgan, Chapter 27) discusses these issues in detail and advocates the need for research at a variety of different scales so that science can both address the need for improved process understanding and the need to assess the practical viability of control/mitigation practices. Rickson asserts that this can be achieved through careful research planning and experimental design. Additional considerations that warrant further consideration by the scientific community are, amongst other things, identifying the drivers for management and involving stakeholders and policy-makers in the decision-making process. These are considered further below.

Drivers for management

In order for scientific research to be effective, it is important to identify the drivers for management and the scale at which they operate (Fig. 28.3; see Salomons and Brils, 2004; Salomons, 2005). At the local level, the main drivers are often site-specific erosion and sediment problems, such as severe erosion on agricultural land, sediment discharge on to roads due to excessive erosion, and dredging of waterways and harbours (Fig. 28.4; see also Morgan, Chapter 27). Thus, local managers (such as land managers, highways authorities, and port and canal authorities) have to deal with analysis, treatment and disposal aspects which are themselves guided by local and national legislation that primarily relate to

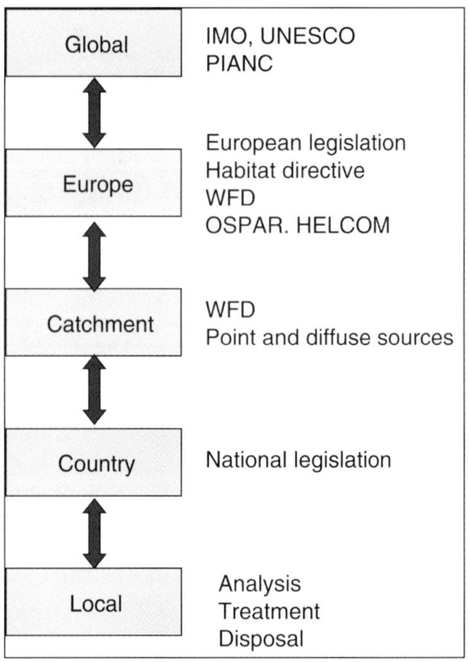

Fig. 28.3. Different spatial scales of drivers for sediment management (from Salomons and Brils, 2004; Salomons, 2005).

Fig. 28.4. (a) Soil erosion on cultivated land, England (photo: P.N. Owens); and (b) sediment dredger, River Elbe, Germany (photo: P.N. Owens). See text for information on amount of sediment dredged and the costs of this.

environmental quality. At the local level, decisions are also made according to socio-economic considerations, with cost and social perception of treatment options (i.e. what to do with dredged material, for example) being particularly important (Bortone *et al.*, 2004).

In the countries of the EU, management decisions are also being increasingly guided by EU (i.e. multi-national) guidelines and legislation. While no dedicated legislation exists for sediment management at the river basin scale in the EU, there are many existing guidelines, recommendations and legislation which relate to sediment, including (Köthe, 2003):

- water legislation (e.g. WFD, Nitrates Directive, Bathing Water Directive);
- soil legislation (see below);
- waste legislation (e.g. Waste Directive, Landfill Directive); and
- other environmental legislation (e.g. Habitats Directive, Fisheries Directive).

The WFD aims to coordinate the application of all EU water-related legislation (as listed above) and to provide a coherent management framework so as to meet the environmental objectives of these instruments as well as the WFD, within a set time frame (i.e. good ecological and chemical status by 2015). The WFD introduces a single system of water management by river basin, instead of according to administrative or political boundaries. In turn, this will enable a coordinated and, if necessary, multi-national approach to achieve the set environmental objectives. As explained above, the WFD does not as yet deal adequately with sediment quantity and quality issues, although this appears to be changing.

Local and national legislation and guidelines for controlling soil erosion are much less developed than those for managing sediment in rivers and the coastal zone, although controlling soil erosion is often identified by its relationship to sediment fluxes in rivers. In England and Wales, for example, soil erosion is not covered by any specific law. The only piece of legislation that could be used directly to prevent erosion from occurring is the Anti-Pollution Works Notices detailed in Section 161 of the Water Resources Act 1990. This legislation gives the Environment Agency for England and Wales the powers to require action to be taken to prevent potential pollution where activities are likely to result in polluting material entering controlled waters (Olmeda-Hodge *et al.*, 2004). There are also additional pieces of legislation that can be used to deal with the impacts of soil erosion once it has happened (Olmeda-Hodge *et al.*, 2004). Other examples of local, district and national policies and legislation that relate to soil erosion and sediment redistribution for Europe, New Zealand and the USA are contained in Owens *et al.* (2005).

In the EU there are plans to develop specific legislation relating to soils within the Thematic Strategy for Soil Protection (TSSP). At the centre of this thematic strategy is the recognition of the main threats to soils in the countries of the EU. The eight main threats have been identified as (Blum and Eswaran, 2004; Van-Camp *et al.*, 2004):

1. Soil erosion (and the delivery of sediments to watercourses).
2. Decline in soil organic matter.
3. Diffuse and local soil contamination.
4. Soil sealing.
5. Soil compaction and other physical soil deterioration.
6. Decline in soil biodiversity.
7. Salinization.
8. Floods and landslides (in rivers).

The threat of soil erosion and sediment fluxes to soil functions and water quality (chemical and ecological) are clearly identified in the EU TSSP (Van-Camp *et al.*, 2004). It is anticipated that this will result in appropriate legislation and policy, both at the level of the EU and also nationally in EU member states. Some countries already have soil protection legislation, such as the Dutch Soil Protection Act, while other member states are in the process of policy consultation and development (e.g. DEFRA, 2004).

There are also some drivers for erosion–sediment management that operate at the global scale, such as measures implemented by global organizations such as UNESCO (e.g. the International Sedimentation Initiative; Spreafico and Bruk, 2004), the International Maritime Organisation (IMO), the Central Dredging Association (CEDA) and the International Navigation Association (PIANC), and include (Köthe, 2003; Salomons and Brils, 2004):

- conventions for the protection of the marine environment (e.g. OSPAR and HELCOM conventions);
- conventions for the trans-boundary movement of hazardous waste (e.g. Basel Convention); and
- international recommendations for the management of dredged material (e.g. CEDA and PIANC recommendations).

Drivers at the global scale also include global trade markets and policies, and food security (Morgan, Chapter 27). In addition, global environmental change is also likely to have profound effects on soil erosion and sediment redistribution in rivers, and will probably influence soil–sediment–water management in catchments (Owens, 2005). In turn, this will probably require global initiatives and legislation to mitigate any detrimental effects.

For any policy and legislative frameworks to be effective in addressing the needs of resource management, in light of the threats and issues identified above, it is clear that an integrated approach to soil–sediment–water systems is required. There is an intrinsic link between the three resources, in terms of the common sources, pathway mechanisms and receptors. It is encouraging that, at least in the EU, there appear to be moves towards this goal.

Stakeholder participation

A key developing issue for soil–sediment–water management at the river basin scale is the identification and involvement of stakeholders in the decision-making process. With soil–sediment–water management increasingly moving towards the scale of the river catchment or basin, it is clear that the complexity of the issues is such that stakeholders must be involved if sustainable solutions are to be developed that balance the needs of society. Gerrits and Edelenbos define stakeholder involvement as

> the early involvement of individual citizens and other organized stakeholders in public policy-making in order to explore policy problems and develop solutions in an open and fair process of debate that has influence on decision-making. Stakeholder involvement differs from traditional public consultation procedures mainly in that stakeholders are involved early enough to influence policy as it is formulated, as opposed to merely being given the opportunity to modify proposals slightly after they have been developed, or not giving them the opportunity at all.
>
> (2004, p. 240)

There are a variety of degrees of stakeholder participation. Table 28.4 lists some of these and also identifies the participation of experts (the scientific community and consultants) and policy-makers. The scientific community should welcome the input of stakeholders in a variety of different ways, including (Gerrits and Edelenbos, 2004):

- helping to identify the key issues of concern for management;
- provision of local knowledge and sources of information;
- provision of support and finance;
- helping to counteract the obstructive impacts of pressure groups; and
- assisting with the development and implementation of appropriate policy and legislation.

A good example of how stakeholder involvement can offer a way forward for resource management at the local and national levels is provided by the Land Care movement in Australia, which brings together farmers and the local community, extension services, and the State and Federal Governments to address local and regional problems of land management (Morgan, Chapter 27). Importantly, they also provide a mechanism for liaison between the local community, erosion and sediment experts, industry and regulators. Morgan suggests that such an approach could, and should, be adopted in other countries, and identifies some important considerations. Similarly, the Watershed Reef Interconnectivity Scientific Study in Belize, described by Nunny *et al.* (Chapter 25), had stakeholder involvement and dissemination of results to stakeholders as important project objectives.

At the multi-national and global level, increased interaction between the scientific community and both stakeholders and policy-makers via common networks is helping to bridge the gap between the various parties, and is serving to develop sustainable solutions to soil–sediment–water resources management at the river basin scale. Examples of such networks include the European Sediment Network, SedNet (Brils, 2004b) and the UNESCO International Sedimentation Initiative (Spreafico and Bruk, 2004).

Conclusion

The previous chapters have presented recent research on issues of soil erosion and sediment redistribution in river catchments. They have covered a variety of spatial (from plot to national) and temporal (from event-based to historical) scales and geographical environments (from semiarid to temperate to tropical to urban), and have addressed soil–sediment redistribution on hillslopes and in rivers, floodplains, lakes and the coastal zone. It is clear that considerable progress is being made in our understanding of the main processes of soil erosion (e.g. rill, interrill, gully) and sediment redistribution (including transport, deposition, remobilization), and the rates and spatial patterns of these processes. It is also clear from these chapters that there is a

Table 28.4. Degrees of participation and influence of stakeholders and their link to experts (scientists and consultants) and policy-makers in soil–sediment–water management (modified from Gerrits and Edelenbos, 2004).

Degrees of stakeholder influence	Governance styles within the scale of participation	Role of the stakeholder	Role of the expert	Role of the policy-maker
Not involved	Closed authoritarian	None	Deliver information to policy-makers on demand; no information to stakeholders	Determine policy; policy process is closed, no information is issued
Informed but remain passive	Open authoritarian	Receive information but do not deliver input to the process	Deliver information to stakeholders on demand of the policy-makers	Determine policy; information is issued to stakeholders
Consulted	Consulting style	Consulted	Deliver information to participants on demand of all parties: experts provide another flow of information to the process, next to the flow of stakeholders	Determine policy and open the process to input by stakeholders, but are not obliged to accept their recommendations
Give advice	Participative style	Advisors to the process	Deliver information to all parties on demand and investigate suggestions from participants on demand of the policy-makers	Policy process is open to input by stakeholders; take account of input but have the right to deviate from it in their decisions
Co-producers	Delegating style	Co-decision makers within the set of preconditions	Treat policy-makers and stakeholders as equal clients; provide advice and knowledge to both types	Take the input of stakeholders into account and adopt if it fits into the set of preconditions
	Cooperative style	Policy-partners on the basis of equivalence	Treat stakeholders as equal knowledge providers; keep open mind to suggestions and ideas from stakeholders	Interact with stakeholders on the basis of equivalence and take stakeholder input seriously
Produce solutions and help in decision process	Facilitating style	Take initiatives and make decision	Supply stakeholders with knowledge and treat as clients; need no approval of policy-makers	Offer support (money, time, civil servants etc.) and leave the production of solutions and decision to the participants

move towards more applied research, in that the research is increasingly focusing on specific management needs in terms of providing an improved knowledge-base (e.g. sediment source identification, magnitude and timing of sediment fluxes, sediment–contaminant interactions) and set of tools (e.g. fingerprinting techniques, more realistic models) for management.

It is apparent from many of the chapters in this book that there is an increased level of integration between those concerned with soil erosion and those concerned with sediment delivery and redistribution in rivers. This integration must, however, be developed much further and at a quicker rate if we are to provide answers to the relevant questions being asked by catchment managers and policy-makers. Most importantly, this requires a greater level of interdisciplinary collaboration both between different scientific disciplines and between scientists and stakeholders. The chapters presented in this book provide a good foundation for this collaboration by integrating research in soil erosion on land and sediment redistribution in rivers, and by considering measurement, modelling and management approaches and requirements. In order to be useful to science and society, this collaboration must both continue and become more inclusive. Blum and Eswaran state:

> Application of mitigation technology alone is seldom satisfactory (for management), particularly over a long time-frame. The solution lies in seeking approaches for a collaborative effort between natural sciences, economic and social sciences as well as cultural sciences, in order to find ways to use soils and sediments in a sustainable way.
>
> (2004, p. 71)

References

Appleby, P.G., Hawarth, E.Y., Michel, H., Short, D.B., Lapte, G. and Pilipsonian, G.T. (2003) The transport and mass balance of fallout radionuclides in Blenham Tarn, Cumbria, UK. *Journal of Paleolimnology* 29, 459–473.

Asselman, N.E.M., Middelkoop, H. and van Dijk, P.M. (2003) The impact of changes in climate and land use on soil erosion, transport and deposition of suspended sediment in the River Rhine. *Hydrological Processes* 17, 3225–3244.

Baker, A. (2002) Fluorescence properties of some farm wastes: implications for water quality monitoring. *Water Research* 36, 189–195.

Barceló, D., Petrovic, M., de Deckere, E., Bakker, J.F., Scrimshaw, M., Stegemann, J., Kowalewska, G., Heininger, P., Vale, C., Hansen, P.D., Segarra, M.J.B., Parker, A., Poulsen, V., Manz, W., Blasco, J. and Casillas, A.D.-V. (2004) Sediment quality and impact assessment: synthesis of SedNet Work Package 3 outcomes. *Journal of Soils and Sediments* 4, 223–224.

Beven, K.J. (1996) Equifinality and uncertainty in geomorphological modelling. In: Rhoads, B.L. and Thorn, C.E. (eds) *The Scientific Nature of Geomorphology*. Wiley, Chichester, UK, pp. 289–313.

Beven, K.J. (2002) Towards an alternative blueprint for a physically-based digitally simulated hydrologic response modelling system. *Hydrological Processes* 16, 189–206.

Blake, W.H., Walling, D.E. and He, Q. (2002) Using cosmogenic beryllium-7 as a tracer in sediment budget investigations. *Geografiska Annaler* 84A, 89–102.

Blum, W.E.H. and Eswaran, H. (2004) Soils and sediments in the anthropocene. *Journal of Soils and Sediments* 4, 71.

Bogen, J., Walling, D.E. and Day, T.J. (eds) (1992) *Erosion and Sediment Transport Monitoring Programmes in Rivers*. IAHS Publcation 210. IAHS Press, Wallingford, UK.

Bogen, J., Fergus, T. and Walling, D.E. (eds) (2003) *Erosion and Sediment Transport Measurements in Rivers: Technological and Methodological Advances*. IAHS Publication 283. IAHS Press, Wallingford, UK.

Bortone, G., Arevalo, E., Deibel, I., Detzner, H.D., de Propris, L., Elskens, F., Giordano, A., Hakstege, P., Hamer, K., Harmsen, J., Hauge, A., Palumbo, L. and van Veen, J. (2004) Sediment and dredged material treatment: synthesis of SedNet Work Package 4 outcomes. *Journal of Soils and Sediments* 4, 225–232.

Brazier, R. (2004) Quantifying soil erosion by water in the UK: a review of monitoring and modelling approaches. *Progress in Physical Geography* 28, 340–365.

Brils, J. (2004a) Sediment monitoring under the EU Water Framework Directive. *Journal of Soils and Sediments* 4, 72–73.

Brils, J. (2004b) Three years of SedNet. *Journal of Soils and Sediments* 4, 216–217.

Carter, J., Walling, D.E., Owens, P.N. and Leeks, G.J.L. (in press) Spatial and temporal variability in the concentration and speciation of metals in suspended sediments transported by the River Aire, Yorkshire, UK. *Hydrological Processes*.

Chambers, B.J. and Garwood, T.W.D. (2000) Monitoring of water erosion on arable farms in England and Wales 1990–1994. *Soil Use and Management* 16, 93–99.

Clark, E.H., Haverkamp, J.A. and Chapman, W. (1985) *Eroding Soils: the Off-farm Impacts*. The Conservation Foundation, Washington, DC.

Collins, A.L. and Walling, D.E. (2004) Documenting catchment suspended sediment sources: problems, approaches and prospects. *Progress in Physical Geography* 28, 159–196.

Collins, A.L., Walling, D.E. and Leeks, G.J.L (1997) Use of the geochemical record preserved in floodplain deposits to reconstruct recent changes in river basin sediment source. *Geomorphology* 19, 151–167.

Coulthard, T.J., Macklin, M.G. and Kirkby, M.J. (2002) A cellular model of Holocene upland river basin and alluvial fan evolution. *Earth Surface Processes and Landforms* 27, 269–288.

Crosson, P. (1997) The on-farm economic costs of soil erosion. In: Lal, R., Blum, W., Valentine, C. and Steward, B. (eds) *Methods for Assessment of Soil Degradation*. CRC Press, Boca Raton, Florida.

Day, T.J. (1992) Network evaluation and planning: Canada's sediment monitoring programme. In: Bogen, J., Walling, D.E. and Day, T.J. (eds) *Erosion and Sediment Transport Monitoring Programmes in River Basins*. IAHS Publication 210. IAHS Press, Wallingford, UK, pp. 337–342.

DEFRA (2004) *First Soil Action Plan for England: 2004–2006*. Department for the Environment, Food and Rural Affairs, London.

Douglas, G., Palmer, M. and Caitcheon, G. (2003) The provenance of sediments in Moreton Bay, Australia: a synthesis of major, trace element and Sr-Nd-Pb isotopic geochemistry, modelling and landscape analysis. *Hydrobiologia* 494, 145–152.

Droppo, I.G. (2001) Rethinking what constitutes suspended sediment. *Hydrological Processes* 15, 1551–1564.

EEA (2003) *Europe's Environment: the Third Assessment*. Environmental Assessment Report No. 10, European Environment Agency. Office for Official Publications of the European Communities, Luxembourg.

Evans, R. (1993) Extent, frequency and rates of rilling on arable land in England and Wales. In: Wicherek, S. (ed.) *Farm Land Erosion: in Temperate Plains Environments and Hills*. Elsevier, Amsterdam, pp. 177–190.

Evans, R.D. and Evans, H.E. (eds) (2001) Proceedings of the 8th International Symposium on Sediment–Water Interactions. *The Science of the Total Environment* 266 (1–3).

Evans, R.D., Wisniewski, J.R. and Wisniewski, J. (eds) (1997) The Interactions Between Sediments and Water: Proceedings of the 7th International Symposium. *Water, Air and Soil Pollution* 99 (1–4).

Foster, I.D.L. (ed.) (2000) *Tracers in Geomorphology*. Wiley, Chichester, UK.

Foster, I.D.L., Dearing, J.A., Simpson, A., Carter, A.D. and Appleby, P.G. (1985) Lake catchment-based studies of erosion and denudation in the Merevale catchment, Warwickshire, UK. *Earth Surface Processes and Landforms* 10, 45–68.

Gerrits, L. and Edelenbos, J. (2004) Management of sediments through stakeholder involvement. *Journal of Soils and Sediments* 4, 239–246.

Golosov, V.N., Ivanova, N.N., Litvin, L.F. and Sidorchuk, A.Yu. (1992) Sediment budget in river basins and small river aggradation. *Geomorphologiya* 14, 62–71 (in Russian).

Gomez, B., Brackley, H.L., Hicks, D.M., Neff, H. and Rogers, K.M. (2004) Organic carbon in floodplain alluvium: signature of historic variations in erosion processes associated with deforestation, Waipaoa River basin, New Zealand. *Journal of Geophysical Research* 109, doi: 10.10129/2004JF000154.

Hagedorn, C., Crozier, J.B., Mentz, K.A., Booth, A.M., Graves, A.K., Nelson, N.J. and Reneau, R.B. Jr (2003) Carbon source utilization profiles as a method to identify sources of faecal pollution in water. *Journal of Applied Microbiology* 94, 792–799.

Hardardóttir, J. and Snorrason, Á. (2002) Sediment monitoring of glacial rivers in Iceland: new data on bed load transport. In: Bogen, J., Fergus, T. and Walling, D.E. (eds) *Erosion and Sediment Transport Measurement in Rivers: Technological and Methodological Advances*. IAHS Publication 283. IAHS Press, Wallingford, UK, pp. 154–163.

Harmon, R.S. and Doe, W.W. III (eds) (2001) *Landscape Erosion and Evolution Modelling*. Kluwer Academic, New York.

He, Q., Walling, D.E. and Owens, P.N. (1996) Interpreting the ^{137}Cs profiles observed in several small lakes and reservoirs in southern England. *Chemical Geology* 129, 115–131.

Heathwaite, L., Haygarth, P., Matthews, R., Preedy, N. and Butler, P. (2005) Evaluating colloidal phosphorus delivery to surface waters from diffuse agricultural sources. *Journal of Environmental Quality* 34, 287–298.

Horowitz, A.J. (2004) Monitoring suspended sediment and associated trace element and nutrient fluxes in large river basins in the USA. In: Golosov, V., Belyaev, V. and Walling, D.E. (eds) *Sediment Transfer through the Fluvial System*. IAHS Publication 288. IAHS Press, Wallingford, UK, pp. 419–428.

Horowitz, A.J. and Walling, D.E. (eds) (2005) *Sediment Budgets, Vol. 2*. IAHS Publication 292. IAHS Press, Wallingford, UK.

Jones, R.J.A., Le Bissonnais, Y., Bazzoffi, P., Díaz, J.S., Düwel, O., Loj, G., Øygarden, L., Prasuhn, V., Rydell, B., Strauss, P., Üveges, J.B., Vandekerckhove, L. and Yordanov, Y. (2004) Nature and extend of soil erosion in Europe. In: Van-Camp, L., Bujarrabal, B., Gentile, A.-R., Jones, R.J.A., Montanarella, L., Olazabal, C. and Selvaradjou, S.-K. (eds) *Reports of the Technical Working Groups Established Under the Thematic Strategy for Soil Protection*. EUR 21319 EN/1. Office for the Official Publications of the European Communities, Luxembourg, pp. 145–185.

Kirkby, M.J. (1998) Modelling across scales: the MEDALUS family of models. In: Boardman, J. and Favis-Mortlock, D. (eds) *Global Change: Modelling Soil Erosion by Water*. Springer, London.

Köthe, H. (2003) Existing sediment management guidelines: an overview. *Journal of Soils and Sediments* 3, 139–143.

Kretzschmar, R., Borkovec, M., Grolimund, D. and Elimelech, M. (1999) Mobile subsurface colloids and their role in contaminant transport. *Advances in Agronomy* 66, 121–193.

Kronvang, B. (ed.) (2003) The Interactions Between Sediments and Water: Proceedings of the 9th International Symposium Between Sediments and Water. *Hydrobiologia* 494 (1–4).

Leithold, E.L., Perkey, D.W., Blair, N.E. and Creamer, T.N. (2005) Sedimentation and carbon burial on the northern California continental shelf: the signatures of land-use change. *Continental Shelf Research* 25, 349–371.

Littlewood, I.G. and Marsh, T.J. (2005) Annual freshwater river mass loads from Great Britain, 1975–1994: estimation algorithm, database and monitoring network issues. *Journal of Hydrology* 304, 221–237.

Martin, Y. and Church, M. (2004) Numerical modelling of landscape evolution: geomorphological perspectives. *Progress in Physical Geography* 28, 317–339.

McHugh, M., Harrod, T. and Morgan, R. (2002a) The extent of soil erosion in upland England and Wales. *Earth Surface Processes and Landforms* 27, 99–107.

McHugh, M., Morgan, R.P.C., Walling, D.E., Wood, G.A., Zhang, Y., Anthony, S. and Hutchins, M. (2002b) *Prediction of Sediment Delivery to Watercourses from Land. Phase II*. Environment Agency R&D Report P2-209. Environment Agency, Bristol, UK.

Morgan, R.P.C. (2005) *Soil Erosion and Conservation*, 3rd edn. Blackwell, Oxford, UK.

Morgenstern, U., Geyh, M.A., Kudrass, H.R., Ditchburn, R.G. and Graham, I.J. (2001) Si-32 dating of marine sediments from Bangladesh. *Radiocarbon* 43 (2B), 909–916.

Netzband, A., Reincke, H. and Bergemann, M. (2002) The River Elbe: a case study for the ecological and economical chain of sediments. *Journal of Soils and Sediments* 2, 112–116.

Nijampurkar, V.N., Roa, D.K., Oldfield, F. and Renberg, I. (1998) The half-life of Si-32: a new estimate based on varved lake sediment. *Earth and Planetary Science Letters* 163, 191–196.

Olley, J.M. (2002) Organic carbon supply to a large lowland river and implications for aquatic ecosystems. In: Dyer, F., Thoms, M.C. and Olley, J.M. (eds) *The Structure, Function and Management Implications of Fluvial Sedimentary Systems*. IAHS Publication 276. IAHS Press, Wallingford, UK, pp. 27–33.

Olmeda-Hodge, T., del Salto, J.F.R., Vandekerckhove, L., Arnoldussen, A., Böken, H., Brahy, V., Düwel, O., Gettermann, J., Jacobsen, O.H., Loj, G., Vilela, A.M., Owens, P.N., Poitrenaud, M., Prasuhn, V., Redfern, P., Rydell, B., Strauss, P., Theocharopoulos, S.P., Yordanov, Y. and Zùquete, A. (2004) Policy options for prevention and remediation. In: Van-Camp, L., Bujarrabal, B., Gentile, A.-R., Jones, R.J.A., Montanarella, L., Olazabal, C. and Selvaradjou, S.-K. (eds) *Reports of the Technical Working Groups Established Under the Thematic Strategy for Soil Protection*. EUR 21319 EN/1. Office for the Official Publications of the European Communities, Luxembourg, pp. 215–239.

Owens, P.N. (2005) Soil erosion and sediment fluxes in rivers: the influence of anthropogenic activities and climate change. In: Lens, P., Grotenhuis, T., Malina, G. and Tabak, H. (eds) *Soil and Sediment Remediation*. IWA Press, London, pp. 418–433.

Owens, P.N. and Walling, D.E. (2002) Changes in sediment sources and floodplain deposition rates in the catchment of the River Tweed, Scotland, over the last 100 years: the impact of climate and land use change. *Earth Surface Processes and Landforms* 27, 403–423.

Owens, P.N. and Walling, D.E. (2003) Temporal changes in the metal and phosphorus content of suspended sediment transported by Yorkshire rivers, U.K. over the last 100 years as recorded by overbank floodplain deposits. *Hydrobiologia* 494, 185–191.

Owens, P.N., Walling, D.E., He, Q., Shanahan, J. and Foster, I.D.L. (1997) The use of caesium-137 measurements to establish a sediment budget for the Start catchment, Devon, UK. *Hydrological Sciences Journal* 42, 405–423.

Owens, P.N., Batalla, R., Collins, A.J., Gomez, B., Hicks, D.M., Horowitz, A.J., Kondolf, G.M., Marden, M., Page, M.J., Peacock, D.H., Petticrew, E.L., Salomons, W. and Trustrum, N.A. (2005) Fine-grained sediment in river systems: environmental significance and management issues. *River Research and Applications* 21, 693–717.

Petticrew, E.L. (2005) The composite nature of suspended and gravel stored fine sediment in streams: a case study of O'Ne-eil Creek, British Columbia, Canada. In: Droppo, I.G., Leppard, G.G., Liss, S.N. and Milligan, T.G. (eds) *Flocculation in Natural and Engineered Environmental Systems*. CRC Press, Boca Raton, Florida, pp. 71–93.

Petticrew, E.L. and Arocena, J.M. (2003) Organic matter composition of gravel-stored sediments from salmon-bearing streams. *Hydrobiologia* 494, 17–24.

Petticrew, E.L. and Droppo, I.G. (2000) The morphology and settling characteristics of fine-grained sediments from a selection of Canadian rivers. In: *Contributions to the International Hydrological Programme-V by Canadian Experts*. IHP-V Technical Documents in Hydrology No. 33. UNESCO, Paris, pp. 111–126.

Phillips, J.M. and Walling, D.E. (2005) Intra-storm and seasonal variations in the effective particle size characteristics and effective particle density of fluvial suspended sediment in the Exe basin, Devon, United Kingdom. In: Droppo, I.G., Leppard, G.G., Liss, S.N. and Milligan, T.G. (eds) *Flocculation in Natural and Engineered Environmental Systems*. CRC Press, Boca Raton, Florida, pp. 47–70.

Phillips, J.M., Webb, B.W., Walling, D.E. and Leeks, G.J.L. (1999) Estimating the suspended sediment loads of rivers in the LOIS study area using infrequent samples. *Hydrological Processes* 13, 1035–1050.

Poesen, J.W.A., Verstraeten, G., Soenens, R. and Seynaeve, L. (2001) Soil losses due to harvesting of chicory roots and sugarbeet: an underrated geomorphic process. *Catena* 43, 35–47.

Polyakov, V.O., Nearing, M.A. and Shipitalo, M.J. (2004) Tracking sediment redistribution in a small watershed: implications for agro-landscape evolution. *Earth Surface Processes and Landforms* 29, 1275–1291.

Rowan, J.S., Goodwill, P. and Franks, S.W. (2000) Uncertainty estimation in fingerprinting suspended sediment sources. In: Foster, I.D.L. (ed.) *Tracers in Geomorphology*. Wiley, Chichester, UK, pp. 279–290.

Ruysschaert, G., Poesen, J., Verstraeten, G. and Govers, G. (2004) Soil loss due to crop harvesting: significance and determining factors. *Progress in Physical Geography* 28, 467–501.

Salomons, W. (2005) Sediments in the catchment-coast continuum. *Journal of Soils and Sediments* 5, 1–7.

Salomons, W. and Brils, J. (eds) (2004) *Contaminated Sediments in European River Basins*. European Sediment Research Network. TNO, The Netherlands.

Small, I.F., Rowan, J.S. and Franks, S.W. (2002) Quantitative sediment fingerprinting using a Bayesian uncertainty estimation framework. In: Dyer, F., Thoms, M.C. and Olley, J.M. (eds) *The Structure, Function and Management Implications of Fluvial Sedimentary Systems*. IAHS Publication 276. IAHS Press, Wallingford, UK, pp. 443–450.

Spreafico, M. and Bruk, S. (2004) UNESCO's International Sedimentation Initiative: background and current status. *Journal of Soils and Sediments* 4, 217–218.

Summer, W. and Walling, D.E. (eds) (2002) *Modelling Erosion, Sediment Transport and Sediment Yield*. Technical Documents in Hydrology No. 60. UNESCO, Paris.

Syversen, N., Øygarden, L. and Salbu, B. (2001) Cesium-134 as a tracer to study particle transport processes within a small catchment with a buffer zone. *Journal of Environmental Quality* 30, 1771–1783.

USEPA (2004) *The Incidence and Severity of Sediment Contamination in Surface Waters of the United States, National Sediment Quality Survey: Second Edition*. US Environmental Protection Agency, Office of Science and Technology, Washington, DC.

Van-Camp, L., Bujarrabal, B., Gentile, A.-R., Jones, R.J.A., Montanarella, L., Olazabal, C. and Selvaradjou, S.-K. (eds) *Reports of the Technical Working Groups Established Under the Thematic Strategy for Soil Protection*. EUR 21319 EN/1. Office for the Official Publications of the European Commission, Luxembourg.

Vandekerckhove, L., Arnoldussen, A., Bazzoffi, P., Böken, H., Castillo, V., Crescimanno, G., Düwel, O., Esteve, J.F., Imeson, A., Jarman, R., Jones, R.J.A., Kobza, J., Lane, M., Le Bissonnais, Y., Loj, G., Owens, P.N., Øygarden, L., Mollenhauer, K., Prasuhn, V., Redfern, P., Diaz, J.S., Strauss, P. and Üveges, J.B. (2004) Monitoring soil erosion in Europe. In: Van-Camp, L., Bujarrabal, B., Gentile, A.-R., Jones, R.J.A.,

Montanarella, L., Olazabal, C. and Selvaradjou, S.-K. (eds) *Reports of the Technical Working Groups Established Under the Thematic Strategy for Soil Protection.* EUR 21319 EN/1. Office for the Official Publications of the European Communities, Luxembourg, pp. 295–309.

Van Rompaey, A.J.J., Govers, G. and Puttemans, C. (2002) Modelling land use changes and their impact on soil erosion and sediment supply to rivers. *Earth Surface Processes and Landforms* 27, 481–494.

Wallbrink, P.J., Murray, A.S., Olley, J.M. and Olive, L.J. (1998) Determining sources and transit times of suspended sediment in the Murrumbidgee River, New South Wales, Australia, using fallout ^{137}Cs and ^{210}Pb. *Water Resources Research* 34, 879–887.

Walling, D.E. and Horowitz, A.J. (eds) (2005) *Sediment Budgets, Vol. 1.* IAHS Publication 291. IAHS Press, Wallingford, UK.

Wasson, R.J. (2002) What approach to the modelling of catchment scale soil erosion and sediment transport should be adopted? In: Summer, W. and Walling, D.E. (eds) (2002) *Modelling Erosion, Sediment Transport and Sediment Yield.* Technical Documents in Hydrology No. 60. UNESCO, Paris, pp. 1–11.

Webb, B.W., Phillips, J.M. and Walling, D.E. (2000) A new approach to deriving 'best-estimate' chemical fluxes for rivers draining the LOIS study area. *The Science of the Total Environment* 251/252, 45–54.

Whiting, P.J., Matisoff, G., Fornes, W, and Soster, F.M. (2005) Suspended sediment sources and transport distances in the Yellowstone River basin. *Geological Society of America Bulletin* 117, 515–529.

Zapata, F. (ed.) (2002) *Handbook for the Assessment of Soil Erosion and Sedimentation Using Environmental Radionuclides.* Kluwer, Dordrecht, The Netherlands.

Zolitschka, B. (1998) A 14,000 year sediment yield record from Western Germany based on annually laminated lake sediments. *Geomorphology* 22, 1–17.

Index

Page numbers in **bold** refer to illustrations and tables

α calculation 166
absolute particle size distribution (APSD) 113, 114, 118
afforestation impacts 79
  see also logging
aggregates
  composition 104, 112
  destruction and redistribution 158
  size 123, 126, 157, 167, 229
  splash redistribution 155
  stability 235
  water-stable 154–155, 163
agriculture
  catchments 132
  land management techniques 287
  land use 18–21, 77–81, 85, 265
  learning from 289–290
  productivity reduction 3
  sediment main sources 288
  sugarcane cultivation 254–263
  see also cultivation effects; farming; land use; tillage
algorithms 168, 197
anhysteretic remanent magnetism (ARM) 64–65
Anoia River basin 85–93
anthropogenic effects 128, 177–184, 267, 269–270, 272
Anti-Pollution Works Notices, Water Resources Act 1990 310
ants, bioturbation 52, **53**, 59
Aquatic Sediment Quality Protection and Management guidelines 276, **281**, **282**
arable land
  catchment 40–50
  lowland 218, 219–220, **221**, 225
  soils 217, 220–222
  use 71, 218
  water erosion 72
audits 291, 292

bank erosion 106–107, 139, 250, 288
bar transects 241
Baru catchment 241, **243**
bedloads 104–106, 107, **298**
Belize 264–271
Belizean Institutions reports 270
beryllium-7 16, 17–18, 19, 20
  see also radionuclides
best management practices **109**, 283, 291
bioavailability, metals 276–283, 284
bioturbation 51–60
Bitterroot Mountains 179–184
Blue Gum Creek 51–60
Breakpoint Climate Data Generator (BPCDG) 188
budget
  calculation **255**
  composition 260–261
  concept 14
  construction 21, 44, **46**, 49
  functioning 24–26
  landscape element identification 254
  processes contribution 40, 125
buffer strips 180, 290
bulk sediment, chemistry 275–283
Bush Integrated Monitoring project 101–110

cadmium 276, **277**, **280**–282
caesium-137
  conversion 31–34, 47–48
  depth profiles **16**–17, **33**, 47
  detection 130
  fallout 15–16
  inventories 20, 29–30, 31, 35
  measurements 24–25, 28
  spatial distribution **19**
  use limitations 132
Cannata watershed, Sicily 187–192
carbon 125, 288
carbon isotopes 115–116, **121–122**, **124**, 125

Cascade Mountains **178**, 179–184
catchment
delivery 70–81
erosion **224**, 241, 242–243
land use **71**–72, 132
management 13–14, 107, 289
river 297–313
scale 239, 242–243, 306
sediment budget 24–25
soil degradation processes 40–50
systems 207–213
yield **245**
Catchment Water Modelling (CATCHMOD) 103
channel
banks 139, 250
erosion **80**, 241–242
length 187
Manning roughness coefficient 189
morphology 159
widths and depths distribution 204
Chemicals, Runoff and Erosion from Agricultural Management Systems (CREAMS) 189, 196
classification, Landsat Thermatic Mapper 220
clay loam 154, 156, **157**, **158**, **159**, 187
climate
Bitterroot Mountains 180–183
Blue Gum Creek 53
Breakpoint Climate Data Generator (BPCDG) 188
Cascade Mountains **178**, 180–183
Danum Valley Conservation Area (DVCA) 239
Stavropol Upland 40
see also precipitation; rainfall
collaboration 313
see also participatory approach
compaction 80
see also mechanical damage; tillage
conductivities, saturated hydraulic 250
conservation
areas 239–252
reef 262, 264–271, 311
soil 287, 289, 291
strategy 5
technique 5
contaminants
barrier reef 265, 271
dilution 269
loading, hydraulic sorting, street sediment 276
monitoring guidelines 301
sinks 23
source area identification 276
urban 272
see also metals; pollution; radionuclides
continuous simulation models 145, 149–151, 152, 306
continuum mechanics 162, 306
control
community-based 290–291
effectiveness, scale affect 233
factor, soil texture 255
measures 21, 288–289, 292
programmes 21
strategies 13
techniques 5
conventions 310
copper 269, **270**, 276, **277**, **280–281**, 282
coral populations, stresses 269
cores 19–20
Coshocton-wheel **298**
costs 4, 217, 288, 291, 307
CREAMS (Chemicals, Runoff and Erosion from Agricultural Management Systems) 189, 196
critical areas 197, 199
critical shear stress 157, **158**, 159, 188, **189**, 204
Critical Source Area 187
critical threshold 197, **200**
crop
cover 255
see also vegetation, cover
cultivation 189
different, impact 79
fields 72–75
increase impact 80
runoff and erosion vulnerability 80–1
type, erosion risk **79**
cropping factor, USLE 231
Cuckney-1 soils 75
cultivation effects 16–17, 49, 189, 254–263, **309**
see also agriculture; tillage
Curry Report 81
Curve Number approach, USDA 174

dams, debris **53**, 59, 248, 250–252
Danum Valley Conservation Area (DVCA) 239–252
dating methods 130–132, 303
see also radionuclides; time series
debris dams **53**, 59, 248, 250–252
decision-support systems 292, 306–307
deconstructing processes 233
degradation 40–50, 153, 229, 264
delivery
catchments 70–81
control 5
cropped fields 72–75
pathways 132
rates prediction **181**
ratios 229, 231
tracing 16–18
transport 49, 50, 139
water courses 73, **74**
yields **75**
DEM (digital elevation model) 196, 197, 199, 200
density estimation 129–130
deposition 210, 211, 254–263
depth profiles
caesium-137 **16**–17, **33**, 47
channel widths and depths distribution 204
depth-age curve estimation 131–132
lead-210 130
radionuclides distributions **16**, **17**
storm runoff 189–191
desertification 229
Desmet and Govers approach 172, 173

detachment
 by flow **163**–164
 incision 196
 probability 162, 168
 rates, double averaging 163–168
 runoff 197
 splash erosion 210, 229, 233
 systems 170–171, 175
 time 165
detector error possibility 35
detention ponds 283–284
diatoms 104
digital elevation model (DEM) 196, 197, 199, 200
discharge 87–88, 91–93, 94–100, 203–204, **266**–267
discrete particle approach 162–168
discretization, watershed morphology 187–188
dispersion 265
distributions
 mapping 273
 Poisson cumulative probability **218**
 spatial **19**, **24**, 239–252, 275–276, **277–279**
disturbances
 human 128, 177–184, 229, 267, 269–270
 logging 239–252
 minimization 290
 natural 177–184, **247**, 248–**249**, 250–252
 see also fire; urban environments
diverter berms 290
double-averaging approach 162–168
drainage
 density 241
 flow increase/decrease 101
 impacts 79, 132, 136, 139, 255
 layout, Cannata watershed **188**
 see also flood events; hydrology; palaeohydrology; pipes
dredged materials recommendations 310
dredger **309**
drivers 81, 288, 301, **308**–311
driving forces, resistance to **149**
drying effect 155–156, 157, 159

educational programmes 291–292
effective particle size distribution (EPSD) 113, 114, 118
EGEM (Ephemeral Gully Erosion Model) 196
elevation changes 57–58
embedding 234
Enborne catchment 211, 212
engineering 288
enhancement, magnetic 62–68
entrainment 210, 211, 275
Environment Agency of England and Wales (EA) 225
environmental radionuclide *defined* 15
ephemeral gullies
 defined 195
 development 41, 43
 erosion volumes estimation 48
 incision 195–205
 surface lowering 47
 transport pathways 49, 50
 volumes estimates **45**
 see also gullies
Ephemeral Gully Erosion Model (EGEM) 196
EPICS (Erosion Productivity Index Calculator) 212
EPSD (effective particle size distribution) 113, 114, 118
erodibility 153–160, 178, 305
erosion
 bridge 54–55, 56–58, 59, 241
 rates and volumes **45**, **183**–184, 254–263
 spatial distribution 239–252
 types occurrence and severity, Europe **300**
Erosion Productivity Index Calculator (EPICS) 212
European Union Directives 4
European Union recommendations 299
European Union Water Framework Directive *see* Water Framework Directive (WFD)
eutrophication 4
extreme events role 252

fallout 15–16, 21, 130, **131**
 see also radionuclides
farming 76, 264–271
 see also agriculture
F_c determination 197
fields
 channel erosion **80**
 cropped 72–75
 cultivated 76–77
 erosion occurrence and rates **77**
 size impact 79
fine sediment
 gravel stored 115, 118–121
 role 13
 size and composition 115
 sources 104, 106–107
 storage 112–126
 transport 112–126
fingerprinting 18, 21–23, 107, 302–303
fire
 downstream impacts 62–68
 erosion rates following **178**, **183**–184
 fuel management sediment yields 179–180
 hillslope soil losses 51–60
 prescribed 180
 suppression 177
first flush phenomenon 283
fish 21–23, 101, 106, 107, 112–126
flocs *see* aggregates
flood events 94–100, 262
floodplains **17**, 23–24, 254–263
flow
 capacity 208
 competence 105, 275
 control 268
 detachment by 163–164
 direct runoff equation 210
 free-surface gravity-driven **167**
 lines 76
 network 197, 201
 overbank 23–**24**, 25, 72, 255, 261, 262

flow *continued*
overland 188, 196–197, 200, 250
rain-impacted 170–171
regime changes 159
saturation 197, 209
strips 234
transport 105–106, 170–171, 175
types, water 261
width, depth and velocity 157
FLUENT hydrodynamic software 166
fluxes 113, 210, 264–271, 299–302
force balance 164–165
force distributions 149
forests 177–184, **240**, 249–250, 264
see also logging
fragmenting processes 233
fuel management 177, 179–180, 184
furrow bed morphology 157, 159

geographical extrapolation 219–220
Geographical Information Systems (GIS) 219–222
geotextiles **231**–233, 233–234
GeoWEPP 186, 187–188, 192
see also Water Erosion Prediction Project (WEPP)
Global Assessment of Human Induced Soil Degradation (GLASOD) 229
government responsibilities 288
grain size
determination 95–96
distribution 96, 99–100, 210, 274
key variable 210–211
potential deposition rate 211
street sediment 274, **275**, 276
transport flow requirements 105–106
see also size
grassland 218, 219–220, **221**, 222
gravels
aggregates storage 123
particle size 104–105, 123
sediment storage 115, 118–121, 125
spawning 21–23
grazing 79, 81, 136
ground cover 28, 255, 257, 259–**260**, 290
ground level change
long-term **247**
lowering 56–57, 243, **246**
monitoring 54, **55**
post-fire rainfall events 59
rises 245
see also surface soils, level change
guidelines 276, **281**, **282**, 301, 309–310
gullies
development, logging 242
environmental change effects prediction 195
erosion **73**
incipient 243–244, 250
models 196
runoff volumes prediction 201–202
rutted systems 245
soil loss 195, 217
volume measurement 48
width equation 198
see also ephemeral gullies
Gully Thermoerosion and Erosion Model (GULTEM) 196

Habitats Directive 13
Hanslope soils 75
Harmonised Monitoring Scheme (HMS) 299, 300, **301**
harvesting 180, 299
see also logging
headlands 254–263
Helley-Smith bedload sampler **298**
Herbert River 255, **256**
hillslopes
erosion causes 180
erosion prediction using runoff as a factor 170–175
farmed areas 264
fire soil losses 51–60
scale 145, 183
sediment generation estimation 217–225
sediment yield **181**
see also slopes
HMS (Harmonised Monitoring Scheme) 299, 300, **301**
Hortonian overland flow 197, 250
hydraulic
conductivity **189**
critical shear 188, **189**
radius 196
simulation, gully 204
soil slope **305**
sorting 274–275, 276, 284
hydrodynamics 166–167
hydrology
annual values 87–90
changes 62–68
connectivity 106
data 86
forcing 207, 213
human intervention 128
model 209, 212
response 101, 103–104
see also drainage; flood events; palaeohydrology
hysteresis loops 96, **97–98**

IAHS (International Association of Hydrological Sciences) 297
IASWS (International Association for Sediment Water Science) 298
image grabbing system, automated 114–115
impacts
assessment 276
multiple 3–4
off-site 4, 229, 287
on-site 229
reduction 252
INCA-Sed (Integrated Catchment Model of Sediment) 207–213

incentives 289
incisions 196, 197, 199, 245
index of vegetation-consumed-by-fire 64
infiltration 59, 118, 120–121, 234
institutional framework provision 291
Integrated Catchment Model of Sediment (INCA-Sed) 207–213
interill erodibility 153, **154**–156, 188, **189**
International Association of Hydrological Sciences (IAHS) 297
International Association for Sediment Water Science (IASWS) 298
iron 62, 63, **278**, **280**, **281**, 282
isothermal remanent magnetization (IRM) measurements 65
isotopes 115–118, **121–122**, **124**, 125

Jorba 87, 88, **89**, 90, 91

Kastanozen 154, **155**, **156**
Kenilworth sewershed 273–274
Kolmogrov-type algorithm, particles failure 168
Kruskal-Wallis tests, surface level change **259**
Kuamut Formation rocks 240

lakes 128–140, 303
Lambourn catchment 211, 212
laminations 130
Land Care Movements 290–292, 311
land cover mapping 220, 225
land management 7–8, 287, 291
land restoration 289
land use
  activities, occurrence frequency 106
  catchment **71**–72, 132
  changes 37, 77–81, 128, **304**
  classification 220
  determination 219–220
  intensification 72, 76, 77, 79, 81
  percentage calculation 208
  policy shifts 77, 79
  systems evaluation 150
  urban 272–284, 308
  water courses pollution 70–81
  yield relation to 70–81
  see also agriculture; arable land; farming
Landsat Thematic Mapper classification 220
landscape elements identification 254
landslides **247**, 248–**249**, 250–252
Large Eddy Simulation (LES) 166–167
lead 276, **278**, **280**, **281**, 282
  see also metals
lead-210 15, 16–18, 19, 130, 132
  see also radionuclides
legislation 4, 309–310
LES (Large Eddy Simulation) 166–167
Limburg Soil Erosion Model (LISEM) 196, 199, 204, 205
litter dams **53**, 59
livestock stocking intensity 79, 136

loads 99–100, 136, 252, 267, **301**
loams 241
Loess soils 204
logging
  impacts 239–252
  see also afforestation impacts
lumped estimates 299

magnetic enhancement, soil water repellency 62–68
maize 80, 81
mammal activity 52, 58
management
  best management practices **109**, 283, 291
  decision-support systems 292, 306–307
  defined 5
  drivers **308**–311
  implication 101–110
  planning tools 217–225
  recommendations 107, 262–263
  strategies improvement 272–284
manganese **279**, **280**, **281**, 282
Manning roughness coefficient 189
mapping 220, 225, 306
maps
  Anoia River basin **86**
  atmospheric fallout, United Kingdom **131**
  Belize study locations **265**
  Blue Gum Creek, Cataract River basin **52**
  Cannata watershed **187**, **188**
  catchment erosion probability percentage **224**
  copper concentrations **270**
  Danum Valley Conservation Area location map **240**
  Herbert River catchment study area location **256**
  Kenilworth sewershed **273**
  landslides, Baru catchment **249**
  Nattai River basin **63**
  River Bush catchment location **102**
  sediment generation **223**
  Segama River **240**
  Stavropol Upland study site **42**
  topographic **200**
  Zagożdżonka catchment location **95**
marine systems, Belize 267–270
mass balance 47, 210, 211
mass calculation 130
measurement 4, 6–7, 34–35, 297–302
mechanical damage 72, 80, 242, 243–244, **247**
  see also tillage
Mediterranean Basin 85–93
MEDRUSH model 234
metals 269, 272–284, 308
  see also contaminants; pollution
Meteorological Office Rainfall and Evaporation Calculation System (MORECS) 209
minerals, magnetic analysis 63, 64–65
Minimum Source Channel Length 187
Ministry for the Environment and Energy (MOEE) guidelines 276, **281**, **282**
mitigation 283–284, 313
mobilization tracing 16–18

modelling 4–5, 7, 303–307
models
  continuous simulation 145, 149–151, 152, 306
  data sets 234, 304–305
  data variability 34, 145, 146–149, 150, 151–152, 306
  development 303–304
  digital elevation 196, 197, 199, 200
  integrated decision-support 292
  MEDRUSH 234
  multicomponent mixing 22
  process information 304–305
  process-based 171
  proportional 47
  scales 305–306
  stochastic 162–168
  un-mixing **139**
  use 290
  see also prediction
MOEE (Ministry for the Environment and Energy) guidelines 276, **281**, **282**
monitoring 37, 291, 297–302
MORECS (Meteorological Office Rainfall and Evaporation Calculation System) 209
morphology 157, 159, 187–188
multiscale approach 234

Nash-Sutcliffe efficiency factor 172
Nash-Sutcliffe values 212
NASQUAN (National Stream Water Quality Accounting Network) monitoring programme 299
national assessment need 217
National Marine Environmental Quality Monitoring Programme 271
National River Flow Archive (NRFA) 299
National Sediment Quality Survey (NSQS) 302
National Soil Inventory (NSI) 217
National Soil Resource Institute (NSRI) Soil Map (NatMap) 220
National Stream Water Quality Accounting Network (NASQUAN) monitoring programme 299
Nattai River **63**
natural adjustment 287–288
Navier-Stokes equations 166
networks 182, 197, 201, 299, 311
nitrogen budgets 125
NRFA (National River Flow Archive) 299
NSI (National Soil Inventory) 217
NSQS (National Sediment Quality Survey) 302
NSRI (National Soil Resource Institute) Soil Map (NatMap) 220
nutrients 23, 112–113, **124**, 125, 126

OFEs (Overland Flow Elements) 188
off-site impacts 4, 229, 287
on-site impacts 229
organic matter 115, 118, **121**–122, 124
organisms, benthic 276, 284
overbank flow 23–**24**, 25, 72, 255, 261, 262
overgrazing impacts 79, 136
overland flow 196–197, 200, 250
  see also runoff, direct
Overland Flow Elements (OFEs) 188

palaeoenvironmental reconstruction methods 139
palaeohydrology 128–140
palaeolimnology 132–136, 303
participatory approach 290, 292, 308
  see also collaboration
particles
  conglomerate 272
  distributions 113, 114–115, 118, 168
  failure algorithms, Kolmogrov-type 168
  populations **121**, 123
  size
    estimation 74, 113–114
    fish resuspended sediments **119**
    gravel 104–105, 123
    transported material composition 94
pathways 5, 59, 132, 287
PCA (Principal Components Analysis) **108**
pedestals **54**, 55, **56**, 58, 59
performance criteria 291
phosphate 4, 70
phosphorus
  budgets 125
  concentrations 132, **135**, 136, **137**
  delivery 287
  loads **138**
pigs, outdoor 80, 81
pin plots 28–37, 255, 256–257, **258**, **259**
pipes 250, 252, 289–290
  see also drainage
policy shifts, land use 77, 79
polluter pays 289, 291
pollution
  directives 13–14, 153
  legislation 310
  payers 289, 291
  water 4, 70
  see also contaminants; metals; radionuclides
precipitation 95–96, **120**
  see also climate; rainfall; snow
precision, spatial 199–201
prediction
  capacity, USLE-M 174, 175
  critical threshold areas **200**
  delivery rates **181**
  environmental change effects, gullies 195
  gullies runoff volumes 201–202
  hillslope erosion using runoff as a factor 170–175
  modelling capabilities and limitations 145–152
  runoff 180, 192, 201–202
  single storm 147
  soil loss **146**, **147**, **148**, **150**, 174, 175
  tools 179
  yield 180, **181**
  see also models; Water Erosion Prediction Project (WEPP)
Principal Components Analysis (PCA) **108**

probability analysis 217–219, 222, 225
probability density function (PDF) of α 166, 167–168

quality, sediment 301
quantification methods 40, **41**

radionuclides 14–21, 23–26
    see also caesium-137; contaminants; lead-210; pollution; time series; tracing
rain-impacted flow 170–171
Raindrop-induced Flow Transport (RIFT) 170–171, 175
rainfall
    data, headland **262**
    experimental 145
    gauges location **52**
    hydraulic sorting, street sediment 274–275
    increase, United Kingdom 132
    kinetic energy level 172
    rate, direct runoff equation 209–210
    seasonality 85, 90, 91–**92**
    tests 154–155
    transport 94–100
    variability 35, **36**, 192
    see also climate; splash erosion; storm
reconstruction 129–132, 136
redd preparation 112, 113, 122–123, 125–126
reef conservation 262, 264–271, 311
reference sites *defined* 16
remediation 5, 283–284
representativeness, temporal variation 34–35, 261
research 217, **230**, 236, 307–308
reservoirs 128–140
resistance distributions 149
resource investment target areas 217
responsibilities 288–289, 307–308
resuspension 115, 118, **119**, 124
revegetation 245, 250
reviews 297
RIFT (Raindrop-induced Flow Transport) 170–171, 175
rills
    channel morphology 159
    creation 261
    erodibility 148–149, 153, **154**, 188, **189**
    erosion **73**, 148, 156–157, 170
    formation 195
    incision 157, 159
    soil loss 217
Ripple Creek catchment 254–263
risk 74–75, **79**, 220–222, 299
river
    basins 5, **265**
    catchments 297–313
    monitoring sites, Belize 265–267
    sampling 86–87
    scale 4
    suspended sediment yields 72–73
    see also maps
River Bush, Northern Ireland 101–110
road
    banks, erosion control 288
    erosion 183
    networks, yield 182
    runoff 76, 245
    sediment yields 178
    unsurfaced logging 245, **247**
roots, exposed 54, 55, **56**
runoff
    coefficient 173, 174
    depth, storm 189–191
    detachment 197
    diluting effect on human impact 267
    direct 208, 209, 210
        see also overland flow
    discharge 203–204
    erosion prediction factor 171–175
    modelling 186–192
    outflow 95
    patterns 156–**157**
    prediction 180, 192, 201–202
    sheetwash 76, 81
    simulated vs observed **191**
    sources 76–77, **182**, 183, 245, **261**, 272
    transport capacity 210, 234
    see also precipitation; rainfall; storm

Sabah Natural Forest Management Area **240**
salinity 267–268
salmon
    carcass nutrients 112–113, 125
    redd preparation 112, 113, 122–123, 125–126
    sources 124
    spawning 21–23, 101, 106, 107, 112–126
sampling
    isotopic carbon 125
    river 86–87
    sample collection 21–22, 64, 274
    samplers 21–22, 86–87, 241, **298**
    schedule, Stuart-Takla Creek **116–117**
sandy loams 241
Sant Sadurní 87, 88–90, 91
saturation 197, 209, 210
scale
    catchment 239, 242–243, 306
    identification 305–306
    issues 228–236
    nested data 234, 305
    research **230**, 236, 308
    river basin 4
    scaling 233
    spatial 228, 305
    studies 299
    temporal 305–306
    watershed, simulation models 186
scatter reduction 90–91
Schomberg clay loam 154, **157**, **158**, **159**
scientific community responsibilities 307–308
seal formation 155, 156
Sealing and Transfer by Runoff and Erosion Related to Agricultural Management (STREAMS) 196

seasonality 85–93, 118
sediment generation **208**, 217–225
sediment production prevention 228–236
Segama River map **240**
sequential extraction procedure 274
sewers **273**–274
sheet erosion **45**, 48–50, 76, 77, 81
SIA (stable isotope analysis) 115–118, 121–122, 124, 125
signatures 63, 67, 121, 122, 125
single storm models 147, 149, 150, 152
sinks 5, 23, 254–263, 270
size
 aggregates 123, 126, 157, 167, 229
 class, trace metal concentrations **281**
 effective particle size distribution (EPSD) 113, 114, 118
 estimation 74, 113–114
 fields 79
 fine sediment 115
 fractions, metals **275**, 276, **281**
 gravel particles 104–105, 123
 suspended sediment characteristics 118
 transported material 94
 see also grain size; particles, size
sizing 74, 114–115, **121**, 274
skid trail, mechanical damage 243, **247**
slopes
 bioturbation quantity and rates **58**
 classes 218, **219**, 220
 disturbed 242, 243–244
 erosion 241–242
 gradients calculation 220
 hydraulically operated **305**
 lengths 191, 233, **233**
 slopewash rates 249–250
 surface soils, magnetic properties **67**
 see also hillslopes
snow 94–100, 182–183
soil
 associations 74–75
 bulk density 80, 222, 224
 composition description 219
 distribution 220
 management recommendations 262–263
 moisture condition 96
 post-fire hillslope 55–58
 quality maintenance 290
 redistribution **20**–21, **43**, 47–48
 resistance 153, 156–157, 159
 strength effect 198
 structure modelling 167–168
 subgroups 221–222
 texture, erosion control factor 255
 thickness variation **44**
 wettability 66, 67
Soil Conservation Districts, USA 291
soil loss
 estimated 44–47, 55–56, 222
 frequency distributions **148**
 observations 228
 observed vs predicted **172**, **173**, **174**
 predictions **146**, **147**, **148**, **150**, 174, 175
 prevention, geotextiles **231**–233, 233–224
 time **157**
 see also Universal Soil Loss Equation (USLE)
soil types
 arable lowlands 220–222
 clay loam 154, 156, **157**, **158**, **159**, 187
 Cuckney-1 soil types 75
 Hanslope 75
 Kastanozen 154, **155**, **156**
 Kuamut Formation rocks 240
 Loess 204
 lowland grassland 222
 sandy loams 241
 subgroups 222
 surface **67**
 topsoils 59, 80, 139
 upland 222
Soil and Water Protection (SOWAP) 228, 234–235
sorting 274–275, 276, 284
Source Channel Length 187
sources
 changes 72, 81, 136, 239–250, **251**
 characterization 64, 107
 differentiation technique 124, 125
 discrimination 139
 hydrological influences 101–110
 identification 25–26, 107, 276
 main 288
 material differentiation 113
 terrestrial 124
 types 21
spatial analysis techniques 308
spatial distribution **19**, **24**, **32**, 239–252, 273, 274
spawning 21–23, 101, 106, 107, 112–126
splash erosion
 aggregates redistribution 155, 158
 cups 229
 defined 170
 detachment 210, 229, 233
 pedestals formation 58
SPM (suspended particulate matter) estimates 113–114
stable isotope analysis (SIA) 115–118, 121–122, 124, 125
stakeholder participation 308, 311, **312**
Stann Creek 264–271
stochastic modelling 162–168
stone lags formation 245
storage 112–126, 302
storm
 design 150, 151
 model, single 147, 149, 150, 152
 peak **245**
 responses 250
 runoff depth 189–191
 yield statistics 191, **192**
 see also rainfall; runoff
stormwater detention ponds 283–284
stream power 208
STREAMS (Sealing and Transfer by Runoff and Erosion Related to Agricultural Management) 196

street sediment 273–**275**, 276, 283, 284
  see also urban environments
Stuart-Takla Creek 113, **116–117**
sugarcane cultivation 254–263
surface soils
  conditions 258
  level change 47, 256, 257, 258–261
    see also ground level change
  magnetic properties **67**
  washoff 273, **275**
suspended particulate matter (SPM) estimates 113–114
suspended sediment
  discharge relationship 91–93, 94–100, **120**
  fish nutrients association 125
  flux 210
  observed vs simulated **212**, **213**
  rating curves **89**, **90**
  seasonality 85–93, 118
  silhouette images **119**
  size characteristics 118
  storm peak **245**
  yield **71**, 72–73
swales, grassed 283

Tarland Catchment Initiative, Scotland 107
techniques
  conservation 5
  land management 287
  measurement 6–7, 297–298
  monitoring 37, 297–298
  participatory 290
  resuspension 115
  sources differentiation 124, 125
  spatial analysis 308
  stable isotope analysis 124
  Water Drop Penetration Time (WDPT) method 64
  see also fingerprinting; models; tracing
Thematic Strategy for Soil Protection (TSSP) 4, 310
tillage 16–17, **45**, 47, 79, 151–152
  see also mechanical damage
timber harvesting 177, 180
  see also logging
time series 128–140, 212, 303
  see also radionuclides
tools
  decision-support systems 292, 306–307
  ground cover establishment 290
  management planning 217–225
  measurement 4, 6–7
  modelling 4–5, 7
  prediction 179
  toolboxes 307
  see also models
topsoils 59, 80, 139
tracing 13–26, 62–68, 106, 107, 139, 303
traffic effects 72, 80, **178**, 242, 243–244, **247**
transfer 70, 196, **209**, 287
translocation **45**, 47, 49, 58, 59
transport
  annual load 243
  chemical 302
  fish influences 112–126
  flow 105–106, 170–171, 175
  management implications 101–110
  non-selective 156
  pathways 49, 50, 139
  process components and linkages **208**
  regulators 250
  spatial and temporal variation 104
  systems 170
  wave energy 268–269
turbidity 268, 269, **269**, 270

un-mixing models **139**
uncertainty analysis 306
United Nations Environment Programme (UNEP) 229
Universal Soil Loss Equation (USLE)
  application 151
  cropping factor 231
  as event model 148, 171
  modified for Russia 44, **45**
  revised 151, 171–172
  soil loss 48–49
  variants 172, 174, 175
  see also soil loss
upland soils **221**
urban environments 272–284, 308
USLE *see* Universal Soil Loss Equation (USLE)

variability 35, **36**, 192, 235
variables, averaging, hydrodynamic 163
vegetation
  cover 28, 255, 257, 259–**260**, 290
  effect 37, 231
  engineering properties 288
  pasture areas 187
  revegetation effects 245, 250
  role 263, 290
  vegetation-consumed-by-fire index 64
velocity-concentration approach 162, 168
Vertisol 154, 156
vulnerability 225, 229

washed-out layer, formation and removal 156, 158
water
  courses 72, 73, **74**, 81, 217
  discharges **266**–267
  erosion 47, 81, 186
  flow types 261
  loss equation 209
  management strategies improvement 272–284
  pollution 70–81
  quality 81, 266, 284
  repellency 62–68
  storage 128–140
  supply, burnt 62–68
Water Drop Penetration Time (WDPT) method 64
Water Erosion Prediction Project (WEPP) 147–151, 153, 179–180, 184, 186–192

Water Framework Directive (WFD) 4, 13, 225, 309–310
Water Resources Act 1990, 310
water stable aggregates (WSA) 154–155, 163
waterfall sections 245, 250
watershed 183, 186–192, 287
Watershed Reef Interconnectivity Scientific Study (WRIScS) 264–271, 311
WDPT (Water Drop Penetration Time) method 64
weather effects **178**, 180, **182**
see also climate; rainfall
WEPP (Water Erosion Prediction Project) 147–151, 153, 179–180, 184, 186–192
wetness index 197
WFD (Water Framework Directive) 4, 13, 225, 309–310
wildfire 62–68, 178, 179–180, **183**–184
see also fire
wind–wave effects 268, 270–271
WRIScS (Watershed Reef Interconnectivity Scientific Study) 264–271, 311
WSA (water stable aggregates) 154–155, 163

yield
changes **135**, **136**, **243**
discrepancies 76
estimation 130
histories, United Kingdom **133**, **134**
measurement 182
monthly, Baru catchment area **245**
prediction 180, **181**
reconstruction 129–132
relation to land use 70–81
sources relative contribution **268**
statistics 191, **192**

Zagożdżonka catchment location **95**
zinc 276, **279**, **280**, **281**